S0-AUH-171

New Developments in Survey Sampling

New Developments in Survey Sampling

A Symposium on the Foundations of Survey Sampling
Held at the University of North Carolina,
Chapel Hill, North Carolina

Norman L. Johnson, Editor
Harry Smith, Jr., Editor

WILEY-INTERSCIENCE
A DIVISION OF JOHN WILEY & SONS
New York · London · Sydney · Toronto

10 9 8 7 6 5 4 3 2 1

Library of Congress Catalogue Card Number: 78-87080

SBN 471 44487 1

Printed in the United States of America

Preface

The impact of sampling procedures throughout the modern world is tremendous. The ability to predict the outcome of elections, to check the reliability of census figures, to estimate the number of illegal abortions, etc., through sample survey techniques is astonishing to the statistically uneducated public. However, there remain many problems in the sample survey field which need to be solved. In particular, the problems of non-sampling errors, of large sampling errors, etc., still exist. In 1967 it became apparent that much progress was being made in the sample survey field, but the results were unavailable to most people. In addition, there was a distinct difference of opinion concerning the problem of "labelling", and it was deemed necessary to take some action to bring these controversial and/or unavailable pieces of research into focus for assessment by the sampling experts of the world.

Under the prodding of Dr. V. P. Godambe, University of Waterloo, Canada, the departments of Biostatistics and Statistics, University of North Carolina, Chapel Hill, North Carolina, agreed to sponsor an International Symposium on Survey Sampling in April, 1968. We were very fortunate in that the National Center of Health Statistics, USPHS, became very interested in the idea and provided us with some of the funds and necessary guidance in making the Symposium possible.

The symposium was planned to bring together experts in the field of "Theory of Sampling" and experts in the practical uses of Survey Sampling. This confrontation was indeed successful. Some thirty experts assembled for a week in Chapel Hill and this volume contains most of the papers given during that week. Unfortunately, we have been unable to include the lively discussions which took place at this meeting in this volume. In particular, we regret that we have been unable to include the comments of the invited discussants, Professors T. V. Hanurav, A. M. Mood, and J. Sedransk. However, some indication of the tone and impact of the discussions that took place can be found in Professor Barnard's summary at the end of the conference. We also give the names of discussants at the end of Table of Contents.

We would like to thank Mr. Jon Kettenring for his invaluable work as Administrative Assistant in preparation and organization of the symposium. We also wish to acknowledge the support of the Departments of Biostatistics and Statistics, and the efficient services of the Extention Service of the University of North Carolina.

Finally, we would like to thank the many graduate students and stenographers who worked so diligently with us to prepare this manuscript for publication.

Chapel Hill, North Carolina
May, 1969

N. L. Johnson
H. Smith, Jr.

Contents

Discussants and Session Chairmen

Session I Chairman: H. SMITH, University of North Carolina

1. Opening: M. H. HANSEN and B. J. TEPPING
V. P. Godambe

2. V. P. GODAMBE, University of Waterloo, Canada
W. E. Deming, J. Durbin, W. A. Ericson, T. V. Hanurav, H. O. Hartley, W. G. Madow, J. N. K. Rao, R. Royall, J. Sedransk

*3. D. B. LAHIRI, Indian Statistical Institute, Calcutta, India
T. Dalenius, T. V. Hanurav, V. M. Joshi

Session II Chairman: H. O. HARTLEY, Texas A and M University

4. F. YATES, Rothamsted Experimental Station, England
W. G. Madow, F. F. Stephan

*5. P. J. McCARTHY, Cornell University
J. Durbin, V. P. Godambe, D. B. Lahiri, W. G. Madow, J. M. K. Rao

6. F. F. STEPHAN, Princeton, University
T. V. Hanurav, W. G. Madow

Session III Chairman: A. L. FINKNER, Research Triangle Institute, Durham, North Carolina

*7. R. SMITH and C. E. ZIMMER, National Center for Air Pollution, Cincinnati, Ohio
D. S. Robson

8. RUTH PATRICK, Acedemy of Natural Sciences of Philadelphia
T. Dalenius

9. D. S. ROBSON, Cornell University
J. Durbin, D. J. Hudson, L. Kish, F. Yates

Session IV Chairman: J. E. GRIZZLE, University of North Carolina

10. H. O. HARTLEY and J. N. K. RAO, Texas A and M University
W. A. Ericson, V. P. Godambe, T. V. Hanurav, S. Koller, W. G. Madow

11. M. R. SAMPFORD, University of Liverpool, England
J. Durbin, T. V. Hanurav, H. O. Hartley, J. N. K. Rao

12. V. M. JOSHI, Maharashtra Government, Bombay, India
T. V. Hanurav, P. Thionet

13. J. N. K. RAO, Texas A and M University
R. Frauendorfes, V. P. Godambe, T. V. Hanurav, H. O. Hartley, J. Sedransk, B. J. Tepping

Session V Chairman: W. E. DEMING, Consultant, Washington, D. C.

14. S. KOLLER, University of Mainz, West Germany
R. L. Anderson, H. O. Hartley, W. R. Simmons

15. D. G. HORVITZ, Research Triangle Institute and G. G. KOCH, University of North Carolina
    V. P. Godambe, B. J. Tepping

16. P. THIONET, University of Poitiers, France
    J. Neter

Session VI Chairman: N. K. NAMBOODIRI, University of North Carolina

17. N. R. DRAPER and I. GUTTMAN, University of Wisconsin
    G. A. Barnard, V. P. Godambe, H. O. Hartley, D. J. Hudson, W. G. Madow

18. W. A. ERICSON, University of Michigan
    H. O. Hartley, W. G. Madow, J. N. K. Rao

19. J. D. KALBFLEISCH and D. A. SPROTT, University of Waterloo, Canada
    T. V. Hanurav, H. O. Hartley, V. M. Joshi

Session VII Chairman: F. YATES, Rothamsted Experimental Station, England

20. T. DALENIUS, University of Stockholm, Sweden
    R. L. Anderson, K. A. Schäffer

21. L. KISH, University of Michigan
    W. E. Deming, H. O. Hartley, D. J. Hudson, J. Sedransk

Session VIII Chairman: W. R. SIMMONS, National Center for Health Statistics, Washington, D. C.

22. L. R. FRANKEL, Audits and Surveys Co., Inc., New York
    T. Dalenius, W. G. Madow

23. R. J. JESSEN, University of California, Los Angeles
    W. E. Deming, W. G. Madow

24. J. NETER, University of Minnesota
D. B. Lahiri

Session IX Chairman: F. F. STEPHAN, Princeton University

25. D. R. COX, Imperial College, University of London, England
R. L. Anderson, W. G. Madow, K. D. Ware

26. P. V. SUKHATME, F.A.O., United Nations, Rome and B. V. SUKHATME, Iowa State University
V. P. Godambe, H. F. Huddleston, L. Kish, D. B. Lahiri

27. N. KEYFITZ, University of Chicago
H. O. Hartley, D. B. Lahiri

28. G. NATHAN, Central Bureau of Statistics and Hebrew University, Isreal

Session X Chairman: N. KEYFITZ, University of Chicago

29. W. R. SIMMONS and JUDY ANN BEAN, National Center for Health Statistics, Washington, D.C.
T. Dalenius, H. O. Hartley, S. Koller, D. B. Lahiri, W. G. Madow, F. F. Stephan, B. J. Tepping

30. J. DURBIN, London School of Economics, England
D. R. Cox, W. E. Deming, D. J. Hudson, W. G. Madow, P. Thionet

31. W. E. DEMING, Consultant, Washington, D. C.
H. O. Hartley, O. Kempthorne, F. F. Stephan

Session XI Chairman: W. G. Madow, Stanford Research Institute, California

32. O. KEMPTHORNE, Iowa State University
E. E. Bryant, W. A. Ericson, V. P. Godambe, T. V. Hanurav, H. O. Hartley, D. J. Hudson, W. G. Madow, R. Royall, P. Thionet

33. G. A. BARNARD, University of Essex, England
    T. V. Hanurav, H. O. Hartley, K. D. Ware

34. Closing: J. NEYMAN, University of California, Berkeley

---

* Papers by P. J. McCarthy, "Pseudo-replication; Half Samples," D. B. Lahiri, "On the Unique Sample: the Surveyed One, and R. Smith and C. E. Zimmer, "Survey Sampling of Air Quality" are not included in this book.

New Developments in Survey Sampling

PROGRESS AND PROBLEMS IN SURVEY METHODS AND THEORY ILLUSTRATED BY THE WORK OF THE UNITED STATES BUREAU OF THE CENSUS

Morris H. Hansen
and
Benjamin J. Tepping

U.S. Bureau of the Census

Opening Address at the Symposium on Foundations of Survey Sampling at the University of North Carolina, Chapel Hill, North Carolina, April 22-26, 1968.

U.S. DEPARTMENT OF COMMERCE/Bureau of the Census

Introduction

This paper is intended to describe some significant developments as well as current important problems in the design and operation of sample surveys. It is illustrated specifically in terms of the survey work of the Bureau of the Census. We feel that this is not a severe restriction since progress and problems in the work of the Bureau of the Census exhibit the advances that have been made, and also emphasize the need for more thought and substantial research in the techniques of designing and conducting surveys.

In the earlier period of the development of the theory of sample surveys, attention was focused primarily on problems of the measurement and control of sampling errors, subject to conditions of cost and other administrative considerations. This included examining alternatives and identifying and making effective use of resources. More recently, there has been increasing emphasis on the effects of nonsampling errors,* variously termed measurement or response errors, reflecting the effects of such factors as variation in the interpretation of questionnaires, inability or unwillingness to provide "good" information on the part of respondents, mistakes in the recording or the coding of data obtained, and so forth. (See, for example, [1].) The problem of designing surveys includes taking account of both sampling and nonsampling aspects, and is essentially concerned with discovering ways in which available resources can best be used to achieve the goals for which surveys are undertaken. In this light, incidentally, a complete census is not to be distinguished from a sample survey, of which it is simply a special case.

Although there is no justification for complacency in the present state of the theory of sampling errors, that theory is in a relatively satisfactory condition.

---

* That these were not entirely neglected is shown by a paper of Deming [2] on sources of error in surveys.

Many of the important problems have been well identified and useful solutions explored. There is a body of effective techniques for attacking new problems as they arise. The literature contains a host of papers on techniques of approximating optimum sample selection and estimation, including the results of empirical studies and bases for advance speculation on the values of needed parameters. For example, Kenneth Brewer has recently prepared a manuscript in which he reviews and compares some 32 alternative techniques which have been published on the problem of selecting sampling units with varying probabilities and preparing estimates based on such samples.

The state of the theory of nonsampling errors is in considerable contrast to this. Some beginnings have been made. Models that attempt to describe the joint structure of sampling and nonsampling errors have been developed and studied. One such model developed by the staff of the Bureau of the Census and others has been employed extensively (see [4], [6], [7]). Large-scale experiments and evaluation studies were incorporated into the Population and Housing Census of 1950 and 1960 for the purpose of estimating parameters of the model (see [10], [14]). Experiments have also been conducted in Canada by the Dominion Bureau of Statistics. The results have had important influence on the substantial changes that were made in the design of the 1960 Census and the even more radical changes in the design of the 1970 Census, plans for which are now being completed. Much work remains to be done in gaining insight into the character of the nonsampling errors, for the purpose of providing control over the magnitudes of the parameters of the model. Also, research needs to be extended beyond the univariate situations that have been its principal concern, to multivariate situations.

A General Model

It may be helpful at this point, to outline the main features of the model just referred to.

We conceive, first, of some set of essential conditions under which a survey is taken, and of independent repetitions of the survey taken under those same essential conditions. The essential conditions are considered to include the design of the questionnaire, survey procedures, the types of qualifications of personnel for various activities, their training, the control operations, etc.

We consider that there is an unknown quantity U which we should like to estimate from our census or survey, and let x denote the survey estimate of U. The measure may be an aggregate, an average, a difference, a ratio, a regression coefficient, or another kind of measure. Thus, we might take a census or a sample survey of a population and prepare from it an estimate of the number of persons that receive incomes within a specified range, or the difference between the average incomes for specified subclasses of the population. A series of surveys may produce various time series.

The mean square error (mse) of a survey estimate can be analyzed in the form

$$
\begin{aligned}
E(x-U)^2 = &\ \text{Sampling variance} \\
&+ \text{nonsampling-error variance} \\
&+ \text{covariance of sampling and nonsampling errors} \\
&+ (\text{Bias})^2
\end{aligned}
$$

We sometimes add another term to the mse, that measures the relevance of the measure, U, to the specific purposes for which we apply the survey results.

The sampling variance term depends on the design of the sample, and can be analyzed into appropriate components of variance. Values for these components, estimated from empirical studies and prior experience, along with the appropriate theory and ingenuity in identifying and using these and various other resources, serve to guide the design of efficient sample surveys. The sampling variance and covariance terms are, of course, zero in the case of a complete census.

The nonsampling-error variance arises from the variations, other than those attributable to sampling, that would occur in repetitions of a survey taken under the same essential conditions. Particularly important are contributions that arise through correlated response errors, for example, within the work of an interviewer, or a supervisor, or a coder. The response variance term has proved to be quite important in some of our work, and its measurement has led to drastic changes in the technique of conducting the Census of Population and Housing.

The covariance of sampling and nonsampling errors is not necessarily zero, but we have usually presumed it to be small and have not given it much attention.*

The bias term has been exceedingly important in many of our studies. It can also be exceedingly difficult to evaluate — at least this is true for that part of the bias arising from measurement errors. The sampling biases may be small (although this is sometimes far from true) and in such case the bias is primarily a result of nonsampling errors. Since the values we seek are usually unknown, and unknowable, the actual bias must be unknown to us. We do not know how to obtain measurements without error. But we can learn something about measurement bias by comparing the average effects of <u>alternative</u> measurement procedures. If we are willing to regard some procedure as providing a standard, we obtain a measure of bias. We often assume that this bias is independent of sample size. However, this is not necessarily true since, for example, the use of a complete census rather than a sample may alter some of the essential conditions. We have found the bias component to be exceedingly important in our surveys, and consequently have given a great deal of attention to measuring

---

* Some work has been done on this component by Ivan Fellegi [4] and further research would be desirable.

and controlling it in some of our work. Much remains to be done on the measurement and control of response bias.

The remainder of this paper is primarily concerned with problems involved in two major activities of the Bureau of the Census: the Decennial Census of Population and Housing and the Current Population Survey. These problems illustrate both advances that have been made and difficulties that remain to be explored and resolved. Although the emphasis of the Bureau has been on the search for solutions to urgent unsettled questions relating to the design and execution of its surveys, this has led to supporting more basic research efforts devoted to attaining the deeper understanding of the problems and their implications required for progress towards answers to the more obvious questions.

The Design of the 1970 Censuses

The methods and procedures of the 1970 Censuses of Population and Housing, and some of the analytical tools that will be made available, represent a revolutionary change from the methods in use in the 1950 and earlier censuses, and a substantial change even from the 1960 Censuses when a number of major steps had already been taken. Some of these changes relate to aspects of the total systems design that illustrate the broad issues considered in survey design, going somewhat beyond issues closely related to sampling. The highlights of these developments will be briefly summarized, and a few immediately related to the principles of survey design will be discussed more fully, in the following remarks.

(1) Role of Sampling. Sampling will be used to collect most of the information in the census [13]. In the Population Census the complete census (100 percent) questions will consist only of a listing of the population by age, sex, race, relationship and marital

status.* Other questions will be collected only for a sample. These questions will include information on place of birth, occupation, industry, work status, education, migration, place of work, fertility, income, many characteristics of housing, and other subjects. Most of these are the more difficult and costly items of information to obtain in a census, and a number of them involve manual coding.

We began with a limited amount of sampling in the 1940 Census and extended it in 1950. Sampling at the full level planned for 1970 was used in the 1960 Censuses. Later we shall discuss the philosophy for the use, and the extent of use, of sampling in the censuses.

(2) Census by Mail. Much of the census will be taken by mail in many parts of the United States. The census questionnaires will be mailed to addresses rather than to names. Householders will be requested to fill out the questionnaires and mail them back. Seventy-five percent of the households will receive the short questionnaire, containing only the "100%" census questions, and 25 percent will receive a much longer questionnaire, containing the sample questions as well as the complete census questions.

A control register of the addresses will be prepared and maintained, and interviewers will visit and collect the information from nonrespondents. They will also edit the returns received by mail and follow-up (usually by telephone) those who return unacceptable forms. On the basis of our tests we expect a mail response rate of about 85 percent for both the long and short forms. Of the forms returned by mail, we expect that about 80 percent of the short forms and perhaps 40 percent of the long forms will be acceptable, although many of the unacceptable forms will have only a few omissions of items that will be followed up by telephone.

---

* Some housing items will also be addressed to all households.

The Post Office Department will be an important participant in the census. The mail carriers, most of whom are well acquainted with their own territory, will attempt to identify and add any missing addresses to the list of addresses prepared by the Census. Thus we will take advantage of a major resource that was untapped in earlier censuses. We have found that with an enumerator canvass a small fraction of residential addresses is completely missed, especially in the larger cities and in rural areas, and that the participation of the Post Office can substantially reduce this source of incompleteness in a census.

The mail census was introduced in part to reduce costs, but the cost reductions are relatively small; the major gains are improvements in quality of coverage, obtained from Post Office participation, and improvement in quality of response obtained from the process of self-enumeration, as distinguished from direct interviewing by enumerators. We can give intuitive reasons for the improved quality of response, but on the other hand, we can also give intuitive reasons why an enumerator canvass should produce better results. Our research generally has shown that intuition and informal experience do not suffice, but that decisions should be guided by research and measurement rather than by intuition or untested inferences from past experiences, or from deductive approaches. Later we shall discuss the principles and empirical studies that pointed to the advantages of self-enumeration.

(3) Evaluation and Research Programs. We have conducted a continuing program of research, evaluation, and experimental studies of census methods as a part of the censuses, and in the inter-censal periods. These are reported, in part, in [3], [8], [10] and [14]. The studies have included small and large-scale experiments with alternative methods, reinterview studies conducted under conditions similar to those of the census, and intensive reinterview studies conducted under as nearly ideal conditions as we can reasonably create through more careful selection, training, and supervision of staff, and by using intensive questionnaires and procedures. They have also included name-by-name matching studies, matching individual census records with prior census returns, with the Current Population Survey returns, and with various administrative records.

The whole program has generally been guided by the error model we summarized earlier. The studies have taken the form of measuring coverage errors and content errors, (i.e., response errors within the returns we have obtained), and attempting to ascertain the principal sources of these. Thus, we have undertaken to estimate major components of the mean square error in the census results.

The measurement and control of bias has proved to be particularly important, as well as difficult. The studies of bias have involved developing summary measures from intensive reinterview studies, comparisons of measurements from a census with measurements from other sources, and extensive analytical studies. Studies of response variance have involved both reinterview and matching studies, and randomized experiments.

Large-scale experiments incorporated in the 1950 Census, and inter-censal studies, had a profound effect on the procedures adopted for the 1960 Census. In one of the 1950 Census experiments, a set of experimental interviewer assignment areas was designated. In each of these areas, each of several interviewers was given two or more randomly selected assignments, in contrast to the procedure employed through the rest of the U.S., where interviewers were assigned on the basis of administrative convenience. These experiments provided a basis for making estimates of variance components, between and within interviewers, and of the intraclass correlation of response errors within the work of interviewers. Let σ_{id}^2 be the variance of repeated independent measurements, under the conditions of the survey, on the i-th elementary unit of the population; and let σ_d^2, the simple response variance, be the average of the σ_{id}^2.

In general, the response errors may be correlated, and in particular, they may be correlated within the work of an interviewer. It is easily shown that if m is the number of enumerators in an area, $\bar{n}$ is the number of persons in each enumerator's assignment, and ρ is the intraclass correlation among response errors within the work of an interviewer, then the response variance component may be written in a form that displays the effect of correlated measurement errors within interviewers, namely

$$\frac{\sigma_d^2}{m\bar{n}} \{1 + \rho(\bar{n}-1)\} \ .$$

Estimates from our evaluation studies have indicated that for many of the more difficult items in a census, items like occupation, industry, work status, income, education, and such, in a complete census taken by an enumerator canvass (the conditions of the 1950 Census) the response variances were such that the variability of the complete census estimates were as large, on the average, as if only a 25-percent sample had been taken in the absence of nonsampling variability. Also, the response bias studies pointed to biases of a substantial order, and the joint contributions of bias and response variance were such that the amount lost by collecting such items for a sample of the order of 25 percent instead of a complete census was far less than one might assume — even for very small areas. For large areas (and consequently large samples, since the sampling fraction was uniformly 25 percent) any sampling variance contribution becomes trivial. Such evidence and arguments, plus the possible economies, and the feasibility of substantially increasing the timeliness of the census results, led to the adoption in 1960 of a 25-percent sample of households as the basic procedure for obtaining the census information for most difficult-to-measure items. The sample estimates were needed for small areas such as census tracts or minor civil divisions — areas that typically have populations of perhaps 1,000 to 5,000 people.

The results of the 1950 experiments led to even further changes in census procedures. For a typical difficult item we estimated that the intraclass correlation (within interviewers) was of the order of .03, which seems small enough, but when the average size of interviewer assignment is approximately $\bar{n}$ = 700 people, the term $\{1 + \rho(\bar{n}-1)\}$ becomes a factor of more than 20. The simple response variance, σ_d^2, itself was moderately large for many of the difficult items, and when multiplied by the factor of 20 the result was the very substantial contribution to variance that we have indicated.

The mathematical model suggested the desirability of a method of data collection which would reduce the effect of the factor $1 + \rho(\bar{n}-1)$. One approach is to use self-enumeration, in which respondent households would independently fill out their own questionnaires. Extensive experimental studies with self-enumeration indicated that response bias would not be increased and response variance could be greatly reduced. The consequence of the extensive studies and research was the introduction of a self-enumeration procedure (with the questionnaires left by the enumerators in the 1960 Censuses instead of delivered by mail) for the sample questions.

Since it was necessary for an interviewer to complete the 1960 Census questionnaire for those households that did not mail a completed form, or whose questionnaires were incomplete or inconsistent, the 1960 procedure also gave rise to an interviewer contribution to the variance. To measure that variance, experiments similar to those used in 1950 were also conducted as a part of the 1960 Census. Analysis of the results of these experiments and those of other intensive evaluation studies indicated that both the response variances and the biases were substantially reduced. On the average, for difficult items, the response variance was estimated to have been reduced to about one-third of its magnitude in 1950. Thus there was a substantial pay-off from the new procedures. Presumably this was the result of a complex of factors, including improved questionnaire design, better quality control, and other factors, but also was accomplished in part as a consequence of a system that involved self-enumeration.

Additional studies, experiments, and analyses during and after the 1960 Census finally led to the plans for taking a census by mail in 1970.

These steps should yield important gains in accuracy of census results, but it must be emphasized that they are not enough. Our evaluation studies indicate serious problems of undercoverage in the census in 1960, even after substantial steps were taken to improve coverage. Similarly, the levels of response variance were non-trivial, and response bias remains a major problem. For 1970 the new methods will yield gains, but still the coverage and measurement problems presumably will be our primary source of concern.

The Canadians have verified and extended some of our results [4]. Much more needs to be done. In part this is an appeal for broader participation in the development of appropriate models, theory, and empirical studies, to advance on the task of providing vastly improved information needed in dealing with the problems of economic and social development with which we are concerned in our respective nations.

The Current Population Survey

The Current Population Survey ([9], [11], [12]) is a monthly survey of the population of the United States based on a rotating sample. About 50,000 households are in the sample each month. The survey obtains responses for the members of the sampled households to a standard set of labor force and related questions, and estimates are prepared. In addition to the labor force information obtained monthly, the survey employs supplemental questions to obtain information of varying content, including more detailed labor force data as well as data on education, income, housing, fertility, recreation, smoking habits, health, and a wide variety of other subjects. It is used, also, for drawing subsamples for special surveys (some intensive and some not) of youth, the aged, and other special groups. More recently a separate sample has been identified from the same primary and second stage units — but a different subsample of households — for a quarterly survey of expenditures for home maintenance and repair, and of consumer inventories, purchases, and intentions to purchase.

Thus, the sample survey design is a multi-purpose one, with some of the important goals identified in advance. In the design, we try to take into account both sampling and nonsampling errors, and costs, as well as the need for rapid collection and publication of results.

The general design is a multi-stage sample, utilizing the principle of defining heterogeneous primary sampling units, grouping these into strata, and sampling them with varying probabilities, with considerations of optimization determining these and other aspects of design. The primary sampling units (psu's) are individual counties, or sometimes two or more adjoining counties, with 449 distinct psu's included in the sample. The 112 largest psu's including the major metropolitan areas of the country, are included in the sample with certainty.

The procedure for sampling within primary sampling units varies depending on the nature of the resources available. However, relatively small areal units of about 200 households are, with certain exceptions, the second-stage sampling units. Clusters of housing units or other living quarters are sampled within the selected areal units. The occupants of the specified set of dwelling units are included in the sample, so that the sample reflects population shifts as they take place. Of course, the design is such that newly created quarters have their appropriate chance of inclusion in the sample as they are created.

The final sampling units selected within the second-stage areal sampling units contain approximately six housing units. We refer to these clusters of approximately six housing units as segments. Only one such segment is included in the sample in a given month from a particular second-stage unit. However, the sample of segments is related over time, with a new segment brought into the sample from the same second-stage unit, as the sample is rotated, to replace a segment that has been in use. Ultimately the second-stage units, too, are replaced.

Sample Rotation. The approximately 9,000 segments (containing about 50,000 households) included in the sample are divided into eight subsamples of approximately 1,100 segments each, referred to as rotation groups or alternatively, as panels. Each rotation group is itself a national sample of more than 6,000 households spread over the 449 primary sampling units. The rotation of the sample involves each rotation group being in the survey for four consecutive months, out of the survey for eight consecutive months, then back in the survey for the same four consecutive calendar months as before, but one year later. Thus, of the eight rotation groups in the sample in any particular month, one is the sample for the first time, a second is in the sample for the second time, and so on through the eighth rotation group which is in the sample for its eighth and last time. The rotation pattern is such that at any given month, six of the current rotation groups were also in the sample the previous month, while two rotation groups were not in the sample in the previous month. Four of the current rotation groups were also in the sample in the same month a year earlier, while the other four were not.

Estimation procedures. Estimates are made from the survey for a great many different items, including estimates of summary measures and detailed interrelations at a point in time as well as monthly time series. Special interest also centers in changes over various intervals of time, and in averages over several months or a year or more.

The estimation procedure involves two initial stages of ratio estimation. At the first stage, certain related supplementary information from the prior census is used. The second-stage ratio estimates make use of independent current estimates of the total population by age, sex, and race. This two-stage ratio estimate process produces an estimate x'_u, for month u, which is incorporated into a composite estimator, x''_u, of the general form

$$x''_u = w[x''_{u-1} + x'_{Iu} - x'_{I(u-1)}] + (1-w)x'_u.$$

In this expression w is a weight satisfying the condition

$$o < w < 1 ,$$

x'_u denotes the two-stage ratio estimate of the number of people having a particular characteristic, based on the whole sample surveyed at time u; x'_{Iu} denotes the same form of two-stage ratio estimate for month u, but based only on those 6 rotation groups that are in the same in both month u and month u-1; $x'_{I(u-1)}$ denotes the same form of estimate for month u-1 based on the same 6 rotation groups; and x''_{u-1} is the composite estimate established for month u-1.

Principles that guide choice of sample design. Many facets of the design are based on considerations of optimum sample survey design, that is, maximizing the information obtained per unit of cost, but subject to fast time schedules and certain other administrative restrictions, and controlling of measurement error and other sources of error, as well as sampling error to the best of our ability. Thus, the method of defining the primary sampling units and the sampling units at subsequent stages, the workload of the interviewer, the size of the ultimate sample from a primary unit, the definition of the primary strata, the selection of primary units with varying probabilities, the estimation procedure, and other elements of the design all involve considerations of optimization. In this paper only a few facets of the design will be selected for very brief discussions of the kinds of considerations involved, and of some of the problems that remain.

We shall mention some of the advances that have been introduced in the development of this design, but primarily we shall focus attention on difficulties and unresolved problems.

A number of new aspects of theory were developed and applied in this design, including the concept of varying the probability of selection of units as a means of reducing sampling variance, certain optimum stratification principles, the sample rotation and composite estimation procedure, and others. Extensive empirical work has been done on choosing the type and size of sampling units at various stages. Much attention and resources have been put on training and supervision of the interviewers in the survey, and into a quality inspection and control system on the work of interviewers. The effectiveness of the survey is so controlled that we regard it as providing more accurate measures of labor force characteristics than we obtain in the decennial census (which lacks the opportunity for selection and training of interviewers and quality controls that are introduced into this survey).

Problems in the use of composite estimators

The sample rotation scheme and the form of the estimator were also chosen with a view to optimization. Unfortunately, there is no adequate theory for the choice of an optimum rotation pattern under the cost and operating conditions of the Current Population Survey. However, we have achieved some reduction in variance of estimates of month-to-month and year-to-year changes by the use of the rotation design that has been introduced, which involves substantial overlap of the samples from month to month and year to year.

The composite estimator, given above, takes explicit advantage of the overlap in sample in adjoining months, but does not do so for the year-to-year overlap. Novertheless the sample rotation pattern itself implicitly gives some worthwhile gains in measuring year-to-year changes.

A uniform value of the weight w is used for all statistics, with $w = 0.5$. This value gives moderate variance reductions for a number of estimates, as compared with the variance of the 2-stage ratio estimator, and little or no increase in variance for others. We have done some work (see [5], [15]) on optimum sample rotation, on adding additional terms to the composite estimate, and on optimum values of w for various estimates, but have not moved to more nearly optimum estimates for two reasons. The first reason is that in the CPS the correlations from year to year and from month to month are low enough that the gains are not great and thus not worth accepting certain disadvantages that go with them. The second is that the nonsampling errors, which we shall discuss shortly, create complications that we have not been able to resolve.*

---

* The situation is quite different in our retail trade survey in which we use a different pattern of monthly rotation of the sample, and in which we do obtain striking gains from the rotation pattern and the composite estimator — reducing the variances of many important items to ½ or ¼ of what they would be for the simpler ratio estimator. See [16].

It should be noted that theory and extensive empirical studies have guided the design and improvement of the CPS over the years. New theory is sometimes added to guide changes in design, and the research and development program has resulted in substantial gains in information per unit of cost. For the estimation of unemployment and some other labor force characteristics, in a 15-year period the information provided per unit of cost (in equivalent dollars) was doubled as the consequence of the introduction of various advances, even though the initial design was relatively sophisticated.

Some additional problems

We have indicated that much is done towards achieving an optimum total approach, and towards a balanced total system, with appropriate expenditure of resources and efforts at various stages. But we should emphasize that we do not have a body of theory to cover the total system, or even many specific aspects of it, even from a sampling viewpoint. Still, only a small part of our current efforts, and of our efforts in recent years, have been directed at extending the sampling theory and developing models that might describe the total systems. Most of our efforts have been directed at what our research and evaluation studies show to be our primary problems, namely the improved identification of sources of response or measurement errors, especially biases, in the total system and the means of controlling them.

We may illustrate by means of two of the many problems in this area.

One of these problems is that the results obtained in a survey often are a function of the exposure of the units in the survey to the survey process itself. Consequently a repetition of a survey on the same units (or a repetition on some of the same units) at different times, may provide seriously biases measures of changes. Similarly, estimates of the level of an aggregate in a particular one-time survey may be seriously affected.

The problem becomes evident (and has additional

facets) when the sample consists of a set of rotating panels. Thus, we find in almost every area where we use rotating panels, each of which is a random sample of the same population of units (people, households, establishments, etc.) that statistics based on different panels have different expected values, depending on the amount of exposure of the panel. We have noted above that in the Current Population Survey eight different panels or samples are used in each monthly survey; one panel is in the survey for the first time, a second panel has been in the survey once before and is included for the second time, a third panel is included for the third time, etc. The levels of unemployment estimated differ among the eight panels that are used in any one monthly survey, in a way that we do not adequately understand. The differences themselves vary over time, so that recently we have observed a range twice as great as the range that had been observed earlier. Also, the pattern varies strikingly for various classes of the population. Research efforts to date have not brought explanations or verifiable hypotheses concerning causal factors. These panel (or rotation group) differences cause serious difficulties in certain types of analyses. Broader participation in research in this area might bring both improved understanding of the phenomenon and more reliable and meaningful measurements.

A second illustration of a problem, which may well be related to the first illustration, is the observed high level of gross errors in presumably well-controlled efforts at measurement of apparently simple facts in censuses or surveys. This is the manifestation of a response or measurement variance, which in itself is not surprising, but its relatively large magnitude in many social science measurements is not well understood. Major questions for discussion are how to understand the essential conditions under which the variance arises, and how to bring such measurements under better control. The response variance may seriously affect and distort certain types of measurements, although it may have relatively little effect on others. For example, turnover among the unemployed, as measured by the CPS, is grossly exaggerated simply because of the variance of response. We have developed models to reflect and attempt to correct for such response variances in simple circumstances, but thus far they seem inadequate in dealing with multiple classifications.

We look for broader participation in research on such problems. Participation should include the extension and confirmation of measurements, as well as developing theory and models and hypotheses, and testing them.

We indicated earlier that the CPS is a multipurpose survey, intended to collect various labor force measurements, and that it provides a basis for both analytical studies and descriptive measures. It has been used for a wide range of special studies for various groups of the population, and on an exceedingly wide variety of subjects. We believe that this survey and other surveys of this type have great utility as vehicles for such <u>multi-purpose</u> use. We have successfully used the survey for limited studies of agriculture, identifying agricultural households within the CPS sample, and supplementing the sample for certain types of households. Similarly, we have successfully used it for certain types of surveys directed at business establishments. But these latter types of surveys, while well served for limited and occasional one-time use, ordinarily call for at least different samples of second-stage units within the same primary units and often for a different general approach to sample survey design.

The exceedingly wide range of uses of the CPS has made it one of the greatest statistical resources of the country. It has great flexibilities in providing samples of special groups that are identified in ongoing surveys, perhaps supplemented in various ways. We often draw samples of a class of the population from administrative records, for example, of the aged from retired and Medicare records, or of disabled veterans from the veterans records, and get general samples of eligible populations from the CPS and supplements for comparative studies. Often exceedingly extensive studies are involved, with mail and personal interviews used jointly, and a variety of other methods.

Also the survey, through its regular rotation of

panels, provides the possibilities of longitudinal studies. It makes possible comparisons for individuals of intentions to buy, for example, with actual purchases in an interview a year later, and measuring month-to-month changes for identicals in labor force participation, or year-to-year changes in income, and related studies. Longer term longitudinal studies have been designed.

These wide ranging uses are of great benefit to a nation in learning things about its programs, their operations, and is increasingly being taken advantage of.

There has been consideration of multi-purpose versus special-purpose surveys in international discussions. We believe this is an illustration of a proven and highly effective demonstration of a multi-purpose (but not of an all-purpose) survey.

Concluding Remarks

In this paper we have indicated some of the significant developments that have taken place in the design and execution of surveys and have pointed to areas in which critical problems demand more attention. The most important of the problems are those involving nonsampling errors. The problems should be considered in the context of the total design of a survey: the identification and effective use of resources, the design of the sample, the structure of the data-collection organization, the formulation of appropriate interviewing instruments, the control of operations, the construction of efficient and useful estimators of population parameters and the components of the mean square errors of estimates, and the exploration of alternatives. The existing theoretical structure for the entire survey system, whether it uses a sample or a census, is far from being adequate for guiding us to an optimum choice of a survey design. In the Appendix we list some problems, of both quite general and fairly specific natures, which present a challenge for research in survey methodology and whose solutions would contribute to the improvement of surveys.

Appendix

Some Important Research Problems in Survey Methodology

1. Survey models. There is a need for theoretical models that will provide a basis for the optimization of the whole survey design and that take into account the sources of error that include the respondent, the observers or interviewers, the organizational structure of the collection machinery, the questionnaire or other recording device, the coding and editing procedures employed, the imputation methods used for faulty or missing data, and the estimators or analytical techniques that are implied, as well as the design of the sample.

2. Standards. How may standards be set for levels of response error, degree of editing, and amount and accuracy of imputation for inconsistencies and non-response, taking into account the needs for data and the available resources?

3. Response error effects on multivariate analysis. In what manner does response error in data affect the various types of analyses that depend upon cross-tabulations and other functions of several statistical variates?

4. Record linkage. Records from several sources may need to be linked for the purpose of constituting a sampling frame, or to assemble information, or for other purposes. Further development of existing models is needed, particularly emphasizing specific goals of linkage.

5. Memory bias. The nature and the means of controlling biases of various types in retrospective surveys need to be studied.

6. Redundancy. The deliberate introduction of redundancy in questionnaires, in coding, and in other operations might be a means for controlling nonsampling errors arising from variation in response, coding errors, and so forth.

7. Estimation on successive occasions. The optimum design of samples and estimators for estimation on successive occasions has received considerable attention, which has however been confined to the control of sampling errors only. This work needs to be extended to take account of nonsampling biases and random errors, such as the phenomenon of rotation group bias.

8. Estimation for units not sampled. A problem that recurs fairly frequently is that of estimating the characteristics of each unit of a population based on direct observation of only a sample of those units plus collateral information available for all. Investigation is needed to suggest new ways of attacking this problem, or sharpening or modifying the approaches that have been suggested.

9. Sample-based choice of estimator. In general, this is the problem of choosing one or another estimator on the basis of statistics calculated from the sample. For example, it may be proposed to use one estimate or another of the population total, depending upon which of the estimators has the smaller estimated mean square error. What are the properties of the resulting estimator?

10. Role of assumptions in sampling. Assumptions may enter into the choice of sample selection or estimation procedures. A common procedure (the "cutoff method") is to confine sampling to a portion of the universe of interest. How sensitive is the usefulness of sample results to the errors in the assumptions? Bayesian theory may have a role here.

11. <u>Utility of statistical information</u>. Considerable research needs to be done on useful ways to measure the value of statistical information and the risks involved in using imperfect information, particularly in the context of manifold uses of the same information.

12. <u>Information theory in statistical processing</u>. Mathematical information theory may be able to make important contributions to the effectiveness and efficiency of data-processing operations, such as the specification of codes, sorting methods, editing, computer programming, and similar operations.

13. <u>Estimation of distributions</u>. The reference here is to problems in which the characteristic according to which the units are to be distributed is estimated from a sample, and either all or a sample of the units are observed. This is closely related to estimating the distribution when the observations are subject to response error.

14. <u>Treatment of non-response</u>. The central problems in this area are those of devising means to control the magnitude of non-response, of developing detailed models that reflect the influence of non-response, and of constructing estimators whose bias is small despite the presence of non-response.

15. <u>Quality control of survey operations</u>. A survey is a chain of operations extending from the collection of information to its publication. Various degrees of control can be exerted on the quality of each link in the chain. Investigation is needed of the optimum balance of effort.

REFERENCES

[1] Dalenius, T. "Recent advances in sample survey theory and methods." Annals of Mathematical Statistics, 33 (1962).

[2] Deming, W. E. "On errors in surveys." American Sociological Review, Vol. 9 (1944).

[3] Eckler, A. R. and Hurwitz, W. N. "Response variance and biases in censuses and surveys." Bull. ISI, Vol. 36, part 2 (1957).

[4] Fellegi, I. P. "Response variance and its estimation." Journal of the American Statistical Association, 59 (1964).

[5] Gurney, M. and Daly, J. F. "A multivariate approach to estimation in periodic sample surveys." Proceedings of the Social Statistics Section of the American Statistical Association (1965).

[6] Hansen, M. H., Hurwitz, W. N., and Bershad, M. A. "Measurement Errors in Censuses and Surveys." Bull. ISI, Vol. 38, part 2 (1961).

[7] Hansen, M. H., Hurwitz, W. N., and Pritzker, L. "The Estimation and Interpretation of Gross Differences and the Simple Response Variance." Contributions to Statistics Presented to Professor P. C. Mahalanobis. Pergamon Press, Oxford (1964).

[8] Taeuber, C. and Hansen, M. H. "A preliminary evaluation of the 1960 Census of Population and Housing." Demography, Vol. 1 (1964).

[9] U. S. Bureau of the Census. Concepts and Methods used in Manpower Statistics from the Current Population Survey. Current Population Reports, Series P-23, No. 22 (1967). Also issued by the Bureau of Labor Statistics as BLS Report No. 313.

[10] U. S. Bureau of the Census. The Post-Enumeration Survey: 1950. Technical Paper No. 4 (1960).

[11] U. S. Bureau of the Census. The Current Population Survey Reinterview Program, Some Notes and Discussion. Technical Paper No. 6 (1963).

[12] U. S. Bureau of the Census. The Current Population Survey — A Report on Methodology. Technical Paper No. 7 (1963).

[13] U. S. Bureau of the Census. Sampling Applications in Censuses of Population and Housing. Technical Paper No. 13 (1965).

[14] U. S. Bureau of the Census, Evaluation and Research Program of the U. S. Censuses of Population and Housing, 1960: Series ER60, Nos. 1-6.

[15] Waksberg, J. and Pearl, R. B. "The Current Population Survey: A Case History in Panel Observations." Proceedings of the Social Statistics Section of the American Statistical Association (1964).

[16] Woodruff, R. S. "The use of rotating samples in the Census Bureau's monthly surveys." Journal of the American Statistical Association, 58 (1963).

SOME ASPECTS OF THE THEORETICAL DEVELOPMENTS IN SURVEY - SAMPLING

V. P. GODAMBE

UNIVERSITY OF WATERLOO

Summary

This, primarily is an expository paper. In sections 1 and 2, is given, a brief historical ac ount of the development of the subject up to and including Newman's contribution and works along his line of thought. The section 3 introduces the new model and explains the conventional U.M.V. estimation. In section 4, the likelihood function for the new model is defined. Its relationship with the "conventional" likelihood function is explained. In sections 5 and 6, the Fiducial and Bayes approaches, in relation to survey-sampling, are investigated. The section 7 gives the assessment of the new model, in relation to the survey-practice. The subsequent sections review some relevant works of Dempster and others.

1. The Back-Ground: The Hypothetical Population Model.

As is commonly understood the theory of probability originated in the 16th century, with the study of the chance phenomena associated with the gambling instruments such as a roulette or a die. The outcome of the random experiment such as turning a roulette or throwing a die was a well defined event. It gradually became apparent that the theory of probability had some basic connection with the possible outcomes of the repetition of the same random experiment, over and over again. For example one might imagine the outcomes (head or tail) of infinite tosses made with the same coin. Thus, speaking very broadly, a hypothetical population of the outcomes obtained by infinite repetition of a random experiment, came to be regarded, implicitly at any rate, a possible object of investigation for the theory of probability. This, obviously one can say even after taking into account the basic differences, which all along persisted, among the probabilists, concerning the definition of probability. Now the concept of hypothetical population and its relationships with the theory of probability played a central role in the development of statistical theory from its very beginning.

It is generally believed that, statistics or statistical theory originated with the investigations of biological and socialogical phenomena such as inheritance and the like, during the last century. Soon the statistical theory was directed towards studying the relationships between different factors that influenced those phenomena and to study the chance mechanism underlying them. For instance the sizes of a dozen human skulls were supposed to have been produced by some kind of chance mechanism operating in the background. It was then but a small step to replace the chance mechanism by a hypothetical population generated by the repeated applications of the chance mechanism. Thus the dozen observed skulls, were supposed to be a random sample from some hypothetical population. It was the job of the statistician to study this hypothetical population on the basis of the given random sample, with the help of the tools provided by the classical theory of probability. This, one may say resulted over years into a mathematical theory of statistics or briefly theoretical statistics. The point to be emphasized here is this: The general theoretical statistics developed by Galton, Pearson, Fisher, etc. took for granted the model of hypothetical* populations which evolved along with the development of the classical theory of probability: The given set of observations (sample) was assumed to have been drawn at random from some hypothetical population.

With the introduction of the artificial randomisation, which implied use of devices like random num-

*Fisher often in his writings, has stressed the hypothetical nature of the populations dealt with in theoretical statistics.

ber tables to draw a random sample, first in the Design of Experiments (by Fisher) and then in Survey-Sampling during 1920's (?) a logically very different situation than the one described in the preceeding paragraph, emerged. The logical difference between the situations about which we would speak shortly, was however understoo very slowly and the failure to recognize this difference explicitly, restricting ourselves to survey-sampling, we would say, caused considerable confusion in the theoretical development of the subject. This was the difference: The artificial randomisation or briefly randomisation is usually carried out by means of the devices such as random number tables. Now any such device could be implemented to draw a random sample from a population only if the individuals of the population are in some way labelled. Thus in contrast with the hypothetical populations presupposed by the general theoretical statistics, as developed by Galton Pearson, Fisher, etc., randomisation presupposed as actual population i.e. a population whose individuals bear respective name tags or labels. Such actual populations would of course be finite. One may note that the survey-populations, (for example, population of all families in a town), are actual populations. The general theoretical statistics often addressed itself to the problem of inference about some unknown parameter of the otherwise (assumed to be) known frequency function of some hypothetical population. In reality, as said before, the problem was of inferring about the possible chance mechanism that produced the given set of observations; and the reference to the (hypothetical) population was only incidental. In contrast when with a specified randomisation device such a random number tables, a random sample is drawn from an actual population with labelled individuals, the problem is not of inferring about the randomisation or chance device, (which is completely specified anyway), but the problem is of inferring about the actual populations on the bas s of the random sample, drawn from it. One may for instance like to infer about the average income of a given population of 10,000 families on the basis of a random sample of say 100 families (and of course their respective incomes) drawn from the population.

Thus the two inference problems, one concerning the actual population, are fundamentally different.

But this fundamental difference, as said earlier was not understood for quite a long time. One for instance would argue as follows: When we try to infer about the chance of "head" turning up on a certain coin being tossed, on the basis of the results of say 10 previous tosses, after knowing how many out of 10 tosses resulted into heads, whether the 5th or the 7th toss resulted into head contained no relevant informa ion. This is quite intuitive. Besides there are some formal arguments in theoretical statistics to support it. If so, while estimating the proportion of white balls in a bag containing black and white balls, on the basis of 10 balls drawn from it at random, after knowing how many of them are white, does the knowledge, which of the balls are actually white, contain any relevant information? Answer: it may. For suppose the balls in the bag are labelled before the sample is drawn. Then in 10 random draws made with replacement, which of the balls, because of the labels we know, have been drawn more than once. Surely now, even intuitively proportion of white balls among the distinct draws provides a better estimate than the proportion among all the draws, for the proportion of white balls in the bag. Actually though both are unbiased estimates, the proportion among the distinct draws has smaller variance (Basu (1958), Des Raj and Khamis (1958)). Now, t e above illustration is defective insofar as it gives the impression that the individual labels are significant only when sampling is done with replacement. That this is not the case, would be evident from the next section where we have reviewed the works on estimation, in survey-sampling. In fact it would be clear from the next section that the search for an unbiased minimum variance estimate for the survey-population mean, for the first time suggested that the classical hypothetical-population model was fundamentally inadquate for survey-sampling which presupposes actual populations.

2. *Neyman's Approach. An Attempt to Fit Survey-Sampling in the Hypothetical-Population Model.*

In the first major work on survey-sampling, Neyman (1934) introduced the then novel technique of Markov's Theorem (subsequently rightly called Gauss-Markov Theorem) to obtain a linear unbiased least variance estimate for the mean of a survey-population. He *apparently* established that for simple random sampling the sample mean was the linear unbiased least variance estimate of the survey-population mean.* He then similarly investigated stratified random sampling, optimum allocation, etc. The success of Neyman's work appeared so overwhelming that for several years to follow sample-survey statisticians often concerned themselves with finding a linear unbiased least variance estimate, with the help of the Gauss-Markov Theorem, for different situations. The situation at times might have been characterised by the prior knowledge of some correlation coefficient, etc. This invariably resulted in claims like the regression estimate, the ratio estimate (allowing for the small bias) ..., etc., were linear unbiased least variance estimates for the mean of the survey-population, provided "such and such" conditions were satisfied. Talk of "most efficient estimates" became common. Ample instances of all this are provided by text books on the subject. Again, elaborate sampling procedures such as sampling with arbitrary probabilities, multi-stage sampling, and so on, were put forward for reducing the variance of estimates. However, as the sample-survey statisticians started employing elaborate sampling procedures, the artificiality, complexity, and occasionally even impossibility, of the Gauss-Markov approach to estimation of survey-sampling became evident.

Horvitz and Thompson(1952) were the first to give

---

* A very interesting reference concerning the history of arithmetic mean is Plackett (1958).

a clear expression to the general uneasiness over Neyman's approach based on the Gauss-Markov Theorem. They constructed three classes of estimates for the mean of a survey-population and raised a general question concerning linear unbiased least variance estimates.

Soon (Godambe (1955)) with appropriate formalisation of the concepts of sampling design and the linear estimator, for survey-sampling, the non-existence of the uniformaly minimum variance estimation, was rigourously demonstrated. It was also shown (Godambe (1955)) that Neyman's (1934) use of the Ganss-Markov Theorem was based on the 'incorrect'* assumption that the problem of estimation in survey-sampling could be fitted into the set-up of the conventional estimation of a parameter of a hypothetical population (refer section 1), which was supposed to have some known or unknown form of the frequency function#.

3. An Appropriate Model for Survey-Sampling. The Problem of Estimation.

The author (1955) proposed a theoretical model within which most of the problems of survey-sampling could be discussed. No doubt it evolved gradually, especially through the works of Horvitz-Thompson (1952) and Midzuno (1952). The model with its suitable extensions and refinements by Koop (1957), Hájek (1958), Godambe (1965), and Godambe and Joshi (1965) incorporating in it the sufficiency-criterion of Basu (1958),

---

* A remark by R. A. Fisher during the discussion on Neyman's (1934) paper suggests that even Fisher granted this assumption

\# Of course Neyman's (1934) classic work on optimal allocation in stratified sampling is free of this criticism.

Takeuchi (1961) and Pathak (1963), is briefly as follows:

Let $i = 1,\ldots,N$ denote the individuals of a given survey-population U. One may say U is the set of integers i, $i=1,\ldots,N$. Further if x is a real variate defined on U, with $x(i) = x_i$ for $i=1,\ldots,N$, then $\underset{\sim}{x} = (x_1,\ldots,x_N)$ is a point in the Euclidean space R_N. Now, for reasons which would be clear later, we call any function of R_N a <u>parametric function</u>. The <u>parametric function</u> T is said to be the <u>population total</u> if,

$$T(\underset{\sim}{x}) = \Sigma_1^N x_i, \quad \text{for all } \underset{\sim}{x} \in R_N, \ldots, \tag{1}$$

Usually our problem would be to estimate T on the basis of individuals i sampled from the population U and the values x_i associated with them i.e. on the basis of $(s, x_i{:}i\varepsilon s)$ where s is a subset of U, $s\subset U$, drawn from U, with a given <u>sampling design</u> p:

DEFINITION 3.1. A <u>sampling design</u> is any function p on S, the set of all possible subsets s of U, such that $p(s) \geq 0$, $\Sigma p(s) = 1$, $s\varepsilon S$. It may not be difficult to see that all the sampling procedures in common use such as simple random sampling, stratification, arbitrary probability sampling etc., are special cases of <u>sampling design</u> defined above.

Now in connection with the estimation of T in (1) we introduce the following

DEFINITION 3.2. Any real function $e(s,\underset{\sim}{x})$ on $R_N x S$, (S as before being the totality of the subsets s of U), which depends on $\underset{\sim}{x}$ only through x_i, $i\varepsilon s$, is an estimator.

DEFINITION 3.3. An estimator $e(s,\underset{\sim}{x})$ is said to be <u>linear</u> if

$$e(s,\underset{\sim}{x}) = \alpha(s) + \sum_{i\varepsilon s} \beta(s,i)x_i$$

where α is a real function on S and β a real function on S×U with $\beta(x,i) = 0$ if $i \notin s$.

Now the non-existence theorem of the author (1955) can be stated as follows: In the class of all linear (Def. 3.3) unbiased estimators of T in (1), none has uniformly (in $\underset{\sim}{x}$) minimum variance, whatever the sampling design (Def. 3.1) may be, except for some unicluster designs. For exceptions we refer to Hanurav (1965). This non-existence theorem was generalized by the author and Joshi (1965) removing the restriction of linearity.

Many estimators that are employed in practice are special cases of what is called Horvitz-Thompson estimator. This is defined as follows: Let for the given sampling design (Def. 3.1)

$$\sum_{s \ni i} p(s) = \pi_i, \quad i=1,\ldots,N \qquad (2)$$

Then π_i, $i=1,\ldots,N$ are called inclusion probabilities for the sampling design p. Next we assume that the sampling design is such that $\pi_i > 0$ for all $i=1,\ldots,N$. Now Horvitz-Thompson estimator is

$$e^*(s,\underset{\sim}{x}) = \sum_{i \varepsilon s} x_i/\pi_i. \qquad \ldots\ldots \qquad (3)$$

The following properties of e* have already been established. (A) In the class of all unbiased estimators for the population total T in (1), e* is admissible with respect to the variance in all the intervals of R_N barring some exceptions, and for any sampling design with $\pi_i > 0$, $i=1,\ldots,N$ (Godambe and Joshi (1965)). For any fixed sample size design having $\pi_i > 0$, $i=1,\ldots,N$, (B) e* is admissible for T, in R_N, with respect to the squared error, in the class of all estimators. (Joshi (1965)), (C) in the class of all polynomial unbiased estimators of T in (1), e* is uniquely hyper-admissible i.e. admissible in every subset of the co-ordinates of R_N, with respect to the variance

(Hanurav (1965)), and (D) in the class of all unbiased estimators for T, in (1), e* has minimum expected variance for a class of prior distributions on R_N, defined in terms of π_i's in (2) (Godambe (1955)), Godambe and Joshi (1965)).

REMARK 3.1. It would be clear from the subsequent discussion, (sections 6, 7) with the sampling theory developed to this day, that the only way one can make some definite sense of different sampling designs (Def. 3.1), i.e. different modes of randomisation, is through the properties (C) and (D) above.

In (D) we can take the class of prior distributions which is consistent with our prior knowledge of the population U. These priors would determine the inclusion probabilities π_i, i=1,...,N, on the basis of which we can choose an appropriate sampling design p, (Def. 2.1).

REMARK 3.2. As a special case of (D) simple random sampling without replacement and sample mean are optimum, under suitable conditions. It is important to distinguish the above approach of finding optimum sampling designs and estimators by means of the property (D) from Bayesian approach (discussed in the section 6). The former approach is based on unbiased estimators[#]; and unbiased estimation has no place in Bayesian approach. In fact as it would be clear later, in section 6, it is difficult to make sense of different sampling designs (Def. 3.1), i.e. different modes of randomisation within the Bayesian frame-work.

The unbiased estimation sometimes can lead to absurd results: Consider a population of two individuals. One of the two individuals (we do not know which one) is known to have x-value 0 and remaining one has x-value non-zero. The unbiased minimum variance estimator, of the population total, based on a sample of one individual drawn with equal probability of selection to both individuals, surely gives an absurd estimate when the non-zero x-value is observed.

4. The Likelihood Function

From the discussion so far it should be clear that the total observation in survey-sampling is

$$(s,\ x_i:\ i\in s) \tag{4}$$

where s is the subset of individuals i drawn from the population U, (consisting of N individuals i = 1,...,N), with the sampling design p and the x-values associated with the different individuals $i\in s$ i.e. x_i: $i\in s$. Now suppose

$$\underset{\sim}{x}' = (x_1',\ldots,x_N'), \tag{5}$$

are the x-values associated with the different individuals i in the population U, then for the given sampling design p, (Def. 3.1), we can write the probability of (4) as

$$\text{Prob.}\ (s,\ x_i:\ i\in s \mid \underset{\sim}{x}') = \begin{cases} p(s) \text{ if } x_i = x_i' \text{ for all } i\in s \\ \\ 0 \quad \text{otherwise.} \end{cases} \tag{6}$$

Now from (6) it would be clear that in our model $\underset{\sim}{x}'$ in (5) plays the role of the <u>parameter</u> with the Euclidean space R_N as the <u>parametric space</u>. (Later on we would discuss the cases when the parametric space consists of a subset of R_N.) From (6) we can write the likelihood function ℓ, (Godambe, (1966)), for the parameter $\underset{\sim}{x}'$ given (s, x_i: $i\in s$) as follows:

$$\ell(\underset{\sim}{x}' \mid s,\ x_i:\ i\in s) = \begin{cases} p(s) \quad \text{if } \underset{\sim}{x}' \in R_N(x_i : i\in s), \\ \\ 0 \qquad \text{if } \underset{\sim}{x}' \notin R_N(x_i : i\in s) \end{cases} \tag{7}$$

where $R_N(x_i:\ i\in s) \subset R_N$ such that $\underset{\sim}{x}' \in R_N(x_i:\ i\in s)$ iff $x_i' = x_i$ for all $i\in s$. Thus (7) attaches equal (greater than zero) likelihood to all points $x' \in R_N(x_i: i\in s)$ and '0' likelihood to the points $x' \notin R_N(x_i: i\in s)$. Thus the

REMARK 4.1. The likelihood function ℓ given by (7) is independent of the sampling design. Hence the likelihood principle (Barnard, et. al (1962)) implies that the inference about $\underset{\sim}{x}$ must be independent of the sampling design.

Now we will see what a 'conventional parametrised frequency' may mean in the model introduced above. Take for instance the example of bag containing N balls, black and white, (denoting black by 1 and white by 0) the proportion of black to N being say θ. In our model this means our parametric space in (6) or (7) is no more the whole of R_N but a subset $R_{N\theta}$, $R_{N\theta} \epsilon R_N$, of $\underset{\sim}{x}'$ in (5) such that,

$$R_{N\theta} = \left\{ \underset{\sim}{x}' : \begin{array}{l} \text{(1) For } i = 1,\ldots,N \; x_i' = 0 \text{ or } 1 \\ \text{(2) (Total number of i's for which } \\ \quad x_i' = 1)/N = \theta \end{array} \right\} \qquad (8)$$

It is easy to see if any point $\underset{\sim}{x} \in R_{N\theta}$ then the point obtained by any permutation of $\underset{\sim}{x}$ would also be in $R_{N\theta}$. Conversely if $\underset{\sim}{x} \notin R_{N\theta}$, no permutation of $\underset{\sim}{x}$ can be in $R_{N\theta}$. Now if $\underset{\sim}{n}$ balls are drawn from the bag at random without replacement we have from (6)

$$\text{Prob.}(s, x_i = i\varepsilon s \mid \underset{\sim}{x}') = \begin{cases} 1/{}^N C_n & \text{if s contains n individuals and if } x_i = x_i' \text{ for all } i\varepsilon s \\ 0 & \text{otherwise} \end{cases} \qquad (9)$$

Now if we eliminate s from (9) the resulting probability would be the same for all $\underset{\sim}{x}' \in R_{N\theta}$ in (8). In fact if n_1 of x_i: $i\varepsilon s$ are equal to 1 and the rest equal to 0, then eliminating s from (9) we get

$$\text{Prob}\ (n_1) = {}^{N_1}C_{n_1} \times {}^{N-N_1}C_{n-n_1} \Big/ {}^{N}C_n, \text{ where } \frac{N_1}{N} = \theta \quad (10)$$

Next, suppose in the above illustration θ is <u>unknown</u> and we want to infer about θ on the basis of $(s, x_i: i\epsilon s)$.

QUESTION. Which of the likelihood functions*, one given by (9) or the one given by (10) is appropriate?

This question we do not answer. However in sections 5 and 7 we would show how even for a situation as symmetrical as the one given by (8) and (9), the individual lables s, play a central role in the statistical inference.

The above illustration would at once suggest how we can <u>directly</u> interprete, in our model, certain <u>distributional assumption</u> about the population. A survey-sampler might for instance assume that, in a certain population of families, income distribution has a form F_θ, where θ is an unknown parameter $(\theta\epsilon\Omega)$, F_θ being <u>completely specified</u> for every $\theta\epsilon\Omega$.# This, in our model means the following assumption: The parameter x' in (6) belongs to the subset of R_N given by

$$\bigcup_{\theta\epsilon\Omega} R_{N\theta} \quad (11)$$

---

* A likelihood function similar to (10) is given by Royall (1967).

It is important to note the approximation involved here. Most distributional assumptions in survey-practice are of <u>continuous</u> distributions such as normal or Pareto, etc., while finite populations can admit of only <u>discrete</u> distributions.

where

$$R_{N\theta} = \left\{ \underset{\sim}{x}' : \begin{array}{l} \text{The proportion of the number of indivi-} \\ \text{duals i (i = 1,...,N) for which} \\ x_i' \leq y = F_\theta(y), \ -\infty < y < \infty \end{array} \right\} \quad (12)$$

Generally we can get a likelihood function directly for θ, when the sampling design (Def. 3.1) is given by simple random sampling without replacement, by eliminating s from Prob$(s, x_i\colon i\epsilon s | \underset{\sim}{x}')$ in (9), using (12), as illustrated in (10).

5. A Fiducial Argument

Let in (12),

$$F_\theta(y) = F(y - \theta), \quad -\infty < \theta < \infty \qquad (13)$$

and

$$\int_{\infty}^{\infty} ydF(y - \theta) = \theta, \ -\infty < \theta < \infty \qquad (14)$$

In short, θ is the location parameter of F, F having it's mean value at θ itself. Actually, because of (14), for every point $\underset{\sim}{x}' \epsilon R_{N\theta}$ in (11),

$$\theta = \Sigma_1^N x_i'/N = \overline{x}_N'. \qquad (15)$$

Hence the problem of inferring about the population means $\overline{x}_N'$, on the basis of the observation $(s, x_i\colon i\epsilon s)$ in (4), is the same as the problem of inferring about the parameter θ of the otherwise known distribution function F_θ. Next we assume sampling design, (Def. 3.1), to be the one given by, simple random sampling without replacement with fixed say n, number of draws. Hence the probability Prob.$(s, x_i\colon i\epsilon s | \underset{\sim}{x}')$ would be the same as given by (9). Now putting

$$\overline{x}_s = \sum_{i\epsilon s} x_i/n \qquad (16)$$

we can write

$$(s, x_i\colon i\epsilon s) \Longleftrightarrow (s, \overline{x}_s, x_i - \overline{x}_s\colon i\epsilon s) \qquad (17)$$

Next the probability distribution of the unordered (i.e. without the individual labels i), set of values $x_i - \bar{x}_s$: $i \epsilon s$ is the same for all the parametric points $\underset{\sim}{x}'$ in (5) such that

$$\underset{\sim}{x}' \in \bigcup_{-\infty<\theta<\infty} R_{N\theta} = R_N^* \text{ say} \tag{18}$$

the latter being defined by (11), (12), (13) and (14). Hence in the usual terminology the unordered set of $x_i - \bar{x}_s$: $i \epsilon s$, which for convenience we would denote by

$$[x_i - \bar{x}_s\colon i \epsilon s] \tag{19}$$

is an ancillary statistic. Thus from (17) and (18) we have the likelihood function (ref. (7)),

$$\ell(s, x_i\colon i \epsilon s | \underset{\sim}{x}') = \text{Prob.}(s, \bar{x}_s | [x_i - \bar{x}_s\colon i \epsilon s], \underset{\sim}{x}') \tag{20}$$

for all $\underset{\sim}{x}' \in R_N^*$ in (18). Now consider the function

$$s, \bar{x}_s - \bar{x}_N' - (\bar{x}_s' - \bar{x}_N') \tag{21}$$

where $\bar{x}_s'$ is the mean of x_i': $i \epsilon s$ and $\bar{x}_N'$ as defined in (15). Clearly the conditional distribution of (21), given $x_i - \bar{x}_s$: $i \epsilon s$ in (19) is,

$$\text{Prob.}(s, \bar{x}_s - \bar{x}_N' - (\bar{x}_s' - \bar{x}_N') \mid [x_i - \bar{x}_s\colon i \epsilon s], \underset{\sim}{x}')$$

$$= 1/{}^{N}C_n \text{ iff, s contains just n individuals i and } \bar{x}_s - \bar{x}_N' - (\bar{x}_s' - \bar{x}_N') = 0 \text{ for all } x' \in R_N^* \text{ in (18).} \tag{22}$$

Thus (22) shows that the function s, $\bar{x}_s - \bar{x}_N' - (\bar{x}_s' - \bar{x}_N')$ is a pivotal (i.e. a function of the random variate and the parameter having its distribution independent of the parameter) conditional on the given $[x_i - \bar{x}_s\colon i \epsilon s]$ in (19). Moreover this pivotal, because of (20) satisfies the Hacking's (1965) criterion of irrelevance conditionally on the given $x_i - \bar{x}_s$: $i \epsilon s$. From (20) it also follows that conditionally on the given $[x_i - \bar{x}_s\colon i \epsilon s]$, s, $\bar{x}_s$ is a sufficient statistic for the parameter $\underset{\sim}{x}'$, $\underset{\sim}{x}' \in R_N^*$ in (18). Hence the pivotal in (21) is based on the conditionally sufficient statistic $(s, \bar{x}_s)$. Further the uniqueness of the pivotal satisfying the above qualities is not difficult to demonstrate in the present case. Hence the

REMARK. We may obtain the fiducial distribution of $\bar{x}_N'$, the population mean in (15), on the basis of the observation $(s, x_i: i \in s)$ in (4), by eliminating from (22), $\bar{x}_s' - \bar{x}_N'$.

The above fiducial distribution can be written more easily with the following notation. In (5), using (15) we write

$$x_i' = \bar{x}_N' + y_i, \qquad i = 1,\ldots,N \tag{23}$$

Hence we may write the population vector $\underset{\sim}{x}$ in (5) as

$$\underset{\sim}{x}' = (\bar{x}_N', \underset{\sim}{y}) \tag{24}$$

where

$$\underset{\sim}{y} = (y_1,\ldots,y_N) \tag{25}$$

as defined by (23). Now from (13), (14) and (15) it follows that for all $\underset{\sim}{x}' \in R_N^*$ in (18) the c.d.f. given by the proportion of the number of individuals i for which $y_i \leq a$ in $\underset{\sim}{y}$ is precisely $F(a)$, $-\infty < a < \infty$, F being the same as in (13). (In short, it means that, all the points $\underset{\sim}{x}' \in R_N^*$ in (18), and only those, are obtained by giving x_N' in (24) all values $(-\infty, \infty)$ and putting for y all the permutations of a fixed vector $(Y_1,\ldots,Y_N)$ determined by the c.d.f. F in (13).) Hence from (22) and the Remark above, the fiducial distribution of $\bar{x}_N'$ can be written, putting $\bar{y}_s$ for the mean of y_i, $i \in s$, as

$$\text{Prob.}(\bar{x}_s - \bar{x}_N' = \bar{y}_s \,|\, [y_i - \bar{y}_s: i \in s], \underset{\sim}{y}) \tag{26}$$

which of course is the *same* for *all* $\underset{\sim}{y}$ such that $(\overline{x}'_N, \underset{\sim}{y}) \in R^*_N$, (i.e. for all $\underset{\sim}{y}$ which are permutations of $(Y_1, \ldots, Y_N)$ referred to above).

In the next section we would obtain the fiducial distribution given by (26) as the *Bayes posterior* distribution for some specific prior distribution. In spite of this there is a *basic* difference between the *fiducial* and *Bayes* approach, as made clear in the *Remark* 6.2, in the next section. For more details we refer to Godambe (1967). Some relevant remarks are given at the end of the section 7.

It may be interesting to note that if the c.d.f. F in (13) is *approximately normal*, then in (26), $\overline{y}_s$ and $y_i - \overline{y}_s$: $i \in s$ are probabilistically independent (Basu (1955)). Hence we have

$$E(\overline{x}_s - \overline{x}'_N | \underset{\sim}{x}') = 0 \tag{27}$$

and

$$V(\overline{x}_s - \overline{x}'_N | \underset{\sim}{x}') = \left(\frac{1}{n} - \frac{1}{N}\right) \sigma^2 \tag{28}$$

for all $\underset{\sim}{x}' \in R^*_N$ in (18), σ^2 being the variance of F. Thus the usual *confidence* intervals for the population mean $\overline{x}'_N$, based on the sample mean $\overline{x}_s$ and its variance $(1/n - 1/N)\sigma^2$, are also *fiducial intervals* provided the distribution of x-values in the population is *approximately normal*.

6. Bayes Approach and the Problem of Randomisation.

Now using the probability function defined in (6), for the observation s, x_i: $i \in s$, in (4), given the parameter $\underset{\sim}{x}'$ in (5), we can write, for any *prior probability distribution* $\xi(\underset{\sim}{x}')$, (which possibly is a good formal description of our prior beliefs) on the parametric space R_N, the *Bayes posterior probability*, (Godambe, (1966)) for $\underset{\sim}{x}'$ on the basis of s, x_i: $i \in s$ as,

$$\text{Prob}\,(\underset{\sim}{x}' | s, x_i : i \in s) = \frac{\text{Prob}(s, x_i : i \in s | \underset{\sim}{x}')\, \xi(\underset{\sim}{x}')}{\int_{R_N(x_i : i \in s)} \text{Prob}(s, x_i : i \in s | \underset{\sim}{x}')\, \xi(\underset{\sim}{x}')\, d\underset{\sim}{x}'} \tag{29}$$

where $R_N(x_i: i \in s)$ is the same subset of R_N as in (7). Now if in (29) we put

$$\text{Prob}(s, x_i : i \in s | \underset{\sim}{x}') = p(s)\ \gamma_s(x_i : i \in s, \underset{\sim}{x}') \qquad (30)$$

where p(s) is the probability of drawing the subset s with the <u>sampling design</u> p (Def. 3.1) and

$$\gamma_s(x_i : i \in s, \underset{\sim}{x}') = \begin{cases} 1 \text{ if } \underset{\sim}{x} \in R_N(x_i : i \in s) \\ 0 \text{ if } \underset{\sim}{x} \notin R_N(x_i : i \in s) \end{cases} \qquad (31)$$

From (29), (30) and (31) we have the

THEOREM. The Bayes posterior probability distribution in (29) is <u>independent</u> of the <u>sampling design</u> p.

Next suppose ϕ is any real function defined by the Bayes posterior distribution in (29), given s, x_i:i$\in$s, $\phi = \phi(s, x_i : i \in s)$, (note because of the above Theorem, ϕ is <u>independent</u> of the sampling design p). Then <u>Bayes</u> apriori expectation of ϕ, given the prior distribution $\xi(\underset{\sim}{x}')$ on R_N and the sampling design p, is

$$\Sigma_S\ p(s) \int_{R_N} \phi(s, x_i : i \in s)\xi(x)dx = \Sigma_S\ p(s)\Phi(s), \text{ say} \qquad (32)$$

Now, for the variations of the sampling designs p over a class of <u>fixed</u> sample size designs p, (i.e. those designs for which the <u>number</u> of individuals i$\in$s=n(s)$\neq$n, a <u>fixed</u> number $\rightarrow$ p(s) = 0 for all s$\in$S, (refer Def. 3.1)), (32) is minimised (or maximised) for a sampling design which attaches probability p(s) = 1 for that subset s for which Φ(s), in (32), is minimised (or maximised). Hence the

REMARK 6.1. The considerations of Bayes prior expectation (as in (32)), of certain desirable (numerical) property, of the Bayes posterior distribution, often leads to <u>purposive</u> (i.e.) <u>non-random</u> selection of some sample

Now it is easy to see that Bayesian approach as summarised in the above Theorem and Remark 6.1 is in direct conflict with a very important aspect of survey sampling practice over years: The survey-statisticians over years, have been employing different sampling designs such as stratification, multistage sampling, arbitrary probability sampling etc. with a view to reduce the variance of their estimates (usually of the population mean). To a Bayesian, this can make sense only if this variance is in some way related to the Bayes posterior distribution. But the above Theorem asserts that the Bayes posterior is independent of the sampling design! This we call the problem of randomisation in survey-sampling. Presumably, randomisation is adopted in practice, with a view to utilise the sampling distribution generated by randomisation. But this is impossible within Bayesian framework.

As an illustration consider the following prior distribution $\xi(\underset{\sim}{x}')$ on R_N or rather on R_N^* in (18). Using the notation (24) we write

$$\xi(\underset{\sim}{x}') = \phi(\overline{x}_N') \; \psi(\underset{\sim}{y}) \tag{33}$$

Now assuming ψ in (33) to be a discrete distribution, using (17) we can obtain, the Bayes posterior distribution of $\overline{x}_N'$ from (29) as

$$\text{Prob}(\overline{x}_N' | \, s, \, x_i : i\epsilon s) = \sum_{\underset{\sim}{y}} \frac{\phi(\overline{x}_N')\psi(\underset{\sim}{y})}{\sum\limits_{(x_N, \underset{\sim}{y})} \int \phi(\overline{x}_N)\psi(\underset{\sim}{y})d\overline{x}_N} \tag{34}$$

where $\underset{\sim}{y}\epsilon A(\overline{x}_N, \, x_i : i\epsilon s)$, $(\overline{x}_N, \underset{\sim}{y})\epsilon B(x_i : i\epsilon s)$

and where $A(\overline{x}_N', \, x_i : i\epsilon s)$ and $B(x_i : i\epsilon s)$ are the subsets of R_N^* in (18) such that, using the notation (24),

$$A(\overline{x}_N, \, x_i : i\epsilon s) = \{\underset{\sim}{y} : y_i = x_i - \overline{x}_N' \text{ for all } i\epsilon s\}. \tag{35}$$

and

$$B(x_i : i\epsilon s) = \{(\overline{x}_N, \, \underset{\sim}{y}) : y_i = x_i - \overline{x}_N \text{ for all } i\epsilon s\}. \tag{36}$$

Now letting $P_\psi($ $)$ to denote the probability of ( ) when ψ in (33^ψ) obtains and noting (17), (35) and (36) we from (34) get,

$$\text{Prob}(\bar{x}'_N | s, x_i : i \epsilon s)$$
$$= \frac{\phi(\bar{x}'_N) P(\bar{y}_s = \bar{x}_s - \bar{x}'_N, y_i - \bar{y}_s = x_i - \bar{x}_s : i \epsilon s)}{P_\psi(y_i - \bar{y}_s = x_i - \bar{x}_s : i \epsilon s)} \tag{37}$$
$$= \phi(\bar{x}'_N) P_\psi(\bar{y}_s = \bar{x}_s - \bar{x}'_N | y_i - \bar{y}_s = x_i - \bar{x}_s : i \epsilon s).$$

Next, substituting in (33),

ϕ ≡ the uniform degenerate distribution on $(-\infty, \infty)$,

ψ ≡ uniform distribution over N! permutations of some fixed vector $(Y_1, \ldots, Y_N)$,

from (37), we get the Bayes posterior distribution of $\bar{x}'_N$, the same as the fiducial distribution of $\bar{x}'_N$ given by (26). But the important thing to note is this:

REMARK 6.2. The Bayes posterior distribution of $\bar{x}'_N$ is independent of the sampling design (ref. Theorem in this section), while the fiducial distribution of $\bar{x}'_N$ is inseparably tied to the simple random sampling, without replacement.

For a detailed discussion of the above point we refer to Godambe (1967).

Now generally about Bayes approach one may say this: In most situations our prior knowledge is so vague that seldom it is formally equivalent to any specific prior distribution. Hence after in principle accepting, that given a prior distribution, the Bayesian posterior distribution represents our entire inference on the basis of the data, in practice what one can do is to adopt

(A) an otherwise reasonable procedure of inference which is also consistant with a Bayes posterior distribution for some prior distribution. A result is the Likelihood-Principle (Barnard et al (1962)).

The implication of the likelihood principle for survey-sampling was discussed in section 4. Or to adopt

(B) an otherwise reasonable procedure of inference which is also consistent with each one of the Bayes posterior distributions obtained from the class of all prior distributions which may be regarded as equivalent to our prior knowledge. The result is the Principle of Bayesian Sufficiency (Godambe, (1966a)).

Thus here we assume that our prior knowledge (of the parameter space) can be characterised by a class C, of the prior distribution ξ, (on the parameter space). In contrast the conventional Bayesian view is that prior distribution is unique at least for the concerned person. Now given the class C of the priors ξ, for any parametric function ϕ, we define the Bayesian sufficient statistic as follows:

DEFINITION 6.1. A function $t_{\phi C}$, on the sample space is said to be Bayesian sufficient statistic for ϕ, with respect to the class C of prior distributions ξ, if the Bayes posterior distribution of ϕ, for every $\xi \in C$, depends on the sample only through $t_{\phi C}$.#

This is a generalisation of Fisherian Sufficiency in the sense that the Bayesian sufficiency is defined for any parametric function while Fisherian Sufficiency is defined only for the entire parameter itself. In other way, Bayesian sufficiency is more restrictive as it is defined for a specified class C of prior

---

A slightly restrictive version of this definition was previously given by Raiffa and Schlaifer (1961). The present author owes this reference to Dr. Ericson.

distributions while Fisherian sufficiency is defined regardless of what the prior distribution is. Thus the directive given by the Principle of Bayesian Sufficiency is this: <u>If the prior knowledge about the parameter, can be represented by a class C of prior distributions ξ, then the inference about the parametric function ϕ, should depend on the data only through the Bayesian sufficient statistic $t_{\phi C}$ defined above.</u>

As an example of the Bayesian sufficient statistic, consider the conventional <u>hypothetical population</u> model. Let the random variate z_1 be distributed as $N(\theta_1, \sigma^2)$ and another z_2 be distributed as $N(\theta_2, \sigma^2)$ where θ_1, θ_2 and σ^2 are unknown. Let further n_1 observations be made on z_1, their mean and standard deviation being denoted by $\bar{z}_1$ and s_1 respectively. Similarly let $\bar{z}_2$ and s_2 denote the mean and standard deviation of n_2 observations on z_2. Now let C be the class of <u>all</u> prior distributions ξ on $(\theta_1, \theta_2, \sigma^2)$ such that $(\theta_1, \theta_2, \sigma^2)$ when distributed as ξ are probabilistically <u>independent</u> θ_1, θ_2 having uniform degenerate distribution over $(-\infty, \infty)$ and σ^2 having <u>any</u> arbitrary distribution over $(0, \infty)$. Now for <u>the</u> class C of the prior distributions so defined, the Bayesian sufficient statistic, for the parametric function $\phi(\theta_1, \theta_2, \sigma^2) = ((\theta_1 - \theta_2), \sigma^2)$, is given by $t_{\phi C} = (\bar{z}_1 - \bar{z}_2, n_1 s_1^2 + n_2 s_2^2)$.

The application of the Principle of Bayesian Sufficiency to our Model of survey-sampling, described in sections 3 and 4, can, under some conditions result in striking reduction of the problem of estimating the parametric function $T(\underset{\sim}{x}') = \Sigma_1^N x_i'$, as in (1), $\underset{\sim}{x}' = (x_1', \ldots, x_N')$ being the parameter as in (5), on the basis of the data $(s, x_i : i \in s)$ in (4). Suppose our <u>prior knowledge</u> about the parameter x implies that the different components $x_1', \ldots, x_N'$ of $\underset{\sim}{x}'$, are in certain, not necessarily very well defined sense of the word, <u>unrelated</u>. We may assume this <u>prior knowledge</u> as equivalent to a <u>class</u> C^*, of <u>prior probability distributions ξ on</u> R_N, <u>which are product measures</u>. [One may here raise the

QUESTION: Under such prior knowledge, characterised by the class C^*, of prior distributions, is it at all meaningful to attempt to estimate $T(\underset{\sim}{x}') = \Sigma_1^N x_i'$ or more specifically, $\Sigma_{i\notin s} x_i'$, on the basis of the data $(s, x_i : i\in s)$?

Answer: Yes. Sometimes the estimation problem may have some structure, like invariance discussed subsequently, which would render meaning to estimating $T(\underset{\sim}{x}')$ on the basis of $(s, x_i : i\in s)$. Now for any prior probability distribution $\xi\in C^*$, the Bayes posterior distribution of the parametric function $T(\underset{\sim}{x}')$, given $(s, x_i : i\in s)$, can be obtained from (29). This Bayes posterior distribution for all $\xi\in C^*$ evidently depends on $(s, x_i : i\in s)$ only through $(s, \sum_{i\in s} x_i)$. Hence we have the

THEOREM 6.1. For any class C^* consisting of product measures on (the parametric space) R_N, a Bayes sufficient statistic, (Def. 6.1), for (the parametric function), the population total $T(\underset{\sim}{x}')$ in (1) is given by $(s, \sum_{i\in s} x_i)$.

Now after reducing the data $(s, x_i : i\in s)$ to Bayesian sufficient statistic $(s, \sum_{i\in s} x_i)$, we try further reduction by considerations of invariance. [This is quite analogous to the routine practice of using sufficiency (conventional) and then invariance for reducing the data, (Hall et al (1965))]. From practical considerations, origin and scale invariant estimation has certain intuitive appeal, in survey- sampling. We therefore introduce,

DEFINITION 6.2. An estimator (Def. 3.2), $e(s, \underset{\sim}{x})$ is said to be origin invariant for the population total $T(\underset{\sim}{x})$ in (1), iff, for any real constant k, $(-\infty < k < \infty)$, $e(s, \underset{\sim}{x} + \underset{\sim}{k}) = e(s, \underset{\sim}{x}) + N.k$ for all $x\in R_N$ where $\underset{\sim}{k} = (k,\ldots,k)$.

DEFINITION 6.3. An estimator (Def. 3.2), $e(s, \underset{\sim}{x})$ is said to be scale invariant for the population total $T(\underset{\sim}{x})$ in (1), iff, for any real constant k, $(-\infty < k < \infty)$,

$e(s, k\underset{\sim}{x}) = k\, e(s, \underset{\sim}{x})$, for all $\underset{\sim}{x} \in R_N$.

Next following Definition 6.1 and Theorem 6.1 we have the

DEFINITION 6.4. An estimator $e(s, \underset{\sim}{x})$, (Def. 3.2), is said to be Bayes sufficient estimator, iff, $e(s, \underset{\sim}{x})$ depends on $\underset{\sim}{x}$ only through $\sum_{i \in s} x_i$.

The following theorems follow easily (for the proofs we refer to Godambe (1966a)) from the above Definitions 6.2, 6.3 and 6.4.

THEOREM 6.2. A Bayes sufficient, origin and scale invariant estimator, for the population total is <u>uniquely</u> given by

$$e(s, \underset{\sim}{x}) = N\bar{x}_s, \tag{38}$$

where as before $\bar{x}_s$ is the mean of $x_i : i \in s$.

In survey-sampling there are occasions when we have the ancillary information in the form of the knowledge of the vector $\underset{\sim}{y}$ which is a realisation of the <u>unknown</u> vector $\underset{\sim}{x}$, (being sampled now), at some previous moment. Suppose further that the sampler feels that the unknown $\underset{\sim}{x}$ is close to $\underset{\sim}{y}$ in R_N. Then following theorems are relevant. Again using the Definitions 6.2, 6.3 and 6.4 we have,

THEOREM 6.3. A Bayes sufficient and <u>origin invariant</u> estimator $e(s, \underset{\sim}{x})$ which is exactly equal to the population total $T(\underset{\sim}{x})$ when $\underset{\sim}{x} = \underset{\sim}{y}$ is <u>uniquely</u> given by

$$e(s, \underset{\sim}{x}) = N(\bar{x}_s - \bar{y}_s) + N\bar{y}_N, \tag{39}$$

where $\bar{y}_s$ is the mean of $y_i : i \in s$, and $\bar{y}_N$ the mean of $y_i : i = 1,\ldots,N$.

THEOREM 6.4. If in the above Theorem 6.3, <u>origin invariance</u> is replaced by <u>scale invariance</u> we have instead of (39),

$$e(s, \underset{\sim}{x}) = \bar{y}_N(\bar{x}_s/\bar{y}_s). \tag{40}$$

It is easy to see that the estimators (39) and (40) are the usual 'ratio' and 'difference' estimators respectively. It would be of interest to note that the estimator (40) was given an independent justification in terms of Distribution-free sufficiency, Unbiasedness and the Principle of censoring by the author (1966) previously. This alternative justification, presumably is also available for the estimator (39). Of course, as a special case of (39) or (40) we can have the estimator (38).

7. Some General Comments concerning the New Model.

Now we propose to discuss, to what extent the new model introduced in section 3 and developed in the subsequent sections, has in general succeeded in explaining and extending the survey-practice.

The survey-practice today largely depends on two types, (D_1) and (D_2), of sampling designs and two types, (E_1) and (E_2), of estimators where,

D_1 = sampling designs obtained by simple random sampling and stratification,
D_2 = sampling designs obtained by arbitrary probability sampling,
E_1 = estimators based on sample mean in (38) and
E_2 = estimators based on HT-estimator in (3).

Now the ratio and difference estimators in (40) and (39) respectively and the conventional estimator of population total in stratified random sampling belong to the E_1 type of estimation above. As we have already seen the new model explains the optimality of these estimators, within the Bayes set-up of prior distributions, assuming the criterion of invariance (section 6). The randomisation as we have seen in section 6 in general poses a problem for the Bayesian. However the above referred to estimators of E_1 type are employed only when some appropriate sampling design of type D_1 obtains. And upto this extent randomisation can be explained, though perhaps not very satisfactorily, from Bayesian view-point. An excellent reference in this direction is Ericson (1967).#

Alternatively, randomisation can be taken care of by introducing along with Bayesian sufficiency, the criterion of unbiasedness (Godambe,(1966)). But this is apparently conflicting. The Bayesian approach implies preferential treatment to different parametric points while unbiasedness means equal treatment, (Hall (1965)). The conflict however is only apparent. To be sure, that we have given just enough preferential treatment as santioned by the prior knowledge, and no more we may adopt the criterion of Bayesian sufficiency together with unbiasedness. This is just a logical extension of the situation when the criterion of sufficiency (conventional) together with unbiasedness are appropriate, (Barnard, (1963)). Actually Bayesian sufficiency, (Godambe (1966a)), or Distribution-Free sufficiency (Godambe (1966)) along with the criteria of unbiasedness and censoring (Pratt (1961)) explains the optimality of Midzuno (1952) estimator, (which is an E_1 type estimator in our above classification), for an appropriate D_2 type sampling design.

Now often when a D_2 type sampling design is used the corresponding estimator is either HT-estimator in (3) or Midzuno's estimator (1952). The domains of the applicability of the two estimators are different. The HT-estimator is appropriate when we have much stronger prior knowledge about the population, namely; (a) we have prior expectation associated with each individual in the population and (b) we know that the variate values associated with different individuals in the population are unrelated. For this prior knowledge the appropriate HT-estimator is uniquely unbiased Bayes estimator (Godambe (1955)), Godambe and Joshi (1965)). On the other hand, the condition (b) of unrelatedness, (i.e. the class C* of the priors

---

Ericson (1967) also gives a thorough going Bayesian account of the survey-sampling theory and practice, in the new model.

referred to in the Theorem 6.1), is enough for the optimality of Midzuno's estimator referred to in the preceeding paragraph. For some interesting results concerning HT-estimator, we refer to Hanurav (1967), Fellegi (1963).

However it is still true that the role, of the randomisation introduced by the D_2 type of sampling designs, in survey-sampling is not at all clear. These do not seem to have any explanation within Bayes set-up. All the discussion above in fact confirms the Remark 3.1 in the section 3. One may therefore feel, that the D_2 type of sampling designs, need be investigated further, theoretically and empirically, to arrive at any definite conclusions concerning them.

Now the above discussion has been independent of the distributional assumption introduced in the section 4 and 5. An additional point, in favour of the new model, when the distributional assumption is introduced is as follows:

With the notation as in (4), consider the distribution of the unordered $x_i : i \in s$ (that is the x-values in the subset s without the individual labels) denoted as $[x_i : i \in s]$. Using (9) and (12) - (15) we can write the distribution of $[x_i : i \in s]$. This would be some symmetrical distribution of dependent random variates $[x_i : i \in s]$ containing the location parameter θ. Let us denote it as

$$(f[x_i : i \in s]) \qquad (-\infty < \theta < \infty) \tag{41}$$

(Note: F in (13) is c.d.f. while f above is the corresponding frequency function). Now, similar to the Question raised in section 4, we ask the

QUESTION. Does the likelihood function for θ given by (41) contain all the relevant information? In other words, given (41), can we ignore the likelihood function given by (9) which in fact generated (41)?

The answer to the above question seems to be negative. For, it is well known that the fiducial (Hora and Buehler (1966)) and Bayes posterior, inferences are exactly identical for the frequency function in (41) while the two methods, fiducial and Bayes, when applied to (9) are distinguishable as explained in the Remark 6.2 in section 6.

8. Some Comments on Dempster's Approach.

In an interesting theory of statistical inference Dempster (1966) has attempted to extend the concept of identifiability of individual, which as shown above has already played a central role in the theoretical developments of survey-sampling, to the sampling of what we have in section 1, called hypothetical populations. However, this attempt of Dempster when worked out in detail (Dempster (1967)) results as I (1968) have shown, into the absurdity: 'Postulate sampling of an hypothetical aggregate of individuals when they do not exist; and ignore the individuals when they actually exist'. Using Dempster's (1967) own illustrative example of sampling finite populations, it has already been shown (Godambe (1968)) that Dempster's theory of inference is seriously self-contradictory.

9. Comments on Royall's (1967) and Hartley and Rao's (1967) Works:

In spite of all that has been said about the relevance of the individual labels in the preceeding sections, under certain conditions of symmetry, an intuitive feeling persists that individuals labels carry no relevant information. This feeling is given a clear expression in an interesting work by Royall (1967). Following Hacking's (1965), line of thought, he puts forward and develops the idea of Best Supported Estimators, (BSE). They are similar to maximum likelihood estimators, but have firmer logical footing. To the extent, Royall's approach goes, namely upto stratified sampling, it is both interesting and useful.

Next, Hartley and Rao (1967), have suggested extensive use of hypergeometric and multinomial models for estimation in sampling finite populations. A similar previous suggestion is due to Wilks (1958). This again, en effect means, ignoring the individual labels, while estimating the population parameters. Such models, as is evident from the survey-practice over years, roughly express practioner's intuition (which of course has roots in the hypothetical-population model, discussed in section 1), when prior knowledge at best could be used upto stratification. The reasons, why such models, have been abandoned in the recent developments of the sampling theory, are by now very well-known. To put it briefly, they are incapable of even expressing the problems of sampling let apart, solving them. On the contrary the new model put forward in section 3 and developed in subsequent sections could accommodate within it most problems of survey-sampling. It could also explain (ref. sections 4 - 7, especially 7) rationale behind considerable portion of survey-practice. Its further use by theoreticians and practitioners would cettainly result in placing survey-sampling on sounder scientific footing.

ACKNOWLEDGEMENTS

The author wishes to thank Mr. J. D. Kalbfleisch, Professors D. A. Sprott, S. W. Dharmadhikari and K. Jogdeo for their comments on the original manuscript. He is also grateful to Mrs. Suerich and Miss Hilgenberg for typing the manuscript.

REFERENCES

Barnard, G. A., Jenkins, G. M. and Winsten, C. B. (1962), "Likelihood inference and time series," J.R. Statist. Soc. A, 125, 321-372.

Barnard, G. A. (1963), "The logic of least squares," J.R. Statist. Ser. B, 25, 124-127.

Basu, D. (1955), "On statistics independent of complete sufficient-statistic," Sankhya, 15, 377-380.

Basu, D. (1958), "On sampling with and without replacement," Sankhya, 20, 287-294.

Dempster, A. P. (1966), "New Methods for reasoning towards posterior distributions based on sample data," Annals of Math. Statist., 37, 355-374.

Dempster, A. P. (1967), "Upper and lower probability inferences based on a sample from finite univeriate population," Biometrika, 54, 515-528.

Des Raj and Khamis, S. H. (1958), "Some remarks on sampling with replacement," Annals Math. Statist. 29, 550-557.

Ericson, W. A. (1967), "Subjective Bayesian models in sampling finite populations, I." Unpublished.

Fellegi, I. (1963), "Sampling with varying probabilities without replacement: Rotating and non-rotating samples," J. Amer. Statist. Asso. 58, 183-201.

Godambe, V. P. (1955), "A unified theory of sampling from finite populations," J. R. Statist. Soc. B, 17, 268-278.

Godambe, V. P. (1965), "A review of the contributions towards a unified theory of sampling from finite populations," Int. Statist. Inst. Rev. 33, 242-258.

Godambe, V. P. (1966), "A new approach to sampling from finite populations, I, II," J. R. Statist. Soc. B, 28, 310-328.

Godambe, V. P. (1966a), "Bayesian sufficiency in survey-sampling," Tech, Report No. 63, Dept. of Stat. J.H.U. (To appear in Annals of Statist. Math. (1969)).

Godambe, V. P. (1967), "A fiducial argument with application to survey-sampling," Tech. Report No. 69, Dept. of Stat. J.H.U. (To appear in J. R. Statist. B, (1969).

Godambe, V. P. (1968), "Discussion of Dempster's paper," J. R. Statist. Soc. B, 30, p. 243.

Godambe, V. P. and Joshi, V. M (1965), "Admissibility and Bayes estimation in sampling from finite populations — I," Ann. Math. Statist., 36, 1707-1722.

Hacking, Ian, (1965), Logic of Statistical Inference Cambridge University Press.

Hâjek, J. (1959), "Optimum strategy and other problems in probability sampling," Casopis Pest. Mat. 84, 387-423.

Hall, W. J. (1965), A personal communication.

Hall, W. J., Wisman, R. A. and Ghosh, J. K. (1965), "The relationship between sufficiency and invariance with applications in sequential analysis," Ann. Math. Stat. 36, 575-614.

Hanurav, T. V. (1965), "Hyper-admissibility and optimum estimators in sampling finite populations," Ph.D. Thesis, Indian Statist. Inst., Calcutta.

Hanurav, T. V. (1967), "Optimum utilization of auxiliary information: πps sampling of two units from a stratum," J. R. Statist. Soc. B, 29, 374-391.

Hartley, H. O. and Rao, J.N.K. (1967), "A new estimation theory for sample surveys," Unpublished Report, Texas A&M University.

Hora, R. B. and Buehler, R. J. (1966), "Fiducial theory and invariant estimation," Ann. Math. Statist., 37, 643-656.

Horvitz, D. G. and Thompson, D. J. (1952), "A generalisation of sampling without replacement from a finite universe," J. Amer. Statist. Assoc. 47, 663-685.

Joshi, V. M. (1965), "Admissibility and Bayes estimation in sampling finite populations - III," Ann. Math. Statist. 36, 1730-1742.

Koop, J. C. (1957), "Contributions to general theory of sampling finite populations without replacement and with unequal probabilities," Inst. Statist. North Carolina Univ. Mimeo. Ser. 296.

Midzuno, H. (1952), "On the sampling system with probability proportional to the sum of the sizes," Ann. Inst. Statist. Math., 3, 99-108.

Pathak, P. K. (1964), "Sufficiency in sampling theory," Ann. Math. Statist. 35, 795-808.

Plackett, R. L. (1958), "Studies in the history of probability and statistics VII. The principle of the arithmetic mean," Biometrika 45, 130-135.

Pratt, J. W. (1961), "Review of Testing statistical Hypotheses by E. L. Lehmann," J. Amer. Statist. Asso. 56, 163-167.

Raiffa, H. and Schlaifer, R. O. (1961), Applied statistical Decision Theory, Boston: Division of Research, Graduate School of Business Administration, Harvard University.

Royall, Richard M. (1967), "An old approach to finite population sampling theory," Technical Report 420, J.H.U.

Tacheuchi, K. (1961), "Some remarks about unbiased estimation in sampling finite populations," Jour. Univ. of Scientists and Engineers, Japan, 8, 35-38.

Wilks, S. S. (1960), "A two-stage scheme for sampling without replacement, _Bull. de l'Inst. Intern. Statist._, 37, 241-248.

THE EVOLUTION OF A SURVEY ANALYSIS PROGRAM

F. Yates

Rothamsted Experimental Station

The main object of this paper is to describe the work done intermittently at Rothamsted over the last ten years on the development of a general computer program for the analysis of surveys of the research type.

In 1954 we acquired the Elliott 401, a drum machine similar to the IBM 650 but paper-tape oriented. A somewhat primitive card reader was added in 1956. The first large-scale survey analysis undertaken was that of the 1957 Survey of Fertiliser Practice. (See Sampling Methods, Sect. 4.23, for a general description of these surveys.) There was pressing need to analyse this survey on the computer, as it was three times the size of any previous Fertiliser Survey, and earlier efforts to replace edge-punched cards by ordinary punched cards, using punched card equipment, had not been rewarding, largely because of the necessarily complicated sampling procedure and associated preliminary numerical work.

The operation was very successful, the analysis being done in record time, but the preliminary programming effort required was very great. It soon became apparent, when we attempted to repeat our success on other surveys, that unless the analysis could be fully specified in advance and ample notice of requirements was given, writing an ad hoc program for each survey - wothout even an autocode - would give rise to delays which were just as serious as those of hand computing.

It was natural, therefore, to consider whether programming dificulties could be reduced by the development of a general program. An early approach of this type was made at Rothamsted by H.R. Simpson who in 1958 completed a program for generating up to four way tables from stratified random samples with variable sampling fractions. This, however, covered only a small part of our needs, as most of the surveys we had to analyse are of a research type with relatively few units but with complicated questionnaires and multi-stage sampling structures.

In 1959 I was asked to prepare a further edition of Sampling Methods, and by good fortune, just as this was becoming really urgent, a three months visit to India gave me the opportunity, free in the evenings from other distractions, to get down to the task of adding a chapter on computers, and the ways in which they might be used for survey analysis. I first went through the book to ascertain all the various analytical procedures that were there described, and then considered the sort of general program that would enable the statistician to perform these operations as required without getting involved in the intricacies of machine programming. This was an uninhibited exercise, to see what could be done on what were then considered large machines. The outcome was a generalised notation which would enable the required operations to be conveniently expressed, and could form the basis of an autocode for survey analysis. The main aspects covered were:

(a) the formation of derived variates from the basic (data) variates, including such operations as grouping for classification purposes;

(b) specification of the counts and tabulations required to form the basic tables;

(c) formation of further derived tables from these, e.g. tables of means from tables of totals and counts, raising by strata raising factors, etc.

At first sight the implementation of such a system appeared to be impracticable on a small computer such as the Rothamsted machine, but somewhat surprisingly we succeeded quite quickly in constructing a general program for the 401 on these lines. Although the autocode form of specification described in Sampling Methods was not followed, virtually all the operations there described were made available in this program. Details are given by Yates and Simpson (1960, 1961).

The program, although inevitably somewhat slow, abundantly proved its worth. After it was completed we never had to write any special survey analysis programs. Demands for 401 machine time were relieved by the presentation by Elliott Automation in 1961 of a 402. This machine was used solely for the formation of tables, with a transcription of Part 1 of the survey program, and served us well until the EMA General Survey Program was available on the Orion.

Early survey programs on the Orion

When the replacement of the 401 and 402 by a Ferranti (now ICT) Orion computer was decided on a start was made on writing a General Survey and Experiments Program (SEP) which it was hoped would provide facilities for analysing both surveys and experiments. In essence the Survey and Experiments Program is a statistical autocode oriented to table manipulation. The autocode devised was a very general and powerful one which on paper was capable of performing all the operations normally required, including the operations of matrix algebra, and with good facilities for providing tables properly annotated by headings etc. Provision was also made for the inclusion of standard routines of SEP instructions so that the more standard types of experiment could be analysed merely by calling in the appropriate routine (Gower, Simpson and Martin, 1967, (a) and (b)).

Because of the need to make a start on the programming before the Orion was available and before details of the promised EMA compiler had been finally settled, it was decided to write the program in machine code. This, it was thought, would also result in a faster program. This decision was undoubtedly a mistake. The Orion EMA compiler compiles very efficient programs, and there is no reason to believe that writing in machine code gave any appreciable increase in speed. Moreover, the fact that the Survey and Experiments Program is in machine code means that it cannot be transferred to another machine without being completely re-written. Undoubtedly also, writing in machine code made it much more difficult to locate and correct programming errors or to make modifications.

Progress on the program was very slow, and when the Orion was delivered it became urgently necessary to provide programs for the analysis of the more standard types of experiment. Programs for this work were therefore written in EMA on much the same lines as the programs we had previously written for the 401, but with much better annotation and layout of tables and with proper provision for covariance analysis. On the 401 we had no good program for analysing general factorial designs with more than two factors, or with confounding. This problem was re-examined and a very powerful general factorial program was devised (Yates and Anderson, 1966) which has since been further improved. With the construction of this program it finally became clear that the Survey and Experiments Program would never be needed for any but the most exceptional experimental designs. The simpler parts of the program were, however, proceeded with in the expectation that they would be useful in survey analysis.

In the meantime it had also become urgently necessary to make available a usable survey analysis program. An EMA program for the analysis of surveys was therefore written (Anderson, 1966). This followed the main lines of the 401 survey program, but because of the need to complete it as speedily as possible, many of the features which made that program convenient for the user, in particular automatic assignment of addresses and automatic construction of the table index, were omitted.

As an autocode was available on the Orion it was decided that the organisation of the reading of the data from cards or magnetic tape and also the derived variate instructions could well be written by the user in EMA for each individual survey. For Part 1 operations, therefore, the user was provided with a standard set of routines, the main one (apart from card reading and magnetic tape routines) being the tabulation routine. This made the required entries for each unit in the tables defined by the tabulation instructions read in at the start. Part

2 was not separated from Part 1, sequences of processing and printing instructions being also obeyed by entry into standard chapters from the user's EMA program.

This use of EMA had the advantages (a) that cycling over hierarchical sets of data could be more general; (b) that the derived variate instructions were compiled instead of being interpreted for each unit, which might be expected to lead to increased speed of processing.

The program was completed in 1965 and has been in use for survey analysis since then. It proved satisfactorily fast. It was found, however, that developing programs for specific surveys was very troublesome and absorbed a good deal of machine time. This was largely due to the defects of the EMA compiler which, although it produces an efficient program, compiles very slowly and has inadequate facilities for speedily incorporating in compiled form parts of a program which are already correct. This, coupled with the fact that Part 1 and Part 2 were not separated, meant that loading the program for each test run was a very lengthy process.

When the program was written it was, of course, envisaged that if it were permanently needed it would have to be re-written so as to be more convenient for the user and less clumsy in operation. When SEP was sufficiently complete to do the simpler types of survey analysis it was found that although analyses could be easily specified and errors of specification eliminated, the program was too slow for use on any but the smallest jobs. The causes of this have never been fully analysed and need not be discussed here. Suffice to say that it soon became apparent that a thorough revision of the EMA program would be necessary. The outcome of this work is described in the following sections.

Revised version of the EMA General Survey Program

In the summer of 1966 we started on the task of improving the EMA General Survey Program (GSP). Most of the details of the modifications and improvements were settled in the autumn of that year and many of the component parts were programmed during the winter. The work then had to set aside because of other tasks but was taken up again in the autumn of 1967 and the program is now (March, 1968) nearly complete.*

One of the most important changes, which was only decided on in the autumn of 1967, was to divide the new program into two independent parts, similar to those of the 401 program. Part 1 deals with the reading of the data, calculation of derived variates and formation of the basic tables of totals and counts, and contains parts written by the user in EMA, but Part 2 is a self-contained program not requiring EMA additions by the user. This has the consequence, which is important on the Orion, that it can be stored on the library tape in bimapped form and can therefore be loaded very rapidly.

Separation of the program into two parts in this manner makes the user's task very much simpler, as the amount of program that has to be compiled or read in basic (a process which may have to be repeated several times for test runs) is limited to Part 1 and is therefore comparatively small. Writing and testing the standard parts of the program was also made easier by the clear division into two parts.

The fact that Part 2 can be rapidly loaded also makes it easy to process and print the tables by stages, thus leading to a more exploratory approach to analyses, thereby avoiding the output of much material subsequently found to be useless.

Apart from this the main improvements in the revised version are:

---

* The revision was made jointly by A. J. B. Anderson and myself, with some help in the early stages from Professor S. Lipton of Sidney, during part of his sabbatical leave.

(a) More convenient and compact specification of the tabulation instructions.

(b) A convenient autocode form for specification of the processing instructions.

(c) Provision for loops, modification, jumps and jumpdowns in the processing instructions.

(d) The inclusion of processing instructions for various operations not available in GSP 1.

(e) Much improved facilities for printing tables, including easily used facilities for headings and numbers of digits to be printed.

(f) Automatic assignment of table addresses.

(g) Better magnetic tape facilities.

The compilers

There is at present a separate compiler for each part of the program. The compilers read in the various lists that are required in the two parts, and also the tabulation instructions for Part 1 and the processing and printing instructions for Part 2. When a compilation is complete, the results are stored on the dumps magnetic tape, and are read back at the start of a run of Part 1 or Part 2, though we may well eliminate this use of magnetic tape in Part 2, by including the Part 2 compiler in the main Part 2 program. The dumps tape is also used for intermediate dumps in Part 1, to give restart facilities.

The tables formed in Part 1 are stored on a separate tables tape at points in the program dictated by the user. Before storage they are expanded, and the marginal totals (or maxima or minima) are inserted. They are read back for processing and printing in Part 2, which also provides for the storage of processed tables.

Tabulation instructions

Each tabulation instruction or set of such insrtuctions is headed by a classification set or sets written explicitly with the variate numbers enclosed in brackets: thus

(1 2 3)

indicates that the variates contained in the tabulation instructions which follow are to be tabulated in tables classified by variates 1, 2, 3, in that order. Up to five-way classifications may be used.

More than one classification set may be written at the head of a set of tabulation instructions, in which case the tables for the first classification set are constructed first, followed by those for the second, etc. (Table numbering in the tabulations is automatic, starting from some specified number.)

A group of classification sets with consecutive numbers for one of the variates may be written in the form

(1 2 3-5)

which is equivalent to writing

(1 2 3) (1 2 4) (1 2 5)

Each tabulation instruction appears on a separate line and contains the following components (or such of them as are required) in order:

No. of variate or variates to be tabulated (C for count).
Packing instruction (P2 = half words, P4 = quarter words)
Associated count (A).
Packing instruction for associated count (as above).
Raising variate (*variate no.)

Restricting variate and nature of restriction (see below).

Thus

21 P2 A P4

indicates tabulation of variate 21 packed in half words, together with an associated count packed in quarter words. Similarly

22 *50

indicates tabulation of variate 22 raised by variate 50. (Raised tables cannot be packed.)

More than one variate may be included in the same tabulation instruction, provided the same conditions regarding the associated counts, packing, raising and restriction apply. Thus

C 24 25 *50

will give a count and tabulations of variates 24 and 25, all raised by variate 50. Similarly

27 28 30 A P4

will give tabulations of variates 27, 28 and 30 (not packed) each with an associated count packed in quarter words.

If a set of consecutive variates requires to be tabulated, this may be written

27-30

which is equivalent to

27 28 29 30

Only one such set may appear in a single instruction, and single variates or C must not ve included in the same instruction.

Entries in a table may be made contingent on some other variate (the restricting variate) satisfying certain conditions. The permissible conditions and form of specification are as follows (restricting variate n; a, b, c, integers between 0 and 31):

Rn = a	Restriction to value a
Rn = a b	Restriction to value a or b
Rn = a b c	Restriction to value a, b or c
Rn = → b	Restriction to values $\leq$ b
Rn = a →	Restriction to values $\geq$ a
Rn = a → b	Restriction to values $\geq$ a and $\leq$ b

The = may be replaced by ≠. This will restrict to values excluded by the above specifications.

Introduction of the word MAX or MIN immediately after the numbers of the variates to be tabulated will cause the max. or min. of the values for each cell to be entered in that cell instead of their total. In the expanded tables the margins will contain the max. or min. of the corresponding component cells. Empty cells will have the value-0. Raising and restricting instructions can follow MAX or MIN.

Provision is made for variates relating to sub-units recorded in parallel fields on the same card. This can best be illustrated by a specific example. Suppose there are 4 pairs of parallel fields containing variates numbered

21, 22; 23, 24; 25, 26; 27, 28

i.e. two variates per sub-unit for up to four units. This grouping is defined under PARALLEL in the compiler.

If we wish to enter the items from each pair of variates in the same table, this is done by specifying the table required for the first pair of variates only, following these variate numbers by +. Thus

(1 21+)

22+

will give a combined table of variate 22 classified by variates 1 and 21, variate 24 classified by variates 1 and 23, etc. Similarly

(1 21+)

C

will give a combined count of the number of entries for each level of variates 21, 23, 25 and 27, classified by variate 1.

The parallel fields facility also provides a means of making entries in the same table for several variates, one classification of the table being by an attribute of these variates, e.g., acreages of different crops with crops as the classifying variate. If, for example, there are 10 crops, 10 dummy variates, permanently set to values from 1 to 10 (not necessarily in order) are defined and these are used as classifying variates for the crop acreage variates.

Processing instructions

One of the weaknesses of GSP1 is that it requires long strings of tabulation and processing instructions which for many jobs are essentially looping, jumps, etc. should be incorporated in the table processing instructions of the revised program. This did not prove difficult and with the standard EMA form of packing (quarter words of 12 digits) did not demand any extra storage for an individual instruction. Each instruction is packed in seven quarter words, four of which are modifiable. The first three digits of a modifiable quarter-word are used to indicate the modifying index and the last 9 digits contain the numerical value which is to be added to the current value of the index concerned, if any.

A conspectus of the actual instructions is given is Appendix 1. This conspectus omits some details - the user instructions must be consulted for these - but should give sufficient indication of the scope and convenience of the available instructions.

Facilities for printing tables

Much improved facilities are provided for printing tables, so as to obtain a good layout and clear headings with minimum specification. Headings can be listed and referred to by number or appended to the instructions which form the tables, or to a print instruction, and the appropriate heading is then reproduced (together with the table number and classifying variates) whenever the table or parts of it are printed. Variates can be given short names, and will then be referred to by name instead of number. Totals, counts, means (totals/counts), percentages, max. and min., are recognised and automatically designated in the heading, as is raising and restriction. Longer headings can be printed where required. Short level names can be specified for the levels of classifying variates, and will then replace level numbers.

The number of digits before and after the decimal point can be specified, either in the instructions, or against variate numbers. For the latter, digits for totals, raised totals and means are separately specified. If digits are not specified the table is examined before printing and suitable digit values are assigned.

The layout of the tables is made as compact as possible, consistent with clarity. The number of columns across the page is determined from the number of digits in each column, and in three- or more-way tables the basic two-way blocks (for the last two factors) are placed side by side if there is room.

Any selected margin of a table can be printed, or all margins of a given order, e.g. all one-way margins. Margins of the parts printed can be omitted if desired. Factors can be re-ordered when printing without re-ordering the table in the store.

Following the print of any table or part table a further auxiliary table which contains all the factors of the main table except the last but one can be printed. Headings are abbreviated in such a table and the format is identical with the line margins of the main table.

This enables us, for example, to print a two-way table of percentages, arranged so that the columns have totals of 100, with the numbers on which each percentage is based beneath each column.

Several tables or one- or more-way margins defined by given factors can be printed in parallel columns. These tables need not all be classified by the same factors; the only restriction (apart from the obvious one that the specified factors must be present in all the tables) is that these factors are consecutive in all the classification sets. If they are classified by the same factors any subset may be taken, with re-ordering, and all margins, or all except those for the last factor, may be omitted.

There are of course occasions when more elaborate forms of print, e.g. interspersed rows or columns from a pair of tables, are useful for final tables that are required for incorporation in a report. Although a table with such interspersed rows or columns could be constructed by the combine tables instruction, it could only be properly printed if the number of digits before and after the decimal point were the same for the two tables. To avoid over-elaboration of the present program it appears better to write a separate print program covering requirements of this kind. Such a program can be designed to call up the required tabular material from GSP magnetic tapes, using the same reference system.

Why no statistical procedures?

Various general survey programs and "report generators" provide for simple statistical procedures on the constructed tables, in particular the calculation of χ^2 for 2-way tables of counts and some provision for the calculation of sums of squares or standard deviations of quantitative data. It may be asked why no such procedures were incorporated in the present program.

In our experience the calculation of χ^2 is scarcely ever necessary. Associations between pairs of variates, if they are of any consequence, are usually obvious without any formal tests of significance. Often also the levels of the classifying variates are ordered quantitatively; in such cases if a test of association is required χ^2 with one degree of freedom rather than (p - 1)(q - 1) degrees of freedom is what is required (Yates, Biometrika, 1947)). A more refined approach, which gives a measure of the degree of association as well as its significance, is to fit constants for rows, columns and linear interaction (quantal data), which can be done by our fitting constants program. For the calculation of sampling errors, and investigations on the efficiency of alternative sampling designs, variances are of course required. No special provision is made for this in GSP, as they can readily be obtained by forming squares of the required variates in Part 1 and tabulating these to give sums of squares (or even more simply by tabulating the variates raised by themselves), and then using the processing instructions of Part 2. An example of the processing instructions required for a simple calculation of this type is given in Appendix 2.

A much more important type of statistical calculation on the completed tables is that of fitting constants, both to quantitative and quantal data. In our opinion such computations are best made by a separate program so constructed that tables provided by the GSP program can be read in as data for the fitting constants program. Our own fitting constants program is organised in this manner.

General survey programs on other computers

I had hoped to review the survey programs available on other computers, but collection of up-to-date particulars proved difficult. The following general comments are based on particulars of a few programs that were readily available.

These programs are designed to deal mainly with sample surveys which require no great statistical finesse for their analysis, such as market research surveys. In most of the programs there is no provision for table processing operations. This I regard as a serious omission. Two operations that are required to deal conveniently with stratified samples are the addition of tables of the same order and the multiplication of tables of different order with subsequent recalculation of the marginal totals, the first when tables are constructed stratum by stratum and totals over the whole population are required, and the second for raising when strata with variable sampling fractions, e.g. size groups, are used for classification. It is then usually more convenient to construct the basic tables in unraised form and raise them subsequently by multiplication by a table of the sampling fractions; if the sampling fractions have to be adjusted to take account of non-response and defective returns, this must be done if preliminary tabulations to determine the adjustments are to be avoided.

Although some of the operations provided in our programs may not often be needed, if any are included there is everything to be said for providing a good selection, as additional functions do not greatly complicate the program, nor do they make it appreciably more difficult for the user to understand.

The program for the ICT 1900 series does contain a good set of table processing operations. These operate only on two-way tables, and no provisions are made for modification, but apart from our POS function provide much the same facilities for handling two-way tables as are provided by our own processing instructions, though the sequence of operations required to attain a given result may be very different.

Another way in which these programs differ from ours is in the use of names instead of numbers for variates and tables. If modification is not required, the use of names or numbers is largely a matter of taste; there would be no difficulty in making names permissible in our program. Numbers have, however, the great advantage that they can be modified in the manner to which everyone acquainted with high-level computer languages is accustomed; and although names may give the beginner an illusion of simplicity they are much more tiresome to write and punch correctly.

There is also no provision for handling three- or more-way tables in any general manner. The ICT program permits the construction of three- or more-way tables by the device of taking all combinations of the levels of two or more of the factors as the levels of the row or column factor, but only provides the margins over all rows and all columns. Thus in a three factor table with combinations of levels of A and B as rows and levels of C as columns, only the AB margin and the C margin are directly available. If other margins are required, these can be obtained by the use of the table processing operations, but only in an intolerably involved manner. Thus, if the "major" row classification is by a factor with p levels, to obtain the three two-way tables, with their margins, no less than 4p such operations are apparently required, most of them involving "sub-tables", the specification of which is tedious. The user will almost certainly opt for the alternative of making additional tabulations.

One curious feature of the ICT program is that it apparently runs through the data afresh for each table formed. If this is really so, the program is likely to be very slow.

References

Anderson, A.J.B. (1966) A note on the construction of a general survey programme in Extended Mercury Autocode. Comput. J., 8, 312-314.

Gower, J.C. (1962) The handling of multiway tables on computers. Comput. J., 4, 280-286.

Gower, J.C., Simpson, H.R. and Martin, A.H. (1967) An outline of a programming language for the analysis of surveys, experiments and multivariate data. Proc. Int. Symp. for Methods of Field Experimentation, Halle,(1965 (Sonderdr. TagBer., 86, 159-180).

Gower, J.C., Simpson, H.R. and Martin A.H. (1967) A statistical programming language. Appl. Statist., 16, 89-99.

Yates, F. (1947) The analysis of contingency tables with groupings based on quantitative characters. Biometrika, 35, 176-181.

Yates, F. (1960) Sampling Methods for Censuses and Surveys. (3rd Edn.) London: Griffin.

Yates, F. and Anderson, A.J.B. (1966) A general computer programme for the analysis of factorial experiments. Biometrics, 22, 503-524.

Yates, F., Gower, J.C. and Simpson, H.R. (1963) A specialized autocode for the analysis of replicated experiments. Comput. J., 5, 313-319.

Yates, F. and Simpson, H.R. (1960) A general programme for the analysis of surveys. Comput. J., 3, 136-140.

Yates, F. and Simpson, H.R. (1961) The analysis of surveys: processing and printing of the basic tables. Comput. J., 4, 20-24.

Appendix 1

GSP2, Part 2. Conspectus of processing instructions

Tables

T1-T399

Classification sets

S1-S98

Specified instructions as (Sn) or explicitly as ($n_1 n_2 \ldots$). Cl. set list includes Pt. 1 sets, compiler list, and those explicitly specified in Pt. 2 instructions. (S0) is the cl. set with no factors. In instructions below, cs denotes that specification of the cl. set is obligatory (unless previously determined for that table), (cs) that there is a default setting if omitted.

Constants

X1-X100

Set by compiler or from paper tape at run time (X0=0).

Indexes

J, K, L, M, N, O, P

Set at run time.

Modification

If a symbol is underlined it may be written as an integer (0-511) or in symbolic form. Symbolic forms comprise index (unbracketed), (index + integer), (integer + index). No other forms are permitted.

Instructions for setting indexes and constants

J = i J = -i J = i $\pm$ i' J = X$\underline{p}$	i and i' are integers (0 - 4095) or indexes (no brackets).
J = REA r X$\underline{p}$ = REA r	Next no. from paper tape on reader r.

Labels

As in EMA (labels 1-50).

Loops, Jumps, etc.

J = $k_1(k_2)k_3$ J = $k_1(k_2)k_3$, K = k'_1 (k'_2)	k_1 etc. are integers (0-511), up to 3 extra indexes, nesting to any depth.
REPEAT	
JUMP n	
JUMP n,s_1=$s_2$etc.	s_1,$s_2$are of form i or i + i' $= \neq > < \geq \leq$ permitted.
JUMPDOWN n	One level only, jump out of routine permitted.
RETURN	

END	
TAG _t_	Set tag to _t_.
STR _s_	Set stratum no. to _s_.
HEA _h_ _n_ HEA: heading: _n_	Print heading no. _h_, or as punched. First: must be on same line as HEA. _n_ gives no. of NL before heading (default setting of 4 if omitted).

Table formation and modification

T_d_ = T_a_ + T_b_ (cs) T_d_ = T_a_ - T_b_ (cs) T_d_ = T_a_ * T_b_ (cs) T_d_ = T_a_ / T_b_ (cs) T_d_ = PCT T_a_/T_b_ (cs)	At least two tables must have same cs. If cs differ variates in smaller set must all be in larger set and in same order. If cs(T_a_)=cs(T_b_), cs(T_d_) must be the same or smaller. If no cs specified, T_d_ has cs of T_a_ and T_b_ or the larger of them.
T_d_ = T_a_ (cs) T_d_ = T_a_ + X_p_ (cs) T_d_ = T_a_ * X_p_ (cs) T_d_ = X_p_ cs	cs(T_d_) may differ from cs(T_a_) provided variates in smaller set are in same order as in larger set. If no cs specified, cs(T_d_)= cs(T_a_).
T_d_ = PCT T_a_ i	Percent — i defines margins required.
MAR T_d_ i	New mar. totals — i defines margins required.
MAR T_d_	New mar. totals (all margins).
T_d_ = POS T_a_ (cs)	1 if value positive, 0 otherwise
T_d_ = SRT T_a_ (cs)	Square root.
T_d_ = REO T_a_ cs	Reorder factors.
T_d_ = CMR T_a_ X_r_ ($z_1 z_2$..)	Combine rows.
T_d_ = CMT T_a_ T_b_ ($z_1 z_1'$..)	Combine tables.

Reading tables from magtape or paper tape

GET d b j	d = deck no., b=1st block no. of job j = jobname (omitting characters 1-3)
//Tn.t/s.C	Current ana. (no GET) table n
// Tn.t/s	Prev. job, same survey tag t, stratum s
Tn = MAG Tn'.t/s cs	Another survey
Tn = REA d b s c cs	From deck d, block b, addr. s.
Tn = REA Tn' r c cs	From paper tape, reader r. c: M margins, C total check, MC both.

Storing tables on magtape

/ prefixed to any instruction forming, modifying, printing (not multiple print**), or reading a table (except //), turn the table.
/ Td store only.

Print table

* Td Aw Z (cs)	If Aw(w=1,2,...) all w way margins are printed, if Z margins of 3 or more way tables are omitted. cs is used for reordering and/or omission of factors (not with Aw).

* prefixed to any instruction forming, modifying or reading a table prints the full table. - or up to 6 + may replace * and gives print of total cell only preceded by up to 6 NL.

Heading and digits

A heading and/or digits (if both in that order) may be added at the end of a print instruction or a table formation or modification instruction (except CMR and CMT), whether or not the instruction is prefixed by *

Forms for heading:

Hh or: heading:

Forms for digits (n before, n' after dec. pt.):

n.n' or an index(= 8n + n'). n, 1-7; n', 0-7.

Print of tables in parallel columns

Up to 3 tables:

$**\underline{n}_1\underline{n}_2\underline{n}_3$ F$\underline{f}$ (cs)	F$\underline{f}$ refers to list of col. headings and digits.

If more than 3 tables precede by one or more instructions

$**\underline{n}_1\underline{n}_2\underline{n}_3\underline{n}_4/$

Core and drum transfers and clearance

CTD table nos.	Core to drum	Up to 4 table nos.
DTC table nos.	Drum to core	$\underline{n}_1\underline{n}_2$... or $\underline{n}_1$-$\underline{n}_2$ (inclusive).
CLC	Clear core.	
CLC table nos.	Clear core except for table nos. shown (any number, including runs $\underline{n}_1$-$\underline{n}_2$, possibly symbolic, /for continuation on new line).	
CLD	Clear drum.	

Appendix 2

Processing instructions for calculating sampling errors per unit in a sample stratified by size groups and districts.

Basic tables from Part 1

T1 count T2 total T3 sum of squares	Classified by size groups and districts

Processing instructions

```
  Get 2 2341 DSG1              Mag tape for Pt. 1 analysis
 //*T1                         T1 from mag tape and print
 //T2                          T2 from mag tape
 //T3                          T3 from mag tape
 *T4 = T2/T1                   Prints table of means
 *T5 = T2*T1:C.F.M.:7.2        c.f.m. for each cell (print)
 *T5 = T3-T5:SS DEV.:7.2       Prints SS of dev.
 MAR T5                        Within cell marginal totals
 T6 = POS T1
 T6 = T1-T6                    D.F.
 MAR T6                        Within cell D.F. with marg.
                               totals
  *T7 = T5/T6: MEAN SQ.
      PER UNIT:4.2             Within cells M.S. and S.D.
  *T8 = SRT T7: ST.DEV.        with marginal means (print)
      PER UNIT:2.2
```

Printing the mean squares and st. dev. in tabular form permits examination of (a) whether any particular cells are excessively variable, (b) whether the mean squares vary from district to district or size group to size group.

If the computation is required for several variates, indexes can be used where necessary for table numbers, and possibly also for print digits, the computation being written in the form of a routine or within a loop.

Tables of counts and means need not have their headings and digits specified, as these are obtained from the compiler lists for the variate concerned.

Note that a few additional processing instructions will provide the sums of squares for a hierarchial analysis of variance by size groups, districts within size groups, and within sub-classes; or vice versa by districts, etc. The analyses can also be neatly printed in the conventional layout.

THREE EXTENSIONS OF SAMPLE SURVEY TECHNIQUE: HYBRID, NEXUS, AND GRADUATED SAMPLING

Frederick F. Stephan

Princeton University

The evolution of modern sample survey theory and technique has been traced in a number of publications (1). They exhibit it as a major accomplishment in the application of the theory of probability to a wide range of practical problems involving the collection of data through extensive programs of observation and the preparation of estimates from such collections of data.

In modern sampling theory and practice the guiding objectives emphasize 1) overcoming practical limitations, 2) producing a product of very high quality, and 3) satisfying theoretical requirements that reach down to the very foundations of statistical theory. The successful reconciling of these diverse objectives stands as a supreme example of the elevation of technical procedures from an intuitive and traditional level to a highly rational and scientific level.

Many of the major features of modern sample survey practice, as thus developed in rational form, can be recognized in the traditional "know-how" of a variety of agricultural and industrial operations as well as in a number of other distinct human activities such as grading grain and fibers, assaying ore, drawing panels of jurors by lot, selecting bonds for redemption, and operating lotteries. Other features of modern sampling, such as the selection of samples with varying probabilities, can be traced to the discovery and correction of biases in commonly accepted methods of assembling data and interpreting them.

Looking back over the past evolution of sample survey practice, one may well ask what direction its further development may take. In the past many advances came about by the successive removal of constraints that had been imposed on selection procedures and estimating functions by their initial conceptualizations as sampling problems. Thus it was customary to expect researchers to examine as thoroughly as they can the material or populations they choose to study, picking out deliberately what appeared to be the most representative sample. This rule has been relaxed, and indeed largely replaced, by a rule that after due preparation, the sample should be obtained by random selection, out of all the material available, of a relatively large number of comparatively small units. This certainly must have appeared at first to be quite an irresponsible and hazardous approach to the problem of getting good data (2). The customary rule that every element of the population should have an equal chance to be selected for the sample has given way to a more sophisticated concept of varying probabilities for individual elements as well as for strata (3). The initial concept that a sample should have a definite size, fixed in advance of the taking of data, has been relaxed to admit the more sophisticated theory of sequential sampling (4). This also must have been somewhat shocking to people who had previously thought that it was wicked to peek at the sample while it was being drawn and to modify the sampling procedure on the basis of what one was getting. Finally, a considerable number of problems and pressures have forced samplers to attempt to reintroduce into their sample selecting procedures more control and greater utilization of information and judgment, through controlled selection and similar procedures (5). The apparent objective of this relaxation of customary rules is to realize, at a higher and more sophisticated level, the objectives of purposive sampling while at the same time retaining the fundamental theoretical basis of all probability sampling, i.e., the essential linking of the sampling process to a suitable mathematical model in which the several possibilities of selecting individual elements and various sub-sets of elements of the population have associated with them specific probabilities.

In this paper three directions of further development will be suggested, all of which involve some relaxation of constraints or extension of concepts. Indeed, they include some merging of previously distinct concepts into hybrids that promise to enrich the options available to the survey sample designer.

A Digression

The future historian of sampling practice is likely to characterize the development of probability sampling as a flight from dependence on judgment and deliberate choice. Perhaps he will trace the relationship of this intellectual development to the development of behavioristic psychology and the predominant emphasis upon objectivity in scientific work that characterized the first half of this century. He will probably consider the adoption of random selection to be a minimax strategy or generalized defense against adverse characteristics of the population that might lead to excessively large sampling error and serious bias in the application of sampling procedures to particular populations. In searching for the fundamental principles that have been followed in the development of modern sampling theory, he will observe that the early emphasis upon simple random sampling put the stress on unbiased selection of the sampling elements. He will then note that this emphasis shifted with the recognition that stratification and variable probability methods could lead to unbiased estimates through an appropriate matching of the bias of selection and the counter-bias of the estimating procedure. He may also find that the strict rule that all bias is to be avoided was softened and modified when it was recognized that some bias can be accepted if it is more than offset by reduction in variance, or, as systems analysts would put it, a favorable trade-off which accepts a modest amount of bias to obtain a greater gain in variance reduction.

The future historian is also likely to observe that simple random sampling was found to be somewhat inefficient in the sense that further control without sacrifice of unbiasedness could frequently produce very substantial reductions in variance. Indeed he would observe that there seemed to be no limit to the possibility of reducing variance through such controls up to complete elimination of sampling variation or reduction of sampling variation to so small a magnitude that it is negligible in the presence of non-sampling variation, including measurement error and the errors of application. Hence he will note as of considerable significance a trend toward utilizing more and more of what is known or knowable about the structure of the population and the logistics of obtaining data from the sample. These logistic processes involve fundamental cost factors and various factors of scheduling and organization as well as other problems of gaining access to the sample.

This attempt to anticipate the diagnosis of our times by a future historian is somewhat of a digression, but it does suggest that we can profitably devote time and resources to exploring ways in which we can take advantage of the structure of the population to a greater extent than we do currently, while preserving our fundamental commitment to rational scientific procedure. The rationality of the procedure essentially involves formulating specific rules and then imposing strict discipline on adherence to the rules in the actual sampling operations. It also involves utilizing for the design of the entire sampling system the results of theoretical tests and empirical comparisons with the performance of alternative sampling schemes.

It should be noted that what is said about the structure of a population can also be extended to the structure of a process when the sampling involves observations of change over time or indeed any aspect of a dynamic system.

Hybrid methods: spaced random sampling and patterned selection

In their discussions of systematic sampling, Bowley, Cochran, Kish, Hansen, Hurwitz, Madow, Quenouille, and

Yates (6) point out the attractive advantages that may be gained from systematic rather than simple random sampling but qualify all this by stern warnings about the dangers that inhere in the possible periodic character of the sequence in which the population is ordered. These dangers are greatest when they are not immediately evident and when the interval used in stepping off the systematic selection coincides with some integral multiple of the fundamental period or periods that are involved.

These writers generally warn the investigator that he alone is responsible for appropriate measures to recognize and cope with this problem but they do not tell him how to discharge that responsibility. If an investigator finds evidence of periodicity in a sequence, it would not seem appropriate in these times to advise him to retreat to simple random sampling. On the contrary, the more that he knows about the periodic structure of the population, the more we might expect him to seek ways to benefit by taking advantage of it, appropriately scaling the interval or intervals used in the systematic selection so as to break step with the periodicity.

Further, as Deming, Hansen, and the author have shown in selecting a mammoth sample of the population of the United States during the 1940 Census (7), a recognizable but somewhat complicated repetitive pattern of variation in the population can be sampled advantageously by an appropriate pattern of selection in which the lengths of the successive steps or intervals are not constant but vary according to a predetermined cycle of specified values. In this case the cyclic pattern necessarily was specified on the basis of very limited information about the structure of the sequence as it would develop in the course of the enumeration of the population.

In more complicated cases, population sequences may show little or no evidence of periodicity but exhibit a high degree of correlation between adjacent units accompanied by correlation that decreases as the distance between a pair of units increases or fluctuates in a damped or possibly irregular manner. Cochran (8) has shown that if the correlogram for pairs of elements with varying spacing in the sequence is concave upwards, systematic selection is advantageous, even relative to a stratified alternative. Since the sampler is often uncertain about periodicity and the strict concavity of the correlogram, he needs other methods of selection that retain some of the advantages of systematic selection while avoiding most or all of the hazards connected with periodicity when the periods involved cannot be determined. Two such essentially hybrid methods may be suggested:

(1) Spaced random selection

(2) Combined random-systematic selection.

In spaced random selection, the equal-step spacing of systematic sampling is introduced partially but otherwise selection is at random. The selection procedure is quite simple. Units are selected one at a time at random from among the remaining eligible units of the population. A unit is not eligible for subsequent steps of selection after it has been selected at some previous step or one or more of a designated subset of neighboring units has been selected. Thus the concept of sampling without replacement is extended to exclude from further exposure to the chance of being selected not only the units already selected but also all the units in specified neighborhoods around these units.

In the combined random-systematic selection, a major component of systematic selection is utilized but random selection is also incorporated into this hybrid procedure. Two simple varieties will be described to illustrate the idea. In the first variety, a systematic sample is drawn to define a sequence of neighborhoods from each of which one unit is then drawn at random. Units drawn to define the original systematic sample are not carried into the final sample unless they also happen to be drawn in the subsequent random selection from their neighborhoods.

In the second variety of this hybrid procedure, after each unit is drawn the sampler makes his next selection by counting further along the sequence of units a constant number of units, k, and then continuing the count to the (k+x)th unit from the immediately preceding selection, where x is a random choice from the integers from 1 to r. Thus he selects the (k+x)th unit following the last unit previously selected. This in turn becomes the starting point for a repetition of the counting with an independent determination of x that leads to the next selection. The procedure starts with a unit drawn at random from the entire population and proceeds along two branches, one going forward and the other backward from this starting point.

Clearly these selection procedures can be elaborated, applied to stratified and multistage sampling, and developed in several directions. For example, in spaced random selection, instead of excluding neighboring units, the sampler can merely reduce their probabilities of selection according to some appropriate rule. In combined random-systematic selection, in place of simple random choice of x from a discrete uniform distribution the choice can be made from some other appropriate distribution. Successive choices of x may be conditioned on previous choices rather than independent of them. In addition, k may be obtained independently at each step by selection from a suitable distribution. The population itself may be arranged in a loop or circle.

Estimating the sampling error variance of estimates produced by these methods of selection is of major importance in their use. The problems are complicated but not different essentially from those encountered in the case of systematic sampling, controlled selection, and selection of one unit from each stratum. To the extent that knowledge of the structure of the population has been used effectively, the error variance should be less than that of equivalent sampling systems using traditional forms of random sampling.

Spaced random selection largely avoids the problems of periodicity. It reduces the effect of high correlation between neighboring units but only partially achieves the benefits of stratification that are implicit in systematic selection.

The first variety of combined random-systematic selection only partly avoids problems of periodicity but controls the high correlation of neighboring units. The second variety of the combined method controls periodicity and the high correlation of neighboring units when k and r are chosen suitably but makes the sample size quite variable. A corrective step can be taken when necessary: oversample and then prune the sample back to the specified size removing at random a sufficient number of the units that had been selected.

Clearly these methods of selection offer options in their specifications that can be used to make them either closer to simple random selection or closer to systematic selection as the sample designer may prefer. They enlarge his region of choice while maintaining a strictly objective formulation of the selection process in terms of a probability model. It may be easier to obtain information about the correlation of neighboring units for the variables that are to be estimated than it is to obtain information about periodicity and more remote relationships in the sequence, or the reverse may be true. Hence the sampler can design the selection procedure using what he knows about the sequence and choosing the degree of mixing of systematic and simple random selection that seems appropriate.

The operating characteristics of these hybrid selection procedures are being developed and will be published in another paper. The probability functions involved in specific forms of the hybrids tend to be somewhat complicated but readily adaptable to computation for specified values of the parameters on modern computers by combinatorial and recursive methods. The spacing feature that is shared by these methods is reflected in the zero (or reduced) probability that pairs of elements, separated in the sequence by k or fewer elements, occur in the same sample. This reduces the variance of sample means and similar statistics when such pairs show a preponderance of positive serial correlation in the correlogram of the sequence. Control or balancing out of periodic variation in the sequence is accomplished to the extent that the probabilities for other pairs approach equality or vary in a manner that does not resonate with the periodic pattern of the sequence. Examples of the pair probability density functions for several forms of hybrid selection are given in Figure 1.

<u>Nexus sampling and configurational analysis</u>

A substantial contrast distinguishes the concepts of population and their equivalents, as used by sample survey statisticians on the one hand, from the concepts as used by biologists and social scientists, on the other. The latter conceive a population to be something of an entity or an organic system characterized by important relations of interaction and structural position of the elementary units that make it up whether they be cells, individual organisms, or social groups. Statisticians think of populations essentially as sets of objects for which any interrelationships that exist can be ignored. For statistical purposes they are considered as if they were simply aggregates or classes or objects, assuming that theelements do, or could just as well, exist in isolation of each other. While statisticians count and measure these objects, assemble then into subsets such as strata, clusters and a variety of sampling units, and, for the analysis of sample data, subdivide the sample into domains and a variety of categories or classifications, each comprising one or more elements, these collections merely create simple bonds of membership in logical assemblages of elements rather than inherent relationships existing apart from the statistical analysis. Thus, the inherent relationships that exist between elements of a human or biological population enter the survey, if at all, merely as characteristics or properties of the elements involved rather than as connections between one element and others.

In <u>nexus</u> sampling these relationships are not ignored. Indeed, they are frequently more important for the subsequent analysis than the elements which they connect. The selection of the sample is necessarily affected by the network of relationships that exists in the population. It is true, of course, that the network of relationships that is recognized in the sampling may be only a minor part of the entire complex of interrelationships that exists among the elements of the population. Hence, nexus sampling may also ignore a substantial number of relationships which, in some other study, would be examined and observed.

This shift in the orientation of the sampler toward the population he is sampling is a second extension of sample survey theory and technique which is likely to be of increasing importance in the future. It will take a great variety of forms. In some of these it may appear to be only a substitution of elementary relationships as units of sampling in place of the elements themselves that ordinarily are taken as the units of sampling. When it is merely this, no very significant extension of theory and practice occurs. Thus a study based on a sampling of divorces differs from a sample of divorced people in that it is concerned with relationships rather than individual human beings as the units of sampling but the techniques of selection and analysis may not be essentially different from those in the more traditional type of sampling. However, very important and fundamental differences are encountered when one proceeds to other types of nexus sampling.

It is not difficult to find a wide range of examples of data in which nexus sampling is involved. They appear in sociometry, the study of interrelationships within human groups, and in anthropological studies of kinship relationships and systems. They arise in studies of traffic and other types of flows within networks of transportation facilities or analogous systems. They occur in neurological studies, in studies of data reported by a system of weather stations, in crystallography, in the record systems of physicians and health agencies that treat some of the same patients, in the analysis of particle tracks, and in studies of political and industrial organizations. In sociology, they are encountered in the propagation of rumor as well as more subtle systems of personal influence and power structures.

Although several articles have been published on the problems of nexus and configurational sampling (9) the foundations of the theory necessary for full exploitation of these types of problems remain to be established. In abstract terms it may be expressed with the aid of graph theory and matrix representation as well as other available systems of describing and analyzing systems of interrelationship. We can start out by defining nexus sampling as a process of extracting from a portion of a system information about the entire system and, in particular, information that is not merely a summary or tabulation of the properties of the elements that make it up. This immediately suggests that there is need for a critical analysis of the types of structure to be found in systems of interrelationships.

The taxonomy of systems is an extensive and deep subject. There is much available from graph theory and its applications (10). For example, Berge defines a graph as 1) a set X and 2) a function Γ mapping X into X. From this one can see that while X is a population, in sampling a graph the mapping function is coordinate in importance. He goes on to explain that, "The parenthood relationships amongst a group of people define a graph, as do the rules of chess, the connections between several pieces of electrical appartus, the victories of competitors in a tournament, etc. ..." Nexus sampling is concerned with the extent, concurrence, redundancy, complexity, and many other features of systems of relationship. The elements thus connected are termed <u>nodes</u> in recognition of their association with relationships.

In this paper there is no room for development of a suitable taxonomy for sampling purposes but some aspects of the anatomy of a system as represented by a graph may be mentioned. Elements of the system are represented by points (vertices). Relationships are represented by arcs (directed lines) connecting pairs of points. Relationships between more than two elements can be represented by drawing all arcs applicable to the relationship. The arcs may have associated with them attributes, as well as various quantities specifying their capacity, cost, and other properties. Paths, circuits, networks, and arborescences are composed of arcs. Edges (undirected lines) make up chains, cycles, trees, and other formal combinations.

From the standpoint of the design of a sampling system, it is important to know how the observer or researcher will have access to the population for the purpose of obtaining data. Here we encounter one of the essential differences between nexus sampling and traditional sampling for we contemplate that the information the researcher is seeking may be available not merely at each node for that particular node, as in the traditional case, but he may find at several nodes parts or all the information involved in the relationship between these nodes as well as information at each node about several other nodes with which it is connected by an arc, an edge, or a path or chain. For example, in the study of divorce, information about the family life of the couple may be obtained from either the former husband or the former wife, as well as from close relatives and other knowledgeable persons.

Clearly an important question arises as to the accuracy of the information that is available in each situs. Setting that problem aside for the moment, one can recognize that there may be a great deal of redundancy in the storage of information within the system. In the extreme, for example, one may find that a single person such as the head-man of a village is able to give information about all the persons resident in the village. Typically, anthropologists obtain data about whole societies through a few informants who were commonly accepted as having relatively full information about genealogies and kinship relations as well as a wide range of customs and cultural traits. Other informants might serve almost as well. In this sense, then, a sample may contain multiple sources of information on the same subjects and particulars.

In other situations the information obtained at each node is supplementary or complementary to information obtained at other nodes and it is necessary to obtain partial information from each of two or more nodes in order to assemble a complete unit of information needed for the research. In contrast to redundancy, therefore, this kind of situation requires the examination or observation of several elements before a single unit of information can be obtained.

Consequently an essential part of the fundamental theory of nexus sampling must include the formulation of the situation with respect to the available means of access to elements, the manner in which the information is stored or lodged in the elements, and the possibility that there is some information that cannot be found at all in the elements but must be obtained directly from an examination of links between them. From a practical standpoint the means of travel from one element to another may become a significant part of the theory. The costs associated with the collection of data will necessarily be involved in the theory when it is directed toward problems of optimal procedure.

Figure 2 presents a simple example of a situation in which information is available at several places or positions. The sampling problem is concerned with the selection of those nodes that will contribute a maximum needed information in some preferable manner within the constraints of cost and other considerations that must be recognized in the development of the sampling plan. The following information would be provided by various samples of two nodes each:

Number of nodes for which information is obtained	Number of distinct samples
2	7
3	11
4	10
5	15
6	2
Total	45

This choice opens at once the possibility of reducing costs by sampling with unequal probabilities. It is further complicated if obtaining redundant information has value in increasing the accuracy of the information obtained through cross-checking or verifying by merging data from more than one source.

Figure 3 exhibits a situation in which the information is complementary or supplementary. It is necessary to have information from all of the nodes that contain it concerning any particular element represented by a capital letter. In this case the following samples are possible when these nodes are selected:

Number of nodes for which complete information is obtained	Number of distinct samples
0	15
1	14
2	5
3	1
Total	35

In some samples, none of the information is complete even for one element. This situation clearly invites not only sampling with unequal probabilities but sampling with conditional probabilities establishing relationships of dependency between the selection of one element and the selection of others associated with it through the sharing of information about the same elements.

Further development of these questions and the techniques appropriate for them as well as the general theory that is applicable will be published in a separate paper. In that paper the use of matrix representation will be compared with other means of developing the sampling techniques and inferring their consequences in terms of the yield of information and the variability of estimates.

One of the methods of selection that has been suggested in previous articles is that of <u>snowball sampling</u>. In this procedure the survey proceeds from an initial sample of elements obtaining from each element information about other elements to which it is connected in the system. The next step is to add to the sample some or all of these related elements, obtaining data from them and also information about still other elements to which they are connected. Thus step-by-step the sampler proceeds from a starting set of elements to a larger set connected with them by one or more links of relationship. This type of sampling has been examined in detail by Leo Goodman (9).

In some situations the sample grows too rapidly. It is then necessary to sub-sample the relationships that are reported at each extension of the snowball process to new elements. Another problem about snowball sampling, of course, is that it is strongly biased toward the examination of nodes that have many interrelationships or are closely coupled to a large number of other nodes that may serve as the starting points of tours that will reach it. In contrast, the probability of selection is least for an isolate because it can only be chosen directly in the initial sample. Similarly, a node with only a few relationships is very unlikely to be selected unless its relationships extend to a subset of nodes that has a high probability of being reached by a random tour that could continue toward the node in question. The seriousness of this problem depends of course upon the purposes of the investigation.

If the means of access to the population includes a frame on which the relationships among elements is designated then the problem of design is quite different than it is when the relationship structure must be discovered and explored during the course of the collection of data. In such instances, for example, there would be no problem about giving isolates an appropriate probability of appearing in the sample. It would also be fairly simple to scale the probabilities in a suitable manner for nodes with differing numbers and relationships of various orders.

On the other hand if the researcher is working with an imperfect or obsolete frame, the deficiency may not be as serious in nexus sampling as it is in traditional sampling since the frame will be utilized only to obtain starting points for random tours that will lead to nodes not represented on the frame providing they are not isolates. Thus the design problems of snowball sampling are intimately involved in questions about the means of access associated with each available frame and the formulation of rules to be followed in laying out random tours.

In contrast with snowball sampling, when the information stored at each node includes information about nodes directly related to it, the sampler will refrain from examining these closely related nodes. He will, of course, proceed to them if his research is directly concerned with relationships between links.

In the classification of system structures it is necessary to take account of the purposes of the research. For some purposes relationships between arcs may be ignored and the arcs themselves will be the only subject of interest. In other investigations it will be quite important to consider not only the separate links but the relationships among links to higher orders of complexity.

Somewhat unique estimation problems are encountered in nexus sampling. For certain questions it is necessary to examine all the nodes in the system in order to obtain answers. For example, if the question is "How many isolates does this system contain?" one can obtain suitable estimates by taking a sample of nodes from a frame and discovering whether or not they are isolates. On the other hand, if the question is "Does there exist in the system one node that is connected to all the other nodes?" and the information can only be obtained by direct examination of this particular node, then almost the entire population must be examined to get an answer to the questions with a reasonable level of confidence.

Nexus sampling will tend to utilize procedures for sampling with variable probabilities. As a consequence the forms of estimation functions that have been developed for use in sampling with variable probabilities are likely to be much more prominent in nexus sampling than in traditional sampling operations.

Interesting questions arise as to the possibility of clustering in nexus sampling. For example, James Coleman suggested in addition to snowball sampling a procedure he called "saturation sampling". This consists of taking one sub-system,such as one community out of a number of communities, or several such subsystems and then interviewing or obtaining data from all the nodes they contain. The procedure is exactly that of cluster sampling in the traditional sense without any further sub-sampling of the clusters that are selected. It leaves open all questions about interrelationships from one cluster to another. In some instances, there may be serious problems arising from restriction of the study to only a few clusters.

Problems of efficiency and optimization can be quite readily solved for some types of nexus sampling but not for others. Not only are there joint costs involved but, as was previously indicated, redundancy in the availability of information at one node and another may make the programming of sample selection and field operations excessively complex. In a formal sense the sampling problems connected with nexus sampling will parallel some of those of controlled selection but there are distinct problems introduced by this extension of traditional procedures that well merit further study and development both by theoreticians and by practitioners.

Graduated sampling: surveys of processes and variable examination of elements

In a number of surveys the objective has been to obtain information about processes or sequences of change occurring among the elements of a population. The elements in the sample are usually surveyed and measured at only one point of time in the interval during which the process is underway. Data may also be obtained at that time about past changes, but in contrast to information obtainable about the current state of the population these data are usually incomplete, inaccurate or especially difficult to acquire. Ordinarily, it is not possible to ascertain what further changes will occur and when the process will terminate. This is true typically of chronic diseases and disabilities but it has many counterparts in life testing of equipment, systems reliability determinations, and a great number of biological, social and demographic research problems. Surveys of a growing crop to provide an estimate of the yield or harvest, traffic surveys to provide a basis for planning new bridges and highways or systems of traffic control, surveys of the development of public opinion and effects of political campaigns, and ecological studies are important examples.

If the population is homogeneous and all elements are undergoing exactly the same metamorphosis or standard process of development, the sampling problems may not be very complicated. However, ordinarily each element has, as it were, its own unique set of parameters and often follows only one of several variants of the process. Usually the rate at which the process is progressing from its initiation to its termination is a variable of primary importance in the research. Estimates of this rate, as well as the actuarial or statistical properties of a population of heterogeneous elements differing in their rates of change and probabilities of the possible outcomes, are uppermost in the research. Estimates are also required, in many instances, projecting from observations of "incomplete intervals" the distribution of the lengths of these intervals when they will have been completed.

In basic survey theory, this phase of sampling may be regarded as cascaded or concatenated, i.e. consisting of a sampling of individual units followed by a sampling of the stages each individual in the sample passes through during the process. The element in the second sampling is one stage, the particular stage the individual has reached at the moment it is observed or measured. Alternatively, the sampling may be viewed as a single step of selecting and observing part of a population of distinct realizations of processes. In either case there are fundamental questions about relative exposure to the risk of being observed or selected since various intervals or stages in each process may be sampled with different probabilities. If the sampling covers the population as it exists at a point of time, the probability that a period or interval will be selected is proportional to its length and that length is not observable at the time the sample is selected (11). Sometimes the situation is more complicated. For instance, the process may not be observable until it has proceeded to a certain stage. The termination of the process may set in motion activities that facilitate or impede taking a sample of the population as each element reaches its terminal state. Clearly there are many possibilities for which the problems of designing a sample procedure are complex and peculiar to the situation in which the sampling is to be done.

Among the fundamental problems for the development of sample survey theory and practice are the relative advantages of 1) a single cross-section survey and 2) continuing to observe the elements in the sample or repeating the observations. In the latter, each element in the sample may be reexamined one or more times or indefinitely until some further stage is reached. Thus the period and intensity of further observation may vary from element to element depending on the stage that has been reached in the process, the elapsed time since that stage was reached, and the rate of change as observed at the initial observations or subsequently. The sampling problems of panel surveys and research programs involving "follow-up" procedures have been relatively neglected in the theory of survey sampling but must be faced in the evaluation of actual survey results.

A closely related problem is that of the "mortality" or attrition of a sample from the time it is selected to the end of the program of data acquisition. Not only may a large proportion of cases be "lost to follow-up" when they refuse to continue to cooperate in a survey or when they can no longer be located, but the losses at the beginning of the data acquisition may not be negligible. Many surveys suffer losses of as much as 5, 15, or even 40 per cent of the sample before the first interview or set of measurements has been obtained. Additional losses are incurred during subsequent observation of the sample. The effects of this attrition are seldom reported and presumably are not known even vaguely in many studies. Their influence on the sampling processes and intervals can be especially serious.

Even less is known about the effects of hidden attrition occurring after the specification of the population upon the choice of a frame, through such deficiencies as inadequate or incorrect addresses and the omission of recent additions to the population.

Attrition is but one form of variation in the extent and intensity of data acquisition from elements of the population. The traditional concept that dominates the theory of survey sampling is that data are taken from all the elements in a sample selected according to a strict and clearly defined procedure but not from the remainder of the population. In contrast the actual performance of sampling operations produces subsets of the population that deviate from the hypothetical sample in ways not often clearly defined or well understood.

Both attrition and inadequate performance of the observing and measuring procedures are constituents of the selective procedures and the actual survey. Why not, then, extend the concept of sample selection to one which allocates the effort of data seeking in differing degrees and forms to every element of the population, obtaining for some elements only the information provided by the frame, for some a single standard set of data but with variations in its actual quantity and quality, and for others an elaborate program of initial and successive steps that develop information about the entire syndrome or process that is the focal point of the research problems of the survey. The concept of variable or graduated data acquisition from all elements of the population generalizes the concept of sampling and makes it more realistic when applied to survey operations. It also strengthens the statistical reasoning by which one proceeds from actual sample data to the inferences that are drawn from them.

Summary

In three general directions, the theory of survey sampling can be freed from some of the traditional constraints implicit in its prevailing conceptualization. This makes it possible to merge and extend certain rational formulations centering in the selection procedures, the specification of the population, and the relations between sampling and data acquisition within dynamic systems and processes of change. These ideas are being examined and further reports are being prepared but the pursuit of some of their implications may well continue to enlarge the scope of sampling theory for years to come. They merit frequent discussion and study by sampling theorists and practitioners, working together as in this Seminar and this Institute of Statistics which are truly an inspiration for cooperation among statisticians and other scientists everywhere.

REFERENCES

(1) Yates, Frank, "A review of recent statistical developments in sampling and sampling surveys." Jour. Royal Stat. Soc., 109 (1946) 12-43.

Mahalanobis, P. C., "Recent experiments in statistical sampling in the Indian Statistical Institute." Jour. Royal Stat. Soc. 109 (1946) 325-370.

Stephan, Frederick F., "History of the uses of modern sampling procedures." Jour. Amer. Stat. Assn., 43 (1948) 12-39.

Seng, You Poh, "Historical survey of the development of sampling theories and practice." Jour. Royal Stat. Soc., Series A, 114 (1951) 214-231.

Moser, C. A., "Recent developments in the sampling of human populations in Great Britain." Jour. Amer. Stat. Assn., 50 (1955) 1195-1214.

Zarkovic, S. S., "Note on the history of sampling methods in Russia." Jour. Royal Stat. Soc., Series A, 119 (1956) 336-338 and the supplement, 125 (1962) 580-582.

Sukhatme, P. V., "Major developments in the theory and application of sampling during the last twenty-five years." Estadistica 17 (1958) 652-679.

Thionet, P., "Développements récents de la théorie des sondages." Jour. Soc. Statist. Paris, 100 (1959) 279-296.

Murthy, M. N., "Some recent advances in sampling theory." Jour. Amer. Stat. Assn., 58 (1963) 737-755.

Sukhatme, P. V., "Major developments in sampling theory and practice." in F. N. David (ed.), <u>Research Papers in Statistics, Festschrift for J. Neyman</u>, Wiley, 1966, pp. 367-409.

Dalenius, Tore, "Recent advances in sample survey theory and methods." Annals Math. Stat., 33 (1962) 325-349.

(2) Bailey, William B. and John Cummings, Statistics, Chicago, A. C. McClurg, 1920, pp. 5-7.

Lynd, Robert S. and Helen M. Lynd, Middletown, Harcourt Brace and World, 1929.

(3) Neyman, Jerzy, "On the two different aspects of the representative method: the method of stratified sampling and the method of purposive selection." Jour. Royal Stat. Soc. 97 (1934) 558-625.

Hansen, M. H., and W. N. Hurwitz, "On the theory of sampling from finite populations." Annals Math. Stat., 14 (1943) 333-362.

(4) Wald, A., "Sequential tests of statistical hypotheses." Annals Math. Stat., 16 (1945) 117-186 and Sequential Analysis, Wiley, 1947.

(5) Goodman, Roe, and Leslie Kish, "Controlled selection — a technique in probability sampling." Jour. Amer. Stat. Assn., 45 (1950) 350-372.

(6) Madow, W. G. and L. H. Madow, "On the theory of systematic sampling." Annals Math. Stat. 15 (1944) 1-24 and 20 (1949) 333-354.

Yates, Frank, "Systematic sampling." Phil. Trans. Royal Soc. A 241 (1948) 348-377.

Quenouille, M. H., "Problems in plane sampling." Annals Math. Stat. 20 (1949) 355-375.

Cochran, W. G., Sampling Techniques, Wiley, 1953.

Kish, Leslie, Survey Sampling, Wiley, 1965.

Hansen, M. H., W. N. Hurwitz, and W. G. Madow, Sample Survey Methods and Theory, Wiley, 1953.

(7) Stephan, Frederick F., W. Edwards Deming, and Morris H. Hansen, "The sampling procedure of the 1940 Population Census." Jour. Amer. Stat. Assn., 35 (1940) 615-630.

(8) Cochran, W. G., Sampling Techniques, 2nd Ed., Wiley, 1963

(9) Mahalanobis, P. C., _op_. _cit_. pp. 47-48, 60.

Coleman, James S., "Relational analysis: the study of social organizations with survey methods." Human Organization, 17 (1958-59) 28-36.

Goodman, Leo, "Snowball sampling." Annals Math. Stat. 32 (1961) 148-170.

(10) Berge, Claude, _The Theory of Graphs and its Applications_, Methuen 1962.

Flamant, Claude, _Applications of Graph Theory to Group Structure_, Prentice-Hall 1963.

Saaty, Thomas L. and Robert G. Busacker, _Finite Graph and Networks_, McGraw Hill 1965.

Harary, Frank. Robert Z. Norman, and Dorwin Cartwright, _Structural Models_: An Introduction to the Theory of Directed Graphs, Wiley 1966.

Barton, D. E., and F. N. David, "The random intersection of two graphs." in F. N. David (ed.), _Research Papers in Statistics, Festschrift for J. Neyman_, Wiley, 1966, pp. 445-460.

(11) Blumenthal, Saul, "Proportional sampling in life-length studies." Technometrics 9 (1967) 197-218.

METHODS OF STUDYING AQUATIC LIFE IN RIVERS

Ruth Patrick

Chairman, Department of Limnology

Academy of Natural Sciences of Philadelphia

Our main problems in studying the aquatic life in rivers are due to the fact that the organisms are highly clumped; that we have an open system; and that we have a very large number of variables determining the structure of the ecosystem. I will briefly describe the method of evaluating the condition of large and small streams as developed by the Limnology Department of the Academy of Natural Sciences, a method based upon studying the principal species other than the decomposers that constitute the aquatic communities in any given area.

Methods of Study of the Ecosystems of a Stream

Where it is desirable to compare various ecosystems or aquatic communities the areas must be carefully selected so that they will be as ecologically similar as possible. The environmental characteristics that are of primary importance are the current, light, composition of river bed, and chemical and physical characteristics of the water. Current patterns must be similar, which means that comparable areas must have slow water areas where sedimentation may take place and areas with varying rates of current passing through them. In shallow streams one usually selects an area long enough to include riffles, slack water, and polls. In deep streams or rivers one selects areas which include shoaling areas as well as cutting areas and, if possible, a backwater.

In rivers, an area extending far enough to include a small meander is usually the most suitable. Such areas should also be selected to include stable debris such as fallen trees where the current is fairly fast. This is because most of the aquatic life lives within the photosynthetic zone, and varies according to the current.

The substrate is also an important consideration. In shallow water streams one wants to include all the common types of substrates. Typically, the riffles will have fairly large rocks and rubble while the slack waters will include rocks, pebbles, and sand. Often on the edges of slack water sandy mud is present. In the pools one finds silt and sandy mud. In deep rivers in the coastal plain most of the bed is sand and it is only around the meanders or in backwaters that sandy mud or silt is found. Most of the bed of such large rivers is often relatively sterile for the sand shifts back and forth and thus stable habitats for aquatic organisms tend to be few. Usually, it is in the more shallow waters at or near the limits of the photosynthetic zone that more stable habitats are found and bottom forms such as mussels occur. On the cutting edge of the river just below the water line the mud banks often form fine habitats for burrowing mayflies. On the lower Savannah River one often finds the banks riddled with their holes. Of course, if the river bed is more stable and not too hard, as in some of the muddy-bed midwestern deep rivers, the bed proper may form good habitats for burrowing mayflies, chironomids, and some worms.

The orientation of the area in relation to light is also important. An east-to-west (or vice versa) flowing section is often more productive than a north-south flowing section. Also a section of stream in open sunlight or filtered sunlight is more productive than one in deep shade.

The physical and chemical characteristics of the water must also be similar. I refer particularly to such physical characteristics as light penetration and temperature, and to such chemical characteristics as hardness, total conductivity, alkalinity, pH, etc.

The study of a selected area is made by a team of scientists, each one a specialist in one or more groups of aquatic life. A typical team consists of a phycologist, protozoologist, invertebrate zoologist, entomologist, and ichthyologist. Also included is a water chemist and a bacteriologist.

Each biologist enters the area of study and tries to collect all organisms of his interest group which seem to have sufficient populations to rate them as established in the area rather than being mere transients. He uses the gear appropriate for collecting his group of organisms. This type of collecting differs from the usual collecting done by the systematist in that all species which seem to be established are collected rather than the collecting being primarily of new species. The collector is careful to collect all life forms of the species and to determine how many stages in the life history of each species are present. He also estimates the abundance of the species as to whether they are very abundant, common, frequent, or rare.

Many attempts have been made to collect quantitatively on a statistically reliable basis. This is almost impossible to accomplish in a large river. In a small stream it can be done, but it is extremely time-consuming. We have done it where the type of study calls for it. However, we find that estimates of abundance by the collector together with relative abundance among the species is in most cases a satisfactory method. The importance of quantitative collecting that is statistically reliable, as with any type of collecting, depends on the questions being asked.

The water chemist also collects samples for determining the general chemical characteristics of the water as these are important parameters in determining the kinds of species present as well as their abundance. Enough samples are taken to obtain a reliable estimate of the mean and the standard error of the mean for the period during which the biological sampling is done. In some studies such analyses are carried out twice a month for all months of the year.

The bacterial determinations are not very extensive. Usually total counts, "coliform type" bacteria, and biochemical oxygen demand tests are done. We realize much more study of the decomposers in the ecosystem should be carried out as they represent one of the most important stages in the transfer of energy; however, due to the meager knowledge of these forms at the present time any thorough study does not seem possible. To me, an understanding of what the decomposers are in a flowing system and what are their ecological requirements is one of our greatest research needs.

The results of several hundred studies of stream areas (Patrick, 1949, 1961; Patrick, Cairns, and Roback, 1967) have shown that in ecologically similar areas of natural streams the number of species of the various groups of aquatic organisms are very similar and that these numbers usually do not vary greatly from season to season, although the particular species present may vary greatly. Furthermore, we have found that most species are represented by relatively small populations, although a few may be common to very common and a few are rare. Our studies have shown that often a single species may not be of much value in indicating ecological conditions, whereas associations of species--particularly if they represent various major groups of organisms--may be very valuable in indicating ecological conditions. Of course, there are exceptions when the occurrence or lack of occurrence of a single species may be indicative, but this is relatively rare.

In our studies of the Savannah River, which began in 1951, we have spent a considerable amount of time with the aid of a computer attempting to determine more of these associations or matrixes of species which occur together more often than chance would indicate in various areas at the same point in time or over time in the same area. To date we have not been able to identify many such matrices at the species level. We can, however, at the genus level in some groups and at the family or order level in other groups identify such associations. Since it is at the species or population level that one usually finds correlation with a given ecological condition, it is not very meaningful to make such correlations at higher systematic levels.

The fact that these matrices of species are so hard to find is probably because the kinds of species are so variable in spite of the fact that numbers of species remain similar. This, we believe, is due to the fact that in continental areas the species pool is much larger than the niches for species occupancy in any one stream. Small changes in the environment which would be very difficult to identify will give, for example, species A the advantage over species B and as a result greatly reduce the population size of B so that it is uncollectable if not eliminated.

If we can identify the groups of species which are performing the same function in the ecosystem and thereby can replace each other in the system, we probably will be able to find matrices of species groups which one can predict will be characteristic of the ecosystem at various stages of energy transfer. That is, we will be able to state, for example, that in the Savannah River at a given point in time one may expect 60 species of a group of 150 species which will function as primary producers and 50 species of a group of 175 species that will function as herbivores, etc.

Effect of Pollution on the Ecosystem

Our studies have shown that perturbation may affect these aquatic communities in various ways. Organic enrichment that is moderate in amount often causes certain species to develop excessively large populations but does not bring about a reduction in species numbers. Indeed, in protozoa, for example, it may bring about an increase in species of ciliates as many of these are bacterial feeders.

More severe organic pollution will bring about a reduction in sensitive species such as many insects, diatoms, and some fish, and the development of large populations of species that usually are relatively rare, such as *Sphaerotilus* and tubificid worms.

Toxic pollution, depending on the amount and type of toxin, restricts the populations of many species and as a result the nutrients in the system flow through those that are tolerant of the existing conditions, and they develop large populations. Mild toxicity often inhibits growth or reproduction but does not kill. As a result, one may find fairly high diversity with very small populations and/or stunted individuals. An example of this is the effect of low pH (4.5 - 5) on diatom communities in a circumneutral stream. In some experiments we carried out in 1967 on communities of diatoms in circumneutral streams we found diversity to be high but division rate to be greatly reduced, and as a result the standing crop was very low. Severe toxic conditions eliminate most species and the food web of a natural stream is completely destroyed.

The first effect of suspended solids is usually to reduce the depth of the photosynthetic zone and the numbers of species and population sizes of algae. More severe pollution of this type greatly reduces species numbers.

In determining the severity of the effect of perturbation in a stream area we compare the condition of the aquatic community in the area studied to that of a natural area of similar type in the same stream, or if not in the same stream, to averages developed from similar streams. Three stages of stream degradation are recognized.

"Semi-healthy" is the condition in which the balance of life has been somewhat disrupted but not destroyed. Often a given species will be represented by a large number of individuals which indicates that something has happened that has destroyed the predator pressure or control of the species. Under other circumstances one group will have many more species present than usually occur, while other groups will be greatly depressed.

"Polluted" is the condition in which the balance of life found in healthy streams is severely upset. However, conditions are favorable for some groups of organisms.

"Very polluted" is a condition which is definitely toxic to plant and animal life. Often many groups are absent, particularly insects and fish. Other groups, such as diatoms and various invertebrates, are greatly reduced.

Special Method for Studying Diatom Communities

Another approach to determine the condition or health of a stream has been developed by studying the structure of diatom communities. Diatoms are single-celled algae which grow in almost all types of water. They have a wall of silica, the characteristics of which form the basis of classification. Because the wall is silica no special method of preservation is needed, and their recognition is in no way impaired by time.

Patrick *et al*. (1954) determined that glass slides placed vertically in an instrument known as a diatometer were suitable for the development of diatom communities whose species composition and diversities were very similar to those growing in the areas in which they were placed. These studies showed that the structure of a diatom community was similar to a normal curve, with many species represented by relatively small populations, a few having large populations, and a few being very small. A sample of this community resulted in a truncated normal curve.

By studying a sufficient number of specimens so that the species in the mode have populations of four to eight specimens, one allows the community to set the size of the sample studied rather than making an arbitrary decision determining the amount of study. We have found that the structure of most natural communities can be determined by counting five to eight thousand specimens, although under certain conditions ten thousand or more specimens are counted (Figure 1).

Further studies have shown that the structure of the curve remains similar from season to season and year to year if no serious change in the environment occurs (Table 1). However, the populations of the kinds of species composing the curve may vary greatly. Furthermore, in always counting sufficient speciments to place the mode in the same interval one is able, by using Bennett and Franklin's normal bivariate distribution function (1954), to develop 95% and 99% confidence intervals for natural streams. This is because we can describe the curve by σ^2 and the height of the mode if we always count sufficient specimens to place the mode in the same interval. We have found these confidence intervals for soft freshwater and brackish water diatom communities (Figure 2).

Various kinds and amounts of pollution have various effects on the structure of the curve. In a natural stream which is typical of other than mountainous country or steep gradient areas of low nutrient value, the curve should cover 9 - 12 intervals. The height of the mode should be 18 or more species.

The first effect of organic pollution is to cause certain species to become excessively common, which causes the curve to have a long tail. If the organic pollution is greater, an additional change in the curve is the reduction of the height of the mode (Figure 3). This same shape of the curve sometimes results from high temperatures that are not severe enough to kill most of the species.

Light toxic pollution sometimes causes tolerant species to become very common, but the height of the mode is definitely reduced.

Heavy toxic pollution results in a curve with a very low mode, and the curve often covers a relatively few intervals, but rarely has a long tail.

Toxic conditions such as low pH that repress cell division may cause the curve to have a very small σ^2.

This method is useful for continually monitoring a stream. Furthermore, this method of developing algal communities on glass slides can be used to determine the relative productivity of various stream sections as measured by standing crop of diatoms or, in some cases, other algae. Since the slides can be placed in the river at the same time and later removed at the same time, standing crops for various streams or different sections of the same stream at a point in time can be compared. Such studies are useful in determining the relative contributions of different sources of nutrients in promoting algal growth.

Further studies have shown that natural aberrent conditions may affect the structure of the diatom community in similar ways as manmade perturbation. I refer to the structure of diatom communities found in acid dystrophic streams in New Jersey (Figure 4). These communities are composed of fewer species, but some of those which are tolerant of these conditions develop very large populations, and the structure of the community is similar to those found in polluted streams.

We have also used various other methods for describing the structure of the diatom community. For some types of study the Shannon-Weaver diversity index is useful. This method is particularly useful for describing the equality of the distribution of the specimens among the species. The limitation of this method is that it is difficult to differentiate between communities that have a heavy dominance of a few species and 20 species that are scarce and a community which has a heavy dominance of a few species and 50 species that are scarce.

Two methods have been used to relate the diversity of a community to a standard or reference community. One method is that described by MacArthur (1965) in which one compares the differences of diversity between the test community and the control community.

The second method is that of Lloyd and Ghelardi (1964) in which they relate the diversity of the community being studied to a theoretical community in which the individuals are distributed among the species according to MacArthur's broken stick model. Thus a community of 50 species with unequal distribution of individuals in the species may have the same diversity as a community of 25 species distributed as in the broken stick model.

Conclusions

In order to understand more thoroughly the structure of an aquatic community and the effects of perturbation of various types upon it, we should know the relationship between diversity and nutrient transfer through the ecosystem. We should also understand how diversity is related to stability and the multiplicity of paths in the food web. To date our knowledge of food preferences and the most efficient use of nutrients is meager for most aquatic organisms. We also do not clearly understand the relationships between the stability of the ecosystem and the numbers of species functioning as primary producers, herbivores, or carnivores. It is hoped that with these types of additional knowledge and the newer computer techniques we will be able to model correctly the structure and functioning of these aquatic ecosystems. From these models we hope to make valid predictions of the effect of perturbation.

TABLE 1

Savannah River
Summary of Catherwood Diatometer Readings at Station 1
October 1953 to January 1958

Date	Specimen No. in modal interval	Species in mode	Species Obs.	Species in theoretical universe
1953				
Oct.	4-8	22	150	178
1954				
Jan.	4-8	19	151	181
Apr.	2-4	24	169	200
July	2-4	23	153	193
Oct.	4-8	21	142	168
1955				
Jan.	4-8	19	132	166
Apr.	2-4	25	165	221
July	2-4	20	132	180
Oct.	2-4	27	171	253
1956				
Jan.	2-4	30	185	229
Apr.	4-8	35	215	252
July	2-4	24	147	185
Oct.	2-4	23	149	206
1957				
Jan.	2-4	29	177	233
Apr.	2-4	21	132	185
July	4-8	29	181	203
Oct.	2-4	25	157	232
1958				
Jan.	2-4	27	152	212
(Apr. 1954-1958 averages)		24	151	194

FIGURE I

RIDLEY CREEK, CHESTER COUNTY, PENNSYLVANIA

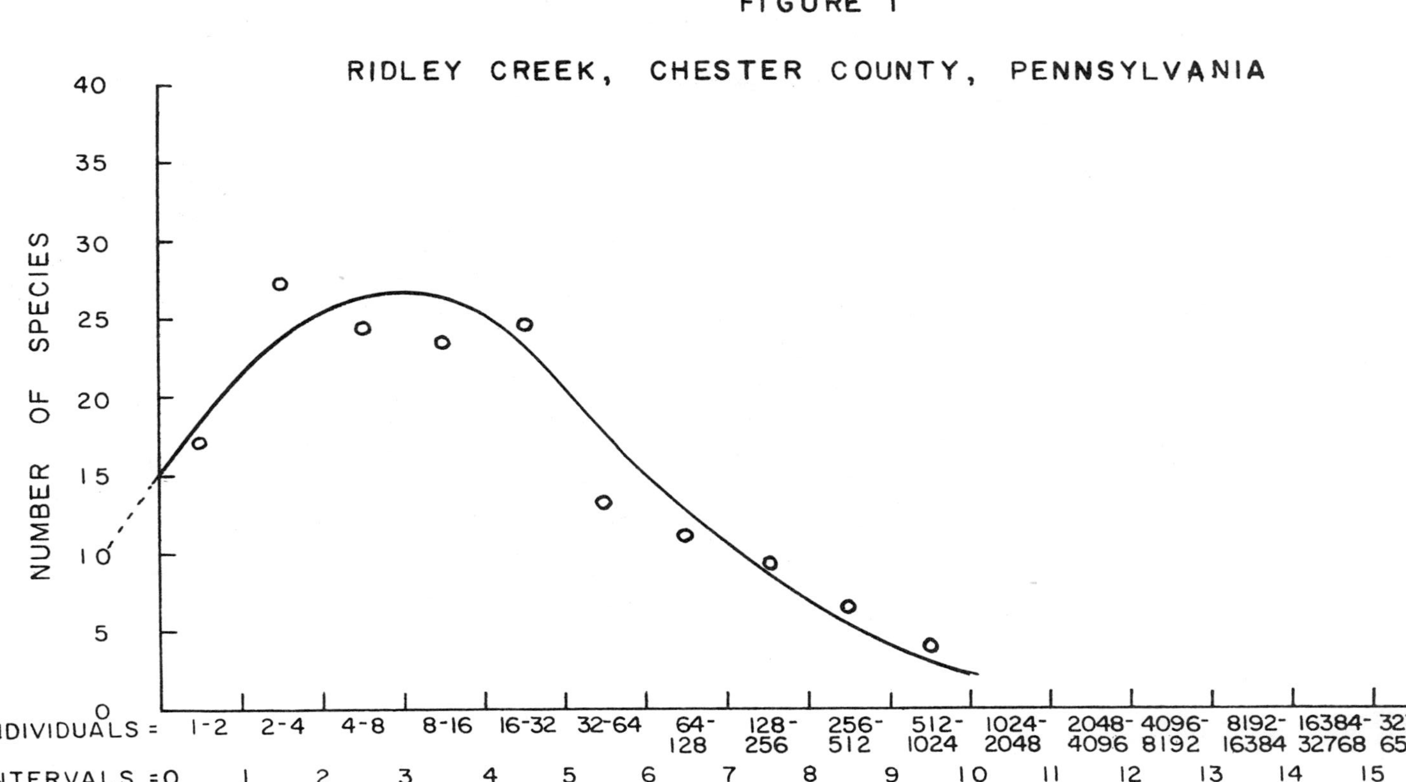

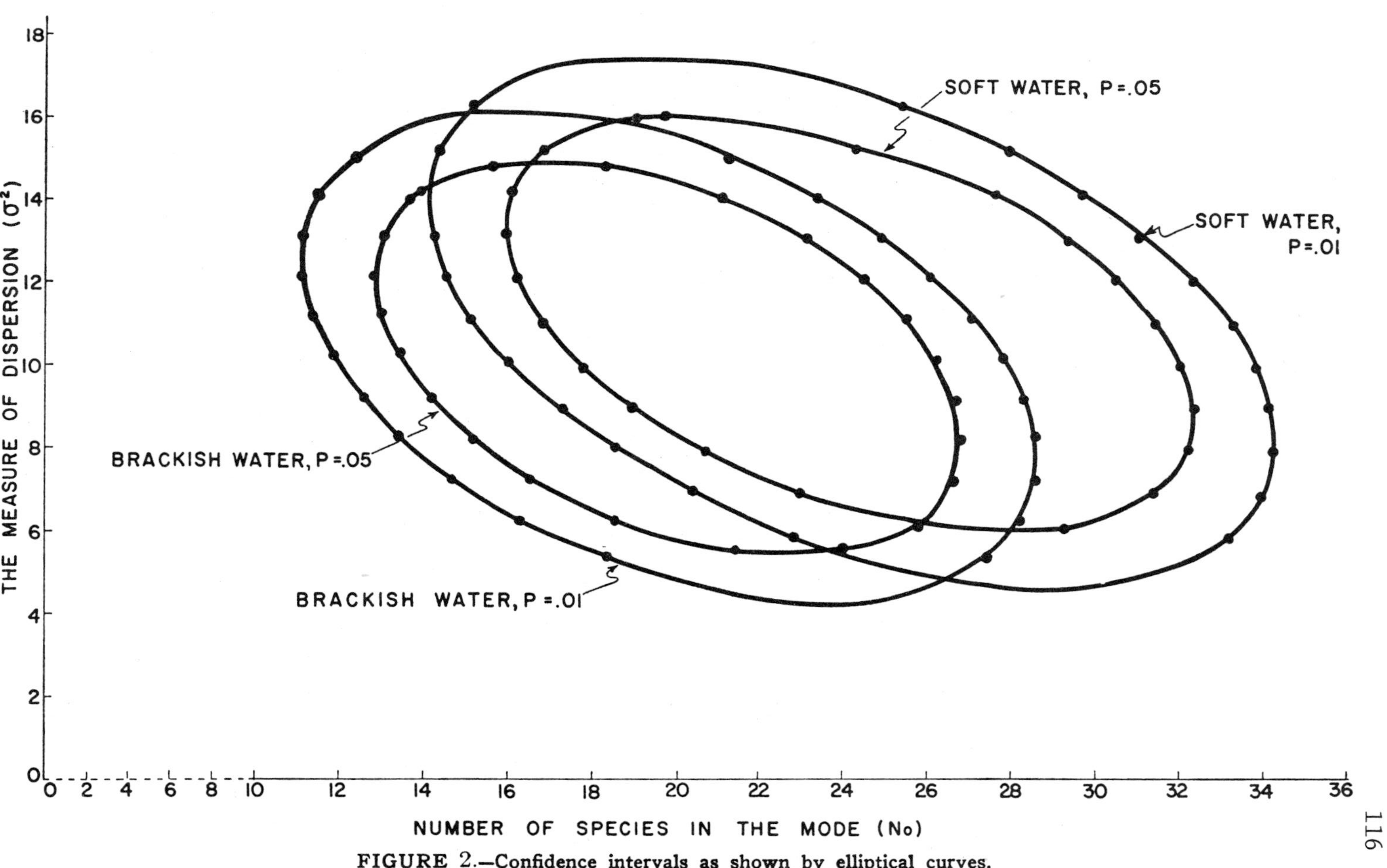

FIGURE 2.—Confidence intervals as shown by elliptical curves.

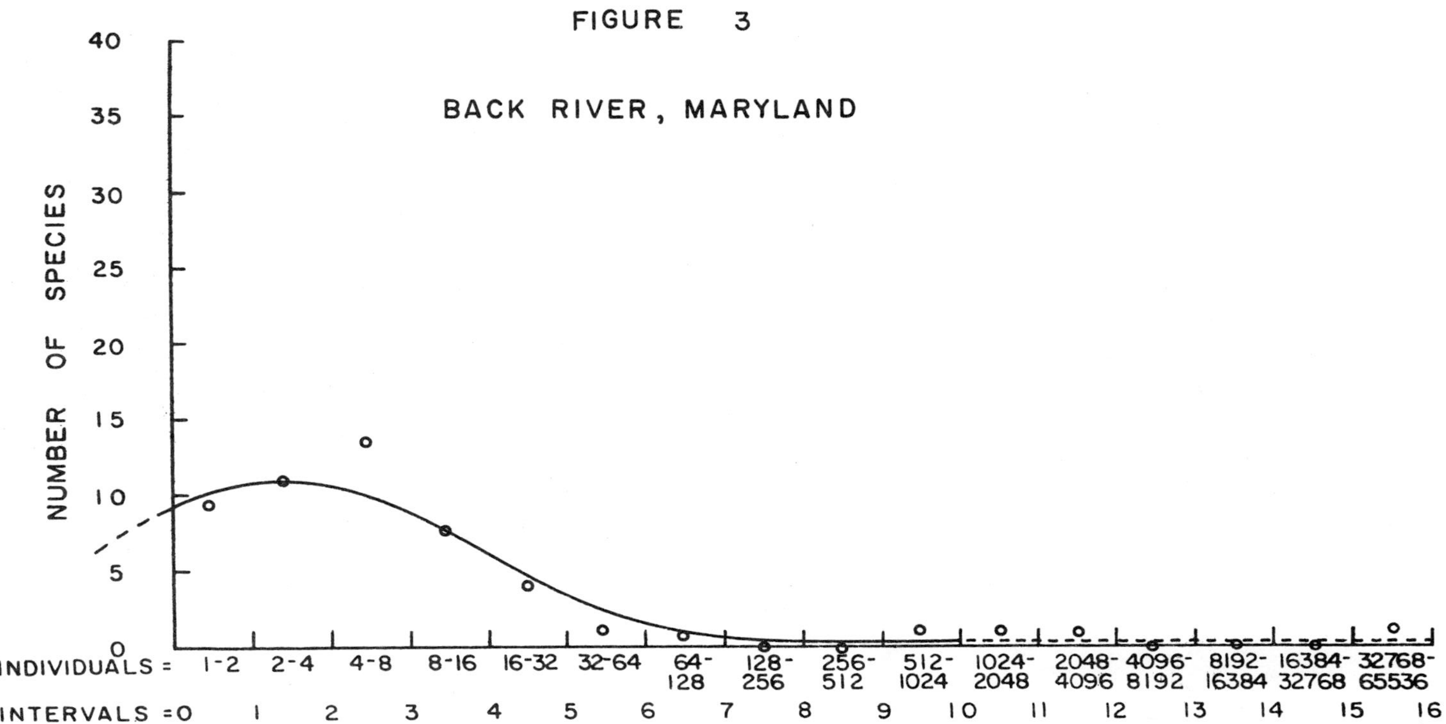

FIGURE 3
BACK RIVER, MARYLAND
NUMBER OF SPECIES
40
35
30
25
20
15
10
5
0
INDIVIDUALS =
1-2
2-4
4-8
8-16
16-32
32-64
64-128
128-256
256-512
512-1024
1024-2048
2048-4096
4096-8192
8192-16384
16384-32768
32768-65536
INTERVALS =0
1
2
3
4
5
6
7
8
9
10
11
12
13
14
15
16

FIGURE 4

EGG HARBOR RIVER, NEW JERSEY

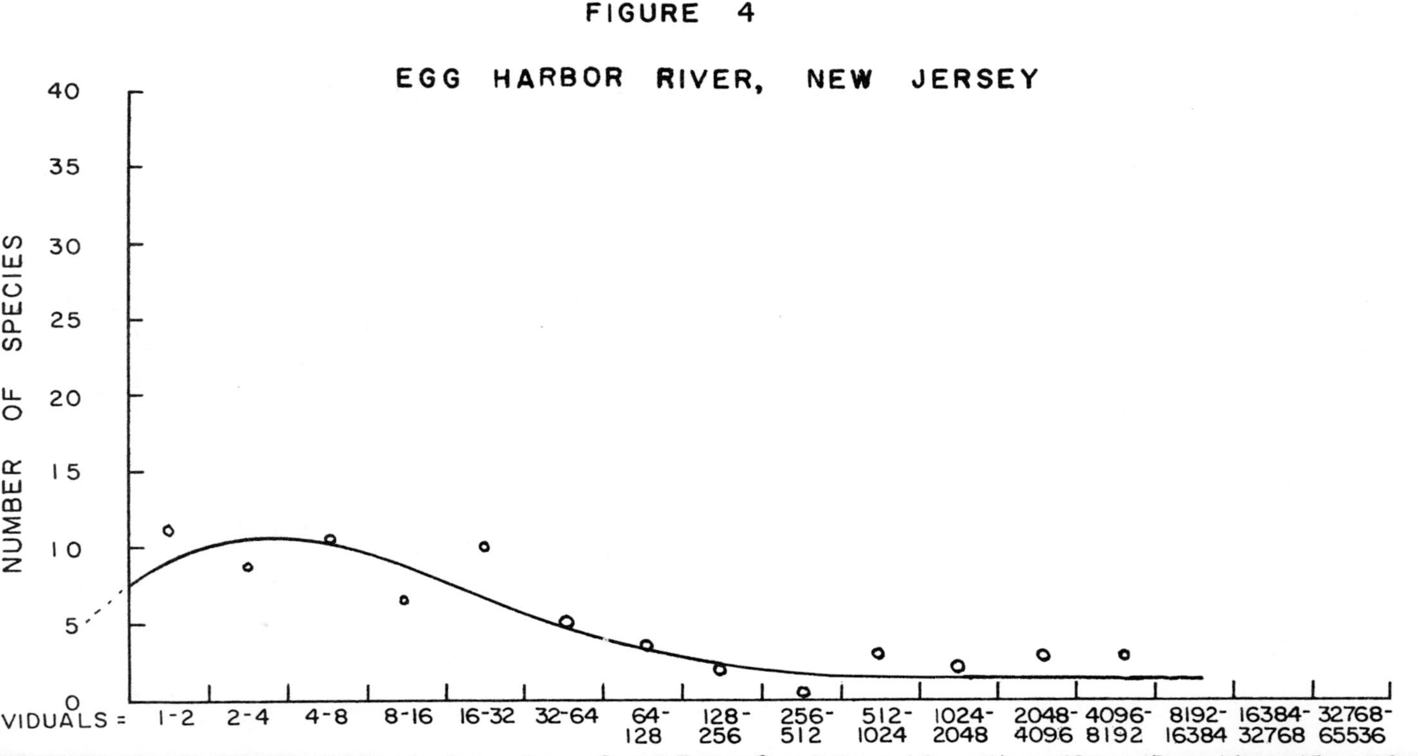

Literature Cited

Bennett, C. and N. Franklin. 1954. Statistical Analyses in chemistry and chemical industry. John Wiley and Sons, Inc., New York.

Lloyd, M. and R. J. Ghelardi. 1964. A table for calculating the equitability component of species diversity. Jour. Animal Ecology, 33:217-226.

MacArthur, R. M. 1965. Patterns of species diversity. Biol. Rev., 40:510-533.

Patrick, R. 1949. A proposed biological measure of stream conditions based on a survey of Conestoga Basin, Lancaster County, Pennsylvania. Proc. Acad. Nat. Sci. Philadelphia, 101:277-341.

_________. 1961. A study of the numbers and kinds of species found in rivers in eastern United States. Proc. Acad. Nat. Sci. Philadelphia, 113(10):215-258.

Patrick, R., M. H. Hohn, and J. H. Wallace. 1954. A new method for determining the pattern of the diatom flora. Notulae Naturae, Acad. Nat. Sci. Philadelphia, No. 259, 12 pp.

Patrick, R., J. Cairns, Jr., and S. S. Roback. 1967. An ecosystematic study of the fauna and flora of the Savannah River. Proc. Acad. Nat. Sci. Philadelphia, 118(5): 109-407.

MARK-RECAPTURE METHODS OF POPULATION ESTIMATION

D. S. Robson

Cornell University

INTRODUCTION

The numerical size of a natural population is often a matter of concern in field biology, and while interest usually centers on the temporal changes in relative abundance, the determination of absolute abundance at some point in time may be essential to the establishment of a meaningful bench mark. Techniques for counting organisms which are mobile and wary of man are still in a relatively primitive state of development and while indices of abundance may be available in a variety of forms the assessment of absolute abundance with any degree of precision generally requires considerable ingenuity and effort. Thus, the number of fish of a given species present in a lake is not an easily accessible parameter; by a judicious selection of times and places to set nets each year, however, the fishery biologist monitors change in relative abundance with little difficulty.

Among the techniques which have been developed for estimating absolute abundance the mark-recapture method is probably the most widely used and is, in fact, routinely incorporated into freshwater fisheries investigation. In its simplest and most commonly applied form the mark-recapture experiment is a two-sample experiment in which the members of the first sample are marked in some recognizable manner and returned to the population. The proportion of marked individuals appearing in the second sample is then regarded as an estimate of the proportion marked in the population. Since the absolute number of marked individuals in the population is known, this reasoning leads directly to an estimate of the absolute size of the population. Thus, if m_1 individuals are marked and released in the first sample and X marked individuals are subsequently recaptured in a sample of size n_2 then the population size N is estimated to be

$$\hat{N} = \frac{m_1 n_2}{X}$$

on the assumption the X/n_2 estimates m_1/N

In more extensive investigations the sampling and marking continues intermittently over a period of time, the unmarked individuals captured on each occasion being marked before being returned with the others to the population. Distinct "batch marks" are sometimes used in such studies; that is, on each sampling occasion all unmarked individuals are given an identical batch-mark but a recognizably different makr is used for each successive batch. More commonly in the multiple sample experiment the "mark" consists of a numbered tag which is attached to the individual and thereafter uniquely identifies him. We shall here consider statistical aspects of the k-sample, tag-recapture experiment.

The k-sample tag-recapture experiment.

Sample size on any one occasion is usually limited by the amount of capture gear and manpower which can be brought into operation at any on time. Under such economic restrictions the only feasible means of increasing the precision of the experiment may be the continuation of sampling on a number of separate occasions, resulting in a k-sample mark-recapture experiment. The number of marked individuals at large in the population may be steadily increased by marking all unmarked individuals captured on each sampling occasion and returning the entire sample to the population.

During the course of this sampling experiment the population itself may undergo change through such processes as mortality, emigration and immigration; and conceivably, the risks associated with these processes may vary with the previous capture history of an individual. In particular, the untagged portion of the population may be subject to different rates of mortality, emigration and immigration than the tagged portion. For example, where immigration into the catchable population results solely from the growth of individuals to catchable size during the course of the experiment then all recruits must enter as untagged individuals.

The situation in which all immigrants are untagged and the remainder of the population is homogenous with respect to the stochastic processes of mortality and emigration has been studied extensively by Darroch (1958, 1959), Seber (1962, 1965) and Jolly (1963, 1965). These authors assume binomial variation in the number captured on any occasion and in the number dying or emigrating during any given epoch, and expressing the likelihood function of the k-sample experiment as a product of conditional binomial probability distributions they have derived maximum likelihood estimators of the corresponding binomial parameters.

A valid objection to their conditional binomial model is that the frequency distribution of the number of organisms captured per unit of sampling effort is typically more closely approximated by a negative binomial rather than a binomial distribution. To avoid such criticism altogether I shall treat the number of individuals surviving any epoch as a fixed but unknown parameter, assuming only that survival during this epoch is independent of the capture history of the individuals prior to the epoch in question. Similarly sample size which is usually subject to some type of chance variation will here be treated conditionally as fixed, observable numbers, thus avoiding the necessity of specifying a probability distribution for sample size.

A further objection to the models of Darroch, Seber and Jolly is the real possibility that the chances of survival during any given epoch may indeed by different for individuals having different histories of prior capture. The authors themselves acknowledge this possibility and outline approximate test procedures for testing homogeneity of survival and capture probabilities. This model testing problem will be given special attention here with the objective of obtaining exact conditional test procedures.

Notation

t_i	= time at which the ith sample is drawn
δ_i	= indicator of capture in the ith sample; $\delta_i = 1$ indicates "captured in the ith sample" and $\delta_i = 0$ indicates "not captured in the ith sample".
h_i	= $(\delta_1, \ldots, \delta_i)$ is used generically as the index variable of capture history during the first i samples.
ϕ_i	= $(\delta_1 = 0, \ldots, \delta_i = 0)$ is the value of the index variable h_i for the specific history "not captured in the first i samples".
H_i	= the range of h_i; H_i consists of the 2^i possible histories $(\delta_1, \ldots, \delta_i)$ where δ is either 0 or 1.
$H_i^{(v)}$	= a subset of H_i, $H_i^{(v)} \subset H_i$.
$N_{h_{i-1}}$	= the number of individuals in the population having the capture history h_{i-1} which are alive and present at the time t_i^- when the ith sample is to be drawn.
$X_{h_{i-1}1}$	= the number of individuals captured in the ith sample and having the previous capture history h_{i-1}.

$m_{h_{i-1}1}$ = the number of individuals out of $X_{h_{i-1}1}$ which are returned to the population at time t_i^+; among the $X_{\phi\, i-1}\,1$ which were untagged on capture, the number $m_{\phi_{i-1}1}$ are tagged and returned to the population.

$N_i^{(v)}$ = $\sum_{h_{i-1} \in H_{i-1}^{(v)}} N_{h_{i-1}}$ is the number alive and present in the population at time t_i^- and having a capture history h_{i-1} belonging to the subset $H_{i-1}^{(v)}$.

$n_i^{(v)}$ = $\sum_{h_{i-1} \in H_{i-1}^{(v)}} X_{h_{i-1}1}$ is the number captured from $N_i^{(v)}$ at time t_i.

$m_{i-1}^{(v)}$ = $\sum_{h_{i-2} | (h_{i-2}1) \varepsilon H_{i-1}^{(v)}} m_{h_{i-2}1}$ is the number released at time t_{i-1}^+ having a capture history in $H_{i-1}^{(v)}$

$S_{i-1}^{(v)}$ = $\sum_{h_{i-2} | (h_{i-2}\, 0)\, \varepsilon\, H_{i-1}^{(v)}} X_{h_{i-2}1}$ is the number captured at time t_{i-1} which would otherwise have entered the class $H_{i-1}^{(v)}$.

$N_{i-1,0}^{(v)} - S_{i-1}^{(v)}$ = $\sum_{h_{i-2} | (h_{i-2};0)\, \varepsilon\, H_{i-1}^{(v)}} (N_{h_{i-2}} - X_{h_{i-2}1})$

is the number alive and present in the population at time t_{i-1}^+ which have a capture history in $H_{i-1}^{(v)}$ and were not captured at time t_{i-1}.

$R_{i-1}^{(v)}$ $= \sum_{h_{i-2} | (h_{i-2}1) \in H_{i-1}^{(v)}} (X_{h_{i-2}11} + X_{h_{i-2}101} + \ldots + X_{h_{i-2}10\ldots01})$ is the number of individuals in class $H_{i-1}^{(v)}$ which are caught in the (i-1)th sample and are subsequently recaptured (at least once).

$T_{i-1}^{(v)}$ $= \sum_{h_{i-1} \in H_{i-1}^{(v)}} (X_{h_{i-1}1} + X_{h_{i-1}01} + \ldots + X_{h_{i-1}0\ldots01})$ is the total number of individuals having a capture history in $H_{i-1}^{(v)}$ during the first (i-1) samples and which are subsequently recaptured (at least once).

$$T_{i-2,0}^{(v)} = S_{i-1}^{(v)} + T_{i-1}^{(v)} - R_{i-1}^{(v)}$$

$$= \sum_{h_{i-2} | (h_{i-2},0) \in H_{i-1}^{(v)}} (X_{h_{i-2}1} + X_{h_{i-2}01} + \ldots + X_{h_{i-2}0\ldots01}).$$

The population model

The changes in composition of a population during the time interval (t_{i-1}, t_i) between the release of the $(i-1)^{th}$ sample and the collection of the ith sample are generally of an unknown stochastic nature, and the object of the mulitple sample tag-recapture experiment is to obtain information concerning this stochastic process. In terms of counts the maximum information obtainable is a complete count and identification of all members of the population, as would result when the sample size n_i is equal to the population size N_i on each occasion, and all releases are tagged with individual identification marks. These complete counts would then constitute direct observation of the chance variables generated by the stochastic processes of the underlying dynamic system and formulation of a mathematical model of the system could then lead to estimation of its parameters. The mathematical models employed to approximate population dynamic systems depend upon the specific circumstances of the population, and the problems of developing such models properly belong to the specialist in quantitative ecology rather than to the statistician. Since complete counts are not available, however, there still remains the preliminary problem of estimating the actual changes being produced by this stochastic process, and it is this first stage of the problem which falls within the domain of finite population sampling theory.

In developing a sampling model for the tag-recapture experiment we are therefore behooved to make as few assumptions as possible concerning the dynamics of the population, so as to allow maximum freedom of choice to the specialist who later uses the data to make inferences concerning population dynamics. To this end we shall here assume only that removals from the population through emigration as well as through death are permanent removals. The desired consequence of this assumption is that all immigrants into the population are unmarked. Further assumptions concerning differential mortality due to tagging, though of interest to the population specialist, are more appropriately classified as features of the sampling model.

The sampling model

At the time t_i^- that the ith sample is to be drawn the population contains individuals of potentially 2^{i-1} different classes with respect to history of capture in the previous i-1 samples. The number $N_{\phi_{i-1}}(t_i^-)$ of untagged elements includes all survivors from the number $N_{\phi_{i-1}}(t_{i-1}^+)$ present at time t_{i-1}^+ plus any immigrants which have entered the population since time t_{i-1}^+ and survived to time t_i^-. The number $N_{h_{i-1}}(t_i^-)$ in any other class $h_{i-1} \neq \phi_{i-1}$ includes only the survivors of those $N_{h_{i-1}}(t_{i-1}^{+})$ present at t_{i-1}^{+}. If all preceding samples were drawn randomly and if tagging has no effect upon survival then the M_i tagged individuals present at time t_i^{-},

$$M_i = \sum_{h_{i-1} \neq \phi_{i-1}} N_{h_{i-1}}(t_i^{-})$$

constitute a random sample of those present at time t_{i-1}^{+}. A solution to the problem of estimating M_i and $N_i - M_i = N_{\phi_{i-1}}$ under these conditions is contained in the work of Jolly [1965], who allows for the added complication of "losses on capture" wherein the number m_i released at time t_i^{+} may be less than the number n_i that were captured.

More generally, however, if tagging or capture does influence an individual's chances for survival then tagged individuals with different capture histories may have different survival rates. In the extreme case each of the 2^{i-1} classes of individuals entering the epoch (t_{i-1},t_i) will have different chances for survival during the ensuing period, and all of the numbers $N_{h_{i-1}}(t_i^{-})$ must then be treated as unknown parameters in the estimation problem. All 2^{i-1} class sizes, however, are not individually identifiable in the tag-recapture model (the number classes 2^{i-1} then exceeds the number of sample counts $2^{i-1} - 1$ in the ith sample); estimability thus requires some degree of homogeneity of survival among the 2^{i-1} different classes entering the epoch (t_{i-1},t_i). We shall assume, therefore, that the collection $H_{i-1} = \{h_{i-1}\}$ may be partitioned into specified subsets $H_{i-1}^{(0)}, H_{i-1}^{(1)}, \ldots, H_{i-1}^{(q_i)}$ such that within a subset all capture histories are equivalent with respect to their effect upon survival. In effect, we assume that if the first i-1 samples were drawn randomly then the $N_{i-1}^{(\nu)}$ survivors in class $H_{i-1}^{(\nu)}$ at time t_i^{-},

$$N_{i-1}^{(\nu)} = \sum_{h_{i-1} \in H_{i-1}^{(\nu)}} N_{h_{i-1}}(t_i^-)$$

constitute a random sample of the members of this subset that were present at time t_{i-1}^+.

A model which allows for recruitment into the class ϕ_{i-1} during (t_{i-1},t_i) must then specify one of these subsets, say $H_{i-1}^{(0)}$, to include only the history ϕ_{i-1},

$$H_{i-1}^{(0)} = \phi_{i-1}$$

and in fact immigration may occur in any subset $H_{i-1}^{(\nu)}$ consisting of a single capture history $h_{i-1}^{(\nu)}$ without distrubing the sampling model. In applications where immigration or recruitment is a physical impossibility there may be no reason for placing untagged individuals in a separate category by themselves, so for purposes of general applicability we shall not insist that $H_{i-1}^{(0)} = \phi_{i-1}$. Two constraints which we do insist upon in the specification of subsets are:

1) If $(\delta_1,\ldots,\delta_{i-1})$ and $(\delta_1',\ldots,\delta_{i-1}')$ are equivalent in H_{i-1} (belonging to the same subset of H_{i-1}) then $(\delta_1,\ldots,\delta_{i-1},0)$ and $(\delta_1',\ldots,\delta_{i-1}',0)$ are equivalent in H_i.

2) For each $i > 2$ there exists a subset $H_{i-1}^{(\nu_i^*)}$ containing both a history of type $(\delta_1,\ldots,\delta_{i-2},0)$ and a history of type $(\delta_1',\ldots,\delta_{i-2}',1)$.

Condition 1) is required for mathematical convenience but is also logically appealing, for if two individuals have the same chances for survival during the epoch (t_{i-1},t_i) and neither is captured at time t_i then they should again have the same chances for survival during the following epoch (t_i,t_{i+1}). Condition 2) is required for identifiability of the parameters $N_i^{(0)}, N_i^{(1)}, \ldots, N_i^{(q_i)}$.

The likelihood function

Under the hypothesis that the sample of size n_k is drawn at random and without replacement from the finite population of size N_k present at time t_k^-, the conditional probability distribution of sample counts $X_{h_{k-1}1}$ given the class sizes $N_{h_{k-1}}$ is multi-hypergeometric,

$$P(\{x_{h_{k-1}1}\}|\{N_{h_{k-1}}\}) = \frac{1}{\binom{N_k}{n_k}} \prod_{\nu=0}^{q_k} \prod_{h_{k-1}\,\epsilon\, H_{k-1}^{(\nu)}} \binom{N_{h_{k-1}}}{X_{h_{k-1}1}} .$$

Multiplying this expression by the conditional probability distribution of $\{N_{h_{k-1}}\}$ given both the subset totals $\{N_k^{(\nu)}\}$ and the results of the preceding sample,

$$P(\{N_{h_{k-1}}\}|\{N_k^{(\nu)}\},\{X_{h_{k-2}1}\},\{m_{h_{k-2}1}\})$$

$$= \prod_{\nu=0}^{q_k} \frac{\displaystyle\prod_{h_{k-2}|(h_{k-2}1)\epsilon H_{k-1}^{(\nu)}} \binom{m_{h_{k-2}1}}{N_{h_{k-2}1}} \prod_{h_{k-2}|(h_{k-2}0)\epsilon H_{k-1}^{(\nu)}} \binom{N_{h_{k-2}}-X_{h_{k-2}1}}{N_{h_{k-2}0}}}{\begin{pmatrix} \displaystyle\sum_{h_{k-2}|(h_{k-2}0)\epsilon H_{k-1}^{(\nu)}} (N_{h_{k-2}}-X_{h_{k-2}1}) + \sum_{h_{k-2}|(h_{k-2}1)\epsilon H_{k-1}^{(\nu)}} m_{h_{k-2}1} \\ \displaystyle\sum_{h_{k-1}\epsilon H_{k-1}^{(\nu)}} N_{h_{k-1}} \end{pmatrix}}$$

and summing over the $N_{h_{k-1}}$ within each subset $H_{k-1}^{(\nu)}$ yields the conditional likelihood of $\{X_{h_{k-1}1}\}$ given both subset totals $N_k^{(\nu)}$ at time t_k^- the results of the preceding sample,

$$P(\{x_{h_{k-1}1}\} | \{N_k^{(\nu)}\}, \{X_{h_{k-2}1}\}, \{m_{h_{k-2}1}\})$$

$$= \frac{1}{\binom{N_k}{n_k}} \prod_{\nu=0}^{q_k} \frac{\binom{N_k^{(\nu)}}{T_{k-1}^{(\nu)}}}{\binom{N_{k-1,0}^{(\nu)} - S_{k-1}^{(\nu)} + m_{k-1}^{(\nu)}}{T_{k-1}^{(\nu)}}}$$

$$\prod_{h_{k-2} | (h_{k-2}1) \in H_{k-1}^{(\nu)}} \binom{m_{h_{k-2}1}}{x_{h_{k-2}1,1}} \prod_{h_{k-2} | (h_{k-2}0) \in H_{k-1}^{(\nu)}} \binom{N_{h_{k-2}} - X_{h_{k-2}1}}{x_{h_{k-2}0,1}}.$$

Continuing this operation back through the successive samples and treating the numbers released $m_{h_{i-1}}$ at t_i^+ as if they were uniquely determined by the numbers captured $X_{h_{i-1}}$ at t_i (i.e., conditioning on $m_{h_{i-1}1}$) yields the likelihood function

$$P = \left\{ \prod_{i=1}^{k} \frac{\prod_{\nu=0}^{q_i} \binom{N_i^{(\nu)}}{T_{i-1}^{(\nu)}}}{\binom{N_i}{n_i}} \prod_{i=2}^{k} \frac{1}{\prod_{\nu=0}^{q_i} \binom{N_{i-1,0}^{(\nu)} - S_{i-1}^{(\nu)} + m_{i-1}^{(\nu)}}{T_{i-1}^{(\nu)}}} \right\} \tag{1}$$

$$\left\{ \frac{T_0!}{x_1!x_{01}!\ldots x_{0\ldots01}!} \prod_{i=2}^{k} \prod_{h_{i-2}} \right.$$

$$\left. \binom{m_{h_{i-2}1}}{X_{h_{i-2}1,1}, X_{h_{i-2}101}, \ldots, X_{h_{i-2}10\ldots01}} \right\}$$

where $T_0 = x_1 + x_{01} + \ldots + x_{0\ldots01}$ is the total number of distinct individuals sampled in the k trials and where

$$\binom{m}{a,b,\ldots,c} = \frac{m!}{a!b!\ldots c!(m-a-b-\ldots-c)!} \quad .$$

Also, note that

$$N_{i-1,0}^{(\nu)} - S_{i-1}^{(\nu)} = 0 \quad \text{if} \quad \{h_{i-2} | (h_{i-2}0) \epsilon H_{i-1}^{(\nu)}\} = \phi$$

$$m_{i-1}^{(\nu)} = 0 \quad \text{if} \ \{h_{i-2} | (h_{i-2}1) \epsilon H_{i-1}^{(\nu)}\} = \phi$$

and, by condition 1),

$$N_{i-1,0}^{(\nu)} = \sum_{\eta_\nu} N_{i-1}^{(\eta_\nu)}$$

where the sum extends over all η_ν such that $H_{i-1}^{(\eta_\nu)} \subset \{h_{i-2} | (h_{i-2}0) \epsilon H_{i-1}^{(\nu)}\}$.

The first $\{\cdot\}$ factor in the likelihood function is thus a function only of the unknown parameters $N_i^{(\nu)}$, the observable parameters n_i, $m_i^{(\nu)}$, and the statistics $S_i^{(\nu)}$, $T_i^{(\nu)}$ while the second factor $\{\cdot\}$ depends only on the observable parameters and the observations $X_{h_i 1}$. The collection $\{S_i^{(\nu)}, T_i^{(\nu)}\}$ thus form a sufficient statistic with respect to the given specification of subsets. Since

$$T_{i-2,0}^{(\nu)} = S_{i-1}^{(\nu)} + T_{i-1}^{(\nu)} - R_{i-1}^{(\nu)} = \sum_{\eta_\nu} T_{i-2}^{(\eta_\nu)}$$

then an equivalent sufficient statistic is $\{R_i^{(\nu)}, T_i^{(\nu)}\}$ or the set of triplets $\{R_i^{(\nu)}, S_i^{(\nu)}, T_i^{(\nu)}\}$. Letting the vector $\underset{\sim}{U}$ denote this sufficient statistic, $\underset{\sim}{N}$ the parameter vector and $\underset{\sim}{X}$ the observation vector, we may then express the likelihood function as a product of the distribution of $\underset{\sim}{U}$ and the conditional distribution of $\underset{\sim}{X}$,

$$\{P(\underset{\sim}{U}|\underset{\sim}{N})\}\{P(\underset{\sim}{X}|\underset{\sim}{U})\} =$$

$$\left\{ \frac{\prod_{\nu=0}^{q_k} \binom{N_k^{(\nu)}}{T_{k-1}^{(\nu)}}}{\binom{N_k}{n_k}} \prod_{i=1}^{k-1} \frac{\prod_{\nu=0}^{q_{i+1}} \binom{N_{i,0}^{(\nu)}}{S_i^{(\nu)}}}{\binom{N_i}{n_i}} \prod_{\nu=0}^{q_{i+1}} \frac{\prod_{\eta_\nu} \binom{N_i^{(\eta_\nu)}}{T_{i-1}^{(\eta_\nu)}}}{\binom{N_{i,0}^{(\nu)}}{T_{i-1,0}^{(\nu)}}} \right.$$

$$\left. \prod_{\nu=0}^{q_{i+1}} \frac{\binom{N_{i,0}^{(\nu)} - S_i^{(\nu)}}{T_i^{(\nu)} - R_i^{(\nu)}} \binom{m_i^{(\nu)}}{R_i^{(\nu)}}}{\binom{N_{i,0}^{(\nu)} - S_i^{(\nu)} + m_i^{(\nu)}}{T_i^{(\nu)}}} \right\} \left\{ \prod_{i=1}^{k-1} \frac{1}{\prod_{i=1}^{q_{1+1}} \binom{m_i^{(\nu)}}{R_i^{(\nu)}} \binom{T_{i-1,0}^{(\nu)}}{S_i^{(\nu)}}} \right.$$

$$\frac{T_0!}{x_1! x_{01}! \ldots x_{0\ldots01}!} \prod_{i=1}^{k-1} \prod_{h_{i-1}}$$

$$\left.\binom{m_{h_{i-1}1}}{x_{h_{i-1}1,1}, x_{h_{i-1}101}, \ldots, x_{h_{i-1}10\ldots01}}\right\}$$

The first factor of this likelihood function, $P(\underset{\sim}{U}|\underset{\sim}{N})$, may be utilized in constructing the maximum likelihood estimator of $\underset{\sim}{N}$; the second factor, $P(\underset{\sim}{X}|\underset{\sim}{U})$, may be utilized in constructing a test of the hypothesis specified by the given partition against some simpler hypothesis specifying a coarser partition. Thus, the given partitions of H_{i-1} into $H_{i-1}^{(0)}, \ldots, H_{i-1}^{(q_i)}$ corresponds to a particular hypothesis concerning the effect of capture and tagging upon future chances for survival. Let $H_{i-1}^{*(0)}, \ldots, H_{i-1}^{*(q_i^*)}$ denote a coarser partition of H_{i-1} where each subset $H_{i-1}^{*(\nu)}$ is a union of subsets from the finer partition, say

$$H_{i-1}^{*(\nu)} = \bigcup_{\eta \epsilon V_i^{(\nu)}} H_{i-1}^{(\eta)} .$$

The sufficient statistic with respect to this simpler hypothesis then satisfies

$$U_{i-1}^{*(\nu)} = \sum_{\eta \epsilon V_i^{(\nu)}} U_{i-1}^{(\eta)}$$

and a (similar) test of this simpler hypothesis against the specified alternative hypothesis may then be developed from the conditional distribution

$$P(\underset{\sim}{U}|\underset{\sim}{U}^*) = \sum_{\underset{\sim}{X}|\underset{\sim}{U}} P(\underset{\sim}{X}|\underset{\sim}{U}^*)$$

$$= \prod_{i=1}^{k-1} \prod_{\nu=0}^{q_{i+1}^{*}} \frac{\prod_{\eta \epsilon V_{i+1}^{(\nu)}} \binom{m_i^{(\eta)}}{R_i^{(\eta)}}}{\binom{m_i^{(\nu)}}{R_i^{(\nu)}}} \frac{\prod_{\eta \epsilon V_{i+1}^{(\nu)}} \binom{T_{i-1,0}^{(\eta)}}{S_i^{(\eta)}}}{\binom{T_{i-1,0}^{(\nu)}}{S_i^{(\nu)}}} \quad . \qquad (2)$$

Example I

We first consider case in which all tagged individuals are homogeneous with respect to survival but may differ from untagged individuals in this respect. We then have

$$H_i^{(0)} = \phi_i \qquad H_i^{(1)} = H_i - \phi_i$$

and the first factor of the likelihood function becomes:

$$\frac{1}{\binom{T_0}{S_1^{(0)}}} \frac{\binom{N_k^{(0)}}{T_{k-1}^{(0)}}\binom{N_k^{(1)}}{T_{k-1}^{(1)}}}{\binom{N_k}{n_k}} \prod_{i=2}^{k-1} \frac{\binom{N_i^{(0)}}{T_{i-1}^{(0)}}\binom{N_i^{(1)}}{T_{i-1}^{(1)}}}{\binom{N_i^{(0)}-S_i^{(0)}}{T_i^{(0)}}\binom{N_i^{(1)}-S_i^{(1)}-m_i^{(1)}}{T_i^{(1)}}\binom{N_i}{n_i}} .$$

Since

$$T_{i-1}^{(0)} = S_i^{(0)} + S_{i+1}^{(0)} + \dots + S_k^{(0)}$$

and

$$S_i^{(1)} = n_i - S_i^{(0)}$$

then a sufficient statistic is $(T_1^{(1)}, \dots, T_{k-1}^{(1)}, S_2^{(1)}, \dots, S_{k-1}^{(1)})$. The maximum likelihood equations

$$\frac{N_i^{(0)}(N_i-n_i)}{N_i(N_i^{(0)}-S_i^{(0)})} = \frac{N_i^{(1)}(N_i^{(1)}-S_i^{(1)}+m_i^{(1)}-T_i^{(1)})(N_i-n_i)}{N_i(N_i^{(1)}-T_{i-1}^{(1)})(N_i^{(1)}-S_i^{(1)}+m_i^{(1)})} = 1$$

are readily seen to have the solution

$$\hat{N}_i^{(1)} = S_i^{(1)} + \frac{m_i^{(1)}(T_i^{(1)}-R_i^{(1)})}{R_i^{(1)}} \qquad \hat{N}_i = \frac{n_i}{S_i^{(1)}}\hat{N}_i^{(1)}$$

for i=2,...,k-1, as given by Jolly [1965].

Example II

In fisheries investigations the operation of attaching a tag is recognized to place the fish under considerable stress and result in increased mortality in the period immediately following release. These short term losses which have been observed among tagged fish held for observation are referred to in the fisheries literature as "Type I losses". We next consider this case where

$$H_i^{(0)} = \phi_i \qquad H_i^{(1)} = (\phi_{i-1},1) \qquad H_i^{(2)} = H_i - \phi_i - (\phi_{i-1},1)$$

which corresponds to the assumption that each epoch (t_i,t_{i+1}) is of sufficient duration to permit complete recovery for individuals surviving to the tagging operation at time t_i.

The first factor of the likelihood function (1) is now:

$$\frac{1}{\binom{T_0}{S_1^{(0)}}} \frac{\binom{N_k^{(0)}}{T_{k-1}^{(0)}}\binom{N_k^{(1)}}{T_{k-1}^{(1)}}\binom{N_k^{(2)}}{T_{k-1}^{(2)}}}{\binom{N_k}{n_k}}$$

$$\prod_{i=2}^{k-1} \frac{\binom{N_i^{(0)}}{T_{i-1}^{(0)}}\binom{N_i^{(1)}}{T_{i-1}^{(1)}}\binom{N_i^{(2)}}{T_{i-1}^{(2)}}}{\binom{N_i^{(0)}-S_i^{(0)}}{T_i^{(0)}}\binom{m_i^{(1)}}{T_i^{(1)}}\binom{N_i^{(1)}+N_i^{(2)}-S_i^{(1)}-S_i^{(2)}+m_i^{(1)}+m_i^{(2)}}{T_i^{(2)}}} .$$

Maximum likelihood estimators are again readily found as functions of the sufficient statistic $(T_1^{(1)},\ldots,T_{k-1}^{(1)},T_1^{(2)},\ldots,T_{k-1}^{(2)},S_2^{(2)},\ldots,S_{k-1}^{(2)})$

$$\hat{N}_i^{(1)} + \hat{N}_i^{(2)} = S_i^{(2)} + \frac{(m_i^{(1)}+m_i^{(2)})(T_i^{(2)}-R_i^{(2)})}{R_i^{(2)}}$$

$$\hat{N}_i = \frac{n_i}{S_i^{(2)}} (\hat{N}_i^{(1)}+\hat{N}_i^{(2)})$$

for i=2,...,k-1, and

$$\hat{N}_i^{(1)} = \frac{T_{i-1}^{(1)}}{T_{i-1}^{(1)}+T_{i-1}^{(2)}} (\hat{N}_i^{(1)}+\hat{N}_i^{(2)})$$

for i=3,...,k-1.

A test of the hypothesis specified in Example I against the alternative specified in the present example may now be developed from formula (2). The statistics of Example I will now be given asterisks to distinguish them from the statistics of the present example. We then have

$$m_i^{*(1)} = m_i^{(1)} + m_i^{(2)} \qquad R_i^{*(1)} = R_i^{(1)} + R_i^{(2)}$$

where

$$R_i^{(1)} = T_i^{(1)} \qquad R_i^{(2)} = S_i^{(2)} + T_i^{(2)} - T_{i-1}^{(2)}$$

and

$$T_i^{*(1)} = T_i^{(2)} \qquad S_i^{*(1)} = S_i^{(2)} \qquad (S_i^{(1)} = 0) \; .$$

Formula (2) thus gives

$$P(\underset{\sim}{U}|\underset{\sim}{U}^*) = \prod_{i=2}^{k-1} \frac{\binom{m_i^{(1)}}{R_i^{(1)}}\binom{m_i^{(2)}}{R_i^{(2)}}}{\binom{m_i^{(1)}+m_i^{(2)}}{R_i^{(1)}+R_i^{(2)}}}$$

leading to k-2 independent chi-square (or exact) tests of the 2x2 tables:

	Captured after t_i	Not captured after t_i
First captured in the ith sample and released	$R_i^{(1)}$	$m_i^{(1)} - R_i^{(1)}$
Recaptured in the ith sample and released	$R_i^{(2)}$	$m_i^{(2)} - R_i^{(2)}$

Example III

As an obvious continuation of the argument followed in Example II we consider the possibility that the stress effect may carry over into the second epoch after the tag is applied. Letting the $H_i^{(2)}$ of Example II now be denoted by $H_i^{*(2)}$, this leads to the partitioning of $H_i^{*(2)}$ into

$$H_i^{(2)} = \{(\phi_{i-2},1,0),(\phi_{i-2},1,1)\} \quad H_i^{(3)} = H_i^{*(2)} - H_i^{(2)} .$$

In this instance the likelihood equations are quadratic and no explicit solutions will be displayed.

Employing formula (2) to develop a test of the hypothesis in Example II against the present hypothesis, we are led to pairs of 2X2 tables:

	Captured at time t_i and	
	Captured after t_i	Not captured after t_i
First caught at time t_{i-1}	$R_i^{(2)}$	$m_i^{(2)} - R_i^{(2)}$
Caught before time t_{i-1}	$R_i^{(3)}$	$m_i^{(3)} - R_i^{(3)}$

	Captured after t_{i-1} and	
	Captured at t_i	Not captured at t_i
First caught at time t_{i-1}	$S_i^{(2)}$	$T_i^{(2)} - R_i^{(2)}$
Caught before time t_{i-1}	$S_i^{(3)}$	$T_i^{(3)} - R_i^{(3)}$

These tests are independent within pairs as well as between pairs.

Discussion

Conspicuously missing items in the preceding examples are variance formulas for the maximum likelihood estimators, or any reference to confidence interval estimation. Jolly [1965] does present variance formulas for the estimators in Example I, but the derivation of variance formulas for a more general case or for the specific Examples II and III remains an unsolved problem. Rather than attempting to derive asymptotic variance (and covariance) formulas for the maximum likelihood estimators, the approach to be recommended is the direct utilization of $P(\underset{\sim}{U}|\underset{\sim}{N})$ to obtain interval estimators of $\underset{\sim}{N}$, as is conventionally done in the k=2 sample case. The sampling distributions of these maximum likelihood estimators are decidedly skewed except for very large samples and asymptotic maximum likelihood theory will usually provide only a very poor approximation for purposes of interval estimation.

Another unsolved problem of practical importance is the development of tests of the hypothesis of equi-catchability, or the hypothesis that the sample of n_i is drawn randomly from the population of N_i. An alternative hypothesis could again be specified in terms of subsets $H_{i-1}^{(\nu)}$; i.e., the $n_i^{(\nu)}$ individuals appearing in the ith sample with capture histories in $H_{i-1}^{(\nu)}$ are a random sample from the corresponding sub-population of $N_i^{(\nu)}$ individuals existing at time t_i^-, the catchability being different for the various sub-populations. Evidently, if survival rates also differ between these same sub-populations then the $N_i^{(\nu)}$ are not identifiable. Seber [1965] has considered a special case of this model testing problem for which he derived an approximate likelihood ratio test, but a general treatment of the problem remains to be developed.

REFERENCES

Darroch, J. N. (1958). The multiple recapture census. I. Estimation of a closed population. Biometrika 45:343-59.

Darroch, J. N. (1959). The multiple recapture census. II. Estimation when there is immigration or death. Biometrika 46:336-51.

Jolly, G. M. (1963). Estimates of population parameters from multiple recapture data with both death and dilution -- deterministic model. Biometrika 50:113-28.

Jolly, G. M. (1965). Explicit estimates from capture-recapture data with both death and immigration -- stochastic model. Biometrika 52:225-47.

Seber, G. A. F. (1962). The multi-sample single recapture census. Biometrika 49:339-49.

Seber, G. A. F. (1965). A note on the multiple-recapture census. Biometrika 52:249-59.

size n_2 then the population size N is estimated to be

$$\hat{N} = \frac{m_1 n_2}{X}$$

on the assumption that X/n_2 estimates m_1/N.

In more extensive investigations the sampling and marking continues intermittently over a period of time, the unmarked individuals captured on each occasion being marked before being returned with the others to the population. Distinct "batch marks" are sometimes used in such studies; that is, on each sampling occasion all unmarked individuals are given an identical batch-mark but a recognizably different mark is used for each successive batch. More commonly in the multiple sample experiment the "mark" consists of a numbered tag which is attached to the individual and thereafter uniquely identifies him. We shall here consider statistical aspects of the k-sample, tag-recapture experiment.

The k-sample tag-recapture experiment.

Sample size on any one occasion is usually limited by the amount of capture gear and manpower which can be brought into operation at any one time. Under such economic restrictions the only feasible means of increasing the precision of the experiment may be the continuation of sampling on a number of separate occasions, resulting in a k-sample mark-recapture experiment. The number of marked individuals at large in the population may be steadily increased by marking all unmarked individuals captured on each sampling occasion and returning the entire sample to the population.

During the course of this sampling experiment the population itself may undergo change through such processes as mortality, emigration and immigration; and conceivably, the risks associated with these processes may vary with the previous capture history of an individual. In particular, the untagged portion of the population may be subject to different rates of mortality, emigration and immigration than the tagged portion. For example, where immigration into the catchable population results solely from the growth of individuals to catchable size during the course of the experiment then all recruits must enter as untagged individuals.

The situation in which all immigrants are untagged and the remainder of the population is homogeneous with respect to the stochastic processes of mortality and emigration has been studied extensively by Darroch [1958, 1959], Seber [1962, 1965] and Jolly [1963, 1965]. These authors assume binomial variation in the number captured on any occasion and in the number dying or emigrating during any given epoch, and expressing the likelihood function of the k-sample experiment as a product of conditional binomial probability distributions they have derived maximum likelihood estimators of the corresponding binomial parameters.

A valid objection to their conditional binomial model is that the frequency distribution of the number of organisms captured per unit of sampling effort is typically more closely approximated by a negative binomial rather than a binomial distribution. To avoid such criticism altogether I shall treat the number of individuals surviving any epoch as a fixed but unknown parameter, assuming only that survival during this epoch is independent of the capture history of the individuals prior to the epoch in question. Similarly, sample size which is usually subject to some type of chance variation will here be treated conditionally as fixed, observable numbers, thus avoiding the necessity of specifying a probability distribution for sample size.

A further objection to the models of Darroch, Seber and Jolly is the real possibility that the chances of survival during any given epoch may indeed be different for individuals having different histories of prior capture. The authors themselves acknowledge this possibility and outline approximate test procedures for testing homogeneity of survival and capture probabilities. This model testing problem will be given special attention here with the objective of obtaining exact conditional test procedures.

Notation

t_i = time at which the ith sample is drawn.

δ_i = indicator of capture in the ith sample; $\delta_i=1$ indicates "captured in the ith sample" and $\delta_i=0$ indicates "not captured in the ith sample".

h_i = $(\delta_1,\ldots,\delta_i)$ is used generically as the index variable of capture history during the first i samples.

ϕ_i = $(\delta_1=0,\ldots,\delta_i=0)$ is the value of the index variable h_i for the specific history "not captured in the first i samples".

H_i = the range of h_i; H_i consists of the 2^i possible histories $(\delta_1,\ldots,\delta_i)$ where δ is either 0 or 1.

$H_i^{(v)}$ = a subset of H_i, $H_i^{(v)} \subset H_i$.

$N_{h_{i-1}}$ = the number of individuals in the population having the capture history h_{i-1} which are alive and present at the time t_i when the ith sample is to be drawn.

$X_{h_{i-1}1}$ = the number of individuals captured in the ith sample and having the previous capture history h_{i-1} .

$m_{h_{i-1}1}$ = the number of individuals out of $X_{h_{i-1}1}$ which are returned to the population at time t_i^+; among the $X_{\phi_{i-1}1}$ which were untagged on capture, the number $m_{\phi_{i-1}1}$ are tagged and returned to the population.

$N_i^{(v)}$ = $\sum_{h_{i-1} \in H_{i-1}^{(v)}} N_{h_{i-1}}$ is the number alive and present in the population at time t_i^- and having a capture history h_{i-1} belonging to the subset $H_{i-1}^{(v)}$.

$n_i^{(v)}$ = $\sum_{h_{i-1} \in H_{i-1}^{(v)}} X_{h_{i-1}1}$ is the number captured from $N_i^{(v)}$ at time t_i.

$m_{i-1}^{(v)}$ = $\sum_{h_{i-2}|(h_{i-2}1) \in H_{i-1}^{(v)}} m_{h_{i-2}1}$ is the number released at time t_{i-1}^+ having a capture history in $H_{i-1}^{(v)}$.

$S_{i-1}^{(v)}$ = $\sum_{h_{i-2}|(h_{i-2},0) \in H_{i-1}^{(v)}} X_{h_{i-2}1}$ is the number captured at time t_{i-1} which would otherwise have entered the class $H_{i-1}^{(v)}$.

$N_{i-1,0}^{(v)}-S_{i-1}^{(v)}$ = $\sum_{h_{i-2}|(h_{i-2},0) \in H_{i-1}^{(v)}} (N_{h_{i-2}}-X_{h_{i-2}1})$ is the number alive and present in the population at time t_{i-1}^+ which have a capture history in $H_{i-1}^{(v)}$ and were not captured at time t_{i-1}.

$R_{i-1}^{(v)} \quad = \sum_{h_{i-2}|(h_{i-2}1) \in H_{i-1}^{(v)}} (X_{h_{i-2}11} + X_{h_{i-2}101} + \ldots + X_{h_{i-2}10\ldots01})$ is the number of individuals in class $H_{i-1}^{(v)}$ which are caught in the (i-1)th sample and are subsequently recaptured (at least once).

$T_{i-1}^{(v)} \quad = \sum_{h_{i-1} \in H_{i-1}^{(v)}} (X_{h_{i-1}1} + X_{h_{i-1}01} + \ldots + X_{h_{i-1}0\ldots01})$ is the total number of individuals having a capture history in $H_{i-1}^{(v)}$ during the first (i-1) samples and which are subsequently recaptured (at least once).

$$T_{i-2,0}^{(v)} = S_{i-1}^{(v)} + T_{i-1}^{(v)} - R_{i-1}^{(v)}$$

$$= \sum_{h_{i-2}|(h_{i-2},0) \in H_{i-1}^{(v)}} (X_{h_{i-2}1} + X_{h_{i-2}01} + \ldots + X_{h_{i-2}0\ldots01}).$$

The population model

The changes in composition of a population during the time interval (t_{i-1}, t_i) between the release of the $(i-1)^{th}$ sample and the collection of the ith sample are generally of an unknown stochastic nature, and the object of the multiple sample tag-recapture experiment is to obtain information concerning this stochastic process. In terms of counts, the maximum information obtainable is a complete count and identification of all members of the population, as would result when the sample size n_i is equal to the population size N_i on each occasion, and all releases are tagged with individual identification marks. These complete counts would then constitute direct observation of the chance variables generated by the stochastic processes of the underlying dynamic system, and formulation of a mathematical model of the system could then lead to estimation of its parameters. The mathematical models employed to approximate population dynamic systems depend upon the specific circumstances of the population, and the problems of developing such models properly belong to the specialist in quantitative ecology rather than to the statistician. Since complete counts are not available, however, there still remains the preliminary problem of estimating the actual changes being produced by this stochastic process, and it is this first stage of the problem which falls within the domain of finite population sampling theory.

In developing a sampling model for the tag-recapture experiment we are therefore behooved to make as few assumptions as possible concerning the dynamics of the population, so as to allow maximum freedom of choice to the specialist who later uses the data to make inferences concerning population dynamics. To this end we shall here assume only that removals from the population through emigration as well as through death are permanent removals The desired consequence of this assumption is that all immigrants into the population are unmarked. Further assumptions concerning differential mortality due to tagging, though of interest to the population specialist, are more appropriately classified as features of the sampling model.

The sampling model

At the time t_i^- that the ith sample is to be drawn the population contains individuals of potentially 2^{i-1} different classes with respect to history of capture in the previous i-1 samples. The number $N_{\phi_{i-1}}(t_i^-)$ of untagged elements includes all survivors from the number $N_{\phi_{i-1}}(t_{i-1}^+)$ present at time t_{i-1}^+ plus any immigrants which have entered the population since time t_{i-1}^+ and survived to time t_i^-. The number $N_{h_{i-1}}(t_i^-)$ in any

A NEW ESTIMATION THEORY FOR SAMPLE SURVEYS, II*

H. O. Hartley and J. N. K. Rao

Texas A&M University, College Station, Texas

1. Introduction

This paper is a sequel to an earlier one (Hartley and Rao, 1968) on the same topic. Accordingly, it will be necessary to briefly recall the basic results of the earlier paper and relate that paper to the present one. Our first paper was predominantly concerned with simple random sampling (with or without replacement) from a finite population. In the present paper we are concerned with examining the relation of our findings to the more complex sampling procedures such as unequal probability sampling as well as stratified and multistage sampling.

The basic feature of our theory was a special parametrisation of a finite population of N units with k characteristics attached to each unit. Denote by $\underset{\sim}{y}_i$ the k-vector attached to the i-th unit. We assume that all elements of the $\underset{\sim}{y}_i$ are measured on discrete scales and that only a finite set of T measurement vectors $\underset{\sim}{y}_t$ $(t = 1,2,\ldots,T)$ are possible for the $\underset{\sim}{y}_i$. Denote then by

$$N_t = \text{no. of units in the population having } \underset{\sim}{y}_t \qquad (1)$$

satisfying the conditions

$$N_t \geq 0 \text{ and } \sum_{t=1}^{T} N_t = N. \qquad (2)$$

Henceforth, sums and products for t are over 1,2,...,T.

* Research supported by U. S. Army Research Office.

The parameters $N_1,\ldots,N_T$ completely describe any finite population. The number T is usually large although sometimes occasions arise when T is small or moderate and the estimation of the N_t is of intrinsic interest, as for example when the N_t represent a frequency distribution such as the number of households in the community comprising t persons. However, in most cases we shall be concerned with the estimation of a few simple parametric functions such as the population moments and not with the separate estimation of the excessively large number of parameters N_t.

Finite population sampling will normally consist of (a) the sample design, i.e., the procedure of drawing a sample of n distinct units (where n may be fixed or random) and with measuring the $\underset{\sim}{y}_t$ for these units, (b) the use of the measured $\underset{\sim}{y}_t$ to compute estimators of the population parameters.

In our previous paper we restricted (a) to simple random sampling and we confined the computation of estimators (b) to what we termed 'scale-load' estimators. These were defined as mathematical functions of the scale vectors $\underset{\sim}{y}_t$ and of their sample loads (frequencies) n_t = no. of units in the sample having $\underset{\sim}{y}_t$. Thus any identifying labels, i, that may be attached to the units may or may not be used for the implementation of the sample design; however, labels are not directly used in the computation of the estimators. Nevertheless, in situations where the labels, i, are observable characteristics of the units and are considered informative observables, the labels may be adjoined to the vectors $\underset{\sim}{y}_i$ as a (k + 1)-th component.

We were able to show that within the class of 'scale-load' estimators many of the estimators in current use possess interesting optimality properties in simple random sampling. Specifically the estimators are either UMV (unbiased minimum variance) or maximum likelihood estimators or both. Some of these results are briefly restated in Section 2. In the remaining sections of the present paper we are concerned with the role these results play in the more complex sampling procedures. Briefly our findings are: (1) The above parametrization of finite populations will continue to yield useful likelihood formulations for sampling designs providing maximum likelihood and Bayesian estimation procedures. UMV property will be the exception rather than the rule. (2) We consider that identifying labels of primary units (or all but the last stage units) will often be available as well as informative. There are, however, situations in which higher stage units are not labelled as is the case, for example, for certain subsets of machine parts produced in bulk, the water supply of water works produced during certain time periods, etc. Certain situations where labels of higher stage units are not informative also exist, for example identifiable subsets of certain lists. Both 'scale-load' and label-dependent estimators are therefore required. As would be expected, there is usually no UMV estimator in the class of label-dependent estimators. (3) A particular problem arises when label dependence of estimators is used in conjunction with Bayesian concepts and separate prior distributions are allowed for the individually identifiable units. The resulting posterior distributions and hence Bayesian inferences do not depend on the survey design which in the frame work of Bayesian theory becomes a randomization procedure irrelevant in making posterior inferences. However, the absurd result that Bayesian theory leads to when applied to simple sampling or ultimate-stage unit sampling (Godambe, 1966) is perhaps our strongest point in favor of examining estimators that do not depend on the labels of the ultimate-stage units.

2. Simple random sampling

If a simple random sample of fixed size n is drawn without replacement from the population of N units, the likelihood of the n_t is given by

$$L(N_1,\ldots,N_T) = \Pi\binom{N_t}{n_t}/\binom{N}{n} \qquad (3)$$

where $n_t \geq 0$ and $\Sigma n_t = n$. We confine ourselves here to the case of a single character y attached to the units (i.e., k = 1). In our previous paper we have shown that any function of the n_t is an UMV estimator of its expectation. Specifically some of the more important parametric functions and their UMV estimators are given below:

Parametric function	UMV estimator	
N_t/N	n_t/n	
$\mu_r' = N^{-1}\Sigma N_t y_t^r$	$m_r' = n^{-1}\Sigma n_t y_t^r$	(4)
$\sigma^2 = \mu_2' - \mu_1'^2$	$\frac{n(N-1)}{N(n-1)} (m_2' - m_1'^2)$	

Notice that the estimators do not depend on T or the non-observed y_t. When N/n is an integer, n_t/n and m_r' are also the maximum likelihood estimators (see the Appendix). When N/n is not integral, the maximization of (3) over the integral grid N_t can be achieved by the algorithm given in the Appendix; however, since UMV estimators exist, the maximum likelihood estimators may not have particular merit for small samples. The possibility of using maximum likelihood estimators of the N_t when T is small and the N_t are parameters of interest is being examined by a Monte Carlo study.

Turning now to Bayesian estimation, we have used in our previous paper the mathematically convenient prior distribution suggested by Hoadley (1968) and given by

$$\varphi(N_1,\ldots,N_T) \propto \Pi \frac{(N_t + \nu_t - 1)!}{N_t!(\nu_t - 1)!}, \quad \nu_t > 0. \tag{5}$$

The 'Bayes estimator' of μ_r' is the posterior expectation of μ_r' and is given by

$$E'(\mu_r') = (1 - \tfrac{n}{N})[wm_r' + (1 - w)M_r'] + \tfrac{n}{N} m_r' \tag{6}$$

where

$$w = n/(n + \nu), \quad \nu = \Sigma\nu_t \tag{7}$$

and

$$M'_r = \nu^{-1}\Sigma\nu_t y_t^r \ . \tag{8}$$

It should be noted that the estimator (6) only requires the knowledge of M'_r (the prior mean of μ'_r) and w, i.e., in the case of r = 1 the knowledge only of the prior mean M'_1 and the relative weight w of the sample and prior information. Moreover, although the ν_t are akin to a priori sample frequencies, the posterior mean is not simply the mean of the pooled 'sample' $\nu_t + n_t$. It duly recognizes the fact that, as $n \to N$, the sample mean m'_1 will tend to μ'_1 and that the prior is ignored.

The expected loss which the decision maker faces bychoosing the 'Bayes estimator' is given by the posterior variance

$$V'(\mu'_r) = \frac{(N-n)(N+\nu)}{N^2(n+\nu+1)}\left[wm'_{2r} + (1-w)M'_{2r} - \left\{wm'_r + (1-w)M'_r\right\}^2\right]. \tag{9}$$

The 'Bayes estimator' of σ^2 is given by

$$E'(\sigma^2) = \frac{(n+\nu-\nu/N)}{(1+\nu/N)} V'(\mu'_1)$$
$$+ \frac{n}{N}\left[m'_2 - m'^2_1 + (1-\frac{n}{N})(1-w)^2(m'_1-M'_1)^2\right] \ . \tag{10}$$

It should be noted that, if the prior information is solely based on a pilot sample, M'_r and ν would roughly represent the r-th sample moment based on the pilot sample and the pilot sample size respectively.

Turning to simple random sampling with replacement, suppose a random sample of fixed size m is drawn with equal probability and with replacement. Let n denote the number of distinct units in the sample and n_t the number of distinct units having the value y_t in the sample. The total likelihood is given by

$$L(N_1,\ldots,N_T) = P(n) \frac{\Pi\binom{N_t}{n_t}}{\binom{N}{n}} \qquad (11)$$

where the probability $P(n)$ is a function only of m and N. For this sample design no UMV exists, but the maximum likelihood estimator of μ_r' is $m_r' = n^{-1}\Sigma n_t y_t^r$ provided N = cx least common multiple of 1, 2,...,m (c = integer). In particular, the maximum likelihood estimator of the population mean μ_1' is the sample mean based only on the distinct units in the sample and it is uniformly more efficient than the customary sample mean based on all the sample draws. With the prior distribution (5), the 'Bayes estimator' of μ_r', the posterior variance of μ_r' and the 'Bayes estimator' of σ^2 are respectively given by (6), (9) and (10), where n and the n_t are as defined above.

3. Estimation with concomitant variables.

In our earlier paper we have considered a situation customarily dealt with by ratio or regression method of estimation in which two variates y and x are attached to each of the units and the population mean $\bar{Y}$ of 'target variate' y is to be estimated utilizing the available information about x. Assuming that only the population $\bar{X}$ of x is known, we have shown that an approximation to the maximum likelihood estimator of $\bar{Y}$ is closely related to the customary regression estimator, provided the sample size n is moderately large. In this section we extend this result to multiple concomitant variables $x_1,\ldots,x_k$, assuming that only the population means $\bar{X}_1,\ldots,\bar{X}_k$ are known. We show that, for moderately large n, an approximation to the maximum likelihood estimator of $\bar{Y}$ is closely related to the customary multiple regression estimator.

As before, we assume that a finite set of T distinct, known values y_t are feasible for y. Likewise, we assume that I_j distinct, known values x_{ji_j} are feasible for $x_j (j = 1,\ldots,k)$. Let $N_{i_1\ldots i_k t}$

denote the number of units in the population which have

$x_{1i_1}, \ldots, x_{ki_k}$ and y_t attached to them. Let $n_{i_1 \ldots i_k t}$ be the number of units in the simple random sample of size n (drawn without replacement) which have $x_{1i_1}, \ldots, x_{ki_k}$ and y_t attached to them.

We consider only the multinomial situation in which $N \to \infty$ and $N_{i_1 \ldots i_k t}/N \to P_{i_1 \ldots i_k t}$ while n is held fixed. The likelihood L is then given by the multinomial distribution with probabilities $P_{i_1 \ldots i_k t}$. The restrictions on the $P_{i_1 \ldots i_k t}$ are given by

$$\underset{\sim}{P} \geq 0, \ \underset{\sim}{P}'\underset{\sim}{i} = 1 \text{ and } \underset{\sim}{P}'\underset{\sim}{Z} = \bar{\underset{\sim}{X}} \tag{12}$$

where $\underset{\sim}{P}'$ is the nx1 vector of the $P_{i_1 \ldots i_k t}$, $\underset{\sim}{i}$ is the 1xn vector of 1's, $\bar{\underset{\sim}{X}}' = (\bar{X}_1, \ldots, \bar{X}_k)$ and $\underset{\sim}{Z} = (\underset{\sim}{x}_1^* | \ldots | \underset{\sim}{x}_k^*)$ where $\underset{\sim}{P}'\underset{\sim}{x}_j^* = \bar{X}_j$ $(j = 1, \ldots, k)$. As in our previous paper, it can be shown that for moderate sample sizes n the global maximum of the multinomial likelihood can only be attained if $P_{i_1 \ldots i_k t} = 0$ for all those variate combinations for which $n_{i_1 \ldots i_k t} = 0$, and $P_{i_1 \ldots i_k t} > 0$ for the remainder. Confining then the maximization to the latter $P_{i_1 \ldots i_k t}$ only and introducing the Lagrangian multipliers λ and $\underset{\sim}{\mu}' = (\mu_1, \ldots \mu_k)$, the maximization of log L subject to (12) is attained for $\underset{\sim}{P} = \hat{\underset{\sim}{P}}$ where

$$\hat{P}_{i_1 \ldots i_k t} = \frac{n_{i_1 \ldots i_k t}}{n} \left[1 + \frac{1}{n} \sum_1^k \mu_j (x_{ji_j} - \bar{X}) \right]^{-1} . \tag{13}$$

Expanding $\hat{\underset{\sim}{P}}'\underset{\sim}{i} = 1$ to first three terms we obtain

$$n(\bar{\underset{\sim}{x}} - \bar{\underset{\sim}{X}})'\underset{\sim}{\mu} = \underset{\sim}{\mu}'\underset{\sim}{X}^{*\prime}\underset{\sim}{X}^*\underset{\sim}{\mu} \tag{14}$$

where $\bar{x}' = (\bar{x}_1, \ldots, \bar{x}_k)$ is the vector of sample means and $\underset{\sim}{X}^{*\prime}\underset{\sim}{X}^{*} = \underset{\sim}{S}^{*} = (s^{*}_{jp})$ where

$$s^{*}_{jp} = n^{-1}\Sigma..\Sigma n_{i_1..i_k t}(x_{ji_j} - \bar{X}_j)(x_{pi_p} - \bar{X}_p).$$

It is readily seen that the solution of (14) is given by

$$\underset{\sim}{\mu} = n(\underset{\sim}{X}^{*\prime}\underset{\sim}{X}^{*})^{-1}(\underset{\sim}{\bar{x}} - \underset{\sim}{\bar{X}}). \tag{15}$$

Now using (15) and expanding (13) to the first two terms we get

$$\underset{\sim}{\hat{P}} = \frac{1}{n}\left[\underset{\sim}{n} + \underset{\sim}{X}^{+}(\underset{\sim}{X}^{*\prime}\underset{\sim}{X}^{*})^{-1}(\underset{\sim}{\bar{X}} - \underset{\sim}{\bar{x}})\right] \tag{16}$$

where $\underset{\sim}{n}$ is the 1xn vector of the $n_{i_1..i_k t}$ and X^{+} is given by $\underset{\sim}{y}'\underset{\sim}{X}^{+} = (s^{*}_{1y}, .., s^{*}_{ky})$ where $\underset{\sim}{P}'\underset{\sim}{y} = \bar{Y}$ and

$$s^{*}_{jy} = n^{-1}\Sigma...\Sigma n_{i_1..i_k t} y_t (x_{ji_j} - \bar{X}) \quad , \; j = 1,..,k.$$

An improved approximation, along the lines of our previous paper, can be obtained by expanding (13) to the first three terms.

Using (16), an approximation to the maximum likelihood estimator of the population mean $\bar{Y} = \underset{\sim}{P}'\underset{\sim}{y}$ is given by

$$\hat{\bar{Y}} = \underset{\sim}{\hat{P}}'\underset{\sim}{y} \doteq \bar{y} + (\underset{\sim}{\bar{X}} - \underset{\sim}{\bar{x}})'\underset{\sim}{S}^{*-1}\underset{\sim}{s}^{*}_{.y} \tag{17}$$

where $\underset{\sim}{s}^{*}_{.y}$ is the k-vector of the s^{*}_{jy}. The customary multiple regression estimator is given by

$$\hat{\bar{Y}}_r = \bar{y} + (\underset{\sim}{\bar{X}} - \underset{\sim}{\bar{x}})'\underset{\sim}{S}^{-1}\underset{\sim}{s}_{.y} \tag{18}$$

where $\underset{\sim}{S} = (s_{jp})$, $\underset{\sim}{s}_{.y}$ is the k-vector of the s_{jy} and

$$s_{jp} = s^{*}_{jp} - (\bar{x}_j - \bar{X}_j)(\bar{x}_p - \bar{X}_p)$$

$$s_{jy} = s^{*}_{jy} - \bar{y}(\bar{x}_j - \bar{X}_j).$$

Although (17) differs slightly from (18), the above development clearly shows that, at least in large samples, the customary multiple regression estimator is essentially the maximum likelihood estimator.

4. Stratified simple random sampling without replacement.

4.1. UMV Estimator

Suppose there are L strata in the population with N_i units in the ith stratum $(i = 1,..,L)$. Denote by N_{it} the number of units in the population belonging to the ith stratum and having the measurement $y_{it}(t = 1,..,T_i)$ so that $\sum_t N_{it} = N_i (\Sigma\Sigma N_{it} = N)$. A stratified simple random sample $(n_1,..,n_L)$ is drawn without replacement, $(\Sigma n_i = n)$, and n_{it} denotes the number of units in the sample belonging to the ith stratum and having the measurement y_{it}, $(\sum_t n_{it} = n_i)$. Now the likelihood of the n_{it} is given by

$$L(N_{11},..,N_{LT_L}) = \prod_1^L \left[\frac{\binom{N_{i1}}{n_{i1}} \cdots \binom{N_{iT_i}}{n_{iT_i}}}{\binom{N_i}{n_i}} \right] . \quad (19)$$

Therefore, the n_{it} are complete sufficient for the N_{it} and, hence, the UMV estimator of the population mean $\bar{Y} = N^{-1}\Sigma\Sigma N_{it}y_{it}$ is the customary estimator

$$\hat{\bar{Y}} = N^{-1}\sum_{it}\sum \hat{N}_{it}y_t = N^{-1}\sum_i N_i\bar{y}_i \quad (20)$$

where $\hat{N}_{it} = (N_i/n_i)n_{it}$ is the UMV estimator of N_{it}. It also follows that the maximum likelihood estimators of the N_{it} and $\bar{Y}$ are the UMV estimators $\hat{N}_{it}$ and $\hat{\bar{Y}}$ respectively, when the N_i/n_i are integral. Notice that each stratum is described by its <u>separate</u> set of parameters, i.e., we have an additional subscript i to index the strata.

An interesting special case occurs when the stratification is according to the size of the units, say x_i. If we assume that x_i is constant within strata and use the allocation proportional to total size, i.e., $n_i = n(N_i x_i / \Sigma N_i x_i) = N_i P_i$ (say) where $\underset{it}{\Sigma\Sigma} P_i = n$, we get

$$\hat{\bar{Y}} = N^{-1} \sum_{it}\sum \hat{N}_{it} y_{it} = N^{-1} \sum_{it}\sum \frac{n_{it} y_{it}}{P_i} \tag{21}$$

which is a 'Horvitz-Thompson' type estimator.

4.2. Bayesian optimization of stratified sampling.

Ericson (1965) has presented a solution to the problem of optimum allocation when prior information in the form of a prior distribution is available. He has, however, assumed: (a) $N_i = \infty$, $i = 1,..,L$, (b) normality of the within stratum populations and (c) known within stratum population variances σ_i^2. Assuming that the within stratum population means μ_i have independent normal priors with means m_i and variances v_{ii}', he has shown that the posterior variance of the population mean $\mu = \Sigma\pi_i\mu_i$ is given by

$$v'' = \sum_i \left[\pi_i^2 \Big/ \left(\frac{1}{v_{ii}'} + \frac{n_i}{\sigma_i^2}\right)\right] \tag{22}$$

where π_i is the known proportion of the population units falling within the ith stratum. Ericson has given a computational algorithm to find $n_i \geq 0$ $(i = 1,..,L)$ such that (22) is minimized subject to the cost constraint

$$\Sigma c_i n_i = C \tag{23}$$

where C is the given budget.

Recently, Draper and Guttman (1968) have relaxed the assumption (c) and presented a sequential allocation scheme which appears simpler than Ericson's algorithm. They have also considered the case of unknown proportions π_i. Using our present approach, one of us (J.N.K. Rao, 1968) has given a solution which is free from the restrictive assumptions (b) and (c). Extension to multiple priors and/or multiple characteristics by the use of convex programming was also considered. In this section we present a complete solution by relaxing the assumption (a) also.

We assume that prior information on the N_{it} is available in the form of (5) for each i and that the priors are independent. Therefore, the prior distribution of $N_{11},..,N_{LT_L}$ is

$$\varphi(N_{11},..,N_{LT_L}) \propto \prod_i \left[\prod_t \frac{(N_{it} + \nu_{it} - 1)!}{N_{it}!(\nu_{it} - 1)!} \right] \tag{24}$$

$$\nu_{it} > 0, \; \sum_t \nu_{it} = \nu_i \quad .$$

Now, since $\bar{Y} = N^{-1}\sum_i N_i \bar{Y}_i$ where $\bar{Y}_i$ is the ith stratum population mean, we get using (6) and (9) the posterior mean of $\bar{Y}$ as

$$E'(\bar{Y}) = N^{-1}\sum_i N_i \left[\left(1 - \frac{n_i}{N_i}\right) \sum_t \frac{n_{it} + \nu_{it}}{n_i + \nu_i} y_{it} + \frac{n_i}{N_i} \sum_t \frac{\nu_{it}}{\nu_i} y_{it} \right] \tag{25}$$

and the posterior variance of $\bar{Y}$ as

$$V'(\bar{Y}) = N^{-2}\sum_i N_i^2 \left(1 - \frac{n_i}{N_i}\right)\left(1 + \frac{\nu_i}{N_i}\right)(n_i + \nu_i + 1)^{-1} .$$

$$\left[\sum_t \frac{n_{it} + \nu_{it}}{n_i + \nu_i} y_{it}^2 - \left(\sum_t \frac{n_{it} + \nu_{it}}{n_i + \nu_i} y_{it} \right) \right] \tag{26}$$

Since the posterior variance (26) depends on the sample values n_{it}, we take the expectation of (26) with respect to the marginal distribution of the n_{it}. It follows from Hoadley (1968) that the marginal distribution of the n_{it} is given by

$$f(n_{11},\ldots,n_{LT_L}) = \prod_i \left[\frac{\prod_t \binom{n_{it}+\nu_{it}-1}{n_{it}}}{\binom{n_i+\nu_i-1}{n_i}} \right] \tag{27}$$

which is identical to that in the case of infinite populations with Dirichlet prior distributions. Therefore, using the results of J.N.K. Rao (1968) it follows from (26) that the expected posterior variance of $\bar{Y}$ is

$$V'(\bar{Y}) = \sum_i \frac{N_i^2}{N^2}\left(1 - \frac{n_i}{N_i}\right)\left(1 + \frac{\nu_i}{N_i}\right)\frac{A_i}{n_i + \nu_i} \tag{28}$$

where

$$\frac{A_i}{\nu_i} = (\nu_i + 1)^{-1} \sum_t \frac{\nu_{it}}{\nu_i}\left[y_{it} - \left(\sum_t \frac{\nu_{it}}{\nu_i} y_{it}\right)\right]^2 . \tag{29}$$

It follows, using (9) and (10), that

$$\text{Prior variance of } \bar{Y}_i = \left(\frac{1}{\nu_i} + \frac{1}{N_i}\right)A_i \tag{30}$$

and

$$\text{Prior mean of } S_i^2 = A_i \tag{31}$$

where $N_i\sigma_i^2 = (N_i - 1)S_i^2$.

Now (28) is a separable convex function in the n_i and, therefore, the values n_i which minimize (28) subject to (23) and $0 \le n_i \le N_i$ ($i = 1,\ldots,L$) can be obtained by convex programming.* It is also possible to develop a sequential allocation procedure analogous to that of Draper and Guttman (1968).

---

* In our original version we ignored the restriction $n_i \le N_i$ and Ericson has pointed this out.

It is important to note that the knowledge of the complete priors is not essential for the optimum allocation — only that of the prior mean of σ_i^2 and prior variance of μ_i is needed. If the priors are solely based on pilot samples within each stratum, then $[(\nu_i+1)/\nu_i]A_i$ and ν_i would roughly represent the pilot-sample variance and the pilot sample size respectively.

The extension of the above results to multiple priors and/or multiple characteristics follows along the lines of J.N.K. Rao (1968) and the optimum allocation is obtained by convex programming.

5. Single-stage unequal probability sampling.

In the preceding sections we have been mainly concerned with sampling procedures in which all the units had an equal chance of selection. The only exception is stratified sampling (Section 4.1) in which strata allocations n_i proportional to the products $N_i x_i$ gave all the N_i units in the ith size stratum an equal inclusion probability of $P_i = n(N_i x_i/\Sigma N_i x_i)$ which was varied from stratum to stratum. While unequal probability sampling by 'size strata' may be satisfactory for many practical purposes, situations often arise in which we desire to vary the inclusion probability from unit to unit. However, this type of unequal probability sampling mainly arises in the selection of primary sampling units in multi-stage sampling which we discuss in Section 6. Here we confine ourselves to the (rare) situations where unequal probability sampling is used in 'single-stage' or 'ultimate-stage' sampling of units which are not necessarily identifiable in advance of sampling.

As an example of p.p.s. sampling of this kind, we may mention here the sampling of farm operators in Iowa counties proportional to the land acreages they operate. If a county map can be covered by a rectangle with dimensions Z by W and $(z_i, w_i), i = 1,..,r (r \geq m)$ denote uniform variables with $0 \leq z_i \leq Z$ and $0 \leq w_i \leq W$, co-ordinates (z_i, w_i) can be pinpointed on the map and the interviewer can be instructed to ascertain (in order of draw) the first m operators whose land acreages contain the pinpointed land marks. This results in p.p.s. sampling with replacement in which $p_i = x_i/X$ (x_i = land acreage of ith operator only known for sampled operators, X = total land acreage known in advance of sampling) where p_i denotes the probability of selection of ith operator at a single draw. This well-known situation of 'multinomial sampling' is the only one discussed in this section. We show that it can be reparametrized in such a way that optimality properties can be formulated for certain estimators.

Let $r_i = y_i/p_i$ and denote by r_t $(t = 1,..,T)$ the set of T discrete scale points feasible for the r_i. Let the score m_i denote the number of times i^{th} unit is included in the sample $(i = 1,..,N; \Sigma m_i = m)$. We now classify the r_i into T groups and denote by

$$p_{it} = \begin{cases} p_i & \text{if for the } i^{th} \text{ unit } r_i = r_t \\ 0 & \text{otherwise} \end{cases}$$

$$y_{it} = \begin{cases} y_i & \text{if for the } i^{th} \text{ unit } r_i = r_t \\ 0 & \text{otherwise} \end{cases}$$

$$m_{it} = \begin{cases} m_i & \text{if for the } i^{th} \text{ unit } r_i = r_t \\ 0 & \text{otherwise} \end{cases}$$

The multinomial distribution of scoring m multinomial scores into N classes with probabilities p_i may then be written in the form

$$L(p_{11},..,p_{NT}) = \frac{m!}{\prod_{i,t} m_{it}!} \prod_{i,t} p_{it}^{m_{it}} \tag{32}$$

and may be reparametrised as follows:

$$\left.\begin{aligned} p_t &= \sum_i p_{it} \\ \gamma_{it} &= \begin{cases} p_{it}/p_t & \text{if } p_t > 0 \\ 1 & \text{if } p_t = 0 \end{cases} \end{aligned}\right\} \tag{33}$$

so that

$$\sum_i \gamma_{it} = 1 \quad \text{if } p_t > 0. \tag{34}$$

Writing $m_t = \sum_i m_{it}$, we may factorize (32) as

$$L(p_{11},\ldots,p_{NT}) = \left[\frac{m!}{\prod_t m_t!} \prod_t p_t^{m_t} \right] \left[\frac{\prod_t m_t!}{\prod_{i,t} m_{it}!} \prod_{i,t} \gamma_{it}^{m_{it}} \right]. \tag{35}$$

Equation (35) shows that the m_t are sufficient for the p_t since the latter are only involved in the marginal distribution of the m_t and not in the conditional distribution of the m_{it} given the m_t.

The maximum likelihood estimators of the p_t are given by the ratios m_t/m and, hence, the maximum likelihood estimator of the population total

$$Y = \sum_i y_i = \sum_{it}\sum y_{it} = \sum_t r_t \sum_i p_{it} = \sum_t r_t p_t \tag{36}$$

is given by

$$\hat{Y} = \sum_t r_t \frac{m_t}{m} = \frac{1}{m} \sum_t r_t \sum_i m_{it} = \frac{1}{m} \sum_i \frac{y_i m_i}{p_i} \tag{37}$$

which is the customary unbiased estimator of Y in p.p.s. sampling with replacement.

Finally, it should be noted that (35) is the likelihood for the scores which do not necessarily represent counts of distinct units in the population. However, it is possible to obtain the likelihood of the number of distinct units in the sample with scale ratio r_t which we denote by n_t. The distribution of the m_t is given by

$$L(p_1,..,p_T) = \frac{m!}{\prod_t m_t!} \prod_t p_t^{m_t} \tag{38}$$

and the conditional distribution of the n_t given the m_t can be obtained in terms of the γ_{it} from formula (4.3) of Kullback (1937). Finally the likelihood of the n_t can be obtained by summation of the product (i.e., the joint distribution) over $m_t = n_t$ to m subject to $\Sigma m_t = m$. We intend to examine this distribution in more detail elsewhere.

Although only one single method of unequal probability sampling is examined in this section and although the method examined is known not to be particularly efficient, the discussion clearly indicates the possibility of deriving concrete likelihoods for other unequal probability sampling methods with the help of our technique of parametrisation.

6. Two-stage sampling.

In order to simplify the discussion we confine ourselves to two-stage sampling in which the primaries are selected with equal or unequal probabilities. Consider then a population consisting of L primary units $i = 1,..,L$ of which ℓ will be sampled and denote by N_i the number of secondary units in the ith primary. Denote by N_{it} the number of units in the ith primary which have the scale value $y_{it}(t = 1,..,T_i)$ so that $\sum_t N_{it} = N_i$. Let $u_i = 1$ if the ith primary is in the sample and zero otherwise. Denote by $P(u_1,..,u_L)$ the joint distribution of the u_i corresponding to the primary sampling procedure adopted and let n_i denote the number of secondary units to be drawn from the

i^{th} primary if it is in the sample. The n_i are all specified apriori for i = 1,..,L. In this paper we only consider equal probability sampling of secondaries without replacement.

If we denote by n_{it} the number of secondaries having scale value y_{it} in the ith sampled primary, then the joint likelihood of the u_i and the n_{it} is given by

$$L(N_{11},..,N_{LT_L}) = P(u_1..u_L)\prod_i\left[\prod_t\binom{N_{it}}{u_i n_{it}}\Big/\binom{N_i}{u_i n_i}\right]. \tag{39}$$

6.1. Maximum likelihood estimation.

We confine ourselves here to the case of $N_i = \infty$, i = 1,..,L. The likelihood (39) reduces to

$$L(p_{11},..,p_{LT_L}) = P(u_1,..,u_L)\prod_i\left[\frac{(n_i u_i)!}{\prod_t u_i n_{it}!}\prod_t p_{it}^{u_i n_{it}}\right]. \tag{40}$$

Maximisation of (40) subject to $\sum_t p_{it} = 1$ for i = 1,..,L leads to

$$\hat{p}_{st} = n_{st}/n_s \text{ (primary s in the sample)} \tag{41}$$

while any values of p_{jt} are permissible for j not in the sample. The maximum likelihood solution will, therefore, in general not be unique. Furthermore, we do not have complete sufficiency here and, hence, no UMV estimator exists. We have not considered here 'scale-load' estimators which do not depend on primary labels.

6.2. Bayesian estimation.

Since the complete likelihood is given by (39), the posterior distribution of the N_{it} is identical to that in the case of stratified sampling (section 4) noting that $n_i = 0$ is allowed for the latter. Therefore, the 'Bayes estimator' of $\bar{Y}$ is given by (25) and it may be recast as

$$E'(\bar{Y}) = N^{-1}\Sigma_1 N_i\left[\left(1 - \frac{n_i}{N_i}\right)\Sigma_t \frac{n_{it} + \nu_{it}}{n_i + \nu_i} y_{it} + \frac{n_i}{N_i}\Sigma_t \frac{\nu_{it}}{\nu_i} y_{it}\right]$$

$$+ N^{-1} \Sigma_2 N_i \Sigma_t \frac{\nu_{it}}{\nu_i} y_{it} \tag{42}$$

where Σ_1 and Σ_2 are respectively the summations over sampled and non-sampled primaries. It should be noted that we must have a prior distribution from each primary. If the prior distribution is solely based on pilot samples, this implies that the pilot sample must include at least one secondary unit from each primary.

The above analysis clearly shows that the sampling procedure adopted for selection of the primaries is entirely irrelevant as far as a full Bayesian analysis is concerned. However, if the likelihood based on a selected estimator is used for a (partial) Bayesian analysis based on insufficient statistics, then the posterior distribution and, hence, the 'Bayes estimator' would depend on the sampling procedure. These are the two alternatives available to the Bayesian analyst and, at this stage, we do not wish to take sides in this issue.

Appendix

Maximum likelihood estimation for simple random sampling without replacement

Case 1. N/n is an integer.

To prove that the $\hat{N}_t = (N/n)n_t$ maximise the likelihood (3) subject to (2), consider a set N_t satisfying (2) and differing from the $\hat{N}_t$. Because the N_t satisfy (2), there is at least one N_θ with $N_\theta > \hat{N}_\theta$ and at least one N_τ with $N_\tau < \hat{N}_\tau$ so that $N_\theta \geq 1$ and $n_\tau \geq 1$.

Define $N'_t = N_\theta - 1$ for $t = \theta$; $N'_t = N_\tau + 1$ for $t = \tau$ and $N'_t = N_t$ for $t \neq \theta, \tau$. Then

$$\frac{L(N'_1,..,N'_T)}{L(N_1,..,N_T)} = \frac{1 - n_\theta/N_\theta}{1 - n_\tau/(N_\tau+1)} . \qquad (A.1)$$

Now $N_\theta/n_\theta > N/n$ and $(N_\tau + 1)/n_\tau = N/n + (1-m)/n_\tau$ with $m \geq 1$ and integral. Therefore, (A.1) is greater than 1. It follows that for any set N_t satisfying (2) and differing from the $\hat{N}_t$, we can always find a set N'_t which increases the likelihood. Hence the $\hat{N}_t$ maximise the likelihood (3) over the finite set of admissible N_t.

Case 2. N/n is not integral.

Write $N/n = R + q$ where R is an integer and $0<q<1$. Then $(N/n)n_t = Rn_t + qn_t$, $N = Rn + qn$ and, hence, qn is an integer. Let $qn_t = R_t + e_t$ where R_t is an integer and $0 \leq e_t < 1$. It follows that $qn = \Sigma R_t + \Sigma e_t$ so that Σe_t is integral. We now define 'approximations' $\tilde{N}_t$ to the maximum likelihood estimators $\hat{N}_t$ as follows: Split the t-indices into θ-set and τ-set such that $\tilde{N}_\theta > (N/n)n_\theta$ and $\tilde{N}_\tau \leq (N/n)n_\tau$. The $\tilde{N}_t$ will now be obtained. First for all t with $n_\tau = 0$ define $\tilde{N}_t = 0$ and place these t in the τ-set. For the remaining t compute the ratios of $(N/n)n_t$ rounded up and then divided by n_t, i.e., the ratios

$$Q_t = R + \frac{R_t + 1}{n_t} \quad . \qquad (A.2)$$

Order the Q_t in order of magnitude (resolving ties arbitrarily) and perform the following steps:

Step 0 (initial step). Select the largest Q_t and define the associated

$$\tilde{N}_t = Rn_t + R_t \qquad (A.3)$$

and the 'cumulative error sum'

$$S = -e_t = \tilde{N}_t - \frac{N}{n} n_t . \qquad (A.4)$$

This means that for this t-index $\tilde{N}_t \leq (N/n)n_t$ so that it is placed in the τ-set and eliminated from further consideration.

Step 1 (rounding-up step). Enter this step whenever the previous $S = S_{old}$ is negative. From among the t-indices still under consideration select the one with the smallest Q_t and define the associated

$$\tilde{N}_t = Rn_t + R_t + 1. \qquad (A.5)$$

This means that for this t-index $\tilde{N}_t > (N/n)n_t$ so that it is placed in the θ-set and eliminated from further consideration. Alter the previous $S = S_{old}$ to

$$S_{new} = S_{old} + (1-e_t). \qquad (A.6)$$

Step 2 (rounding-down step). Enter this step whenever the previous $S = S_{old} \geq 0$. From the t-indices still under consideration select the one with the largest Q_t and define the associated

$$\tilde{N}_t = Rn_t + R_t \ . \qquad (A.7)$$

This means that for this t-index $\tilde{N}_t \leq (N/n)n_t$ so that it is placed in the τ-set and eliminated from further consideration. Alter the previous $S = S_{old}$ to

$$S_{new} = S_{old} - e_t \; . \qquad (A.8)$$

Continue with either step 1 (if $S_{old} < 0$) or step 2 (if $S_{old} \geq 0$) until all indices t are placed in either the θ-set or τ-set.

It is clear that throughout this process $|S| < 1$ and for the final value of S we have

$$S_f = \text{no. of steps of type 2} - \Sigma e_t = \text{integral.} \qquad (A.9)$$

Therefore $S_f = 0$ and $\Sigma\tilde{N}_t = N$. Note also that $Q_\theta \leq Q_\tau$ for any θ and τ in the respective sets. It will be apparent from the analysis below that normally the $\tilde{N}_t$ will be the $\hat{N}_t$. Specifically, it can be shown (the proof is somewhat lengthy) that $\hat{N}_\tau = \tilde{N}_\tau$ for all indices in the τ-set.

We now prove the following theorem: If for a set N^*_θ with $\Sigma N^*_\theta = \Sigma\tilde{N}_\theta$ we have

$$\frac{N^*_{\theta_1} + 1}{n_{\theta_1}} > \frac{N^*_{\theta_2}}{n_{\theta_2}} \qquad (A.9)$$

for any $\theta_1 \neq \theta_2$ in the θ-set, then $\hat{N}_t = N^*_\theta$ for t in the θ-set and $\hat{N}_t = \tilde{N}_\tau$ for t in the τ-set. To prove this theorem assume that a global maximum has a θ-index (say θ_1) with $\hat{N}_{\theta_1} > N^*_{\theta_1}$ and hence a θ_2 with $\hat{N}_{\theta_2} < N^*_{\theta_2}$. Then from (A.9)

$$\frac{\hat{N}_{\theta_1}}{n_{\theta_1}} \geq \frac{N^*_{\theta_1} + 1}{n_{\theta_1}} > \frac{N^*_{\theta_2}}{n_{\theta_2}} \geq \frac{\hat{N}_{\theta_2} + 1}{n_{\theta_2}} \; . \qquad (A.10)$$

Thus if we define $N''_t = \hat{N}_{\theta_1} - 1$ for $t = \theta_1$, $N''_t = \hat{N}_{\theta_2} + 1$ and $N''_t = \hat{N}_t$ for $t \neq \theta_1, \theta_2$ we would have $L(N''_1,..,N''_T) > L(\hat{N}_1,..,\hat{N}_T)$ leading to a contradiction.

The above theorem permits us to obtain the $\hat{N}_t$ from the $\tilde{N}_t$ by a series of 'interchanges'. Starting with $N_t^{(0)} = \tilde{N}_t$ we step from $N_t^{(i-1)}$ to $N_t^{(i)}$ by the i-th interchange defined as follows: Let θ_{i1} be the θ-index for which $N_\theta^{(i-1)}/n_\theta$ is a maximum and θ_{i2} be the θ-index for which $(N_\theta^{(i-1)}+1) n_\theta$ is a minimum. Then if

$$(N_{\theta_{i2}}^{(i-1)} + 1)/n_{\theta_{i2}} < N_{\theta_{i1}}^{(i-1)}/n_{\theta_{i1}} \qquad \text{(A.11)}$$

we clearly have $\theta_{i1} \neq \theta_{i2}$ and define $N_t^{(i)} = N_{\theta_{i1}}^{(i-1)} - 1$ for $t=\theta_{i1}$, $N_t^{(i)} = N_{\theta_{i2}}^{(i-1)} + 1$ for $t = \theta_{i2}$ and $N_t^{(i)} = N_t^{(i-1)}$ for $t \neq \theta_{i1}, \theta_{i2}$. If (A.11) is satisfied, then $L(N_1^{(i)},.., N_T^{(i)}) > L(N_1^{(i-1)},..,N_T^{(i-1)})$. The process is repeated until for some i the condition (A.11) is violated when, by the above theorem, the $N_t^{(i-1)}$ provide the global maximum. Because the likelihood increases at each interchange, the process cannot cycle and must come to a close as $L(N_1,..,N_T)$ is defined over a finite set.

It will be noted that we could have started the process from any set $N_t^{(0)}$ satisfying $\Sigma N_t^{(0)} = N$. However, the computation of the $\tilde{N}_t$ as a starting point will be a convenient way of satisfying this condition and usually provide $\hat{N}_t = \tilde{N}_t$ since often (A.11) will be violated for the $\tilde{N}_\theta$. For example if all the $n_\theta = 1$ this will certainly be the case. We also state (without proof) that for the $\hat{N}_t$ derived by the above process, $-1 < \hat{N}_t - (N/n)n_t < n_t$.

REFERENCES

Draper, N. R. and Guttman, I., "Some Bayesian stratified two-phase sampling results", Biometrika, 55 (1968), in press.

Ericson, W. A., "Optimum stratified sampling using prior information", J. Amer. Statist. Assoc., 60 (1965), 750-71.

Godambe, V. P., "A new approach to sampling from finite populations. II.", J. Roy. Statist. Soc., B, 28 (1966), 320-8.

Hartley, H. O. and Rao, J.N.K., "A new estimation theory for sample surveys", Biometrika, 55 (1968), in press.

Hoadley, A. B., "Properties of the compound multinomial distribution useful in a Bayesian analysis of categorical data from finite populations", Submitted for publication (manuscript seen by courtesy of the author).

Kullback, S., "On certain distributions derived from the multinomial distribution", Ann. Math. Statist., 8 (1937), 127-44.

Rao, J.N.K., "Bayesian optimisation of stratified sampling", submitted for publication.

A COMPARISON OF SOME POSSIBLE METHODS OF SAMPLING FROM SMALLISH POPULATIONS, WITH UNITS OF UNEQUAL SIZE

M. R. Sampford

Department of Computational and Statistical Science

The University, Liverpool 3, England

1. Notation

The quantity

$$\eta = \Sigma z_i = \Sigma \alpha_i y_i$$

is to be estimated, where Σ denotes summation over all (N) units of the population or stratum, α_i is in some sense a measure of unit size, and is known for all units ($A = \Sigma\alpha_i$; $\bar{A} = A/N$), y_i is to be determined on the units of a sample of size n, and α (and hence z) is expected to have an appreciably larger coefficient of variation than y (e.g. α_i = area, y_i = yield per unit area, on plantation i).

2. A selection of available methods of sampling and estimation

I. Equal probability sampling without replacement

(a) Estimator $Y_1 = A\bar{y}_{uw} = A\ Sy_i/n$ (1)

(where S, here and elsewhere, denotes summation over the units of the sample).

(b) $$Y_2 = N\bar{z} = NS\alpha_i y_i/n \ . \qquad (2)$$

(c) [The 'classical' ratio estimator]

$$Y_3 = Ar = A\ Sz_i/S\alpha_i \ . \qquad (3)$$

Approximation to sampling variance:

$$V(Y_3) = \frac{N^2(1-f)}{n} \frac{\Sigma\alpha_i^2(y_i-R)^2}{N-1} \; ; \; R = \eta/A \; ; \; f = n/N \; . \quad (4)$$

Customary variance estimator:

$$v(Y_3) = \frac{N^2(1-f)}{n} \frac{S\alpha_i^2(y_i-r)^2}{n-1} \; . \quad (5)$$

(d) [Regression estimator]

$$Y_4 = N[\bar{z}+b(\bar{A}-\bar{\alpha})] \; , \quad \bar{\alpha} = S\alpha_i/n, \quad (6)$$

where $b = \dfrac{S(z-\bar{z})(\alpha-\bar{\alpha})}{S(\alpha-\bar{\alpha})^2}$.

Approximation to sampling variance:

$$V(Y_4) \simeq \frac{N^2S_e^2}{n}(1-f); \text{ where} \quad (7)$$

$$S_e^2 = \frac{\Sigma(z-\xi)^2 - B^2\Sigma(\alpha-\bar{A})^2}{N-1} \; ; \; \xi = \frac{\eta}{N} \; ;$$

$$B = \frac{\Sigma(z-\xi)(\alpha-\bar{A})}{\Sigma(\alpha-\bar{A})^2} \; .$$

Customary estimate:

$$v(Y_4) = \frac{N^2s_e^2}{n}(1-f); \text{ where} \quad (8)$$

$$s_e^2 = \frac{S(z-\bar{z})^2 - b^2S(\alpha-\bar{\alpha})^2}{n-2} \; .$$

A further factor (n+1)/n may be included in both terms.(Cochran [1963].)

Of these four estimators Y_1 may be seriously biased, and Y_2 though unbiased is relatively imprecise: they are included only as bases for comparison. Exact expectations and variances, and unbiased variance estimators, are available from standard formulae.

For reasonably large samples, from large populations, Y_3 and Y_4 are nearly unbiased, and their variance formulae are satisfactory: for smallish populations (e.g. some strata), however, they are less so. Bias is not a serious problem: a formula for the bias of the ratio estimator is available (Hartley and Ross [1954]) and less biased modifications are available (Ibid.; Quenouille [1956]; Durbin [1959]), while the regression estimate is unlikely to be seriously biased: however $v(Y_3)$ and $v(Y_4)$ are biased estimators of $V(Y_3)$ and $V(Y_4)$, which are themselves only approximations, unreliable for small n and N.

II. Selection with sample probability proportional to total 'size'

Lahiri [1951] and Midzuno [1951] considered methods of selection such that the probability of selection of any sample is equal to the corresponding $S\alpha_i$. Midzuno's method is to make the first drawing with probability proportional to α_i, and all subsequent drawings with equal probabilities, and without replacement. The standard form of the ratio estimator,

$$Y_{3M} = ASz_i/S\alpha_i \ ,$$

is then unbiased. No simple exact variance formula exists. Formulae proposed as approximations are:

$$V(Y_{3M}) = V(Y_3) \qquad \text{(formula (4))}$$

$$v(Y_{3M}) = \left(\frac{n-1}{N-1}\right) \frac{A}{S(\alpha_i)} v(Y_3) \qquad (9)$$

(where $v(Y_3)$ is as defined by formula (5)).

III. Selection with replacement, with probability proportional to size

(a) Number of drawings (n) fixed

Let r_i be the number of appearances of unit i in the population. Then (writing $p_i = \alpha_i/A$)

$$Y_{1R} = A\bar{y} = A\ \Sigma r_i y_i/n\ ; \tag{10}$$

$$V(Y_{1R}) = \frac{A^2}{n}\ \Sigma p_i (y_i - R)^2\ ; \tag{11}$$

$$v(Y_{1R}) = \frac{A^2}{n}\ \frac{\Sigma r_i (y_i - \bar{y})^2}{n-1}\ . \tag{12}$$

Y_{1R} and $v(Y_{1R})$ are unbiased estimators.

(b) Number of distinct units (n) fixed (Inverse sampling)

Selection continues until the (n+1)st distinct unit appears: this is rejected. If r is the number of drawings made up to but excluding the rejected drawing, and r_i the number of appearances of unit i, then formulae Y_{1RI}, $v(Y_{1RI})$, obtained by replacing n by r in Y_{1R}, $v(Y_{1R})$, provide unbiased estimators: the exact variance formula is complicated (Sampford [1961]).

Both Y_{1R} and Y_{1RI} are inadmissible estimators (Basu [1958]; Pathak [1964]); the corresponding admissible estimators, in which the r_i are replaced by their expectations, conditional on the set of units appearing in the sample, have smaller variances, but are tedious to compute except for extremely small samples. The loss in efficiency caused by sampling with replacement, particularly with the inverse method, is not severe for small sampling fractions, and the sampling and estimation procedures are attractively simple. The method is also the only practicable way of allowing for variation in the α_i when they are not numerically defined — when, for example, units of varying areas are selected from a map by siting random points.

IV. Selection without replacement, with probability proportional (or approximately proportional) to 'size'

Provided $\pi_i = np_i < 1$ for all units, the most satisfactory method (in theory) is to select by some procedure which makes the probability of inclusion of unit i in the sample equal to π_i, with a non-zero probability π_{ij} of the inclusion of both units i and j $(1 \leq i < j \leq N)$. Then

$$Y_{1WR} = A\bar{y}_{uw} \tag{13}$$

is unbiased (Horvitz and Thompson [1952]), its exact variance is

$$V(Y_{1WR}) = (A/n)^2 \sum_{i=1}^{N-1} \sum_{j=i+1}^{N} (\pi_i\pi_j-\pi_{ij})(y_i-y_j)^2 \tag{14}$$

and an unbiased estimator of this is

$$v(Y_{1WR}) = (A/n)^2 \mathop{S}_{i} \mathop{S}_{j>i} \left(\frac{\pi_i\pi_j}{\pi_{ij}} - 1\right)(y_i-y_j)^2 , \tag{15}$$

where summation is over all pairs of units in the sample (Yates and Grundy [1953]).

The difficulty here is to find a method that will yield the required π_i, such that the π_{ij} are all non-zero and can be calculated without excessive labour for all pairs of units in the sample. It is also desirable that $\pi_{ij} < \pi_i\pi_j$, all i, j, in order that all terms in (15) shall be positive. Many methods have been advanced for obtaining the correct π_i, but except for very small samples and populations most of them require a substantial amount of tedious arithmetic either during the selection process itself, or in the calculation (if attempted) of the π_{ij}. Since it is hard to see why any one method should have a consistently higher or lower variance than another, I shall consider in detail only two exact methods, and one approximate, while observing that alternatives have been published by, among others, Fellegi [1963], Hajék [1964] and Hanurav [1966; 1967]. The two exact methods are that modified by Goodman and Kish [19] from a systematic method due to Madow [1949] (and discussed at length by Hartley and Rao [1962]), because it is the simplest and most elegant method to yield exactly the required probabilities, and a new method of my own [1967]. The approximate method is that of Rao, Hartley and Cochran [1962], which is easily performed, and yields both an exact variance formula and an unbiased variance estimator.

IV. (a) The Madow (Goodman and Kish) method

The units are listed in random order, each followed by its appropriate π_i. The π_i are then cumulated, term by term, from the top of the list, yielding a column of values $C_1, C_2, \ldots, C_N$ (=n): C_0 is defined to be zero. A random number h $(0 \le h < 1)$ is obtained, and the set of numbers

$$h, h+1, h+2, \ldots, h+n-1$$

calculated. If h' is any number of this list, and

$$C_{k-1} \le h' < C_k ,$$

the kth unit in the list is accepted into the sample. Since all the π_i are less than 1, no unit can correspond to more than one h' in a particular set, and it is obvious that probability of selection is proportional to size. (These properties are true for any single ordering of the units: the initial random ordering is necessary to ensure non-zero values of the π_{ij}.)

Estimation is by formula (13), and approximate variance formulae (Hartley and Rao [1962]) are

$$V_M(Y_{1WR}) \simeq \frac{1}{n} \{A\Sigma\alpha_i y_i^2 - \eta^2 - (n-1)\Sigma\alpha_i^2(y_i - R)^2\} \quad (16)$$

$$v_m(Y_{1WR}) = \frac{1}{n(n-1)} \underset{i}{S} \underset{j<i}{S} \{\sum_1^N \alpha_i^2 - A(\alpha_i + \alpha_j) + \frac{A^2}{n}\} (y_i - y_j)^2. \quad (17)$$

Exact formulae for the π_{ij} are not obtainable, and their empirical determination requires consideration of all possible permutations of the list of units (strictly, of only 1 in 2N of them, since the π_{ij} are unaffected by reversal or cyclic permutation of the list, but even this task is impossibly tedious except when n and N are small).

(b) The Rao-Hartley-Cochran method

The N units are divided at random into n groups, containing as nearly as possible equal numbers of units. One unit is selected from each group, with probability proportional to size. This does not give exactly the required probabilities of selection, so that modified formulae are necessary: if P_i is the sum of the p_i over all units in group i,

$$Y_{RHC} = A \underset{\text{groups}}{S} (P_i y_i)$$

is an unbiased estimator of η. The exact variance formula is related to that for sampling with replacement: if all groups contain either M or M+1 units, and there are k groups of M+1

$$V(Y_{RHC}) = \{1 - \frac{n-1}{N+1} + \frac{k(n-k)}{N(N-1)}\} V(Y_{1R}) , \qquad (18)$$

and

$$v(Y_{RHC}) = \frac{N^2+k(n-k)-N_n}{N^2(n-1)-k(n-k)} \underset{\text{groups}}{S} P_i(Ay_i - Y_{RHC})^2 \qquad (19)$$

is an unbiased (and essentially non-negative) estimator of $V(Y_{RHC})$. This method is said (by the authors) to give slightly higher variances than the previous one.

(c) A new method

This is a rejective method, in which a sample is chosen with replacement, and rejected if it does not contain n distinct units. Since $\pi_i = np_i < 1$, we can define, for each unit i,

$$\lambda_i = \frac{p_i}{1-np_i} . \qquad (20)$$

The sampling system is as follows: select the first unit with probabilities p_i; then, sampling with replacement, make all subsequent drawings with probabilities proportional to the λ_i. If the first n units drawn are all different, accept the sample; otherwise reject the sample as soon as the first duplicate appears, and start again.

This method has the advantage over many others that, while the drawing can be made without extensive recalculation of selection probabilities, the π_{ij} can be evaluated quite easily. Define

$$L_0 = 1;\ L_m = \sum_{S(m)} \lambda_{i_1} \lambda_{i_2} \cdots \lambda_{i_m} \quad (1 \le m \le N) \tag{21}$$

where summation is over all possible sets of $\underline{m}$ units from the population; define $L_m(\bar{i}\ \bar{j})$ $(0 \le m \le N-2)$ similarly, for the modified population obtained by omitting units i and j. Then

$$\pi_{ij} = K_n \lambda_i \lambda_j \sum_{t=2}^{n} \frac{\{t-n(p_i+p_j)\}\ L_{n-t}(\bar{i}\ \bar{j})}{n^{t-2}}, \tag{22}$$

where

$$K_n^{-1} = \sum_{t=1}^{n} \frac{tL_{n-t}}{n^t}. \tag{23}$$

It would be idle to claim that these calculations are trivial, but they are <u>relatively</u> trivial compared with those involved in many other methods. The method is quite practicable for such tasks as selecting 5 - 10 units from 20 - 40: such values of n and N are out of the question for most other methods. For values of N of this order, the L's may be calculated systematically without serious round-off errors from the formulae

$$L_m = \frac{1}{m} \sum_{r=1}^{m} (-1)^{r-1} R_r L_{m-r}, \tag{24}$$

where

$$R_r = \sum_{i=1}^{N} \lambda_i^r \quad ,$$

and

$$L_m(\bar{i}\ \bar{j}) = L_m - (\lambda_i+\lambda_j)\ L_{m-1}(\bar{i}\ \bar{j}) - \lambda_i\lambda_j\ L_{m-2}(\bar{i}\ \bar{j}). \quad (25)$$

For example, suppose $n = 5$. Then we require $L_0(=1)$, $L_1(=R_1)$, L_2, L_3 and L_4 to calculate K_5, and $L_0(\bar{i}\ \bar{j})(=1)$, $L_1(\bar{i}\ \bar{j})(=R_1-\lambda_i-\lambda_j)$, $L_2(\bar{i}\ \bar{j})$, $L_3(\bar{i}\ \bar{j})$ for each of the 10 pairs of units actually appearing in the sample.

Once R_1, R_2, R_3 and R_4 are calculated, all that remains is to calculate

$$L_2 = \frac{1}{2}\{R_1L_1 - R_2\}$$

$$L_3 = \frac{1}{3}\{R_1L_2 - R_2L_1 + R_3\}$$

$$L_4 = \frac{1}{4}\{R_1L_3 - R_2L_2 + R_3L_1 - R_4\} \ ,$$

and hence K_5, and for each of the 10 pairs of i, j,

$$L_1(\bar{i}\ \bar{j}) = L_1 - (\lambda_i+\lambda_j)$$

$$L_2(\bar{i}\ \bar{j}) = L_2 - (\lambda_i+\lambda_j)\ L_1(\bar{i}\ \bar{j}) - \lambda_i\lambda_j$$

$$L_3(\bar{i}\ \bar{j}) = L_3 - (\lambda_i+\lambda_j)\ L_2(\bar{i}\ \bar{j}) - \lambda_i\lambda_j\ L_1(\bar{i}\ \bar{j}) \ ,$$

and hence π_{ij}. The whole operation of deriving the L's from the R's requires merely 46 multiplications, 66 additions or subtractions, and 3 divisions: a far from formidable task even on a desk machine.

3. The specimen population and the investigations

The 'population' on which I have compared these methods is that of 35 Scottish farms, appearing as Table 5.1 in Sampford [1962]. This is not a genuine population, but the units, and the x and y values are genuine: the 35 farms are in fact the members of a sample drawn from a rather larger population stratified by size, so that the 'population' as presented is over-weighted with large farms, making the choice of an appropriate estimator rather more critical than it would have been in the parent population. The investigations reported here were confined to considering a sample of 12 units from this population, in two forms: an unrestricted choice of 12 units from 35, and a stratified scheme in which 4 units are chosen from each of the three strata containing farms 1 - 12, 13 - 24, and 25 - 35 (this is of course not an optimal allocation for any method, but with such a small sample, equal allocation to strata seems the only practicable procedure if variance estimates are required).

The properties of the estimators discussed were determined in various ways

Y_1, Y_2, Y_{1R}, Y_{1RI}, Y_{RHC}, Y_{1WR} (Sampford): exact variances and bias of Y_2 obtained from formulae;

Y_3, Y_4, Y_{3M}: bias of Y_3 and approximate variances of Y_3 and Y_4 from formulae;

bias of Y_4, actual variances and expectations of estimated variances (a) by complete enumeration for strata (b estimated by sampling for the whole population.

Y_{1WR} (Madow): the π_{ij} were estimated by sampling (usin 2000 permutations of the unit list for each stratu and for the total). The expectation of the Hartle Rao approximate estimator (17) was obtained similarl

Computations were made on the English Electric KDF 9 Computer at Liverpool University using programs written in ALGOL by the author. Random numbers were generated by an algorithm due to Roberts and Downham [1967].

V. Data and results

Table I shows the population, and Table II the results obtained.

TABLE I

Recorded acreage of crops and grass for 1947, and acreage under oats in 1957, for 35 farms in Orkney

Farm number	Recorded crops + grass (x)	Oats 1957 (z)
1	50	17
2	50	17
3	52	10
4	58	16
5	60	6
6	60	15
7	62	20
8	65	18
9	65	14
10	68	20
11	71	24
12	74	18
13	78	23
14	90	0
15	91	27
16	92	34
17	96	25
18	110	24
19	140	43
20	140	48
21	156	44
22	156	45
23	190	60
24	198	63
25	209	70
26	240	28
27	274	62
28	300	59
29	303	66
30	311	58
31	324	128
32	330	38
33	356	69
34	410	72
35	430	103
Total	5759	1384

TABLE II

Properties of estimators tested on population of Table I (η = 1384)

Estimator	E(Y)	Bias	V(Y)	Standard approx. to V(y)	E(estimated variance)
(a) Stratified sample with 4 units per stratum					
Y_1	1377·4	-6·6	17567	-	UB
Y_2	UB	0	22641	-	UB
Y_3	1382·2	-1·8	14930	14461	14795
Y_{3M}	UB	0	14636	(as Y_3)	
Y_4	1377·5	-6·5	26144	13969 (17461)*	16956 (21195)*
Y_{1R}	UB	0	22180	-	UB
Y_{1RI}	UB	0	18099	-	UB
Y_{1WR}(Madow)	UB	0	15951	16659	16190
Y_{RHC}	UB	0	17206	-	UB
Y_{1WR}(Sampford)	UB	0	16197	-	UB

TABLE II (continued)

Properties of estimators tested on population of Table I (η = 1384)

Estimator	E(Y)	Bias	V(Y)	Standard approx. to V(y)	E(estimated variance)
(b) Unstratified sample of 12 units					
Y_1	1461·3	+77·3	13147	-	UB
Y_2	UB	0	53463	-	UB
Y_3	1390·4	+6.4	18248	17183	16255
Y_{3M}	UB	0	17515	(as Y_3)	
Y_4	1392·4	+8·4	21161	15911 (17236)*	15689 (16996)*
Y_{1R}	UB	0	18618	-	UB
Y_{1RI}	UB	0	13390	-	UB
Y_{1WR}(Madow)	UB	0	10060	11226	10156
Y_{RHC}	UB	0	13101	-	UB
Y_{1WR}(Sampford)	UB	0	10009	-	UB

*Including factor(s) (n+1)/n.

Perhaps the least expected result was the excessively bad performance of the regression estimator (Y4), both in the stratified and, to a lesser extent, in the unstratified sample. In view of the fact that this is far from being a 'pathological' population, one is tempted to suggest that this method is best avoided except for fairly large samples. (The bias in the variance estimate was also considerable, even when the factor (n+1)/n was included.)

The ratio estimate (assuming unpooled ratios) performed very well in the stratified sample, both with equiprobable selection (Y_3) and with sample probability proportional to total size (Y_{3M}). The bias of Y_3 was trivial (but the population points nearly all lie near a straight line through the origin, so this was to be expected), and both the customary variance approximation and the expectation of the usual estimate were very close to the true variances. Performance was relatively less good in the population as a whole.

The two without-replacement, pps, methods were almost equally efficient, and had by far the smallest variances when selection was made from the whole population. Bias in the Hartley-Rao approximation to, and estimate of, the variance for the Madow estimate was not serious.

The effect of the use of non-optimal allocation was larger than I had expected: only the ratio estimates and (naturally) Y_2 obtained any advantage from stratification, and both estimates Y_{1WR} lost greatly in efficiency. This last feature appears to reflect the fact that y was appreciably more variable in stratum 3 than in the remaining two strata, and that selection from the population with probability proportional to size had the effect of increasing allocation to this stratum.

REFERENCES

Basu, D. [1958]. Sampling with and without replacement. Sankhyā, 20, 287-94.

Cochran, W. G. [1963]. Sampling techniques, 2nd Edn. John Wiley & Sons, Inc., New York.

Downham, D. Y. and Roberts, F. D. K. [1967]. Multiplicative congruential pseudo-random number generators. Computer Journal, 10, 74-7.

Durbin, J. [1959]. A note on the application of Quenouille's method of bias reduction to the estimation of ratios. Biometrika, 46, 477-80.

Durbin, J. [1967]. Estimation of sampling error in multistage surveys. Appl. Statist., 16, 152-64.

Fellegi, I. [1963]. Sampling with varying probabilities without replacement: rotating and non-rotating samples. J. Amer. Statist. Assoc., 58, 183-201.

Goodman, L. A. and Kish, L. [1950]. Controlled selection a technique in probability sampling. J. Amer. Statist. Assoc., 45, 350-72.

Hajék, J. [1964]. Asymptotic theory of rejective sampling with varying probabilities from a finite population Ann. Math. Statist., 35, 1491-523.

Hanurav, T. V. [1966]. An optimum πps sampling. ABSTRACT Ann. Math. Statist., 37, 1859.

Hanurav, T. V. [1967]. Optimum utilization of auxiliary information: πps sampling of two units from a stratum. J. Roy. Statist. Soc. B 29, 374-391.

Hartley, H. O. and Rao, J. N. K. [1962]. Sampling with unequal probabilities and without replacement. Ann. Math. Statist., 33, 350-74.

Hartley, H. O. and Ross, A. [1954]. Unbiased ratio estimates. Nature, 174, 270-1.

Horvitz, D. G. and Thompson, D. J. [1952]. A generalization of sampling without replacement from a finite universe. J. Amer. Statist. Assoc., 47, 663-85.

Lahiri, D. B. [1951]. A method for sample selection providing unbiased ratio estimates. Int. Stat. Assoc. Bull., 33, 2, 133-40.

Madow, W. G. [1949]. On the theory of systematic sampling. II. Ann. Math. Statist., 20, 333-54.

Midzuno, H. [1951]. On the sampling system with probability proportionate to sum of sizes. Ann. Inst. Stat. Math., 2, 99-108.

Pathak, P. K. [1964]. On inverse sampling with unequal probabilities. Biometrika, 51, 185-94.

Quenouille, M. H. [1956]. Notes on bias in estimation. Biometrika, 43, 353-60.

Rao, J. N. K., Hartley, H. O. and Cochran, W. G. [1962]. On a simple procedure of unequal probability sampling without replacement. J. Roy. Statist. Soc. B 24, 482-91.

Sampford, M. R. [1961]. Methods of cluster sampling with and without replacement for clusters of unequal sizes. Biometrika 49, 27-40.

Sampford, M. R. [1962]. An introduction to sampling theory. Oliver and Boyd, Edinburgh.

Sampford, M. R. [1967]. On sampling without replacement with unequal probabilities of selection. Biometrika 54, 499-513.

Yates, F. and Grundy, P. M. [1953]. Selection without replacement from within strata with probability proportional to size. J. Roy. Statist. Soc. B 15, 235-61.

ADMISSIBILITY OF ESTIMATES OF THE MEAN OF A FINITE POPULATION

V. M. Joshi

Secretary, Maharashtra Government, Bombay

1. Introduction.

Though admissibility is a basic requirement for statistical estimates, little was known about the admissibility under general conditions, even of the estimates in common use in survey sampling. The only general result known was that due to Godambe (1960) and Roy and Chakravarti (1960), relating to the estimate of Horvitz and Thomson for the population mean viz. that with the squared error as loss function the Horvitz-Thomson estimate is admissible, whatever be the sampling design, if the estimates are restricted to the class of linear and unbiassed estimates. The result was thus of limited scope.

Removing the restrictions on the class of estimates, general results have been obtained by the author, which hold for the entire class of all estimates, linear and non-linear, biassed and non-biassed. They relate to the admissibility as estimates of the population mean of the Horvitz-Thomson estimate, (H. T. estimate for short) as also a ratio estimate which includes the sample mean, as a particular case. These results are as follows: (out of them (i) and (ii) are due jointly to Godambe and the author).

(i) The H. T. estimate is always admissible in the class of all unbiassed estimates, linear and non-linear. (1965-I, Section 4).

(ii) In the entire class of estimates, biassed and unbiassed, the H. T. estimate is inadmissible if the sampling design is not of fixed sample size (1965-I, Section 9).

(iii) In the entire class of estimates, the H. T. estimate is admissible if the sampling design is of fixed sample size (1965-III).

(iv) In the entire class of all estimates the sample mean is always admissible as estimate of the population mean, (1965-II).

(v) The usual confidence intervals based on the sample mean and the sample standard deviation, and more generally confidence intervals based on a ratio estimate and a generalized version of the sample standard deviation, are always admissible, (1967).

The definition of admissibility in (v) does not involve a loss function, but (i) through (iv) are based on the squared error as the loss function. In respect of the sample mean and the ratio estimate a much wider result has been obtained, viz:

(vi) If the loss function V(t), where t is the numerical difference between the estimated and true values, satisfies only

(a) V(t) is non-decreasing in $[0,\infty)$,

(b) for every $K > 0$

$$\int_0^\infty V(t) \exp(-\frac{Kt^2}{2})dt < \infty,$$

then the sample mean, and more generally the ratio estimate is always admissible as estimate of the population mean.

However (iv) is not strictly a special case of (vi), because the latter result holds subject to the restriction of Lebsgue measurability of the estimates, while (iv) holds without such restriction. The measurability restriction also applies in the case of (v), and in case of (iii) if the fixed sample size is 3 or more.

The proofs of all these results excepting (ii) are lengthy. Here, we shall only give the statistical frame work and the definitions on which the results are based, an outline of the proof of the most general result, viz: (vi), with an indication of the proofs of (iii) and (v) which follow the same lines, and discuss the conditions for the validity of the results.

2. Notation and definitions.

A common notation has been used in all the author's papers, referred to in Section 1. U is a population of units u_i, i=1,2,...,N. A sample s means any subset of U. S is the set of all possible samples s. With the unit u_i, i=1,2.3,...,N is associated a variate value x_i. $x = \{x_1, x_2, \dots, x_N\}$ is a point in the N-dimensional Euclidean sample space R_N. On S is defined a probability p,

$$p(s) \geq o, \sum_{s \in S} p(s) = 1 \tag{1}$$

d = (S,p) is the sampling design. Every possible design of random sampling is a special case of d = (S,p).

The sample size n(s) means the number of units in s. A sampling design is said to be of fixed sample size m, if p(s) = o whenever n(s) $\neq$ m.

π_i, i=1,2,...,N denotes the total probability that the unit u_i is included in the sample for a given sampling design d.

An estimate is defined as:

DEFINITION 2.1. An estimate e(s,x) is a function e defined on $S \times R_N$, which depends on x, through only those x_i for which $u_i \in s$.

Given the sample, the value of the estimate is thus independent of the units not included in the sample 2.1 is, however, not the most general definition possible. It does not, for example, include the case where the estimate depends on the order in which the units are observed. It is, however, easily shown that given any estimate $g(s,x)$ of this more general type, it is always possible to construct from it, by applying the Rao-Blackwellization process, another estimate $e(s,x)$ which satisfies definition 2.1 and which is uniformly superior to $g(s,x)$. Therefore, the generality of our results is not affected by the restriction in the definition of estimates.

The population mean $\bar{X}_N$ is given by

$$\bar{X}_N = \frac{1}{N} \sum_{i=1}^{N} x_i \ . \qquad (2)$$

The sample mean and the sample variance are given by

$$\bar{\bar{x}}_s = \frac{1}{n(s)} \sum_{i \in s} x_i , \qquad (3)$$

and

$$v_s^2(x) = \frac{1}{n(s)} \sum_{i \in s} (x_i - \bar{\bar{x}}_s)^2$$

where $i \in s$, is written for short for $u_i \in s$.

The ratio estimate is given by

$$\bar{x}_s = \frac{Y}{N} \frac{1}{y(s)} \sum_{i \in s} x_i , \qquad (4)$$

where,

$$Y = \sum_{i=1}^{N} y_i ,$$

$$y(s) = \sum_{i \in s} y_i ,$$

and $y_i > o$, $i=1,2,\ldots,N$, are arbitrary fixed numbers. Thus the sample mean in (3), is a particular case of the ratio estimate where all the y_i, $i=1,2,\ldots,N$ have the same value. The generalization of the sample variance which enters into the confidence intervals based on the ratio estimate is given by

$$v'^2(s,x) = \frac{1}{y(s)} \sum_{i \in s} y_i \left(\frac{x_i}{y_i} - \bar{x}_s \frac{N}{Y}\right)^2. \tag{5}$$

The generalized variance has a significance which we shall discuss later.

The Horvitz-Thomson estimate of the population mean is

$$\bar{e}(s,x) = \frac{1}{N} \sum_{i \in s} \frac{x_i}{\pi_i} . \tag{6}$$

Next, let V(t) be the assumed loss function. The admissibility of estimates of $\bar{X}_N$ is defined as:

DEFINITION 2.2. For a given sampling design d, an estimate e(s,x) is admissible for the population mean $\bar{X}_N$, if there exists no other estimate e'(s,x), such that

$$\sum_{s \in S} p(s)V(|e'(s,x) - \bar{X}_N|) \leq \sum_{s \in S} p(s)V(|e(s,x) - \bar{X}_N|) \tag{7}$$

for all $x \in R_N$, and the strict inequality in (7) holds for at least one $x \in R_N$.

A confidence interval is determined by two estimates $e_1(s,x)$ and $e_2(s,x)$ such that

$$e_1(s,x) \leq e_2(s,x), \text{ for all } s \in \bar{S} \text{ and all } x \in R_N.$$

Here $\bar{S}$ denotes the subset of S, consisting of all those samples s, for which for the given sampling design d, $p(s) \neq 0$.

For each x, let $\bar{S}_{e_1,e_2,x}$ denote the subset of $\bar{S}$, consisting of all those samples s, for which

$$e_1(s,x) \leq \bar{X}_N \leq e_2(s,x) \quad (9)$$

holds. $\bar{S}_{e_1',e_2',x}$ is defined similarly.

Then we define admissibility of confidence intervals as:

DEFINITION 2.3. The set of confidence intervals $[e_1(s,x),\ e_2(s,x)]$ is admissible, if there exists no other set of confidence intervals $[e_1'(s,x),\ e_2'(s,x)]$, such that

$$e_2'(s,x) - e_1'(s,x) \leq e_2(s,x) - e_1(s,x) \quad (10)$$

$$\text{for all } x \in R_N \text{ and all } s \in \bar{S},$$

and,

$$\sum_{s \in \bar{S}_{e_1',e_2',x}} p(s) \geq \sum_{s \in \bar{S}_{e_1,e_2,x}} p(s) \quad \text{for all } x \in R_N, \quad (11)$$

the strict inequality in (11) holding for at least one $x \in R_N$.

The rationale of the definition becomes obvious on noting that the quantities on the left and right hand sides of (11) are simply the inclusion probabilities of the two sets of confidence intervals.

Then as an intermediate step in the development of the argument, it is necessary to define a weaker type of admissibility, which may be loosely described as a.e. admissibility. In place of definition 2.2, we say that the estimate $e(s,x)$ is weakly admissible, if there exists no alternative $e'(s,x)$, for which (7) holds for almost all $x \in R_N$, and the strict inequality in (7) holds on a non-null set of R_N. (Lebesgue measure). In place of definition 2.3, we say, that the set of confidence intervals $[e_1(s,x), e_2(s,x)]$ is weakly admissible, if there exists no alternative set $[e_1'(s,x), e_2'(s,x)]$, which satisfies (10), and for which (11) holds for almost all $x \in R_N$ (Lebesgue measure), the strict inequality in (11) holding on a non-null set of R_N.

To distinguish from weak admissibility the admissibility defined by Definitions 2.2 and 2.3 is called strict admissibility.

3. Outline of the proofs.

The proofs of results (iii), (v) and (vi) in section 2, follow a common pattern. We shall outline first the proof of (vi), which is the most general of these results.

Result (vi). This relates to the admissibility, with a general loss function V(t), which is subject only to the conditions (a) and (b), in (vi) of section 2, of the ratio estimate $\bar{x}_s$ in (4), which includes the sample mean as a particular case. The first step in the proof is to show that $\bar{x}_s$ is weakly admissible. We consider a prior distribution on R_N, such that the x_i, $i=1,2,\ldots,N$ are distributed independently and normally with mean θy_i and variance y_i, θ itself being distributed normally with mean zero and variance τ^2. Using only the condition (a), that V(t) is non-decreasing in $[o,\infty)$ it is then shown that the Bayes estimate is

$$b_\tau(s,x) = \frac{g}{g_s}\bar{x}_s, \tag{12}$$

where

$$g = (1 + Y\tau^2)^{-1}$$

$$g_s = (1 + y(s).\tau^2)^{-1} \quad .$$

We observe here that $\bar{x}_s$ is the limit of the Bayes estimates $b_\tau(s,x)$ when $\tau \to \infty$, so that $\bar{x}_s$ is the Bayes solution with respect to the improper uniform density of θ on $(-\infty,\infty)$.

Next an upper bound is obtained for the excess of the risk of $\bar{x}_s$ over the Bayes risk. Let $\bar{A}_{\tau,s}$ be the risk for given s of $\bar{x}_s$ and $B_{\tau,s}$ that of the Bayes estimate. Then

$$\bar{A}_{\tau,s} - B_{\tau,s} < \frac{C}{\tau^2} \quad \text{for all } \tau \geq \text{some } \tau_o, \tag{13}$$

where C is a positive constant independent of s.

Next suppose $\bar{x}_s$ is not weakly admissible. Then by the definition of weak admissibility there exists an estimate e'(s,x) for which the strict inequality in (7) holds on a non-null subset of R_N, it being understood that in the r.h.s. of (7) e(s,x) is replaced by $\bar{x}_s$. It follows that there exists at least one sample s_o with

$$p(s_o) \neq 0, \tag{14}$$

and such that,

$$h(s_o,x) = e'(s_o,x) - \bar{x}_{s_o} \neq 0 \tag{15}$$

on a non-null subset of R_N.

Let $A'_{\tau,s}$ be the risk of e'(s,x) for given s. We define a subset T_a of R_N by $x \epsilon T_a$, if and only if

$$|x_i| \leq a. \tag{16}$$

We take a sufficiently large, so that $h(s_o,x) \neq 0$ on a non-null subset of T_a. The difference in risks $A'_{\tau,s_o} - \bar{A}_{\tau,s_o}$ can be expressed as the sum of a part P_1 arising from integration on the set T_a and a part P_2 from integration on the complementary set T_a^c.

Since $\bar{x}_{s_o}$ is the Bayes estimate for the limiting prior distribution as $\tau \to \infty$, P_1 becomes positive for sufficiently large τ. Also because of the factors

$$\frac{1}{\sqrt{2\pi}.\tau}$$

in the density, $P_1=0(\frac{1}{\tau})$. More precisely it is shown, that for all sufficiently large τ,

$$P_1 \geq \frac{\beta}{2\sqrt{2\pi}\tau} \quad \text{where } \beta > 0. \tag{17}$$

Since the improvement in risk over the set T_a^c, due to e'(s,x) must be less than that due to the Bayes estimate, and the latter in turn must be less than the improvement over the whole space R_N, we have by (13),

$$P_2 \geq -\frac{C}{\tau^2} \ . \tag{18}$$

Hence combining (17), and (18)

$$A'_{\tau,s_o} - \bar{A}_{\tau,s_o} \geq \frac{\beta}{2\sqrt{2\pi}\tau} - \frac{C}{\tau^2} \ . \tag{19}$$

Also for all other samples s, by the property of the Bayes estimate,

$$A'_{\tau,s} - \bar{A}_{\tau,s} \geq -\frac{C}{\tau^2} \quad \text{for } \tau \geq \tau_o. \tag{20}$$

Multiply (19) by $p(s_o)$, (20) by p(s), and sum over all $s\epsilon\bar{S}$. Denoting the total risks of e'(s,x) and $\bar{x}_s$, by A'_τ, $\bar{A}_\tau$ respectively and noting that $\sum_{s\epsilon\bar{S}} p(s)=1$, we get

$$A'_\tau - \bar{A}_\tau \geq \frac{\beta . p(s_o)}{2\sqrt{2\pi}.\tau} - \frac{C}{\tau^2} \text{ for } \tau \geq \tau_1 \text{ say.} \tag{21}$$

The r.h.s. of (21) becomes positive for sufficiently large τ. But since by (7) for almost all $x\epsilon R_N$, the expected loss of e'(s,x) $\leq$ the expected loss of $\bar{x}_s$, we must have, on taking expectation with respect to the prior distribution,

$$A'_\tau - \bar{A}_\tau \leq 0 \text{ for every } \tau. \tag{22}$$

The supposition that $\bar{x}_s$ is not weakly admissible thus leads to a contradiction, thus proving its weak admissibility.

The next step in the proof is to extend the weak admissibility to weak admissibility in a class of hyperplanes in R_N, on which some of the variates have fixed values. Let Q^{α}_{N-k} be the hyperplane on which say the last k variates x_{N-k+t}, t=1,2,...,k have fixed values α_{N-k+t} respectively. Let $\bar{S}_k$ be the subset of $\bar{S}$ consisting of all those samples s, which includes each of the last k units u_{N-k+t}, t=1,2,...,k. Then the extended theorem (Theorem 4.1, Joshi (1968)) is:

If

$$\sum_{s\in\bar{S}_k} p(s)\ V(|e'(s,x) - \bar{X}_N|) \leq \sum_{s\in\bar{S}_k} p(s)V(|\bar{x}_s-\bar{X}_N|) \quad (23)$$

for almost all $(\mu_{N-k})x\epsilon Q^{\alpha}_{N-k}$, then $h(s,x)=e(s,x)-\bar{x}_s=0$, for all $s\epsilon\bar{S}_k$, and almost all $(\mu_{N-k})x\epsilon Q^{\alpha}_{N-k}$.

Here (μ_{N-k}) denotes the (N-k) dimensional Lebesgue measure on Q^{α}_{N-k}.

The theorem is proved by establishing a 1-1 correspondence between the points $x\epsilon Q^{\alpha}_{N-k}$ and the points $x'\epsilon Q^{\alpha'}_{N-k}$, where $Q^{\alpha'}_{N-k}$ is a parallel hyperplane on which the variates x_{N-k+t}, assume another set of values α'_{N-k+t}, t=1,2,...,k.

We put,

$$x'_r = x_r + hy_r, \quad r=1,2,\ldots,N-k \quad (24)$$

where

$$h = \sum_{r=N-k+1}^{N} \alpha'_r - \sum_{r=N-k+1}^{N} \alpha_r \Big/ \sum_{r=N-k+1}^{N} y_r\ .$$

This 1-1 correspondence ensures that (see (5.10) to (5.12), Joshi (1966)),

$$\bar{x}_s(x') - \bar{X}_N(x') = \bar{x}_s(x) - \bar{X}_N(x). \quad (25)$$

In (23), $e'(s,x)$ is defined on the hyperplane Q^{α}_{N-k}. We extend its definition to the hyperplane $Q^{\alpha'}_{N-k}$ by setting

$$e'(s,x') = e'(s,x) + \bar{x}_s(x') - \bar{x}_s(x), \quad (26)$$

so that,

$$e'(s,x') - \bar{X}_N(x') = e'(s,x) - \bar{X}_N(x). \quad (27)$$

Thus the inequality (23) holds for almost all $(\mu_{N-k})x\epsilon Q^{\alpha'}_{N-k}$ and hence for almost all $(\mu_N), x\epsilon R_N$. Further supposing the extended theorem to be false, corresponding to the non-null (μ_{N-k}) subset of Q^{α}_{N-k} on which $h(s,x) = e'(s,x) - \bar{x}_s \neq 0$, there exists by the 1-1 correspondence a similar non-null (μ_{N-k}) subset in every $Q^{\alpha'}_{N-k}$, and all these sets together form a non-null (μ_N) subset of R_N on which $h(s,x) \neq 0$.

So far we have defined the alternative estimate $e'(s,x)$ for $s\epsilon \bar{S}_k$. For $s\notin \bar{S}_k$, we put

$$e'(s,x) = \bar{x}_s. \quad (28)$$

Then for the estimate $e'(s,x)$ defined by (23), (26) and (28) and $\bar{x}_s$, the inequality (7) holds for almost all $x\epsilon R_N$, and $h(s,x) \neq 0$ on a non-null subset of R_N. This contradicts the weak admissibility of $\bar{x}_s$ proved in the first step. The extended theorem is thus proved.

We now come to the last step in the proof. Suppose $\bar{x}_s$ is not strictly admissible. Then there exists at least one sample s_o, with $p(s_o) \neq 0$ and one point $a = (a_1, a_2, \ldots, a_N) \epsilon R_N$ such that,

$$h(s_o, a) = h_o \neq 0. \quad (29)$$

Without loss of generality, we take s_o to consist

of the first m units. Then let P^a_{N-m} be the hyperplane defined by

$$x \epsilon P^a_{N-m}, \text{ if, and only if,}$$
$$x_i = a_i, \ i=1,2,\ldots,m \ . \tag{30}$$

It is next shown that whether h_o is positive or negative, we can determine the half Q^a_{N-m} of the hyperplane P^a_{N-m}, such that for $x \epsilon Q^a_{N-m}$,

$$e'(s_o,x) - \bar{X}_N \geq \bar{x}_{s_o} - \bar{X}_N + h_o . \tag{31}$$

We now exclude the case in which V(t) is a constant on ∞ as in that case every estimate, including $\bar{x}_s$ is admissible. Excluding this case, it can be shown (Lemma 3.2, of Joshi (1968)) that (31) implies, that there exists a non-null (μ_{N-m}) subset E^a_{N-m} of P^a_{N-m}, such that for every $x \epsilon E^a_{N-m}$,

$$V(|e'(s_o,x) - \bar{X}_N|) > V(|\bar{x}_{s_o} - \bar{X}_N|). \tag{32}$$

Hence to maintain the inequality (23), we must have for every $x \epsilon E^a_{N-m}$, at least one other sample $s \neq s_o$, $s \epsilon \bar{S}_k$, such that

$$h(s,x) \neq 0. \tag{33}$$

Thus starting with the single point a, we get a set E^a_{N-m} of (N-m) dimensional positive measure, such that at each point of E^a_{N-m}, $h(s,x) \neq 0$ for at least one $s \epsilon \bar{S}_k$. The process is then repeated with each point of E^a_{N-m}. The dimensionability of the set is thus successively increased till we either reach a non-null (μ_N) subset, which contradicts the theorem of weak admissibility or a non-null (μ_k) subset of a hyperplane which contradicts the extended theorem of weak admissibility in a hyperplane. The proof is thus completed.

Result (v). This relates to the admissibility of the confidence intervals for the population mean based on the sample mean and sample variance viz.

$$[\bar{x}_s - k_s v_s(x),\ \bar{x}_s + k_s v_s(x)]$$

and more generally of the confidence intervals based on the ratio estimate and a generalization of the sample variance viz.

$$[\bar{x}_s - k_s v_s'(x),\ \bar{x}_s + k_s v_s'(x)].$$

Here $k_s > 0$ are positive constants which may vary with the sample s.

The proof proceeds on lines exactly similar to that of (vi). The same prior distribution on R_N is assumed. The only special point arises in the second step in the proof.

The 1-1 correspondence between points $x \epsilon Q_{N-k}^{\alpha}$ and $x' \epsilon Q_{N-k}^{\alpha'}$ is established as before.

But then to preserve the inclusion probability of the confidence intervals, the confidence intervals at the point x' are taken to be

$$[\bar{x}_s(x') - k_s \overset{*}{v}(x'),\ \bar{x}_s(x') + k_s \overset{*}{v}(x')]$$

where the new function $\overset{*}{v}(x')$ is defined by

$$\overset{*}{v}(x') = v'(x) \ . \tag{34}$$

It is then necessary to show that the theorems of weak admissibility and the extended theorem of weak admissibility in a hyperplane hold good for the confidence intervals based on $\overset{*}{v}(x)$. This is done by showing that the only properties of the function $v'(x)$ which enter into the proof of admissibility are that $v'(x)$ is non-negative and an estimate, and that under the assumed prior distribution $v'(x)$ and θ are distributed independently and that all these properties also pertain to the new function $\overset{*}{v}(x)$.

Result (iii). This relates to the admissibility for a fixed sample size design of the H.T. estimate defined in (6). By a well-known result for the fixed sample size design with size m the inclusion probabilities satisfy

$$\sum_{i=1}^{N} \pi_i = m \ . \tag{35}$$

In the first step in the proof, we consider a more general estimate given by

$$\bar{e}(s,x) = \sum_{r \in s} b_r x_r \tag{36}$$

where $b_r > 1$ and $\sum_{r=1}^{N} \frac{1}{b_r} = m$.

The prior distribution assumed in this case is such that all the x_i are distributed independently and

$$b_r x_r \text{ is } N(\theta, \sigma_r^2)$$

where

$$\sigma_r^2 = k(1 - \frac{1}{b_r}), \text{ k being some positive constant.}$$

The estimate $\bar{e}(s,x)$ in (36) is not now the limit of a sequence of Bayes estimates. Hence in this case the theorem of weak admissibility is proved by a different method involving an application of the Cramer-Rao lower bound for the variance of an estimate. This raises the question of regularity restrictions on estimates which is discussed later. The rest of the proof proceeds on lines identical with that of (vi).

Results (i) and (iv) are proved by combinatorial methods, using mathematical induction, and are thus valid without the restriction of measurability of estimates. Similarly in the case of result (iii), for the special cases m = 1 or 2, the admissibility of the H.T. estimate is proved by purely algebraic methods and so is not subject to measurability restrictions.

The proof of result (ii) is quite elementary. We take,

$$e'(s,x) = (1-\lambda)\bar{e}(s,x) \tag{37}$$

where $0 < \lambda < 1$. Then if the sampling design is not of fixed sample size, a value of λ can be found, for which $e'(s,x)$ is uniformly superior to $\bar{e}(s,x)$.

4. Generalization of sample variance.

In result (v), the confidence intervals are based on the ratio estimate and the generalization of the sample variance given by (5). This generalization has a significance. The ratio estimate is superior to the sample mean as estimate of the population mean, if prior knowledge exists that

$$x_i \text{ is } N(\theta y_i, \sigma^2 y_i), \quad i=1,2,\ldots,N \tag{38}$$

all x_i being distributed independently.

Set

$$\bar{X}_{N-n(s)} = \frac{Y}{N}\frac{1}{Y-y(s)} \sum_{i \notin s} x_i \; . \tag{39}$$

Then $(\bar{X}_{N-n(s)} - \bar{x}_s)$ is $N(0,\sigma^2 Y^3/N^2[Y-y(s)].y(s))$.

Hence the best confidence intervals for $\bar{X}_{N-n(s)}$ i.e. equivalently for $\bar{X}_N$ are based on a function which gives an estimate of σ^2. It is seen that the estimate of σ^2 is given, not by the sample standard deviation but by the function $v'(s,x)$ in (5) as $E_\theta\{v'^2(s,x)\} \alpha\, \sigma^2$ for all θ.

A small point may be clarified here. The formulae for $\bar{x}_s, \bar{X}_N$ and $v'(s,x)$ in the present paper, differ slightly from those given previously (Joshi (1967), equation (120)). The difference arises because in the previous paper the confidence intervals were estimates of

$$\frac{1}{Y} \sum_{i=1}^{N} x_i ,$$

while here we have taken them as estimates of

$$\frac{1}{N} \sum_{i=1}^{N} x_i .$$

Aparc from the trivial difference of a constant $\frac{Y}{N}$, the two sets of formulae are identical.

5. Generalized loss function.

The conditions (a) and (b) on the generalized loss function assumed in result (vi) are satisfied by any loss function of the form

$$V(t) = \sum_{r=1}^{n} a_r t^{c_r}$$

where a_r and c_r are arbitrary positive constants. Again according to the axioms of Neumann and Morganstern, the loss function must be bounded and every bounded function automatically satisfies the condition (b).

Thus both the conditions (a) and (b) would be satisfied by most loss functions considered in practice. The admissibility of the sample mean is thus established on a very general basis.

6. Regularity restriction.

The Cramer-Rao inequality is used in the proof of result (iii). This raises the question whether any regularity restrictions are necessary on the estimates. A similar point had already been investigated by Hodges and Lehmann in their proof of the admissibility of the sample mean as estimate of the mean of a normal population. By using the Wolfowitz (1947) conditions, which are sufficient for the Cramer-Rao inequality, it was shown by them (Problem 4, Hodges and Lehmann (1951)) that no regularity restrictions on the class of estimates are necessary. By a similar reasoning the regularity restrictions are shown to be unnecessary for result (iii) also (Joshi (1966), section 3)).

7. Measurability.

Result (iv) and the special cases of result (iii) of sample size = 1 or 2, hold without the restriction of measurability while the remaining results hold only for the class of measurable estimates. This raises the question as to how far the measurability restriction is important.

An example of a non-measurable set is given by Rogosinski ((1952), p. 73)). The set V is formed by taking from each set V_ξ of all numbers $\xi + r$, where ξ is an irrational number and r runs through the set of all rationals, some number, say the first which falls in $[0, \frac{1}{2}]$ according to a given enumeration of the rationals. A non-measurable function based on the set may be defined by putting $\emptyset(x) = 1$ if $x \in V$ and $\emptyset(x) = 0$ if $x \notin V$. Though the function $\emptyset$ is thus formally defined, we cannot state its value for any given point say $x = \frac{1}{\sqrt{3}}$, as whether this point belongs to V or not depends on which of the irrational numbers $(\frac{1}{\sqrt{3}} + r)$ has been taken as the first number ξ, in defining V_ξ. As stated by Rogosinski, all known examples of non-measurable sets are based on the axiom of choice. It thus appears that though non-measurable sets and functions may for theoretical purposes be defined by using the axiom of choice, a non-measurable function cannot be used as estimate in practice.

8. Minimaxity.

We shall conclude by indicating in this and the next section two other properties of the sample mean which have been proved in these investigations.

Previously Aggarwal (1959) had shown that with the squared error as loss function, the sample mean $\bar{x}_s$ in (3) is a minimax estimate of $\bar{X}_N$, on the subset D_o of R_N, defined by

$$D_o = [x: \sum_{r=1}^{N} (x_r - \bar{X}_N)^2 \leq \text{const} = N\sigma_o^2 \text{ say}] \tag{40}$$

if the sampling design is of simple random sampling with fixed sample size.

Using the fact that our proof of the result (iv) is valid for any subset of R_N, which is symmetrical in all the coordinates, Aggarwal's result is easily generalized to the case when the sampling design is not of fixed sample size, but for each size m, $1 \leq m \leq N$, the probability P_m is distributed equally between all possible samples of size m. For consider the set

$$D = [x: \sum_{r=1}^{N} (x_r - \bar{X}_N)^2 = N\sigma^2]. \tag{41}$$

It is easily verified that for $x \epsilon D$,

$$\text{M.S.E. of } \bar{\bar{x}}_s = \sum_{m=1}^{N} P_m \left(\frac{\sigma^2}{m}\right) \left[1 - \frac{m-1}{N-1}\right] = \text{a constant.} \tag{42}$$

Hence the admissibility of $\bar{x}_s$ on D in (41) implies its minimaxity on D and hence on the set D_o in (40).

The result is further easily extended to any subset M of R_N which is symmetrical in all the coordinates and on which the M.S.E. of $\bar{\bar{x}}_s$ is bounded. For suppose $x \epsilon M$, $\sup_{x \epsilon S}$ M.S.E. of $\bar{\bar{x}}_s = \sigma^2$. Then consider the subset M_1 of M on which

$$[x: \text{M.S.E. of } \bar{\bar{x}}_s = \sigma^2 - \epsilon]. \tag{43}$$

As x_s is minimax on the set defined in (43),

$$\text{M.S.E. of any other estimate } e_1(s,x) \geq \sigma^2 - \epsilon \text{ for some } x \epsilon M_1. \tag{44}$$

Hence, since ϵ can be taken arbitrarily small

$$\sup_{x \epsilon M} \text{M.S.E. of } e_1(s,x) \geq \sigma^2. \tag{45}$$

9. Uniform admissibility.

So far we have considered the admissibility of estimates, the sampling design d being considered as given. However, the sampling design is generally under the experimenter's control subject to certain conditions. It is, therefore, desirable to define a notion of admissibility jointly for an estimate and a sampling design. Let V(e,d) denote the mean squared error of an estimate e(s,x) for a sampling design d, i.e.

$$V(e,d) = \sum_{s \in S} p(s) [e(s,x) - \bar{X}_N]^2. \tag{46}$$

If there exists another pair $[e_1(s,x), d_1)$ such that

$$V(e_1,d_1) \leq V(e,d) \tag{47}$$

for all $x \in R_N$, with strict inequality for some x, then we should use the pair (e_1,d_1) in preference to the pair (e ,d).

If there are no limitations on d, then obviously the only admissible sampling design will be the one which assigns probability 1 to the sample consisting of the whole population. But in practice through considerations of cost and time, the class of alternative designs d_1 is limited by one of the following conditions: viz. (a) the average (i.e. expected) sample size does not exceed a certain limit or (b) the average cost of sampling does not exceed a certain limit. (b) is equivalent to (a) except when the cost of an observation varies with the unit observed. In practical situations condition (a) is perhaps more commonly met with than (b). Adopting condition (a), a notion of joint admissibility called uniform admissibility is defined as follows:

DEFINITION 8.1. An estimate e(s,x) together with a sampling design d are uniformly admissible, if there exists no other estimate $e_1(s,x)$ and sampling design d_1 such that expected sample size for $d_1 \leq$ expected sample size for d (48)

and,

$$V(e_1, d_1) \leq V(e,d) \text{ for all } x \epsilon R_N, \tag{49}$$

the strict inequality holding either in (48), or in (49) for at least one $x \epsilon R_N$.

The sample mean $\overline{\overline{x}}_s$ together with any sampling design of fixed sample size are shown to be uniformly admissible according to the Definition 8.1 (Joshi (1966) section 7)).

Appendix

According to our results, the Horvitz-Thomson estimate is admissible as estimate of the population mean or equivalently the population total, in the entire class of all estimates biassed and unbiassed, provided the sampling design is of fixed sample size. During informal discussion, a conjecture was suggested that any arbitrary function whatsoever of the sample values is admissible for the population total in case of simple random sampling if the sampling design is of fixed sample size. This conjecture is, however, not true.

Let $\bar{e}(s,x)$ be a linear, unbiassed estimate of the population total $T(x)$, $e_1(s,x) = (1-\lambda)\bar{e}(s,x)$, $V(x)$ = variance of $\bar{e}(s,x)$ and

$$K = \min_{x \in R_N} \frac{V(x)T^2(x)}{1+V(x)T^2(x)} .$$

It is seen by an elementary calculation that the loss due to the estimate e_1 is uniformly less than that due to $\bar{e}$ if $\lambda < \frac{2K}{1+K}$. Hence $\bar{e}$ is inadmissible if K is positive, which in turn is satisfied if the variance $V(x)$ does not vanish anywhere else except the origin. Since

$$V(x) = \sum_{s \in \bar{S}} p(s) \left[\bar{e}(s,x) - T(x)\right]^2,$$

$V(x)$ vanishes at a point x only if its coordinates $x_1, \ldots, x_N$ satisfy the linear simultaneous equations,

$$\bar{e}(s,x) = T(x) \text{ for all } s \in \bar{S}, \quad \text{i.e. for all } s \text{ for which } p(s) > o. \tag{1}$$

Suppose the sampling design assigns positive probabilitie to M samples. Because of the identical relation

$\sum_{s \in \bar{S}} p(s)\bar{e}(s,x) = T(x)$, (1) reduces to (M-1)

linear independent equations. Hence if $M > N + 1$, these equations in general have no solution and the estimate $\bar{e}$ is inadmissible. The following is a simple numerical example

$$\text{Population } U = \{u_1, u_2, u_3, u_4\}.$$

The sampling design is of fixed sample size 2, all the samples being drawn with the same probability $\frac{1}{6}$. Let

$$s_1 = (u_1, u_2);\; s_2 = (u_1, u_3);\; s_3 = (u_1, u_4)$$

$$s_4 = (u_2, u_3);\; s_5 = (u_2, u_4);\; s_6 = (u_3, u_4)$$

and let $\bar{e}(s,x)$ be given by

$$\bar{e}(s_1, x) = 2(x_1+x_2) + 2(x_2-x_3);$$

$$\bar{e}(s_2, x) = 2(x_1+x_3) - 2(x_2-x_3);$$

$$\bar{e}(s_3, x) = 2(x_1+x_3);\; \bar{e}(s_4, x) = 2(x_2+x_3);$$

$$\bar{e}(s_5, x) = 2(x_2+x_4);\; \bar{e}(s_6, x) = 2(x_3+x_4).$$

Then

$$S^2(x) = \sum_{i=1}^{6} p(s_i)\bar{e}^2(s_i, x)$$

$$= \frac{2}{3}\{7x_2^2+3(x_1^2+x_3^2+x_4^2)$$

$$+ 2(x_1x_2+x_1x_3+x_1x_4+x_2x_4+x_3x_4)+x_2x_3\}.$$

By the method of Lagrange's multipliers, it is easily shown that for given $T(x) = k$, $S^2(x)$ and hence $V(x) = S^2(x) - T^2(x)$ are minimized for $x_2=o$,

$x_1=x_3=x_4=k/3$. The minimum value of $S^2(x)$ comes to $\frac{10}{9}k^2$, so that

$$K = \min \frac{V(x)T^2(x)}{1+V(x)T^2(x)} = \frac{1}{10} .$$

Hence $e_1(s,x)$ is obtained by assigning to λ any positive value $< \frac{1}{5}$.

An example of a biassed estimate which is inadmissible can similarly be constructed by taking $\bar{e}(s,x) = a.T(x)$ where $a > 1$.

References

Aggarwal, O. P. (1959) Bayes and Minimax procedures in sampling from finite and infinite populations, Ann. Math. Statist. 30, 206-218.

Godambe, V. P. (1960) An admissible estimate for any sampling design, Sankhya 22, 285-288.

Godambe, V. P. and Joshi, V. M. (1965-I) Admissibility and Bayes estimation in sampling finite populations, Ann. Math. Statist. 36, 1707-1722.

Hodges, J. L. Jr. and Lehmann, E. L. (1951) Some applications of the Cramer-Rao inequality. Proceedings of the second Berkeley Symposium, Math. Statist. Probability 13-22, University of California Press.

Joshi, V. M. (1965-II) Admissibility and Bayes estimation in sampling finite populations,-II, Ann. Math. Statist. 36, 1723-1729.

Joshi, V. M. (1965-III) Admissibility and Bayes estimation on sampling finite populations,-III. Ann. Math. Statist. 36, 1730-1742.

Joshi, V. M. (1966) Admissibility and Bayes estimation in sampling finite populations,-IV. Ann. Math. Statist. 37, 1658-1670.

Joshi, V. M. (1967) Confidence intervals for the mean of a finite population, Ann. Math. Statist. 38, 1180-1207.

Joshi, V. M. (1968) Admissibility of the sample mean as estimate of the mean of a finite population. Ann. Math. Statist. 39, under publication.

Neumann, Von J. and Morganstern. (1947) Theory of Games and Economic behavior, 2nd Edition, Princeton University Press, Princeton, N. J. p. 641.

Rogosinski, Werner W. (1952) Volume and Integral p. 73.

Roy, J. and Chakravarti, I. M. (1960) Estimating the mean of a finite population, Ann. Math. Statist. 31, 392-398.

Wolfowitz, J. (1947) The efficiency of sequential estimates and Wald's equation for sequential process, Ann. Math. Statist. 18, 215-230.

RATIO AND REGRESSION ESTIMATORS

J. N. K. Rao

Texas A&M University, College Station, Texas

1. Introduction

Ratio estimators are often employed in sample surveys for estimating the population mean $\bar{Y}$ of a character of interest 'y' or the population ratio $\bar{Y}/\bar{X}$, utilizing a supplementary character 'x' that is positively correlated with 'y'. Although the classical ratio estimator is biased, the bias in most practical situations is usually negligible provided 'combined' ratio estimators are used. (Considerable empirical evidence has been provided by Kish *et al.*, 1962.) The bias (relative to standard error) may, however, become considerable in surveys with many strata and small or moderate samples within strata if it is considered appropriate to use 'separate' ratio estimators. Unusual situations may also exist which lead to large values of the coefficient of variation of x, C_x, and then the possibility of large bias can arise. In such situations, the use of wholly unbiased or approximately unbiased (i.e., estimators with a smaller bias than the classical ratio estimator) may be of great advantage. Therefore, in recent years, considerable attention has been given to the development of wholly unbiased or approximately unbiased ratio estimators. Before we decide to use one of these estimators, it is, however, necessary to consider the performances of these estimators relative to the classical ratio estimator in small or moderate samples.

In Section 2 we review recent work on some of these ratio estimators and also on some unbiased regression estimators. We give, in Section 3, some new empirical results on the performances of these estimators and four variance estimators, using several sets of live data which represent a wide variety of populations. Attention is given to small and moderate samples since these are the cases in which freedom from bias is important.

2. Recent work on some ratio and regression estimators

We consider only simple random sampling without replacement from a population of N units and confine ourselves, without loss of generality, to estimators of $R = \bar{Y}/\bar{X}$ (assuming $\bar{X}$ is known). If a simple random sample of n units is drawn, the classical ratio estimator of R is

$$r = \bar{y}/\bar{x} \tag{1}$$

where $\bar{y}$ and $\bar{x}$ are the sample means of 'y' and 'x' respectively. The well-known unbiased ratio estimator of Hartley-Ross is given by

$$t_1 = \bar{r} + \frac{n(N-1)}{N(n-1)} \frac{(\bar{y}-\bar{r}\bar{x})}{\bar{X}} \tag{2}$$

where $\bar{r} = n^{-1}\sum_1^n y_i/x_i$.

Three ratio estimators are based on splitting up the sample at random into g groups each of size m, where n = mg. First, following a general method of Mickey (1959), an unbiased estimator of R is

$$t_2 = \bar{r}_g' + \frac{(N-n+m)g}{N\bar{X}} (\bar{y}-\bar{r}_g'\bar{x}) \tag{3}$$

where $\bar{r}_g' = \sum_1^g r_j'/g$ and r_j' is the classical ratio estimator computed from the sample after omitting the jth group. t_2 reduces to t_1 if n = 2. Applying Quenouille's method of bias reduction we get

$$t_3 = \omega r-(\omega-1)\bar{r}_g' \tag{4}$$

where $\omega = g[1-(n-m)/N]$. The asymptotic bias of t_3 does not contain terms of order n^{-1} and order N^{-1} (Jones, 1963). Cochran (1963) has used t_3 with g in place of ω and, therefore, terms of order N^{-1} appear in the asymptotic bias. A ratio estimator based on the group sample means $\bar{y}_j$ and $\bar{x}_j$ is

$$t_4 = \omega^* r - (\omega^* - 1)\bar{r}_g \tag{5}$$

where $\bar{r}_g = g^{-1}\sum_1^g(\bar{y}_j/\bar{x}_j)$ and $\omega^* = [(N-m)g]/[N(g-1)]$. Tin (1965) has used t_4 with $g/(g-1)$ in place of ω^* and, therefore, terms of order N^{-1} appear in the asymptotic bias. t_4 reduces to t_3 if $n = 2$.

Tin (1965) has also considered Beale's estimator

$$t_5 = r[1+(\frac{1}{n} - \frac{1}{N})\frac{s_{xy}}{\bar{x}\bar{y}}]/[1+(\frac{1}{n} - \frac{1}{N})\frac{s_x^2}{\bar{x}^2}] \tag{6}$$

and

$$t_6 = r[1+(\frac{1}{n} - \frac{1}{N})(\frac{s_{xy}}{\bar{x}\bar{y}} - \frac{s_x^2}{\bar{x}^2})] \tag{7}$$

where $s_x^2 = (n-1)^{-1}\Sigma(x_i-\bar{x})^2$ and $s_{xy} = (n-1)^{-1}\Sigma(x_i-\bar{x})(y_i-\bar{y})$. The asymptotic biases of t_5 and t_6 do not contain terms of order n^{-1} and order N^{-1}. By approximately estimating the bias of r, Pascual (1961) has proposed

$$t_7 = r + \frac{(N-1)}{N(n-1)}\frac{(\bar{y}-\bar{r}\bar{x})}{\bar{X}} . \tag{8}$$

Regression estimators are not widely used in sample surveys. Nevertheless, it is important to study their performances in small and moderate samples. The classical regression estimator of R is

$$t_8 = \bar{X}^{-1}[\bar{y}+b(\bar{X}-\bar{x})] \tag{9}$$

where $b = s_{xy}/s_x^2$ is the sample regression coefficient. following Mickey (1959) an unbiased regression estimator based on splitting up the sample into g groups is

$$t_9 = \bar{X}^{-1}[\bar{y}+\bar{b}_g'(\bar{X}-\bar{x}) + \frac{N-n}{Ng}(\sum_1^g \bar{x}_j b_j' - n\bar{x}\bar{b}_g')] \tag{10}$$

where $\bar{b}_g' = g^{-1}\sum_1^g b_j'$ and b_j' is the sample regression coefficient computed from the sample after omitting the jth group. Williams (1963) has proposed the unbiased regression estimator

$$t_{10} = \bar{X}^{-1}[\bar{y}+\bar{b}_g(\bar{X}-\bar{x}) + \frac{N-n}{Ng}\frac{1}{(g-1)}(\sum_1^g \bar{x}_j b_j - n\bar{x}\bar{b}_g)] \tag{11}$$

where $\bar{b}_g = g^{-1}\sum_1^g b_j$ and b_j is the sample regression coefficient computed from the jth group (m>1).

Turning to variance estimators, the classical variance estimator of r is

$$v_1 = (\frac{1}{n} - \frac{1}{N})\frac{1}{\bar{X}^2}(s_y^2 - 2rs_{xy} + r^2 s_x^2). \tag{12}$$

Mickey (1959) has given a general method of constructing non-negative variance estimators of t_1, t_2 and t_9. However, it appears formidable to investigate the performances of these variance estimators and, therefore, we have not included them in our empirical study. Tukey's 'jack-knife' variance estimator of t_3 (or r) is

$$v_2 = (1 - \frac{n}{N})\frac{g-1}{g}\sum_1^g (r_j' - \bar{r}_g')^2 \tag{13}$$

and it does not depend on ω. No variance estimators of t_4, t_5, t_6 and t_7 are available in the literature; however, one could use v_1 or v_2.

The variance estimator of t_8 is

$$v_3 = \left(\frac{1}{n} - \frac{1}{N}\right) \frac{1}{(n-2)\bar{X}^2} (s_y^2 - bs_{xy}). \tag{14}$$

Williams has given an unbiased variance estimator of t_{10}; however, it can take negative values and, hence, is useless in practice. For comparison, we have included the unbiased variance estimator of $\bar{y}/\bar{X}$:

$$v_4 = \left(\frac{1}{n} - \frac{1}{N}\right) \frac{s_y^2}{\bar{X}^2} . \tag{15}$$

We now briefly review some important results on the performances of these estimators and variance estimators that are available in the literature. First we review some asymptotic results which are, perhaps, not very useful in practice. Assuming linear regression of y on x and x normally distributed (Durbin's (1959) model 1), Rao (1965, 67) has shown that g = n is the optimum choice for t_2 and t_3 and that the asymptotic variance of t_2 with g = n is slightly smaller than that of r, but slightly larger than the m.s.e. of t_3 with g = n. Tin (1965) has compared r, t_3 (with g = 2), t_5 and t_6 for large n without assuming any model. His results indicate that t_5 and t_6 are better than the others with regard to bias, precision and approach to normality. Assuming a bivariate normal distribution, Chakrabarty and Rao (1967) have shown that in large samples

$$CV^2[v_1] = CV^2[v_4] + \frac{4(\text{Bias } r)^2}{V(r)} . \tag{16}$$

where CV denotes the coefficient of variation. Therefore, CV of v_1 will be considerably larger than that of v_4 if the bias ratio is substantial.

Turning to exact results for any n, Rao and Webster (1966) and Rao (1967) have shown that, if the regression of y on x is linear and x has a gamma distribution (Durbin's model 2), the following results are true: (1) $g = n$ is the optimum choice for t_2 and t_3, (2) t_2 is considerably better than t_1 for $n \geq 2$ and slightly better than r for $n \geq 8$, (3) t_3 (with $g = n$) and t_6 are better than t_1, t_2, t_4, t_7 and r. Rao and Beegle (1967) have made a Monte Carlo study of the eight ratio estimators r, $t_1, \ldots, t_7$ and the three variance estimators v_1, v_2 and v_4 for small and moderate samples, using Durbin's model 1. Their study essentially supports the above results. Further, they have shown: (1) if $\alpha=0$(i.e., regression through the origin) and C_x is small, the performances of the eight ratio estimators are about the same, $g = n$ is the optimum choice for v_2 and the stabilities of v_1, v_2 (with $g = n$) and v_4 are essentially equal, (2) if $\alpha \neq 0$ and C_x is not small, $g = n$ is again the optimum choice for v_2; the stabilities of v_2 (with $g = n$) and v_1 are essentially equal, but both are considerably less stable than v_4. The Monte Carlo study of Lauh and Williams (1964) indicates that, if $\alpha = 0$, C_x is small and x has a truncated exponential distribution, v_2 (with $g = n$) is much better than v_1 for estimating $V(r)$. Chakrabarty and Rao (1967) have investigated the performance of v_2 as an estimator of $V(r)$ under Durbin's model 2. They have shown that the bias of v_2 is minimum when $g = n$ and the m.s.e. of v_2 (with $g = 2$) is considerably larger than that of v_1. Chakrabarty (1968) has also worked out the formula for m.s.e. of v_2 for general g, but unfortunately no numerical results are as yet available due to some formidable difficulties in numerical integration.

No results are available on the unbiased regression estimator t_9. Under Durbin's model 1, Williams (1963) has shown that $g = (n/3)^{1/2}$ is the optimum choice for his estimator and that it compares favorably with t_8

(provided $m > 3$). His asymptotic results for certain non-normal x-distributions also show that t_{10} compares favorably with t_8 provided the x-distribution is platykurtic.

The above literature review clearly indicates the need for an extensive empirical study of these estimators and variance estimators, using live data.

3. Empirical Study

We have chosen 16 natural populations for the empirical study. Table 1 gives the source, nature of y and x, coefficient of variation of x and the correlation ρ. It is clear from Table 1 that we have a wide variety of populations with N ranging from 7 to 270, C_x from 0.14 to 1.19 and ρ from 0.535 to 0.995. The populations 2 and 4 are of special interest since the ratio $r_i = y_i/x_i$ for city no. 17 is considerably larger than the next largest ratio. It will be seen later that the estimators based on the individual ratios (i.e., t_1, t_4 and t_7) give very poor results for these populations.

A computer program has been written by Mr. William Vaughn to draw all the $\binom{N}{n}$ possible samples from a given set of data. Due to limitations on available computer time we, however, adopted the following procedure which appears quite satisfactory for our present purpose: If $\binom{N}{n} \leq 2000$, draw all the $\binom{N}{n}$ possible samples, otherwise draw 2000 independent samples each of size n; the units in each sample are, however, drawn without replacement. Moreover, we have confined ourselves to the case $g = n$ while investigating the performances of the estimators and the variance estimator based on split-samples (the Williams estimator t_{10} is, therefore, excluded from the present study). Since the theoretical results in Section 2 indicate that $g = n$ is a good choice, we feel this is a reasonable strategy.

If $N \leq 20$ the values are computed only for $n = 2$ and 4; if $20 < N \leq 30$ the values for $n = 2$, 4 and 6 are computed; if $N > 30$ the values for $n = 2, 4, 6, 8$ and 12 are computed.

Table 1. Description of the Populations

Pop. no.	Source	y	x	N	C_x	ρ
1	Sukhatme (1954) p. 165	mean no. livestock	mean agric. area	7	1.01	.995
2	Cochran (1963) p. 156	size of city in U.S. in 1930	size of city in U.S. in 1920	49	1.01	.982
3	Cochran (1963) Cities 25-49	size of city in U.S. in 1930	size of city in U.S. in 1920	25	.87	.958
4	Cochran (1963) Cities 1-24	size of city in U.S. in 1930	size of city in U.S. in 1920	24	1.06	.991
5	Cochran (1963) p. 183	no. of polio cases in the not inoculated group	no. of 'not inoculated' children	34	1.03	.720
6	Cochran (1963) p. 183	no. of polio cases in the placebo group	no. of 'placebo' children	34	1.03	.732
7	Cochran (1963) p. 204	actual weight of peaches	eye-estimated wt. of peaches	10	.17	.974
8	Cochran (1963) p. 325	no. of persons in a block	no. of rooms in a block	10	.14	.652
9	Sukhatme (1954) p. 183	Area under wheat in 1937	Area under wheat in 1936	34	.77	.930
10	Sukhatme (1954) p. 183	Area under wheat in 1936	Total cultivated area in 1931	34	.62	.829

Table 1. (Continued)

11	Sukhatme (1954) p. 279	Area under wheat in a circle	no. of villages in a circle	89	.60	.643
12	Sampford (1962) p. 61	acreage under oats in 1957	acreage of crops and grass for 1947	35	.71	.838
13	Kish (1965) p. 625	dwellings occupied by renters	no. of dwellings	270	1.00	.970
14	Kish (1965) p. 42	dwellings occupied by renters	no. of dwellings	20	1.19	.969
15	Yates (1960) p. 163	measured vol. of timber	eye-estimated vol. of timber	25	.47	.535
16	Yates (1960) p. 159	total no. of persons in a kraal	no. of persons absent from a kraal	43	.47	.647

3.1 Results based on mean square error (m.s.e.)

The following tentative conclusions may be drawn from Tables 2-5 with regard to m.s.e.: (1) Mickey's unbiased estimator t_2 is clearly superior to the Hartley-Ross estimator t_1 when $n > 2$; the gains are considerable for some populations. Mickey's estimator t_2 compares favorably with the classical estimator r, although the latter appears slightly better (particularly for small n). (2) The estimator t_3 is slightly better than t_3^* with $\omega = g$, but the gains are very small. (3) Tin's estimator t_4 is better than t_7. It compares favorably with r, t_3, t_5 and t_6 excepting for populations 2 and 4 where the performance is very poor. In so far as one would like to have robustness in performance, the estimators r, t_3, t_5 and t_6 are clearly preferable to t_4 and t_7. (4) t_5 and t_6 are about same and better than t_3 for $n = 4$. It is not clear-cut between r and $t_5(t_6)$, although t_5 and t_6 appear slightly better. (5) The classical regression estimator appears inferior to the ratio estimators, particularly for n = 4 and 6 (excepting pops. 15 and 1). Mickey's unbiased regression estimator t_9 is definitely inferior to t_8, especially for $n \leq 6$.

3.2 Results based on skewness and kurtosis

We have also computed the skewness and kurtosis of these estimators, but the numerical results are not given here. An examination of these results, however, lead to the following tentative conclusions with regard to kurtosis: (1) t_1 and t_7 are inferior to other ratio estimators. (2) t_8 has often a larger kurtosis than the ratio estimators (excepting t_1 and t_7). t_9 is definitely inferior to the other estimators. (3) The differences between the other estimators are small and, in any case, not clear-cut. With regard to skewness, the differences are small and not clear-cut, excepting that t_9 appears inferior.

Table 2. The m.s.e. of r, $t_1 = t_2$, $t_3 = t_4$, t_5, t_6, t_7 $t_3^* = t_3$ (with $\omega = g$) for $n = 2$.

Est. / Pop. no.	r	t_1	t_3	t_5	t_6	t_7	t_3^*
1	.0006	.0019	.0007	.0005	.0005	.0012	.0009
2	.143	5.796	4.047	.073	.076	2.741	4.420
3	.122	.215	.122	.107	.108	.160	.125
4	.098	11.814	7.764	.031	.039	5.555	9.335
5	.052	.091	.070	.060	.062	.076	.072
6	.156	.265	.212	.178	.185	.226	.218
7	.0009	.0009	.0009	.0009	.0009	.0009	.0009
8	.018	.017	.017	.017	.017	.017	.017
9	.017	.042	.026	.019	.020	.033	.027
10	.0084	.0097	.0118	.0099	.0103	.0107	.0121
11	24081	32194	24558	21303	21607	27484	24764
12	.0041	.0065	.0058	.0048	.0051	.0061	.0059
13	.049	.122	.070	.049	.049	.088	.071
14	.049	.117	.071	.048	.049	.087	.076
15	.198	.326	.255	.082	.075	.225	.305
16	.0086	.0099	.0087	.0082	.0082	.0091	.0088

3.3. Results based on $|\text{bias}|/(\text{m.s.e.})^{1/2}$

Table 6 gives the values of percent $|\text{bias}|/(\text{m.s.e.})^{1/2}$: for $n \geq 4$ the values are given only for those populations where the bias ratio of r is substantial (the values for t_3, t_5 and t_6 are, however, smaller than the bias ratio of r for all the populations (excepting pop. 4) when $n \geq 4$). The following tentative conclusions may be drawn from Table 6 with regard to bias ratio: (1) t_4 is inferior to t_3 and its performance is very poor on populations 2, 4 and 15 even relative to r. (2) t_3 for $n > 2$ is slightly better than t_5 and t_6 and it leads to substantial reductions over r and t_8 (excepting pop. 4). The estimators t_5 and t_6 are also better than r but, for some populations, the bias ratio remains substantial.

Combining the results of Sections 3.1 and 3.3, we tentatively conclude that t_3 (for $n > 2$), t_5 and t_6 appear promising. If, however, an unbiased estimator is needed, Mickey's unbiased ratio estimator is promising. For $n = 2$, it is probably advisable to use t_1 because the bias ratios of the approximately unbiased estimators could be substantial (note that $t_1 = t_2$ and $t_3 = t_4$ in this case).

3.4. Results on variance estimators

We have computed the biases of the classical variance estimator v_1 (as an estimator of m.s.e. of r), the jack-knife variance estimator v_2 (as an estimator of m.s.e. of r or t_3) and v_3, and the numerical results are given in Table 7. These results clearly indicate that the bias of v_2 is often considerably smaller in absolute value than that of v_1. Moreover, v_1 consistently underestimates the m.s.e. of r. The bias of v_3 (as an estimator of m.s.e. of t_8) is considerably large in absolute value. These results also hold good for estimating the variances.

Tables 8 and 9 give the stabilities of the variance estimators as measured by the square of the coefficient of variation. Here $(\text{C.V.})_1^2 = E(v_1 - \text{m.s.e.r})^2/$

Table 3. The m.s.e. of r, t_1, .., t_9, $t_3^* = t_3$ (with $\omega = g$) for $n = 4$

Est. / Pop. no.	r	t_1	t_2	t_3	t_4	t_5	t_6	t_7	t_8	t_9	t_3^*
1	.00024	.00047	.00034	.00031	.00026	.00026	.00027	.00035	.00020	.00029	.00046
2	.034	1.120	.034	.028	.241	.027	.027	.248	.087	.208	.028
3	.043	.060	.044	.039	.038	.038	.038	.039	.154	.340	.040
4	.014	2.036	.013	.011	.356	.010	.010	.464	.035	.080	.011
5	.023	.025	.027	.028	.026	.026	.026	.026	.059	.248	.029
6	.066	.068	.070	.075	.073	.071	.072	.071	.255	.565	.078
7	.0003	.0003	.0003	.0003	.0003	.0003	.0003	.0003	.0005	.0016	.0003
8	.0067	.0065	.0065	.0066	.0066	.0065	.0065	.0065	.0108	.0190	.0065
9	.011	.018	.019	.017	.014	.014	.014	.015	.015	.070	.018
10	.0036	.0035	.0036	.0039	.0040	.0038	.0038	.0039	.0075	.0708	.0039
11	9403	11284	9297	8692	8565	8674	8637	9069	8752	14976	8693
12	.0022	.0026	.0026	.0027	.0026	.0025	.0026	.0026	.0047	.0291	.0028
13	.020	.044	.030	.027	.023	.020	.021	.026	.019	.046	.027
14	.015	.030	.019	.020	.017	.015	.015	.019	.014	.026	.023
15	.048	.078	.038	.032	.027	.034	.033	.033	.037	.077	.030
16	.0037	.0039	.0038	.0038	.0036	.0037	.0037	.0037	.0058	.0189	.0038

Table 4. The m.s.e. of $r, t_1, \ldots, t_9$ and $t_3^* = t_3$ (with $\omega = g$) for $n = 6$

Est. / Pop. No.	r	t_1	t_2	t_3	t_4	t_5	t_6	t_7	t_8	t_9	t_3^*
2	.0153	.4257	.0135	.0124	.0577	.0128	.0126	.0674	.0257	.0488	.0122
3	.0210	.0309	.0204	.0192	.0194	.0191	.0191	.0207	.0386	.0566	.0192
4	.0060	.6490	.0051	.0049	.0683	.0052	.0051	.1103	.0193	.0268	.0048
5	.0146	.0146	.0152	.0159	.0155	.0154	.0155	.0154	.0314	.0427	.0163
6	.0392	.0374	.0376	.0396	.0409	.0399	.0399	.0402	.0861	.1025	.0401
9	.0077	.0105	.0110	.0106	.0089	.0093	.0096	.0094	.0069	.0154	.0114
10	.0024	.0023	.0023	.0024	.0025	.0024	.0024	.0025	.0039	.0034	.0024
11	5416	6581	5329	5152	5029	5146	5136	5211	4813	6880	5148
12	.0014	.0015	.0015	.0015	.0015	.0015	.0015	.0015	.0019	.0022	.0016
13	.0114	.0260	.0147	.0138	.0124	.0116	.0118	.0138	.0099	.0151	.0140
15	.0363	.0675	.0311	.0286	.0217	.0292	.0288	.0240	.0230	.0299	.0265
16	.0025	.0026	.0025	.0025	.0024	.0025	.0025	.0025	.0033	.0042	.0025

Table 5. The m.s.e. of r, $t_1, \ldots, t_9$ for n = 8 and 12

Est. / Pop. No.	r	t_1	t_2	t_3	t_4	t_5	t_6	t_7	t_8	t_9
	n=8									
2	.0096	.2392	.0084	.0080	.0218	.0083	.0082	.0304	.0143	.0194
5	.0106	.0102	.0106	.0109	.0111	.0109	.0109	.0110	.0197	.0224
6	.0264	.0247	.0248	.0255	.0269	.0261	.0260	.0266	.0522	.0531
9	.0060	.0073	.0077	.0077	.0066	.0071	.0071	.0069	.0056	.0105
10	.0016	.0016	.0016	.0016	.0017	.0016	.0016	.0017	.0022	.0018
11	3990	4863	3951	3864	3776	3864	3859	3868	3498	4606
12	.0009	.0010	.0010	.0010	.0010	.0010	.0010	.0010	.0012	.0013
13	.0085	.0186	.0098	.0093	.0088	.0086	.0087	.0096	.0070	.0093
16	.0017	.0017	.0017	.0017	.0017	.0017	.0017	.0017	.0019	.0023
	n=12									
2	.0053	.0987	.0048	.0046	.0070	.0048	.0047	.0101	.0068	.0077
5	.0051	.0049	.0050	.0050	.0052	.0051	.0051	.0052	.0076	.0075
6	.0133	.0125	.0125	.0126	.0133	.0128	.0128	.0133	.0189	.0176
9	.0037	.0041	.0042	.0042	.0039	.0041	.0041	.0040	.0039	.0057
10	.0009	.0009	.0009	.0009	.0009	.0009	.0009	.0009	.0010	.0009
12	.0006	.0006	.0006	.0006	.0006	.0006	.0006	.0006	.0006	.0006
13	.0055	.0119	.0059	.0057	.0056	.0055	.0056	.0059	.0043	.0052
16	.0010	.0010	.0010	.0010	.0010	.0010	.0010	.0010	.0011	.0012

Table 6

Absolute values of % bias/(m.s.e.)$^{\frac{1}{2}}$ of r, t_3, t_4, t_6 and t_8 for n = 2, 4, 6 and 8

Pop. no. \ Est.	r	t_3	t_4	t_5	t_6	t_8	r	t_3	t_4	t_5	t_6	t_8
	n=2						n=4					
1	44	3	3	32	28		26	5	7	15	14	45
2	22	21	21	6	5		17	5	31	4	2	34
3	25	4	4	15	12		23	3	6	12	10	11
4	16	30	30	10	31		7*	13*	45*	6*	9*	47*
5	8	8	8	9	9							
6	4	3	3	4	4							
7	4	1	1	1	1							
8	3	0	0	0	0							
9	10	3	3	8	8							
10	2	2	2	1	0							
11	19	3	3	7	4		14	1	2	4	3	10
12	13	5	5	9	8		7	1	1	2	1	14
13	37	6	6	28	24		30	1	4	17	13	27
14	42	2	2	32	28		25	8	9	13	9	7
15	15	27	27	4	14		12	3	32	1	2	24
	n=6						n=8					
2	13	5	38	1	1	36	10	4	42	2	1	31
3	18	1	3	6	5	15						
11	11	1	2	3	2	10	8	1	4	0	0	10
13	23	3	6	9	6	19	23	1	3	9	7	18
15	11	3	15	3	3	21						

* Based on all possible samples as 2000 samples did not give accurate results on bias.

$(m.s.e.r)^2$, $(C.F.)_2^2 = E(v_2 - m.s.e.r)^2/(m.s.e.r)^2$, $(C.V.)_{2*}^2 = E(v_2 - m.s.e.t_3)^2/(m.s.e.t_3)^2$, $(C.V.)_3^2 = E(v_3 - m.s.e.t_8)^2/(m.s.e.t_8)^2$ and $(C.V.)_4^2 = E[v_4 - V(\bar{y}/\bar{X})]^2/V^2(\bar{y}/\bar{X})$. It is clear from Tables 8 and 9 that the stability of the jack-knife variance estimator is very poor compared to that of v_1 - this is rather unfortunate in view of the results on bias. The choice between v_1 and v_2 is, therefore, not clear cut and one has to decide on other grounds. The jack-knife variance estimator is computationally attractive if we decide to use t_3 because $t_3 = \Sigma r_{Qj}/n$ and

$$v_2 = (1 - \frac{n}{N})\frac{(n-1)}{n(\omega-1)^2}\sum_1^n (r_{Qj}-t_3)^2$$

where $r_{Qj} = \omega r - (\omega-1)r_j'$.

The stability of v_3 compares favorably with that of v_1, although v_1 is slightly better for n = 4 and 6. Excepting populations 9 and 10, the stabilities of v_1 and v_3 compare favorably with that of v_4; in fact considerably better for some of the populations.

Although our empirical study has provided some guidelines on the overall performances of the estimators and variance estimators, it is necessary to continue theoretical and Monte Carlo investigations using realistic models to gain better understanding of the methods.

This research is supported by the National Science Foundation, under Grant GP-6400. My thanks are due to Mr. William Vaughn for the programming help.

Table 7. Biases of v_1, the classical variance estimator of r and v_2, the jack-knife variance estimator of r (denoted by B_1 and B_2 respectively)

	n=4		n=6		n=8		n=12	
Pop. no.	B_1	B_2	B_1	B_2	B_1	B_2	B_1	B_2
1	-.00014	-.00006						
2	-.0181	.0109	-.0056	.0039	-.0029	.0012	-.0014	-.0001
3	-.0174	.0095	-.0062	.0036				
4	-.0085*	.0082*	-.0018	.0018				
5	-.0078	.0033	-.0042	.0011	-.0032	.0001	-.0008	.0003
6	-.0254	.0110	-.0130	.0039	-.0073	.0019	-.0021	.0009
7	-.000009	-.000002						
8	-.000037	.000033						
9	-.0027	-.0011	-.00112	-.00003	-.00098	-.00001	-.00055	-.00007
10	-.00073	.00029	-.00032	.00013	-.00021	.00003	-.00007	-.00001
11	-2581	916	-857	420	-624	-11		
12	-.0052	.00002	-.00021	.00007	-.00009	.00008	-.00009	-.00002
13	-.0113	.0028	-.00451	.00311	-.00297	.00123	-.00158	.00021
14	-.0101	.0057						
15	-.0061	.0139	-.0133	-.0057				
16	-.00036	.00021	-.00035	-.00011	-.00013	-.00002	-.00008	-.00004

*Based on all possible samples.

Table 8. $(C.\ V.)_1^2$, $(C.\ V.)_2^2$, $(C.\ V.)_{2*}^2$, $(C.\ V.)_3^2$ and $(C.\ V.)_4^2$ for n = 4 and 6

	n=4					n=6				
Pop. no.	$(C.V.)_1^2$	$(C.V.)_2^2$	$(C.V.)_{2*}^2$	$(C.V.)_3^2$	$(C.V.)_4^2$	$(C.V.)_1^2$	$(C.V.)_2^2$	$(C.V.)_{2*}^2$	$(C.V.)_3^2$	$(C.V.)_4^2$
1	.50	.47	.42	.63	.38					
2	.61	7.92	11.92	.81	2.00	.50	4.06	6.48	.60	.22
3	.71	3.73	4.56	.82	.80	.49	1.96	2.42	.60	.47
4	.45	16.38	27.74	.77	1.31	.28	3.03	4.67	.70	.79
5	.66	1.52	1.04	.66	1.19	.41	.87	.72	.55	.69
6	1.02	3.18	2.42	.78	.95	.68	1.99	1.94	.61	.56
7	.26	.31	.32	.41	.26					
8	.28	.32	.33	.30	.34					
9	2.83	2.21	1.12	2.17	.76	2.12	2.36	1.31	2.55	.43
10	2.24	4.69	4.09	1.23	.68	1.41	2.61	2.67	.93	.39
11	.59	2.45	2.89	.68	1.07	.49	1.27	1.41	.48	.69
12	.94	1.04	.72	.79	.92	.61	.80	.64	.63	.51
13	.66	2.96	1.67	.81	1.97	.56	3.55	2.38	.65	1.20
14	.59	5.15	2.94	.70	.87					
15	1.29	5.61	13.40	.56	.60	.65	2.44	3.91	.41	.36
16	.66	1.03	1.01	.54	1.47	.43	.54	.54	.41	.97

Table 9. $(C.\ V.)_1^2$, $(C.\ V.)_2^2$, $(C.\ V.)_{2*}^2$ $(C.\ V.)_3^2$ and $(C.\ V.)_4^2$ for n = 8 and 12

	n=8					n=12				
Pop. no.	$(C.V.)_1^2$	$(C.V.)_2^2$	$(C.V.)_{2*}^2$	$(C.V.)_3^2$	$(C.V.)_4^2$	$(C.V.)_1^2$	$(C.V.)_2^2$	$(C.V.)_{2*}^2$	$(C.V.)_3^2$	$(C.V.)_4^2$
2	.39	2.08	3.13	.49	.89	.27	.89	1.18	.37	.53
5	.31	.55	.52	.48	.49	.19	.39	.40	.28	.27
6	.51	1.47	1.58	.52	.39	.34	.82	.92	.32	.20
9	1.43	1.87	1.19	1.40	.29	.86	1.10	.85	.72	.15
10	1.02	1.69	1.78	.83	.26	.62	.63	.87	.59	.14
11	.37	.73	.78	.35	.48					
12	.44	.58	.51	.47	.36	.24	.26	.25	.30	.20
13	.46	2.01	1.65	.55	.88	.33	.89	.82	.37	.58
16	.29	.34	.33	.28	.62	.18	.19	.19	.19	.37

REFERENCES

Chakrabarty, R. P. (1968). _Contributions to the theory of ratio-type estimators_. Ph.D. Thesis, Texas A & M University.

Chakrabarty, R. P., and Rao, J. N. K. (1967). "The bias and stability of jack-knife variance estimator in ratio estimation." _Proc. Amer. Statist. Assoc._ (Social Statistics Section), in press.

Cochran, W. G. (1963). _Sampling Techniques_, 2nd Ed. New York: Wiley.

Durbin, J. (1959). "A note on the application of Quenouille's method of bias reduction to the estimation of ratios". _Biometrika_, _46_, 477-80.

Jones, H. L. (1963). "The jack-knife method". _Proc. IBM Scientific Computing Symposium on Statistics_, IBM.

Kish, L. (1965). _Survey Sampling_. New York: Wiley.

Kish, L., Namboodiri, N. K., and Pillai, R. K. (1962). "The ratio bias in surveys". _J. Amer. Statist. Assoc._, _57_, 863-76.

Lauh, E., and Williams, W. H. (1963). "Some small sample results for the variance of a ratio". _Proc. Amer. Statistist. Assoc._ (Social Statistics Section), 278-83.

Mickey, M. R. (1959). "Some finite population unbiased ratio and regression estimators". _J. Amer. Statist. Assoc._, _54_, 594-612

Pascual, J. N. (1961). "Unbiased ratio estimators in stratified sampling". _J. Amer. Statist. Assoc._, _56_, 70-87.

Rao, J. N. K. (1965). "A note on the estimation of ratios by Quenouille's method, Biometrika, 52, 647-9.

_____ (1967). "Precision of Mickey's unbiased ratio estimator". Biometrika, 54, 321-4.

Rao, J. N. K. and Webster, J. T. (1966). "On two methods of bias reduction in the estimation of ratios". Biometrika, 53, 571-7.

Rao, J. N. K. and Beegle, L. D. (1967). "A Monte Carlo study of some ratio estimators". Sankhya.

Tin, M. (1965). "Comparison of some ratio estimators", J. Amer. Statist. Assoc., 60, 294-307.

USE OF NON-REPRESENTATIVE SURVEYS FOR ETIOLOGICAL PROBLEMS

S. Koller

University of Mainz, West Germany

The basis of all generalizing scientific inferences from empirical material is the assumption that a series of observations is a representative sample from a population whose characteristics can be found out by the sample. In analytical statistics in medicine one seldom starts with drawing a sample from a population. In most cases there is a series of observations resultant from numerous external conditions which has been obtained without any application of sampling procedures. For instance, all patients of a hospital suffering from certain diseases are examined to determine whether there are any specified substances in their serum. Primarily the results describe the status of the examined patients. If they are to be generalized a relation between population and sample must be established. In the present case this can be only reached by searching for that population from which the survey in question can be a representative sample.

In the terminology of sampling theory a survey of the described type is at most a single cluster in a possible two-stage sample from a comprehensive population. The variance between the first-stage sampling units and the variance within these units, i.e. between the second-stage sampling units enter into the error formula of the estimated mean value in the population.

The observed variance within the cluster can be only used for the estimation of the variance of the general mean, if the true first-stage variance is 0 and the variance within all clusters is equal. This cannot be assumed generally. For a confidence region, one confines onself to the formulation, that the mean value would be stabilized in this region, of many further observations of the same kind and under the same conditions of material and observation would be made. This formulation gives a possibility of the generalization of medical findings. A fictive population is defined from which the series of observations is a sample by definition.

For a further generalization with regard to other places, other observers, other groups of patients with the same diagnosis an estimate of the between-cluster variance is at least necessary from the point of view of sampling methods. But if this between-cluster variance proves significant in the analysis of variance, the interest changes. Now the causes of the differences between the clusters are examined; the interest turns to the separation of the groups or the elimination of the factors of influence. An over-all estimate is not made in most cases.

A similar situation arises in therapeutic problems. Mostly a therapeutic effect is tested by examining the patients present in a hospital by a controlled clinical trial. The evaluation may end in establishing a confidence region of a difference in therapeutic response. This region is related to the difference of the mean values in the case of increasing numbers of observations under the same conditions of observation and with the same structure of patients. This generalization leads again to a fictive population. At the same time one examines whether there are subgroups showing differing effects; furthermore, there are other observers who examine whether the findings are also valid in other places, at different times and in other groups of patients with the same disease. If there are no significant differences we generalize from the different groups to their respective fictive population and possibly to a super-population consisting of them. If there exist differences, the analysis of differences is emphasized, whereas the estimate of the therapeutic mean effect in the general population comprising all groups of patients in all places is no longer of interest.

Example: When the effect of strophantin (ouabain) on cardiac insufficiency is tested, it is not meaningful to estimate the average therapeutic effect for the total of cases of cardiac failure, for those patients already treated with digitalis respond badly to strophantin. It is more important to find out if there are contraindications than to estimate the structure of frequencies of the heterogeneous subgroups and by this enumerate a general mean of the therapeutic criterion.

In etiological questions, however, there is another kind of problem. It is a complicated process which shall be examined. This process corresponds for instance to the natural history of chronic diseases under the influence of exogeneous factors such as dust, cigarette smoking, nutrition habits, stress, furthermore of drugs, other diseases, psychical experiences etc. Frequencies or mean values are not the object of the statistical study, but the association between factor and disease. Which are now the conditions for drawing a sample which is to be representative for the associations to be discovered leading to the genesis of diseases in the population under the influence of factors? In terms of what we learned yesterday from Professor Stephan it is a kind of nexus statistics in which the connections themselves cannot be counted or measured but the statistics have to start from the nodes. An etiological hypothesis mostly occurs in the form of a contingency table and thus can be tested.

Total enumeration

		Disease D +	Disease D -	total
Factor A	+			
	-			
total				

Let us begin with a situation where no difficulties of observation exist. I call this case a vitreous population.

In a vitreous population where every person with regard to every attribute can be truly observed this fourfold table could be set up for the whole population. The conditions of observation must be equal. The association between A and D has to be determined. But even in this case the four arrays may not be comparable, e.g. because of structural differences by age and sex, by expositions or by feed back.

As an example let us take coal miners disease (Anthraco-Silicosis) as D and the working in a coal mine underground as A. But if we would enumerate it for association we would possibly find very few D^+ working underground, because workers were transferred to surface work when they got the disease.

What we take as factor A is largely arbitrary. The choice of factor A does not only depend on the etiological hypothesis. Often, however, the hypothesis is only a preliminary draft in relation to general complex situations such as, for instance, the fat content of the food and can be substituted approximately in various ways by measurable variables, e.g. the calories of the visible fat content of the food of the last week, the highly unsaturated fatty acids in the food, the lipid content of the serum, and so on. Thus the choice of the variables is to a great extent arbitrary.

In order to make a comparison between different factors or their substitutes the table should be enlarged:

Factor			Disease D		
A_1	A_2	A_r	+	-	total
+	+	+			
+	+	-			
+	-	+			
-	+	+			
-	+	-			
.	. ....	.			
-	-	-			
	total				

This is the extended form of a (r + 1)-dimensional contingency table. The relative importance of the several factors could be derived from the analysis of this table.

Often we do not test a single hypothesis. In most cases we collect a variety of data for screening and selecting variables or combinations giving strongest associations. But here the interpretation is liable to the common errors. A cause-effect relation can only be accepted per exclusionem after having excluded all other possible sources of associations. Even in total enumeration of a vitreous population a bias is possible due to selective mortality prior to the date of recording.

<u>Samples</u>

If one has obtained a <u>representative sample of the population</u>, the sampling variable of which is neither associated with D nor with A_1 ... A_r (for example, birth-day, first letter of the name), the contingency tables of the sample are proportional to those of the population — with the exception of random deviations. The analysis with its problems is similar.

A representative sample, however, is seldom available. Occasionally there is a cluster sample the selection attributes of which are not associated with the variables of the table (Regional sample such as in Framingham). But for the most part one renounces a representative sample from the population and decides on the retrospective or prospective design.

In the <u>retrospective</u> design which is especially similar to the medical method of observation and gaining of experience the sample from the population is divided into two strata with different sampling ratios. The stratum of the patients is recorded totally or with high sampling ratio in a region or in hospital, the stratum of the healthy persons with low sampling ratio. Variables to be examined are the potential influencing factors in the past, the frequencies of which are compared between the sick and healthy persons.

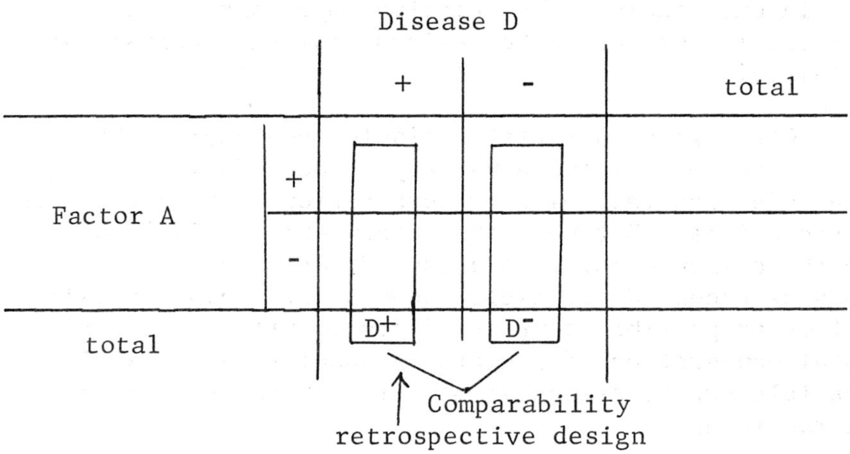

Instead of the single factor A several factors $A_1 \ldots A_r$ must be examined in most cases.

Contrary to sample surveys the sampling ratios and the estimation of population parameters are not of great importance here, but the detection of differences in the frequency of A^+ among the D^+ compared to the D^-. In order to find out true differences the two series of the D^+ and D^- must be <u>comparable</u>. Two series are comparable, if they are

1. equal in structure (for example by age distribution, choice of the hospital, further diseases, selection for a post-mortem examination)

2. equal in observation (for example by examination techniques, interview situation)

3. equal in representativeness, for example by regional origin, profession)

The "comparability" is a necessary set of conditions.

For the most part the group of the sick persons D^+, in the sense of sampling techniques, corresponds to a cluster with some unique features. The group D^- becomes the "control group" which sometimes is obtained separately, but which must be comparable to the group D^+ and consequently, analogous to D^+, must have the same unique features (except D^+). In some situations the realization of equality of observation is most difficult.

A quantitative inference to the general population of all D^+ and D^- is not possible from the single cluster or pair of clusters. In most cases the problem does not require such an inference.

As for sampling techniques the generalization is carried out first of all for a fictive population which would arise, if under the same conditions the series would be infinitely large. The real expert generalization depends on the fact that the requirements of the sampling process have been replaced by the maintenance of the conditions of comparability. The selection factors of the cluster are examined which had to be analysed anyway in order to obtain the restrictions for the control series; the generalization is carried out for that fictive population from which the observed cluster could have arisen by selection factors which are not correlated with the disease and the factors.

In addition to this somewhat speculative, but indispensable procedure a further generalization on the basis of a large material can be obtained by the splitting up into homogeneous subgroups corresponding to blocks. If the associations in the subgroups are preserved, one can possibly take for granted that they are independent of the selection factors of the subgroups (for example hospitals in cooperative studies). We could generalize regarding all populations the samples of which could give — under the same conditions of comparability — the observed contingency tables.

Comparability and capability of generalization are opposite.

For the comparability restrictions are necessary which restrain the generalization. For a wide generalization they must be dropped.

> Example: In statistics of autopsy for example the association between cancer and arteriosclerosis is often examined. In most cases the incorrect comparison shown in the table a) is carried out. One compares the frequency of arteriosclerosis among the patients with cancer with that of the patients without cancer. This control group, however, contains the heterogeneous remaining cases including a large number of myocardial infractions and other diseases connected with arteriosclerosis. Necessarily there must also be an accumulation of serious arteriosclerosis. The negative association which is found out by this comparison only results from a lack of comparability. The bias disappears, if the persons dying by accidents are taken as control series which in respect to arteriosclerosis can be considered as a random sample from the population (table b). In this case there is no association.

a) Cancer

		+	-	
Serious arterio-sclerosis	+	56	192	248
	-	167	122	289
total		223	314	537

$\chi^2 = 66,4$

b) Cancer — Accidents (without cancer)

		+	-	
Serious arterio-sclerosis	+	56	6	62
	-	167	17	184
		223	23	246

$\chi^2 = 0,01$

Serious arteriosclerosis among cancer deaths compared with

a) all remaining autopsies (not comparable)
b) accidents (comparable)

Males, age-group 60-69 years

Pathol. Institute, Mainz (Lange, Hempel, Müller, 1966)

Here, formerly, the confirmation of the negative association was given by several authors — all made the same error. One author (Sv. Juhl) was suspicious about it. He compared the distribution of causes of death in the heterogeneous remainder with that of the vital statistics in the population and found an acceptable agreement.

But this kind of representativeness is not valid for comparability. Representativeness, however, is not necessary for deaths, but for the living population with respect to post-mortem findings. The question for the control group is: What would be the percentage of arteriosclerosis in persons of the same age and sex if they would not have dies from cancer. Then they would live; but the method of observation is the autopsy. Therefore only deaths by accidents which come to autopsy can be used for control.

When this association is examined in subgroups, a positive association between lung cancer and arteriosclerosis is then found.

In the prospective series one starts from groups with factor and without factor. The groups can be obtained separately if we are informed before the examination of the patients whether they belong to A^+ or A^-. Otherwise a uniform method of selection is used; the division into A^+ and A^- is only carried out according to the results of the basic examination (for example in differentiating the habits of smoking, intake of fat or the use of drugs during pregnancy).

The advantage of the prospective approach is that the observation of the factor A is not influenced by the disease.

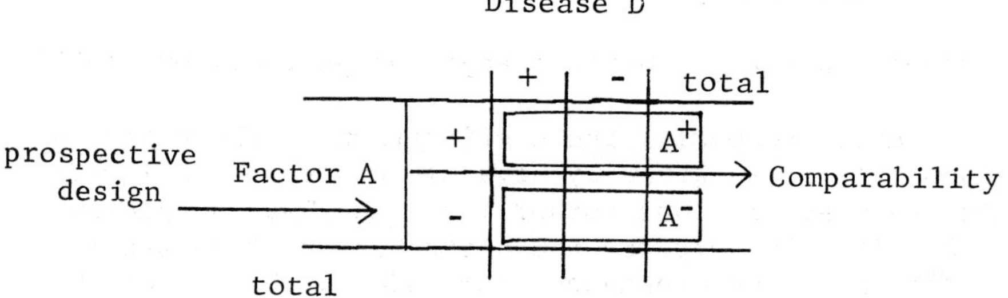

Here, too, the criterion of practicability is only the comparability of the A^+ and A^-. One of the inevitable selection factors of a prospective study is the participation of the persons. For example, it often happens that sick persons are not included; therefore the mortality is lower than that of the general population. This, however, is only a reinforcement of the general selection by survival, for persons who died earlier never participate in a prospective study. In spite of the possible selection there is no other way than to confine the study to the survivors, but there may be questions which cannot be answered retrospectively after a long time because of the important selection by high mortality (for example the consequences of periods of extreme hunger). To the comparability also belongs the equality of representation, not the general representativeness with regard to population. If it is not completely given, one has to form subclusters as blocks or strata. Therefore, in cooperative studies of several hospitals many comparisons must be carried out primarily within the material of each hospital and may be summarized afterwards with some appropriate method.

The generalization beyond the cluster restrictions can be brought about by the examination of other clusters or by the division into subclusters in order to remove the effect of restriction. If the associations of the several groups are different, the analysis of the differences is again more interesting than an inference related to the general population. Examples for a high agreement between different studies with different restrictions can be found in prospective studies on Smoking and Health (studies on veterans; on physicians, on persons participating as volunteers).

We come to the conclusion: In etiological surveys the examination of a representative sample from the population the selection of which does not depend on the disease in question and the factors of influence is desirable, but not necessary. It is sufficient to compare pairs or other multiples of biassed cluster samples in which only the demands for comparability must be strictly observed. In the first instance the results — associations between factor of influence and disease — are only valid for the fictive population from which the series could be a random sample. The generalization beyond the cluster restrictions necessitates investigations of other clusters or the examination of the homogeneity when there are subgroups.

Each restriction concerns a factor able to falsify the test of an etiological hypothesis. Generalization beyond restriction means that groups differing in the restriction factor do not differ in the tested association. Although this inference is only possible in a preliminary, but not definitive way, we go on step by step. The setting up and breaking down of restrictions represent a multidimensional up-and-down strategy which allows generalizing conclusions.

Representativeness with regard to the general population is a requirement existing in the fackground: Since etiological studies should change the population with respect to factors and diseases by preventive medicine the representativeness could not be related only to present but also to future populations. This could not be realized anyway. For an etiological survey representativeness of the total population is not an important criterion to distinguish between good and bad studies. Comparability of the compared groups is the prevailing principle of design and can even provide certain kinds of generalization beyond the observed groups.

THE EFFECT OF RESPONSE ERRORS ON MEASURES OF ASSOCIATION*

G. G. Koch and D. G. Horvitz

University of North Carolina

Research Triangle Institute

I. PRELIMINARY CONSIDERATIONS

1. Introduction. It is now generally recognized both by statisticians and by users of data that non-sampling or measurement error components quite often can dominate the total error of sample survey estimates. During the past twenty years there has been increasing emphasis on the need to measure the nature and extent of non-sampling errors occurring in surveys and on the development of response error models. Most of the significant work in the development of models has been carried on at the U. S. Bureau of the Census. The response error models developed by Morris Hansen and his colleagues have led to new concepts and methods for comparing survey errors and for assessing the relative importance of components of the total errors obtained with alternative survey procedures. These include such measures as the "gross-difference rate," "net-difference rate" and "index of inconsistency." Despite the obvious value of these models and measures, they do not appear, as yet, to have received wide usage. It is, of course, much too early to make judgements concerning their eventual impact.

We would like to suggest that fairly rapid progress toward general recognition and adoption of appropriate techniques for controlling and assessing non-sampling errors by survey practitioners and in the proper inter-

* This research was supported in part by the U. S. Bureau of the Census Center for Measurement Research under Contract No. Cco-9260 and by the National Institutes of Health, Institute of General Medical Sciences, Grant No. GM-12868-04.

pretation of survey data by users might be achieved by adopting simple and yet meaningful accuracy criteria, criteria that are somewhat more closely aligned with classical sampling and estimation theory. The complexity that we feel prevails stems in part from a reluctance on the part of statisticians who have contributed to survey sampling theory and methods to assign a specific functional form to the frequency distributions of the variables under study. Thus, dating at least from the introduction of strictly finite population models, survey statisticians have had to invoke large sample theory in order to provide a basis for the construction of confidence intervals and other inferential statements. This same philosophy has prevailed in the development of response error models. Yet, a willingness to assign a particular distributional form and shape to each of the components in linear additive response models leads quickly to satisfactory measures of the impact of response errors. For example, if each of the error parameters in the model is assumed to follow an independent normal distribution, we can compute its effect on the confidence interval probability appropriate to those limits which apply when there are no response errors. Thus, the "robustness," if you will, of standard confidence interval estimation procedures, in the face of response errors of various types and magnitude, can be measured. The effect of response errors on significance levels and on the power of traditional tests can also be readily determined. The models adopted in this paper, for studying the effect of response errors on the correlation coefficient, stem to a large extent from these considerations.

We shall be concerned with data arising from either a survey or some other experimental situation which may be regarded as conceptually repeatable. In addition, we shall presume that the phenomena being studied is reasonably static over time and is uninfluenced by the measurement process.

The purpose of a survey is to estimate some parameter of a population. Here, we shall be interested in the association between two continuous variables. The estimation of this association will be affected by two sources of error. One will be "sampling error" which reflects the fact that different individuals will tend to perform differently. The other will be "response error" which arises because the measurement process causes the performance of the same individual to be recorded differently over a series of repetitions of the survey. The presence of response errors may simultaneously arise from a number of sources; e.g., respondent variability associated with recall, interviewer variance, processing errors, structural weakness or vagueness in the definition of the phenomena being measured, or other difficulties in reproducing the same conditions of measurement.

The purpose of this report is to consider a bivariate model of measurements which contain a component reflecting sampling error and a component reflecting response error. Within the structure of this model, attention will be focused on the problems of estimating the covariance or the correlation of the two random variables. In particular, the variability of such estimates is partitioned into components related to the various sources of error. Finally, the use of the information from re-surveys is studied. This leads to adjusted estimates of association and the formulation of some tests of significance for the hypothesis of no association.

2. <u>The model</u>. Let $\underline{y}_{ij} = (y_{ij}^{(1)}, y_{ij}^{(2)}, \ldots, y_{ij}^{(p)})'$ denote a p-variate response vector associated with the j-th measurement on the i-th individual where $j=1,2,\ldots,m$ and $i=1,2,\ldots,n$. We shall assume that the $\underline{y}_{ij}$ follow the model

$$\underline{y}_{ij} = \underline{\mu} + \underline{a}_i + \underline{e}_{ij} \tag{1.2.1}$$

where $\mu = (\mu^{(1)}, \mu^{(2)}, \ldots, \mu^{(p)})'$ is a vector of fixed constants reflecting the mean response for the population, $\underline{a}_i = (a_i^{(1)}, a_i^{(2)}, \ldots, a_i^{(p)})'$ is a random vector reflecting the deviation of the mean response for the i-th individual from the population mean, and $\underline{e}_{ij} = (e_{ij}^{(1)}, e_{ij}^{(2)}, \ldots, e_{ij}^{(p)})'$ is a random vector reflecting the deviation of the j-th measurement on the i-th individual from the corresponding mean. It is assumed that the $\{\underline{a}_i\}$ are independently distributed as $N(\underline{0}_p, V_a)$, the $\{\underline{e}_{ij}\}$ are independently distributed as $N(\underline{0}_p, V_e)$, and the $\{\underline{a}_i\}$ and $\{\underline{e}_{ij}\}$ are independent of each other. In this context $\underline{0}_p$ is the p-variate null vector and V_a and V_e are positive definite p X p covariance matrices.

It follows from the previously indicated assumptions that the $\underline{y}_{ij}$ are each normally distributed as $N(\underline{\mu}, V_a + V_e)$ and $Cov(\underline{y}_{ij}, \underline{y}_{ij'}) = V_a$ for $j \neq j'$. In this sense, V_a represents the intra-class covariance matrix.

Within the structure of the model (1.2.1) the $\underline{a}_i$ represent the sampling component of the response. Their variability is characterized by the matrix V_a. On the other hand, the $\underline{e}_{ij}$ represent the non-sampling (response error) component. The variability arising from this source is characterized by the matrix V_e. In this report, the parameters which will be of primary interest to us from the point of view of inference and estimation are those associated with $\underline{\mu}$ and V_a. However, the parameters associated with V_e must be considered because of the effect of the presence of response errors on the distribution of the observations $\underline{y}_{ij}$ and any statistics obtained from them.

3. Inference about $\underline{\mu}$. Let

$$\bar{\underline{y}} = \frac{1}{mn} \sum_{i=1}^{n} \sum_{j=1}^{m} \underline{y}_{ij} .$$

It follows from the model that $\bar{\underline{y}}$ is distributed as

$$N(\underline{\mu}, \frac{1}{n}(V_a + \frac{1}{m} V_e)) .$$

Here $\frac{1}{n} V_a$ represents the sampling component of the variability of $\bar{\underline{y}}$ while $\frac{1}{mn} V_e$ represents the non-sampling component.

Let

$$S_a = m \sum_{i=1}^{n} (\underline{\bar{y}}_i - \underline{\bar{y}})(\underline{\bar{y}}_i - \underline{\bar{y}})'$$

where

$$\underline{\bar{y}}_i = \frac{1}{m} \sum_{j=1}^{m} \underline{y}_{ij} \quad .$$

The matrix S_a has the Wishart distribution $W(S_a; mV_a + V_e, (n-1), p)$. Hence, the statistic

$$F = \frac{mn(n-p)}{p} (\underline{\bar{y}} - \underline{\mu})' S_a^{-1} (\underline{\bar{y}} - \underline{\mu}) \qquad (1.3.1)$$

has the F(p, n-p) distribution. The result in (1.3.1) may be used to make the following types of simultaneous inference statements.

(i) All values $\underline{\xi}$ such that $mn(\underline{\xi} - \bar{y})' S_a^{-1} (\underline{\xi} - \bar{y}) \leq \frac{p}{n-p} F_{1-\alpha}(p,n-p)$ lie in a $100(1-\alpha)\%$ confidence region for $\underline{\mu}$ centered at $\underline{\bar{y}}$.

(ii) For all $\underline{c} \neq \underline{0}$, $100(1-\alpha)\%$ simultaneous confidence intervals for all $\underline{c}'\underline{\mu}$ are given by

$$\underline{c}'\underline{\mu} = \underline{c}' \, \underline{\bar{y}} \pm \sqrt{\frac{p}{n-p} \left(\frac{\underline{c}' S_a \underline{c}}{mn}\right) F_{1-\alpha}(p, n-p)}$$

where $\Pr\{F(p, n-p) \leq F_{1-\alpha}(p, n-p)\} = 1-\alpha$.

When $m \geq 2$, it is possible to estimate the nonsampling component of the variability of $\underline{\bar{y}}$ directly. Namely,

$$S_e = \sum_{i=1}^{n} \sum_{j=1}^{m} (\underline{y}_{ij} - \underline{\bar{y}}_i)(\underline{y}_{ij} - \underline{\bar{y}}_i)'$$ has the Wishart distribution $W(S_e; V_e, n(m-1), p)$. Hence, $\{S_e/n(m-1)\}$ is an unbiased estimate of V_e.

For the univariate case p = 1, Hansen, Hurwitz, and Pritzker [1964] defined the index of inconsistency as

$$I = \sigma_e^2/(\sigma_e^2 + \sigma_a^2) \text{ where } d = \sigma_a^2/\sigma_e^2. \tag{1.3.2}$$

Since S_a and S_e are independent, it follows that

$$(S_e/S_a)\left(\frac{\sigma_e^2 + m\,\sigma_a^2}{\sigma_e^2}\right)\left(\frac{n-1}{(m-1)n}\right)$$

has the F(n(m-1), (n-1)) distribution. Thus, if $\tilde{F} = (n-1)S_e/n(m-1)S_a$, then

$$\left\{\frac{1}{m}\left(\frac{F_{1-\alpha/2}(n(m-1),(n-1))}{\tilde{F}} - 1\right) + 1\right\}^{-1}$$

$$\le I \le \left\{\frac{1}{m}\left(\frac{F_{\alpha/2}(n(m-1),(n-1))}{\tilde{F}} - 1\right) + 1\right\}^{-1} \tag{1.3.3}$$

is a 100(1-α)% confidence interval for I.

In the multivariate case, there may be several appropriate definitions for the index of inconsistency. For example,

$$I_1 = |V_e|/|V_a + V_e| = \prod_{h=1}^{p} (1/(1+d_h))$$

$$I_2 = tr(V_e)/tr(V_e + V_a)$$

$$\tilde{I}_2 = \frac{1}{p}tr(V_e(V_e + V_a)^{-1}) = \frac{1}{p}\sum_{h=1}^{p} (1/(1+d_h))$$

$$I_3 = Ch_{max}(V_e)/Ch_{max}(V_e + V_a)$$

$$\tilde{I}_3 = Ch_{max}(V_e(V_e + V_a)^{-1}) = \max_h(1/(1+d_h))$$

where the d_h are the respective elements of a diagonal matrix D with D being obtained from that non-singular matrix C such that $C\ V_e\ C' = I$ and $C\ V_a\ C' = D$.

Inferences concerning the above indices would depend upon the characteristic roots of the matrices S_e, S_a and $S_e S_a^{-1}$ or $S_a S_e^{-1}$.

4. Inference about V_a. From the remarks made in the preceding section, it is clear that

$$\hat{V}_a = \{S_a/(n-1) - S_e/n(m-1)\}/m \qquad (1.4.1)$$

is an unbiased estimate of V_a. Moreover the hypothesis

$$H_0:\ V_a = 0 \qquad (1.4.2)$$

may be tested by noting that it is identical to the hypothesis

$$H_0:\ m\ V_a + V_e = V_e\ . \qquad (1.4.3)$$

A statistical test of (1.4.3) may be based on the characteristic roots of $S_a S_e^{-1}$ since in the null case, S_a and S_e have the same associated matrix V_e. For further details concerning such tests, the reader is referred to Anderson [1958], Roy [1953], and Schatzoff [1966].

II. THE MEASUREMENT OF ASSOCIATION

1. The bivariate case. From the structure of the model specified in (1.2.1), it follows that the marginal distribution of any pair of variates $(y_{ij}^{(h)}, y_{ij}^{(h')})$ depends only on appropriate corresponding parameters from $\underline{\mu}$, V_a, and V_e; namely $(\mu^{(h)}, \mu^{(h')})$, $(v_{a;hh}, v_{a;hh'}, v_{a;h'h'})$, and $(v_{e;hh}, v_{e;hh'}, v_{e;h'h'})$.

As a result, in a consideration of the association

between any pair of variables, we may restrict attention to the bivariate case p = 2 without any loss of generality.

In the above context, we now arrange all the observations on the first variable into one vector and all the observations on the second variable into another: i.e.,

let $\underline{y}^{(1)} = (y_{11}^{(1)}, \ldots, y_{1m}^{(1)}, \ldots, y_{n1}^{(1)}, \ldots, y_{nm}^{(1)})'$

and $\underline{y}^{(2)} = (y_{11}^{(2)}, \ldots, y_{1m}^{(2)}, \ldots, y_{n1}^{(2)}, \ldots, y_{nm}^{(2)})'$.

If we then write the vector of observations as

$$\underline{y} = \begin{bmatrix} \underline{y}^{(1)} \\ \underline{y}^{(2)} \end{bmatrix}$$

it follows that

$$E(\underline{y}) = \begin{bmatrix} \mu^{(1)} & \underline{j}_{mn} \\ \mu^{(2)} & \underline{j}_{mn} \end{bmatrix} ,$$

$$\text{Var}(\underline{y}) \equiv V = \left[\begin{array}{c|c} (v_{a;11} H_{mn} + v_{e;11} I_{mn}) & (v_{a;12} H_{mn} + v_{e;12} I_{mn}) \\ \hline (v_{a;12} H_{mn} + v_{e;12} I_{mn}) & (v_{a;22} H_{mn} + v_{e;22} I_{mn}) \end{array}\right]$$

where $\underline{j}_{mn}$ is an $(mn \times 1)$ vector of 1's, I_{mn} is the $(mn \times mn)$ identity matrix, and H_{mn} is an $(mn \times mn)$ matrix in which the non-zero elements lie in n diagonal blocks of $(m \times m)$ matrices of 1's.

2. Some variances and covariances. Given the above notation, the elements of the matrices S_a and S_e can be written as quadratic forms

$$S_{a;hh'} = \underline{y}' \, Q_{a;hh'} \, \underline{y}, \quad S_{e;hh'} = \underline{y}' \, Q_{e;hh'} \, \underline{y} \text{ for } h,h' = 1,2$$

In particular, we have

$$Q_{a;11} = \begin{bmatrix} \frac{1}{m} H_{mn} - \frac{1}{mn} J_{mn} & O_{mn} \\ O_{mn} & O_{mn} \end{bmatrix} \quad Q_{e;11} = \begin{bmatrix} I_{mn} - \frac{1}{m} H_{mn} & O_{mn} \\ O_{mn} & O_{mn} \end{bmatrix}$$

$$Q_{a;12} = \frac{1}{2} \begin{bmatrix} O_{mn} & \frac{1}{m} H_{mn} - \frac{1}{mn} J_{mn} \\ \frac{1}{m} H_{mn} - \frac{1}{mn} J_{mn} & O_{mn} \end{bmatrix}$$

$$Q_{e;12} = \frac{1}{2} \begin{bmatrix} O_{mn} & I_{mn} - \frac{1}{m} H_{mn} \\ I_{mn} - \frac{1}{m} H_{mn} & O_{mn} \end{bmatrix}$$

$$Q_{a;22} = \begin{bmatrix} O_{mn} & O_{mn} \\ O_{mn} & \frac{1}{m} H_{mn} - \frac{1}{mn} J_{mn} \end{bmatrix} \quad Q_{e;22} = \begin{bmatrix} O_{mn} & O_{mn} \\ O_{mn} & I_{mn} - \frac{1}{m} H_{mn} \end{bmatrix}$$

where O_{mn} is an $(mn \times mn)$ matrix of 0's and J_{mn} is an $(mn \times mn)$ matrix of 1's.

The above formulation can now be applied to calculate the expectations, variances, and covariances of the elements in S_a and S_e. The method to be used is based on the well-known results that if $\underline{y}$ is $N(\underline{0}, V)$ and if $S_1 = \underline{y}' Q_1 \underline{y}$ and $S_2 = \underline{y}' Q_2 \underline{y}$ are two quadratic forms, then

$$E\{S_h\} = tr(V Q_h), \; Var\{S_h\} = 2tr(V Q_h V Q_h) \qquad h = 1, 2$$

$$Cov\{S_1, S_2\} = 2tr(V Q_1 V Q_2)$$

Because S_a and S_e do not depend on $\underline{\mu}$, we may assume here that $\underline{\mu} = 0$.

Hence, we have that

$$E\{S_{a;11}\} = (n-1)(m\ v_{a;11} + v_{e;11}),\ E\{S_{e;11}\} = n(m-1)v_{e;11}$$

$$E\{S_{a;12}\} = (n-1)(m\ v_{a;12} + v_{e;12}),\ E\{S_{e;12}\} = n(m-1)v_{e;12}$$

$$E\{S_{a;22}\} = (n-1)(m\ v_{a;22} + v_{e;22}),\ E\{S_{e;22}\} = n(m-1)v_{e;22}$$

$$\mathrm{Var}\{S_{a;11}\} = 2(n-1)(m\ v_{a;11} + v_{e;11})^2,$$

$$\mathrm{Var}\{S_{e;11}\} = 2n(m-1)v_{e;11}^2$$

$$\mathrm{Var}\{S_{a;12}\} = (n-1)\{(m\ v_{a;12} + v_{e;12})^2$$

$$+ (m\ v_{a;11} + v_{e;11})(m\ v_{a;22} + v_{e;22})\}$$

$$\mathrm{Var}\{S_{e;12}\} = n(m-1)(v_{e;12}^2 + v_{e;11}\ v_{e;22})$$

$$\mathrm{Var}\{S_{a;22}\} = 2(n-1)(m\ v_{a;22} + v_{e;22})^2,$$

$$\mathrm{Var}\{S_{e;22}\} = 2n(m-1)\ v_{e;22}^2$$

$$\mathrm{Cov}\{S_{a;11}, S_{a;12}\} = 2(n-1)(m\ v_{a;11} + v_{e;11})(m\ v_{a;12} + v_{e;12})$$

$$\mathrm{Cov}\{S_{e;11}, S_{e;12}\} = 2n(m-1)v_{e;11}\ v_{e;12},$$

$$\mathrm{Cov}\{S_{a;22}, S_{e;22}\} = 2(n-1)(m\ v_{a;22} + v_{e;22})(m\ v_{a;12} + v_{e;12})$$

$$\mathrm{Cov}\{S_{e;22}, S_{e;12}\} = 2n(m-1)v_{e;22}\ v_{e;12}$$

$$\mathrm{Cov}\{S_{a;11}, S_{a;22}\} = 2(n-1)(m\ v_{a;12} + v_{e;12})^2,$$

$$\mathrm{Cov}\{S_{e;11}, S_{e;22}\} = 2n(m-1)v_{e;12}^2$$

The above may be verified by calculating the traces of the appropriate matrix products. In these, one exploits the fact that $H_{mn}^2 = m\, H_{mn}$, $J_{mn}^2 = mn\, J_{mn}$, and $H_{mn}\, J_{mn} = m\, J_{mn}$. Finally, because S_a and S_e are independent the covariance of any element from S_a with any element from S_e is zero.

If we let

$$\widetilde{S}_{a;hh'} = S_{a;hh'} - \frac{(n-1)}{n(m-1)} S_{e;hh'}$$

for h, h' = 1, 2 then it follows that

$$E\{\widetilde{S}_{a;11}\} = m(n-1)v_{a;11}$$

$$E\{\widetilde{S}_{a;12}\} = m(n-1)v_{a;12}$$

$$E\{\widetilde{S}_{a;22}\} = m(n-1)v_{a;22}$$

$$\mathrm{Var}\{\widetilde{S}_{a;11}\} = 2(n-1)(m\, v_{a;11} + v_{e;11})^2 + \frac{2(n-1)^2}{n(m-1)} v_{e;11}^2$$

$$\mathrm{Var}\{\widetilde{S}_{a;12}\} = (n-1)\ \{(m\, v_{a;12} + v_{e;12})^2$$

$$+ (m\, v_{a;11} + v_{e;11})(v_{a;22} + v_{e;22})\}$$

$$+ \frac{(n-1)^2}{n(m-1)} (v_{e;12}^2 + v_{e;11}\, v_{e;22})$$

$$\mathrm{Var}\{\widetilde{S}_{a;22}\} = 2(n-1)(m\, v_{a;22} + v_{e;22})^2 + \frac{2(n-1)^2}{n(m-1)} v_{e;22}^2$$

$$\mathrm{Cov}\{\widetilde{S}_{a;11}, \widetilde{S}_{a;12}\} = 2(n-1)(m\, v_{a;11} + v_{e;11})$$

$$(m\, v_{a;12} + v_{e;12}) + \frac{2(n-1)^2}{n(m-1)} v_{e;11} v_{e;12}$$

$$\mathrm{Cov}\{\tilde{S}_{a;22}, \tilde{S}_{a;12}\} = 2(n-1)(m\, v_{a;22} + v_{e;22})$$

$$(m\, v_{a;12} + v_{e;12}) + \frac{2(n-1)^2}{n(m-1)} v_{e;22} + v_{e;12})$$

$$\mathrm{Cov}\{\tilde{S}_{a;11}, \tilde{S}_{a;22}\} = 2(n-1)(m\, v_{a;12} + v_{e;12})^2$$

$$+ \frac{2(n-1)^2}{n(m-1)} v_{e;12}^2$$

Hence, even though the expectations of the elements of the adjusted between individuals matrix $\tilde{S}_a$ do not depend on the "response error" parameters $v_{e;11}$, $v_{e;12}$, $v_{e;22}$, the corresponding variances and covariances do depend on them.

3. The case m = 1. When there is only one observation on each individual, then S_e is not defined. Hence, the estimation of the association between the two variables is necessarily based on S_a.

3.1 Estimation of $v_{a;12}$. In particular, it follows from the results of the preceding sub-section that

$$E\{S_{a;12}/(n-1)\} = v_{a;12} + v_{e;12}$$

$$\mathrm{Var}\{S_{a;12}/(n-1)\} = \frac{1}{n-1} \{(v_{a;12} + v_{e;12})^2$$

$$+ (v_{a;11} + v_{e;11})(v_{a;22} + v_{e;22})\}$$

$$= \frac{1}{n-1} \{(v_{a;12}^2 + v_{a;11}\, v_{a;22})$$

$$+ (v_{a;11}\, v_{e;22} + 2v_{a;12}\, v_{e;12} + v_{a;22}\, v_{e;11})$$

$$+ (v_{e;12}^2 + v_{e;11}\, v_{e;22})\}$$

$$= \frac{1}{n-1} \psi_{12} .$$

Hence, $s_{a;12} = S_{a;12}/(n-1)$ is a biased estimate of $v_{a;12}$; the bias squared is given by $B_{12}^2 = v_{e;12}^2$. As a result, the mean square error of $s_{a;12}$ as an estimate of $v_{a;12}$ is

$$\text{MSE}(s_{a;12}) = \frac{1}{n-1} \psi_{12} + v_{e;12}^2$$

It follows that if $v_{e;12} \neq 0$, then $s_{a;12}$ is not a consistent estimate of $v_{a;12}$. In this case, any real inference about $v_{a;12}$ is not possible independent of the number of individuals. If $v_{a;12}$ is of definite interest, then more than one observation should be taken on each individual so that $v_{e;12}$ can be estimated.

Suppose now that $v_{e;12} = 0$. This assumption may be interpreted as saying that the measurement error on one variable is independent of the measurement error on the other variable. This is not unreasonable if the interviewer (or observer) is able to make separate distinct judgments concerning the two responses. In any event, if $v_{e;12} = 0$, then $s_{a;12}$ is an unbiased estimate of $v_{a;12}$. However, the presence of "non-sampling errors" still has a strong influence on the variance of $s_{a;12}$ through the parameters $v_{e;11}$ and $v_{e;22}$. In particular, we may decompose the variance of $s_{a;12}$ into several components. First of all, let

$$\psi_{a;12} = (v_{a;12}^2 + v_{a;11}\, v_{a;22}).$$

This quantity reflects the effect of the sampling variance component of the total variance. The motivation behind this arises from the relation

$$\psi_{a;12} = \text{Var}\{a_i^{(1)} a_i^{(2)}\} = \text{Var}\{E(y_{ij}^{(1)} - \mu^{(1)} | i) E(y_{ij}^{(2)} - \mu^{(2)} | i)\}$$

where $E(y_{ij}^{(h)} | i)$ denotes the relevant conditional expectation of the i-th individual.

Similarly, we may let $\psi_{e;12} = (v_{e;12}^2 + v_{e;11}\, v_{e;22})$.

This quantity reflects the effect of the non-sampling variance component of the total variance. This follows by noting that

$$\psi_{e;12} = \text{Var}\{e_{ij}^{(1)}\ e_{ij}^{(2)}\}$$

$$= \text{Var}\{[y_{ij}^{(1)} - \mu^{(1)} - E(y_{ij}^{(1)}|i)][y_{ij}^{(2)} - \mu^{(2)} - E(y_{ij}^{(2)}|i)]\}$$

Finally, let

$$\psi_{ae;12} = v_{a;11}\ v_{e;22} + 2v_{a;12}\ v_{e;12} + v_{a;22}\ v_{e;11}.$$

Then we have

$$\psi_{ae;12} = \text{Var}\{a_i^{(1)}\ e_{ij}^{(2)} + a_i^{(2)}\ e_{ij}^{(1)}\}$$

In this sense, $\psi_{ae;12}$ reflects a "cross-variance" component of the total variance.

In all, we have that

$$\psi_{12} = \text{Var}\{(y_{ij}^{(1)} - \mu^{(1)})(y_{ij}^{(2)} - \mu^{(2)})\}$$

$$= \psi_{a;12} + \psi_{ae;12} + \psi_{e;12}$$

Hence, it follows that

$$\text{Var}(s_{a;12}) = \frac{1}{n-1}(\psi_{a;12} + \psi_{ae;12} + \psi_{e;12}).$$

The indicated decomposition of the variance of $s_{a;12}$ applies regardless of whether $v_{e;12} = 0$ or not. However, when $v_{e;12} = 0$, we have that $\psi_{e;12}$ and $\psi_{ae;12}$ simplify to

$$\psi_{e;12} = v_{e;11}\ v_{e;22}$$

$$\psi_{ae;12} = v_{a;11}\ v_{e;22} + v_{a;22}\ v_{e;11}$$

The preceding analysis calls attention to two different indices of inconsistency with respect to the estimation of $v_{a;12}$ by $s_{a;12}$. These are

$$I_e = \psi_{e;12}/\psi_{12},\ I_{te} = (\psi_{e,12} + \psi_{ae;12})/\psi_{12}$$

The index I_e reflects the proportional effect of the non-sampling variance component while the index I_{te} reflects the total proportional effect due to the presence of response (measurement) errors. From some of the preceding considerations, it follows that in the case of re-surveys ($m \geq 2$), then I_e and I_{te} may be estimated.

3.2. Estimation of the variance of $\hat{v}_{a;12} = \{S_{a;12}/(n-1)\}$. Since we have that

$$E\{S^2_{a;12}\} = (n-1)\{(v_{a;12} + v_{e;12})^2$$

$$+ (v_{a;11} + v_{e;11})(v_{a;22} + v_{e;22})\}$$

$$+ (n-1)^2 (v_{a;12} + v_{e;12})^2$$

$$E\{S_{a;11}\ S_{a;22}\} = 2(n-1)(v_{a;12} + v_{e;12})^2$$

$$+ (n-1)^2(v_{a;11} + v_{e;11})(v_{a;22} + v_{e;22})$$

It follows that

$$E\{\frac{(n-1)\{S^2_{a;12} + S_{a;11}\ S_{a;22}\} - 2S_{a;12}}{(n-1)(n+1)(n-2)}\} = (v_{a;12} + v_{e;12})^2$$

$$+ (v_{a;11} + v_{e;11})(v_{a;22} + v_{e;22}) = \psi_{12}\ .$$

Hence, we may use

$$\frac{1}{n-1}\hat{\psi}_{12} = \frac{(n-1)\{S^2_{a;12} + S_{a;11}\ S_{a;22}\} - 2S_{a;12}}{(n-1)^2(n+1)(n-2)}$$

as an unbiased estimate of the variance of $\hat{v}_{a;12}$. If n is large, then the approximation

$$\frac{1}{n-1}\hat{\psi}_{12} \approx \frac{1}{(n-1)^3}\{S^2_{a;12} + S_{a;11}\,S_{a;22}\}$$

may be used. Of course, one must remember that both of the above only estimate the variance of $\hat{v}_{a;12}$ and not the mean square error. To deal with the latter, we require an estimate of $\sigma_{e;12}$ which is only possible in the case of $m > 1$ (i.e., the use of re-surveys).

3.3. Inference concerning the between individuals correlation coefficient $\rho_a = v_{a;12}/(v_{a;11}\,v_{a;22})^{\frac{1}{2}}$.

Let

$$r_{a;12} = S_{a;12}/(S_{a;11}\,S_{a;22})^{\frac{1}{2}}$$

By a series of well-known transformations, it is possible to write $r_{a;12}$ as

$$r_{a;12} = (w_1 + \delta)/\{w_1 + \delta)^2 + \sum_{\xi=2}^{n-1} w_\xi^2\}^{\frac{1}{2}}$$

with

$$\delta^2 = \{\rho^2/(1-\rho^2)\}\sum_{\eta=1}^{n-1} x_\eta^2$$

and

$$\rho^2 = (v_{a;12} + v_{e;12})^2/(v_{a;11} + v_{e;11})(v_{a;22} + v_{e;22})$$

where $w_1, w_2, \ldots, w_{n-1}$ and $x_1, x_2, \ldots, x_{n-1}$ are mutually independent sets of independent N(0,1) random variables. As a result, it follows that for fixed values of δ^2 (which is distributed as $\{\rho^2/(1-\rho^2)\}\,\chi^2(n-1)$ where $\chi^2(n-1)$ denotes the chi square distribution with (n-1) d.f.), the random variable

$$t_x = \{(n-2)\, r^2_{a;12}/(1 - r^2_{a;12})\}^{\frac{1}{2}}$$

has the non-central t-distribution $t(n-2, \delta^2)$. We can simulate the distribution of $r_{a;12}$ by generating $w_1, w_2, \ldots, w_{n-1}$ and $x_1, x_2, \ldots, x_{n-1}$ for various values of ρ^2 and then forming $r_{a;12}$ as indicated above. The parameter ρ^2 may be re-written as

$$\rho^2 = \frac{\left\{\left(\dfrac{v_{a;11}\, v_{a;22}}{v_{e;11}\, v_{e;22}}\right)^{\frac{1}{2}} \left(\dfrac{v_{a;12}}{(v_{a;11}\, v_{a;22})^{\frac{1}{2}}}\right) + \dfrac{v_{e;12}}{(v_{e;11} v_{e;22})^{\frac{1}{2}}}\right\}}{\left\{\left(1 + \dfrac{v_{a;11}}{v_{e;11}}\right)\left(1 + \dfrac{v_{a;22}}{v_{e;22}}\right)\right\}}$$

$$= \left\{(\theta_1\, \theta_2)^{\frac{1}{2}}\, \rho_a + \rho_e\right\}^2/(1 + \theta_1)(1 + \theta_2)$$

where

$$\theta_1 = (v_{a;11}/v_{e;11}),\ \theta_2 = (v_{a;22}/v_{e;22}),$$

$$\rho_e = (v_{e;12}/(v_{e;11}\, v_{e;22})^{\frac{1}{2}}).$$

Let us now consider the effect of response errors on two basic methods which use $r_{a;12}$ as a basis of statistical inference concerning $\rho_{a;12}$. First of all, there is the t-test of no association. In particular if $\rho = 0$, then $t_x \equiv t$ and has unconditionally, the t-distribution $t(n-2)$. Thus, this statistic may be used to form a valid test of the hypothesis H_0: $\rho = 0$ with specified type I error α. However, the real hypothesis of interest is H_{0a}: $\rho_a = 0$. Unfortunately, when $m = 1$, there is no direct way of constructing a valid test of H_{0a} unless one is willing to make the assumption $\rho_e = 0$. In this case H_{0a} and H_0 are identical and the previously cited t-test may be used. If this t-test is applied when $\rho_e \neq 0$, then the type I error will be different from α. Hence, one becomes interested in how both the significance level and the power of this traditional test of no association are influenced by the presence of response errors; i.e., in how $\alpha_a = \Pr\{|t| \geq t_{1-(\alpha/2)}(n-2)\}$ depends on θ_1, θ_2 and ρ_e for fixed values of α and n.

When ρ is different from zero and n is, in some sense, large, then one may use the asymptotic result that

$$Z = \frac{1}{2}\log\left(\frac{1+r}{1-r}\right)$$

has approximately the normal distribution

$$N\left(\frac{1}{2}\log\left(\frac{1+\rho}{1-\rho}\right), \frac{1}{n-3}\right) .$$

Hence, the following 100(2B-1) percent confidence interval for ρ may be formed

$$\rho_L = (ge^{-\xi} - 1)/(ge^{-\xi} + 1) \leq \rho \leq (ge^{\xi} - 1)/(ge^{\xi} + 1) = \rho_U$$

where

$$g = \left(\frac{1+r}{1-r}\right) \text{ and } \xi = \frac{2}{\sqrt{n-3}}\Phi_B,$$

Φ_B being the 100 B-th percent point of the N(0,1) distribution. If ρ is replaced by ρ_a in the above confidence interval, then interest is focused on how the response error parameters affect the true confidence coefficient; i.e., one is interested in how $B_a = \Pr\{\rho_L \leq \rho_a \leq \rho_U\}$ depends on θ_1, θ_2, and ρ_e for fixed values of B and n.

To answer the previously formulated questions, we have undertaken a simulation study of the distribution of $r_{a;12}$. The essence of this research is as follows.

1. For fixed n = 12, 20, 32,64, 128 10,000 samples of the vector

$$\left[w_1 = N(0,1),\ \sum_{\xi=2}^{n-1} w_\xi^2 = \chi^2(n-2),\ \sum_{\eta=1}^{n-1} \chi_\eta^2 = \chi_\eta^2(n-1)\right]$$

were generated.

2. For B - .975 and each fixed combination of

$$\rho_a = .00,\ .10,\ .25,\ .50,\ .75,\ .90$$

$$\rho_e = .00,\ \pm .08,\ \pm .16,\ \pm .32,\ \pm .48,\ \pm .64$$

$$\theta_1, \theta_2 = .10,\ .25,\ 1.00,\ 4.00,\ 10.00$$

the following quantities were obtained from each sampled vector

$$\theta^2 = \left\{(\theta_1\ \theta_2)^{\frac{1}{2}}\ \rho_a + \rho_e\right\}^2 / (1+\theta_1)(1+\theta_2)$$

$$\delta^2 = \left\{\rho^2/(1-\rho^2\right\}\left\{\sum_{\eta=1}^{n-1} \chi_\eta^2\right\}$$

$$r = (w_1 + \delta)/\left\{(w_1 + \delta)^2 + \sum_{\xi=2}^{n-1} w_\xi^2\right\}$$

$$|t| = |\{(n-2)r^2/(1-r^2)\}^{\frac{1}{2}}|$$

$$g = (1+r)/(1-r),\ \xi = \left(\frac{2}{\sqrt{n-3}}\right)\ \Phi_r$$

$$\rho_L = (ge^{-\xi} - 1)/(ge^{\xi} + 1),\ \rho_U = (ge^{\xi} - 1)/(ge^{\xi} + 1)$$

3. Then the counter functions

$$c_1(r) = \begin{cases} 1 & \text{if } |t| \geq t_r(n-2) \\ 0 & \text{otherwise} \end{cases}$$

$$c_2(r) = \begin{cases} 1 & \text{if } \rho_L \leq \rho_a \leq \rho_U \\ 0 & \text{otherwise} \end{cases}$$

were evaluated.

4. Finally, the averages of $c_1(r)$, $c_2(r)$, $c_3(r)$, $c_4(r)$ over the 10,000 samples were computed. The average of $c_1(r)$ will be essentially α_a while the average of $c_2(r)$ will be essentially B_a.

The values of α_a and B_α as a function of θ_1, θ_2, and ρ_e may be compared for fixed values of n, B, and ρ_e. Any evaluation made can be further enhanced by noting that for fixed n, the results obtained for different combinations of θ_1, θ_2, ρ_a, ρ_e and B are based on the same sample. This matching essentially means that the derived averages of $c_1(r)$ and $c_2(r)$ were obtained from the same experimental conditions of variability. Hence any differences observed among them is essentially due to the differences in the corresponding assigned parameters.

The results of the simulation study are summarized in the tables in Appendix A. It seems clear that the non-sampling errors included in the model can have a considerable effect on the probability that the conventional 95 percent confidence interval will cover ρ_a and on the power of the test that $\rho_a = 0$. The effects are smallest for large values of θ_1 and θ_2 (i.e. sampling error variance component large relative to the response error variance component) and largest for small θ_1 and θ_2. When ρ_e is different from ρ_a, the effects on the criteria are dramatic, though modulated by increasing values of θ_1 and θ_2. For example, the conventional 95 percent confidence coefficient drops to 0.45 for $\rho_e = 0.64$, θ_1 and θ_2 each equal to 0.10, n = 12. The corresponding value, when both θ_1 and θ_2 are equal to 10.00, increases to 0.947.

The tables show that the conventional confidence interval procedure and traditional test of hypothesis are modestly robust when ρ_a is small. They are quite seriously affected, however, when ρ_a is large. The affect of increasing the sample size in the presence of non-sampling errors is that the wrong parameters are then estimated with ever greater precision.

III. SOMEWHAT MORE GENERAL MODELS

1. <u>A basic model for the case of correlated sampling and non-sampling errors.</u> Let us now consider the model

$$\underline{y}_{ij} = \underline{\mu} + \underline{a}_i + \underline{e}_{ij} \qquad \begin{array}{l} i = 1, 2, \ldots, n \\ j = 1, 2, \ldots, m \end{array}$$

under the same conditions as (1.2.1) except for the fact that the $\underline{e}_{ij}$ are allowed to be equally correlated with the corresponding $\underline{a}_i$. In other words, we have that the $\underline{a}_i$ are independent and identically distributed as $N(\underline{0}, V_a)$, the $\underline{e}_{ij}$ are independent and identically distributed as $N(0, V_e)$, and the $\underline{a}_k$ and $\underline{e}_{ij}$ have joint normal distributions such that

$$E\,\{\underline{a}_k \underline{e}_{ij}\} = \mathrm{Cov}\,\{\underline{a}_k, \underline{e}_{ij}\} = \begin{cases} C \text{ if } i = k \\ 0 \text{ if } i \neq k \end{cases}$$

where C satisfies the property

$$\underline{u}' V_a \underline{u} \geq \underline{u}' C V_e^{-1} C' \underline{u} \text{ and } \underline{u}' V_e \underline{u} \geq \underline{u}' C' V_a^{-1} C \underline{u}.$$

Hence we have that each $\underline{y}_{ij}$ is distributed as

$N(\underline{\mu}, V_a + C + C' + V_e)$ and

$$\text{Cov}\{\underline{y}_{k\ell}, \underline{y}_{ij}\} = \begin{cases} V_a + C + C' & \text{if } i = k,\ j \neq \ell \\ 0 & \text{if } i \neq k \end{cases} .$$

When the matrix C is non-null, then the sampling and non-sampling sources of variability are no longer statistical independent. As a result, we shall say that there is interaction between them.

2. <u>The bivariate case p = 2</u>. Let $\underline{y}^{(1)}$ and $\underline{y}^{(2)}$ be defined as in section II.1. Again let $\underline{y}$ be defined by $\underline{y}' = [\underline{y}^{(1)\prime}, \underline{y}^{(2)\prime}]$. Then

$$E(\underline{y}) = \begin{bmatrix} \mu^{(1)}\underline{j}_{mn} \\ \mu^{(2)}\underline{j}_{mn} \end{bmatrix} \text{ and}$$

$$\text{Var}(\underline{y}) = \left[\begin{array}{c|c} r & t \\ \hline t & s \end{array}\right]$$

where

$$r = (v_{a;11} + 2c_{11})H_{mn} + v_{e;11}I_{mn}$$

$$s = (v_{a;22} + 2c_{22})H_{mn} + v_{e;22}I_{mn}$$

$$t = (v_{a;12} + c_{12} + c_{21})H_{mn} + v_{e;12}I_{mn}$$

where $\underline{j}_{mn}, H_{mn}$, and I_{mn} have been defined previously and $c_{11}, c_{12}, c_{21}, c_{22}$ are elements of C.

It can be seen from this structure that $(v_{a;11}+2c_{11})$, $(v_{a;12}+c_{12}+c_{21})$, $(v_{a;22}+2c_{22})$ play the same roles as $v_{a;11}$, $v_{a;12}$, $v_{a;22}$ respectively in the results given in section II.2. with respect to the original model. The expectations and covariances of the elements of S_e remain the same as given in section II.2.; moreover, the elements of S_e and S_a are again statistically independent. The proof of this follows by noting that the structure of Var($\underline{y}$) here with respect to V_e is the same as given in section II.1.

If $\tilde{s}_a$ is again defined by the relation

$$\tilde{s}_a = \{S_a - \frac{n-1}{n(m-1)} S_e\}/m(n-1),$$

then

$$E\{\tilde{s}_{a;11}\} = (v_{a;11}+2c_{11}),\ E\{\tilde{s}_{a;22}\} = (v_{a;22}+2c_{22})$$

$$E\{\tilde{s}_{a;12}\} = (v_{a;12}+c_{12}+c_{21}).$$

Hence, even for the case m≥2, it is no longer possible here to obtain unbiased estimators for $v_{a;11}$, $v_{a;12}$, $v_{a;22}$. This bias is due to the interaction between the two sources of error as expressed by C.

3. <u>The case m = 1</u>. As before, in this case S_e is not defined; hence inference will be based on S_a only.

The parameter $v_{a;12}$ can be estimated unbiasedly by $\{S_{a;12}/(n-1)\}$ only if both the conditions

$$v_{e;12} = 0,\ c_{12}+c_{21} = 0 \text{ i.e. } E(a_1e_2) = -E(a_2e_1)$$

the second of the above holds if $c_{12} = c_{21} = 0$.

The variance of $\{S_{a;12}/(n-1)\}$ can be partitioned as the sum of

$$\Psi_{12} = \Psi_{a;12} + \Psi_{ae;12} + \Psi_{e;12} = (v_{a;12} + v_{e;12})^2$$

$$+ (v_{a;11} + v_{e;11})(v_{a;22} + v_{e;22}),$$

$$\Psi_{c;12} = (c_{12} + c_{21})^2 + 4c_{11}c_{22}, \quad \text{and}$$

$$\Psi_{aec;12} = 2c_{11}(v_{a;22}+ v_{e;22}) + 2(c_{12}+ c_{21})(v_{a;12}+ v_{e;12})$$

$$+ 2c_{22}(v_{a;11} + v_{e;11}).$$

The component $\Psi_{c;12}$ represents the "pure" effect of interaction while $\Psi_{aec;12}$ represents the mixed effect. The total effect (increase <u>or</u> decrease) is represented by $\Psi_{aec;12} + \Psi_{c;12}$. The index of inconsistency due to interaction could be defined as

$$I_c = (\Psi_{aec;12} + \Psi_{c;12})/(\Psi_{12} + \Psi_{c;12} + \Psi_{aec;12}).$$

The index of inconsistency due to the presence of non-sampling errors is

$$I_{tec} = 1 - \Psi_{a;12}/(\Psi_{12} + \Psi_{c;12} + \Psi_{aec;12}).$$

Even though $S_{a;12}/(n-1)$ is a biased estimate of $v_{a;12}$, we are able to estimate its variance. This is given by the same quantity as before; namely,

$$\frac{(n-1)\{s^2_{a;12}+s_{a;11}s_{a;22}\}-2s_{a;12}}{(n-1)^2(n+1)(n-2)} .$$

As was pointed out in section III.3.3, the inference regarding the correlation coefficient is more complex. Here, $r_{a;12}$ is directed at the parameter

$$\rho = \frac{(v_{a;12}+c_{12}+c_{21}+v_{e;12})}{\{(v_{a;11}+2c_{11}+v_{e;11})(v_{a;22}+2c_{22}+v_{e;22})\}^{\frac{1}{2}}}$$

The effect of the various aspects of non-sampling errors on the t-test of no association and the confidence interval based on the Z transformation can be studied in a fashion similar to what has been indicated before. In particular, essentially the same simulation routine can be employed to examine how the true error rates, power, confidence probability, etc. are influenced by non-null C and V_e.

5. <u>The generalized model allowing within source correlated errors.</u> The model

$$\underline{y}_{ij} = \underline{\mu} + a_i + \underline{e}_{ij} \qquad \begin{array}{l} i = 1,2,\ldots,n \\ j = 1,2,\ldots,m \end{array}$$

is assumed to satisfy the same conditions given in section III.1. except now the $\underline{a}_i$'s are allowed to be equally correlated with each other and the $\underline{e}_{ij}$'s are allowed to be correlated with each other; i.e., we shall assume

$$E(\underline{a}_i\underline{a}_k') = \begin{cases} V_a & \text{if } i = k \\ \Lambda_a & \text{if } i \neq k \end{cases}$$

$$E(\underline{e}_{ij}\underline{e}_{k\ell}') = \begin{cases} V_e & \text{if } i=k,\ j=\ell \\ \Lambda_{e1} & \text{if } i=k,\ j\neq\ell \\ \Lambda_{e2} & \text{if } i \neq k \end{cases}$$

$$E(\underline{a}_i\underline{e}_{kj}') = \begin{cases} C & \text{if } i = k \\ \Lambda_{ae} & \text{if } i \neq k \end{cases}$$

In the above, the matrices Λ_a, Λ_{ae}, Λ_{e1}, Λ_{e2} must satisfy regularity conditions similar to those placed on C before The reason for these conditions is to insure that all linear combinations of a's and e's have positive variance

From the above remarks, it follows that the $\underline{y}_{ij}$ are normally distributed with mean vectors $\underline{\mu}$ and covariance structure

$$\begin{aligned} \mathrm{Cov}\{\underline{y}_{ij},\underline{y}_{kl}\} &= V_a+C+C'+V_e \quad \text{if } i=k,\ j=1 \\ &= V_a+C+C'+\Lambda_{e1} \quad \text{if } i=k,\ j\neq 1 \\ &= \Lambda_a+\Lambda_{ae}+\Lambda_{ae}'+\Lambda_{e2} \quad \text{if } i=k \quad . \end{aligned}$$

When the matrices Λ_a,Λ_{ae},Λ_{e2} are non-null, then the errors on different subjects are no longer statistically independent. Similarly, when Λ_{e1} is non-null, then the measurement errors on the same subject are no longer statistically independent. Both of these statements indicate a complex interaction between the subjects, the measurement process, the sampling process, etc.

For the bivariate case p = 2, we have

$$E(\underline{y}) = E\begin{bmatrix} \underline{y}^{(1)} \\ \underline{y}^{(2)} \end{bmatrix} = \begin{bmatrix} \mu^{(1)}\underline{j}_{mn} \\ \mu^{(2)}\underline{j}_{mn} \end{bmatrix} \quad \text{and}$$

$$\text{Var}(\underline{y}^{(1)}) = [(v_{a;11}+2c_{11}+\lambda_{e1;11}-\lambda_{11})H_{mn} + \lambda_{11}J_{mn}+(v_{e;11}-\lambda_{e1;11})I_{mn}]$$

$$\text{Cov}(\underline{y}^{(1)},\underline{y}^{(2)}) = [(v_{a;12}+c_{12}+c_{21}+\lambda_{e1;12}-\lambda_{12})H_{mn}+\lambda_{12}J_{mn} +(v_{e;12}-\lambda_{e1;12})I_{mn}]$$

$$\text{Var}(\underline{y}^{(2)}) = [(v_{a;22}+2c_{22}+\lambda_{e1;22}-\lambda_{22})H_{mn} + \lambda_{22}J_{mn}+(v_{e;22}-\lambda_{e1;22})I_{mn}]$$

where

$$\lambda_{11} = \lambda_{a;11} + 2\lambda_{ae;11} + \lambda_{e2;11},$$

$$\lambda_{22} = \lambda_{a;22} + 2\lambda_{ae;22} + \lambda_{e2;22},$$

$$\lambda_{12} = \lambda_{a;12} + \lambda_{ae;12} + \lambda_{ae;21} + \lambda_{e2;12} .$$

Let

$$\tilde{v}^{*}_{a;11} = v_{a;11} + 2c_{11} + \lambda_{e1;11} - \lambda_{11}$$

$$\tilde{v}^{*}_{e;11} = v_{e;11} - \lambda_{e1;11}$$

$$\tilde{v}^*_{a;12} = v_{a;12} + c_{12} + c_{21} + \lambda_{e1;12} - \lambda_{12}$$

$$\tilde{v}^*_{e;12} = v_{e;12} - \lambda_{e1;12}$$

$$\tilde{v}^*_{a;22} = v_{a;22} + 2c_{22} + \lambda_{e1;22} - \lambda_{22}$$

$$\tilde{v}^*_{e;22} = v_{e;22} - \lambda_{e1;22} \quad .$$

The above parameters play the same roles in the covarianc of the elements of S_a and S_e as the original ones did in section II.2.

From what has been indicated here, it is reasonably apparent that any further study of this model would essentially follow the same lines as given in the precedi sections except for the fact that it would pertain to the parameters $\tilde{v}^*_{a;hh'}$ and $\tilde{v}^*_{e;hh'}$. The relationship between this type of inference and that pertaining to V_a and V_e directly will not be discussed any further at present. However, one could always assume the second philosophy of section III.4. and claim that V_a and V_e no longer have any individual meaning and that the parameters of real interest are precisely those in $\tilde{V}^*_a$ and $\tilde{V}^*_e$.

References

Anderson, T. W. (1958). An Introduction to Multivariate Statistical Analysis, John Wiley and Sons, Inc., New York.

Hansen, Morris H., Hurwitz, William N. and Pritzker, Leon. (1964). "The Estimation and Interpretation of Gross Differences and the Simple Response Variance". U.S. Bureau of the Census, Mimeo. Prepared for publication in the Seventieth Birthday Volume in Honor of Professor P. C. Mahalanobis.

Roy, S. N. (1953). "On a Heuristic Method of Test Construction and Its Use in Multivariate Analysis." Annals of Mathematical Statistics, 24, pp. 220-238.

Schatzoff, M. (1966). "Sensitivity Comparisons Among Tests of the General Linear Hypothesis", Journal of the American Statistical Association, 61, pp. 415-435.

Acknowledgement

The authors are grateful to Dr. Tom G. Donnelly, University of North Carolina, who programmed the simulation study for the IBM 360, Model 75 computer.

N = 20 BETA = .97500 RHO-A = 0.0								
PROBABILITY OF 95% CONFIDENCE INTERVAL COVERING TRUE VALUE OF RHO-A								
TH1	TH2	-.640	-.320	-.160	.0	.160	.320	.640
0.10	0.10	.197	.758	.904	.948	.906	.756	.190
0.10	0.25	.270	.783	.910	.948	.912	.780	.260
0.10	1.00	.505	.850	.926	.948	.927	.849	.503
0.10	4.00	.783	.910	.939	.948	.941	.912	.780
0.10	10.00	.875	.933	.945	.948	.944	.931	.877
0.25	0.25	.334	.806	.914	.948	.916	.804	.330
0.25	1.00	.559	.861	.930	.948	.929	.861	.555
0.25	4.00	.806	.914	.941	.948	.942	.916	.804
0.25	10.00	.884	.935	.945	.948	.944	.933	.887
1.00	1.00	.715	.894	.936	.948	.935	.895	.708
1.00	4.00	.861	.930	.944	.948	.943	.929	.861
1.00	10.00	.910	.939	.947	.948	.945	.941	.912
4.00	4.00	.914	.941	.947	.948	.945	.942	.916
4.00	10.00	.935	.945	.948	.948	.946	.944	.933
10.00	10.00	.943	.947	.948	.948	.946	.946	.942

PROBABILITY OF REJECTING HYPOTHESIS RHO-A=0 (TYPE I ERROR)								
0.10	0.10	.800	.239	.093	.051	.092	.241	.807
0.10	0.25	.727	.215	.087	.051	.086	.217	.737
0.10	1.00	.492	.148	.072	.051	.072	.149	.494
0.10	4.00	.215	.087	.059	.051	.057	.086	.217
0.10	10.00	.123	.065	.054	.051	.055	.067	.120
0.25	0.25	.662	.191	.083	.051	.082	.194	.667
0.25	1.00	.438	.137	.068	.051	.070	.136	.440
0.25	4.00	.191	.083	.058	.051	.057	.082	.194
0.25	10.00	.114	.063	.053	.051	.054	.065	.111
1.00	1.00	.282	.103	.062	.051	.062	.103	.289
1.00	4.00	.137	.068	.054	.051	.055	.070	.136
1.00	10.00	.087	.059	.052	.051	.053	.057	.086
4.00	4.00	.083	.058	.051	.051	.053	.057	.082
4.00	10.00	.053	.053	.051	.051	.052	.054	.065
10.00	10.00	.055	.051	.050	.051	.052	.053	.056

N = 64		BETA = .97500			RHO-A = 0.0			
TH1	TH2	-.640	-.320	-.160	.0	.160	.320	.640
PROBABILITY OF 95% CONFIDENCE INTERVAL COVERING TRUE VALUE OF RHO-A								
0.10	0.10	.000	.345	.792	.951	.791	.344	.000
0.10	0.25	.001	.404	.809	.951	.313	.403	.001
0.10	1.00	.047	.599	.863	.951	.864	.588	.046
0.10	4.00	.404	.809	.917	.951	.919	.818	.403
0.10	10.00	.690	.888	.936	.951	.939	.888	.688
0.25	0.25	.008	.458	.827	.951	.830	.461	.005
0.25	1.00	.081	.640	.874	.951	.874	.630	.076
0.25	4.00	.458	.827	.922	.951	.923	.830	.461
0.25	10.00	.723	.895	.937	.951	.941	.895	.721
1.00	1.00	.258	.757	.903	.951	.902	.754	.255
1.00	4.00	.640	.874	.933	.951	.935	.874	.630
1.00	10.00	.809	.917	.941	.951	.946	.919	.818
4.00	4.00	.827	.922	.942	.951	.946	.923	.830
4.00	10.00	.895	.937	.948	.951	.949	.941	.895
10.00	10.00	.927	.943	.950	.951	.950	.946	.928

PROBABILITY OF REJECTING HYPOTHESIS RHO-A=0 (TYPE I ERROR)								
0.10	0.10	.999	.653	.207	.048	.208	.654	.999
0.10	0.25	.998	.595	.189	.048	.186	.596	.998
0.10	1.00	.952	.399	.136	.048	.135	.410	.952
0.10	4.00	.595	.189	.082	.048	.080	.186	.596
0.10	10.00	.309	.111	.063	.048	.060	.111	.310
0.25	0.25	.991	.540	.172	.048	.169	.537	.994
0.25	1.00	.918	.359	.124	.048	.125	.368	.923
0.25	4.00	.540	.172	.077	.048	.076	.169	.537
0.25	10.00	.275	.104	.062	.048	.058	.103	.278
1.00	1.00	.741	.241	.096	.048	.096	.243	.744
1.00	4.00	.359	.124	.065	.048	.064	.125	.368
1.00	10.00	.189	.082	.058	.048	.054	.080	.186
4.00	4.00	.172	.077	.057	.048	.053	.076	.169
4.00	10.00	.104	.062	.051	.048	.050	.058	.103
10.00	10.00	.071	.056	.049	.048	.049	.053	.071

N = 20	BETA = .97500	RHO-A = 0.25						
TH1	TH2	-.640	-.320	-.160	.0	.160	.320	.640
PROBABILITY OF 95% CONFIDENCE INTERVAL COVERING TRUE VALUE OF RHO-A								
0.10	0.10	.036	.400	.652	.344	.936	.939	.521
0.10	0.25	.063	.452	.681	.854	.938	.942	.586
0.10	1.00	.214	.586	.752	.873	.934	.949	.782
0.10	4.00	.514	.733	.818	.883	.924	.948	.928
0.10	10.00	.667	.795	.847	.887	.915	.936	.951
0.25	0.25	.110	.513	.720	.870	.941	.942	.628
0.25	1.00	.299	.652	.793	.891	.941	.947	.783
0.25	4.00	.597	.786	.857	.903	.936	.951	.919
0.25	10.00	.736	.841	.880	908	.930	.945	.947
1.00	1.00	.531	.778	.866	.920	.948	.947	.839
1.00	4.00	.766	.876	.908	.933	.949	.951	.916
1.00	10.00	.853	.903	.923	.937	.948	.951	.943
4.00	4.00	.887	.924	.936	.947	.951	.951	.934
4.00	10.00	.921	.938	.945	.950	.951	.951	.944
10.00	10.00	.937	.947	.949	.951	.951	.951	.946

PROBABILITY OF REJECTING HYPOTHESIS RHO-A=0 (POWER OF TEST)								
0.10	0.10	.757	.208	.080	.052	.108	.278	.847
0.10	0.25	.663	.172	.069	.053	.109	.266	.802
0.10	1.00	.384	.106	.055	.055	.104	.211	.605
0.10	4.00	.140	.059	.050	.057	.086	.138	.320
0.10	10.00	.074	.050	.053	.059	.076	.105	.194
0.25	0.25	.558	.141	.062	.055	.115	.265	.768
0.25	1.00	.291	.080	.051	.062	.117	.228	.604
0.25	4.00	.100	.052	.054	.069	.105	.161	.347
0.25	10.00	.058	.052	.057	.071	.094	.128	.225
1.00	1.00	.131	.051	.054	.080	.139	.233	.523
1.00	4.00	.054	.056	.072	.101	.142	.198	.357
1.00	10.00	.053	.069	.085	.108	.137	.171	.263
4.00	4.00	.059	.086	.108	.135	.165	.204	.300
4.00	10.00	.081	.109	.127	.146	.169	.195	.255
10.00	10.00	.108	.133	.146	.160	.177	.195	.233

		N = 64		BETA = .97500		RHO-A = 0.25		
TH1	TH2	-.640	-.320	-.160	.0	.160	.320	.640
		PROBABILITY OF 95% CONFIDENCE INTERVAL COVERING TRUE VALUE OF RHO-A						
0.10	0.10	.0	.014	.154	.560	.900	.916	.060
0.10	0.25	.0	.025	.198	.593	.902	.925	.101
0.10	1.00	.001	.092	.322	.650	.892	.948	.416
0.10	4.00	.047	.281	.487	.692	.853	.936	.877
0.10	10.00	.176	.424	.568	.704	.816	.895	.952
0.25	0.25	.0	.046	.257	.641	.911	.925	.142
0.25	1.00	.003	.153	.417	.724	.913	.944	.421
0.25	4.00	.102	.401	.599	.775	.895	.948	.848
0.25	10.00	.290	.552	.681	.791	.875	.926	.945
1.00	1.00	.055	.379	.627	.832	.936	.942	.575
1.00	4.00	.351	.660	.794	.887	.939	.951	.839
1.00	10.00	.587	.776	.851	.901	.935	.950	.926
4.00	4.00	.704	.853	.900	.933	.949	.950	.897
4.00	10.00	.834	.902	.924	.941	.950	.952	.932
10.00	10.00	.901	.931	.941	.947	.951	.952	.941

		PROBABILITY OF REJECTING HYPOTHESIS RHO-A=0 (POWER OF TEST)						
0.10	0.10	.998	.580	.162	.050	.265	.726	.999
0.10	0.25	.991	.477	.127	.053	.270	.705	.999
0.10	1.00	.877	.247	.068	.066	.248	.583	.986
0.10	4.00	.368	.083	.048	.079	.185	.372	.799
0.10	10.00	.143	.051	.050	.084	.149	.254	.537
0.25	0.25	.975	.370	.093	.063	.293	.704	.999
0.25	1.00	.757	.163	.052	.095	.300	.624	.986
0.25	4.00	.231	.055	.054	.124	.254	.446	.839
0.25	10.00	.079	.050	.075	.133	.219	.338	.617
1.00	1.00	.339	.058	.060	.163	.375	.635	.965
1.00	4.00	.062	.070	.135	.239	.387	.550	.848
1.00	10.00	.053	.124	.183	.265	.369	.472	.701
4.00	4.00	.085	.185	.264	.360	.457	.567	.768
4.00	10.00	.167	.270	.333	.401	.467	.539	.683
10.00	10.00	.267	.354	.399	.444	.491	.539	.636

N = 20 BETA = .97500 RHO-A = 0.75

TH1	TH2	-.640	-.320	-.160	.0	.160	.320	.640
PROBABILITY OF 95% CONFIDENCE INTERVAL COVERING TRUE VALUE OF RHO-A								
0.10	0.10	.0	.000	.008	.040	.127	.323	.891
0.10	0.25	.0	.002	.014	.054	.148	.348	.886
0.10	1.00	.000	.011	.035	.083	.184	.351	.798
0.10	4.00	.008	.039	.067	.115	.187	.289	.553
0.10	10.00	.028	.061	.089	.128	.175	.237	.396
0.25	0.25	.000	.006	.026	.077	.196	.407	.905
0.25	1.00	.003	.029	.068	.148	.289	.476	.880
0.25	4.00	.032	.091	.146	.226	.331	.453	.727
0.25	10.00	.075	.144	.194	.257	.329	.407	.588
1.00	1.00	.036	.128	.219	.350	.510	.682	.935
1.00	4.00	.189	.347	.445	.551	.661	.768	.921
1.00	10.00	.340	.472	.545	.618	.693	.765	.880
4.00	4.00	.546	.685	.754	.816	.866	.906	.951
4.00	10.00	.720	.809	.847	.876	.904	.924	.951
10.00	10.00	.854	.894	.910	.925	.935	.945	.952

TH1	TH2	-.640	-.320	-.160	.0	.160	.320	.640
PROBABILITY OF REJECTING HYPOTHESIS RHO-A=0 (POWER OF TEST)								
0.10	0.10	.666	.155	.061	.057	.146	.353	.908
0.10	0.25	.520	.113	.052	.070	.172	.382	.905
0.10	1.00	.213	.055	.055	.102	.209	.386	.825
0.10	4.00	.060	.057	.085	.136	.213	.319	.585
0.10	10.00	.053	.079	.108	.148	.201	.268	.428
0.25	0.25	.356	.071	.052	.094	.224	.443	.923
0.25	1.00	.110	.053	.086	.171	.318	.510	.898
0.25	4.00	.054	.110	.170	.256	.364	.486	.758
0.25	10.00	.093	.167	.222	.288	.362	.443	.623
1.00	1.00	.055	.148	.247	.385	.544	.713	.953
1.00	4.00	.215	.380	.478	.584	.693	.796	.938
1.00	10.00	.373	.506	.579	.652	.723	.792	.898
4.00	4.00	.580	.715	.780	.839	.885	.924	.975
4.00	10.00	.753	.833	.866	.896	.922	.943	.974
10.00	10.00	.874	.911	.928	.942	.953	.965	.981

N = 64		BETA = .97500			RHO-A = 0.75			
TH1	TH2	-.640	-.320	-.160	.0	.160	.320	.640
		PROBABILITY OF 95% CONFIDENCE INTERVAL COVERING TRUE VALUE OF RHO-A						
0.10	0.10	.0	.0	.0	.0	.000	.003	.677
0.10	0.25	.0	.0	.0	.0	.000	.005	.668
0.10	1.00	.0	.0	.0	.0	.000	.005	.381
0.10	4.00	.0	.0	.0	.000	.000	.002	.054
0.10	10.00	.0	.0	.0	.000	.000	.000	.009
0.25	0.25	.0	.0	.0	.0	.000	.011	.738
0.25	1.00	.0	.0	.0	.000	.002	.027	.636
0.25	4.00	.0	.0	.000	.000	.004	.020	.233
0.25	10.00	.0	.000	.000	.001	.004	.011	.078
1.00	1.00	.0	.000	.000	.005	.036	.167	.868
1.00	4.00	.000	.005	.018	.053	.140	.308	.805
1.00	10.00	.005	.025	.050	.101	.180	.302	.637
4.00	4.00	.050	.170	.278	.420	.585	.747	.938
4.00	10.00	.224	.404	.511	.624	.734	.827	.934
10.00	10.00	.542	.690	.763	.824	.871	.909	.948

		PROBABILITY OF REJECTING HYPOTHESIS RHO-A=0 (POWER OF TEST)						
0.10	0.10	.992	.421	.092	.080	.402	.844	1.000
0.10	0.25	.966	.272	.058	.125	.475	.874	1.000
0.10	1.00	.590	.070	.065	.243	.578	.878	.999
0.10	4.00	.085	.077	.181	.368	.587	.795	.981
0.10	10.00	.053	.161	.268	.410	.558	.709	.916
0.25	0.25	.848	.133	.050	.220	.613	.925	1.000
0.25	1.00	.262	.053	.185	.474	.795	.960	.999
0.25	4.00	.059	.276	.470	.684	.854	.949	.998
0.25	10.00	.213	.462	.609	.743	.853	.925	.989
1.00	1.00	.068	.408	.669	.876	.972	.997	1.000
1.00	4.00	.593	.872	.945	.981	.996	.999	1.000
1.00	10.00	.865	.959	.980	.992	.997	.999	.999
4.00	4.00	.980	.997	.999	.999	.999	1.000	1.000
4.00	10.00	.998	.999	.999	.999	1.000	1.000	1.000
10.00	10.00	.999	.000	1.000	1.000	1.000	1.000	1.000

ITEM ANALYSIS AND REWEIGHTING

P. Thionet

University of Poitiers, France

1. Introduction

The object of the present paper is not to build some new theory of sampling. In several books and other papers we have shown how probability sampling remains the favorite topic of our researches. However, since 1955, we have had practically no more occasion to select randomly a sample for I.N.S.E.E.* surveys.

As a statistician in the French Ministry of Finances, we had some opportunities to be consulted on sampling problems. And on several occasions, there were great difficulties in applying sampling theory: we ran the risk of solving problems having little (or nothing) to do with the real problem, as it was put by our questioner ("error of 3rd kind").

For instance he was wanting too detailed results, not available by sampling; or his problem had to be translated in terms of Theory of Games, with only a secondary sampling side. (On the struggle against fraud, we gave some years ago a little paper at Aix Meeting on Operational Research.)

Most frequently, however, our interlocutor disposed of some sample, a by-product of administration for example. The sample was not representative at all, because distorted; but individual data were supposed to be excellent, deep verifications being possible. It was desired to make these data comparable with the whole population, from which this sample was

---

* I.N.S.E.E. Institut National de la Statistique et des Études Économiques.

coming. The comparison of sets not having the same composition is a general problem; for demographic statistics, the technique of standard populations is used; for prices, foreign trade, etc. we are accustomed to use index numbers.

The technique presented here is called reweighting. It is used moderately with probability samples, when more or less distorted by non-responses; in this case, reweighted statistics remain random variates, with their own mathematical expectations, biasses, variances,... On the contrary, we were requested to reweight non-random samples; our computations had nothing to do with the Probability Calculus, we were "escaping" from sampling theories.

Let us observe that reweighting (as a tool for comparisons) is by no means esoteric; the layman finds it natural to agree that some units in the sample count for 10 and others for 50. Only negative weights look foolish.

The writer has to work in a country where (notwithstanding its Statistical Society, one hundred years old) trivial concepts relating to Statistical Methods are often misunderstood, at any level of education. And it is difficult to say to our "inspecteurs des Finances" that they are wrong and have to learn sampling, when in fact they are very clever, very powerful and favorable to applied research.

2. Mathematical Problem:

Let us say at once that we have as yet limited experience of the subject: we have had only two occasions to realize computations, and those without powerful means; first with a desk computer and a skillful clerk, at the Ministry of Finances; secondly as a professor at the Institut de Statistique de l'Université de Paris, directing M. POURNIN, an advanced student, in his researches on Agricultural Statistics for the Diplôme de Statisticien.

Both samples were "bad samples", i.e. too beautiful samples. That is always the case when data are obtained from firms having genuine accounts (well-managed firms), from farms having accounts (accepting the help of some agricultural expert). These firms and farms are by no means representative. In fact the first sample (firms) was not too bad; and only the second one was capable of demonstrating our major difficulty, which was to obtain non-negative weights.

Initially we were searching for unknown real numbers, named the weights, one for each unit of the sample, and required to satisfy a system of linear constraints (equations), together with some other way of defining an "optimal" choice.

Optimization will be obtained by way of some quadratic function of the weights to be minimized. Heavy experimental computations were carried out with our 106 firms' data, to investigate several functions, linear combinations of squared weights. Only limited results were published in a Communication, at the Belgrade Session (1965) of the International Statistical Institute. Briefly, we obtained very different weights but almost the same weighted means, when the quadratic function was modified. This is an explanation (not a justification) of our behavior during Pournin's computations on his 39 farms' data. They were restricted to the simplest function: the sum of squared weights. Research was now centered on the elimination of negative weights.

During the first trials, these weights remained very few; and it seemed realistic (though incorrect) to give them a value of zero, when we considered how small they were in modulus and how many errors were possible. But the second experiment, realized under better conditions (only 39 units, with an electronic computer to invert the matrix) gave to one weight in three a definitely negative value. Here was the rub.

Now our problem was to be identified with a (very special) <u>quadratic program</u>. Only the "objective" or "economic function" of weights (to be minimized) was quadratic. Constraints were linear, but of two kinds: equations and inequalities.

<u>Equations</u>: In both trials, they were of two sorts, but it is not possible to assert that this is the most general situation. Our population was divided into some sections (or strata) whose size was known; consequently, the total of weights in each section should take a given value (constraint No. 1).

On the other hand, for the population as a whole, the total (or mean) value of several "bench-mark" variates is supposed to be known. It is desired that the weights be chosen in such a way as to give to weighted means of these variates, computed with sample data, these required values (constraint No. 2).

It would be possible, in further situations, to find bench-mark variates with totals known section by section, or for subsets of sections. But this would modify but little our formulation.

<u>Inequalities</u>: Any sample unit represents itself (at least) and consequently its weight has to be at least 1, $w - 1 \geq 0$. Putting $w - 1 = p$, we should have $p \geq 0$.

It is convenient to translate all constraints into linear equations among positive p (unknown). The p-total is known and consequently we have to minimize Σp^2 instead of Σw^2.

<u>The quadratic function</u>: Σp^2.

We give in Appendix 1 some explanations touching our non-linear choice. Finally it is possible that such a choice be regretted by people having an electronic computer at hand (not our case). However, linear programming would be useful, even with a quadratic objective function, to explore the convex set defined by our linear constraints and, above all, to discover cases (very important in practice) where this set is void, the initial program being too ambitious.

The dual program has a somewhat modest size, if the constraints taking the form of an equation are not too many; and consequently a giant-computer is not necessary.

However, linear programming should be able to give us a feasible program, that is a point of the convex set, supposed non-void. An algorithm giving the optimal program remains to be found.

Engaged in an experimental research with M. POURNIN, our main purpose was to understand where the difficulties were located. We employed geometric methods, only possible in simple cases. We succeeded in visualizing our convex set projected on two-dimentional spaces. The corresponding general theory is given in Appendix 2.

Another very important point is studied in Appendix 4, after having proven (a trivial result) that the economic or objective function (once chosen) takes its minimal value only at one point (appendix 3). It is always possible to neglect at the beginning the constraints $p \geq 0$ as in our BELGRADE paper. The weights have an analytic expression (changing with the objective function); after an inversion of matrix their computation is practicable (by hand if weights are not too numerous). There is no reason why any research should not begin with these computations.

Now suppose we find all p's positive; the work is finished. But it is reasonable to expect some negative p's. It is not possible to approach the optimal program from that imperfect one without having to start off again from zero.

In other words, the objective function is too much reduced; we have to increase its value to obtain a feasible program. Consequently, we have not to substitute 0 for negative weights, decreasing the objective function we are going far away from the set of admissible points. However, it looks reasonable to try to rejoin this set by reducing only some big positive weights, and distinguishing two classes of negative weights, those becoming positive, and those becoming 0. When negative weights are rather few (and others rather many), it is possible to hope to obtain by this way, not only a feasible, but almost an optimal program.

In the next section we complete this rather too general statement by a concrete description of our experiments.

3. An experiment on 39 farms:

Our student was a young "ingenieur agricole" employed by Calvados Farmers Office and having access to all official and private agricultural data, with the possibility of controlling them in the field if necessary. For various practical reasons, the work was limited to only one of the four agricultural regions, the BESSIN. It is sheer coincidence if this country saw your soldiers on the 6th of June, 1944.

Agricultural statistics are published for the whole "département", so that very few variates remain to give linear constraints. The cattle (chief wealth of the country) were known by a special survey (See LENCO, Journal de la Société de Statistique de Paris, 107, (1966), 164-83).

Corn trade, in the hands of a Public Office, is well-known and, of course, acreage. We had difficulties with two other variates: tractors and wage-earners (in farms).

Tractors: data were good, but the sample was too motorized. One farm had 3 and 5 farms had 2 tractors. Each of the others except one, had 1 tractor. In the region, however, there were 1832 tractors for 2101 farms. So it was impossible to keep the number of tractors as a bench-mark variate (the farm without a tractor would have obtained too high a weight).

Wage-earners: This variate had in fact little meaning, the manpower in many cases consisting mostly of the farmer's families. A great part of the wage-earners are only part-time. So it was convenient to substitute wages (translated into work-units) for wage-earners; but there were several divergent estimates of the regional mass of rural wages. This was an elastic bench-mark, if any.

Sections or strata: The sample was sorted into 3 sections:

1) from 10 to 20 hectares (ha):
2) " 20 to 50 ha : (1 ha = 2.5 acres)
3) more than 50 ha :

Consequently farms under 10 ha were absent from the sample, and so bench-mark totals needed to be corrected (with some uncertainty).

	Acreage (ha)	Cattle (head)	Wheat (qtx)	Wage-earners (units)
Section 1 Farm				
1	13.7	18	57	0.03
2	16.0	23	0	0
3	18.5	25	0	0.13
4	18.2	34	0	1.01
Section 2 Farm				
5	21.6	31	24	0.02
6	23.0	25	0	0.34
7	24.0	38	0	0.33
8	25.6	37	0	1.16
9	26.0	45	37	0.92
10	26.9	49	0	1.98
11	28.0	50	0	0.48
12	26.3	45	173	0.70
13	30.0	35	232	1.48
14	30.0	39	44	1.10
15	30.0	41	0	0.38
16	31.7	39	115	1.00
17	31.5	45	236	1.96

(continued)

	Acreage (ha)	Cattle (head)	Wheat (qtx)	Wage-earners (units)
18	34.0	39	212	1.33
19	34.9	42	50	0.66
20	35.0	45	121	1.85
21	35.0	52	117	1.65
22	38.5	40	42	2.73
23	43.0	57	0	1.85
24	41.7	38	357	1.76
Section 3 Farm				
25	50.0	67	0	1.69
26	50.3	52	264	3.24
27	52.0	67	275	1.54
28	52.6	88	0	1.32
29	55.0	72	184	2.44
30	55.7	52	151	2.42
31	60.0	67	409	4.06
32	61.8	57	165	3.88
33	65.0	88	76	3.31
34	69.3	82	0	2.64
35	77.5	91	99	3.64
36	90.8	136	163	5.14
37	97.0	137	350	3.86
38	105.8	966	368	4.21
39	181.7	232	164	7.44

The number of tractors (finally not used), was 1 in most cases and the income (result of a complete normalized system of accounts) were also available for every farm in the sample.

<u>The program</u>: Put $w_i - 1 = p_i$

$w_i \geq 1,\ i=1,2,\ldots,39 \iff p_i \geq 0$

$\Sigma w_i = 810 \quad i=1,2,3,4 \iff \Sigma p_i = 806$

$\Sigma w_i = 1010 \quad i=5,6,\ldots,24 \iff \Sigma p_i = 990$

$\Sigma w_i = 320 \quad i=25,26,\ldots,39 \iff \Sigma p_i = 305$

Variates:	acreage	h_i	(hectares)
	cattle	b_i	(bétail)
	wheat	q_i	(quintaux)
	wage-earners	s_i	(salaries)

(i quintal ≒ 2 cwt or 224 pounds)

$\Sigma w_i h_i = 65{,}415$	$i=1,2,\ldots,39$	$\Sigma p_i h_i = 63{,}608$
$\Sigma w_i b_i = 75{,}340$	" "	$\Sigma p_i b_i = 73{,}024$
$\Sigma w_i q_i = 61{,}836$	" "	$\Sigma p_i q_i = 57{,}351$
$\Sigma w_i s_i = 2{,}290$	" "	$\Sigma p_i s_i = 2{,}214$

Tractor data have been discarded. Wage-earners were finally reduced from 2,290 to 2,230. (In the last constraint, 2,214 became 2,154.)

Admissible program:

$$w_2 = 807;\ w_6 = 337,\ w_{19} = 247,\ w_{22} = 409;$$

$$w_{30} = 186,\ w_{34} = 121.$$

All other $w_i = 1$. $\Sigma w_i^2 = 461{,}335$.

A somewhat better program:

It appeared possible to lower slightly Σw_i^2 (-7,607) (all constraints being maintained) as follows:

$$w_6 \ \&\ w_{19}\text{: } -10;\ w_{22} + 6;\ w_8\text{: } +2;\ w_{15}\text{: } +10;\ w_{14} \ \&\ w_{20}\text{; } +1$$

$$w_{30} \ \&\ w_{34}\text{: } - 2;\ w_{25},\ w_{32},\ w_{33},\ w_{35}\text{: } +1$$

Matricial formulation of the program

Let M designate the following matrix:

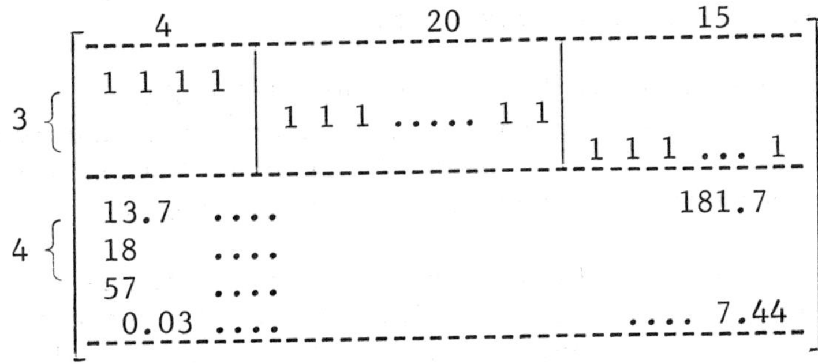

Let C be the vector

$$C = \begin{bmatrix} 806 \\ 990 \\ 305 \\ 63,608 \\ 73,024 \\ 57,351 \\ 2,214 \end{bmatrix}$$

More generally: $M = \begin{bmatrix} J \\ A \end{bmatrix}$; $C = \begin{bmatrix} N \\ S \end{bmatrix}$

with

$$J = \left[\begin{array}{c|c|c|c} IIIIIII & & & \\ & IIIIIIIIIII & & \\ & & IIIIIIIIIII & \\ & & & IIIIIII \end{array}\right]$$

Dimensions: n_i $\quad n_2 \quad n_3 \quad \ldots \quad \Sigma n_j = n$

Unknown vector: $P' = [P'_i \;\; P'_2 \;\; P'_3 \;\; \ldots\;]$

Sign constraints: $p_i \geq 0$, $i=1,2,\ldots,n$.

(contracted) $M \cdot P = C$ or $\begin{bmatrix} J \\ A \end{bmatrix} \cdot P = \begin{bmatrix} N \\ S \end{bmatrix}$

(de-contracted) $\begin{cases} J \cdot P = N \\ A \cdot P = S \end{cases}$

4. Conclusion

Presently in applied mathematics, there are some very powerful methods, which we are tempted to introduce everywhere. Examples are Poisson processes, Markov chains, Linear programming.

The attraction can become irresistible when we have only to put data in an electronic computer and look for the results. For our own part, we have for mathematical programming a pedagogical interest only.

It is to note that, in many situations, mathematical programming takes the place of sampling theory, not immediately but soon. Sampling theory is useful to understand a problem, but finally something else is necessary to obtain the solution. For example, we know the variance has to be reduced, by stratification; but we have to constitute the strata by some means or other.

Before computers or punched carts are employed, some place may be remaining for mathematical programming. In Appendix 5, some further interesting cases are mentioned.

I think the problem is not the same now as in our BELGRADE paper; and it is possible the evolution continues in a more practical direction. But some difficulties appear. In France, my friend THEODORE was afraid these researches might be dangerous, as an incentive to compute farmers' income without doing statistical surveys. He feared evaluations of this income, in this way, computed for the 500 little regions by federations of farmers, with awful political consequences. It is impossible to build a program without"bench-marks" obtained by field work; and finally there is nothing in the results that was not in the data. Our aim is only to obtain a clear presentation of heterogeneous data.

On the other hand, in many caşes,it is impossible to obtain other data, we can only destroy information, a strange procedure. Reweighting appears to be a valuable solution.

APPENDIX 1

Reference to a fictitious random sampling scheme

A "bad sample" is so constituted that we have a very small probability of obtaining the same random by drawing sampling units independently with equal probabilities.

But, (if not too bad) it is not absurd to essay the following hypothesis: Such a sample might have been selected with unequal probabilities of drawing.

In such a case, we know that unbiased estimators are obtained by reweighting sample data proportionally to the inverses of the probabilities.

In fact we do not know these probabilities when somebody gives us a "bad sample". But we are tempted to equate our weights with the inverses of probabilities. (It is difficult to know, however, how to define the distribution over the whole population and we are very far from fiducial probabilities.)

If it was stratified sampling with probabilities of inclusion Π_{hi}, we should use for $\Sigma_h \Sigma_i x_{hi}$ the estimator $\Sigma_h S_i w_{hi} x_{hi}$. An unbiased estimator of the variance is

$$\Sigma_h S_i \; x_{hi}^2 w_{hi}^2 / n_h^2$$

minus a term not depending on w_{hi}.

It would be natural to search for values (of w_{hi}'s) for functions of leading variates, in view to obtain small variances for estimation of other variates (in high correlation with the leading ones. First we attempted to minimize $\Sigma_h \lambda_h S_i w_{hi}^2$, λ_h varying with h. But there is no reason to suppose, for instance, an optimal allocation of the sample into strata (n_h proportional to $N_h . \sigma_h$). Finally this whole background of sampling theory without random selection remains here only to explain why we were interested by quadratic objective function.

We think the theory is not presently too strong; the choice of a certain objective function remains the weak point.

APPENDIX 2

Some Geometric Developments

1. A very simple allocation program:

Let x, y, z, t, be real non-negative unknown numbers under the following linear constraints:

$$\begin{aligned} x + y + z + t &= m \\ 14x + 16y + 19z + 18t &= a \\ 18x + 23y + 25z + 34t &= b \end{aligned}$$

(x,y,z,t) is said to be a point in the primal space R^4. In such a case, usually, dual space means the space R^3 with the following 4 vectors or points:

$$\begin{aligned} u &= 1 \quad 1 \quad 1 \quad 1 \\ v &= 14 \quad 16 \quad 19 \quad 18 \\ w &= 18 \quad 23 \quad 25 \quad 34 \end{aligned}$$

Consequently we shall define here a subdual space, R^2, with the following 4 vectors or points:

$$\begin{aligned} v &= 14 \quad 16 \quad 19 \quad 18 \\ w &= 18 \quad 23 \quad 25 \quad 34 \end{aligned}$$

We shall say that a total weight m is distributed over these 4 points, respectively a weight x for the first point, y for the second, z for the third and t for the last one. And the 2nd and 3rd equations have a physical significance; i.e., the center of gravity of the 4 weighted points has to be

$$\frac{a}{m}, \frac{b}{m}$$

Now, let us consider the convex hull of the 4 points; actually this is a triangle, but it may be a quadrangle, or a segment. Place on the same graph that convex hull (Γ) and the point $G(a/m, b/m)$.

1) G is not a point of Γ: no solution at all for x, y, z, t.
2) G is some vertex of Γ: only one solution (say, $x = m$, $y = z = t = 0$)
3) G is in Γ or on the boundary (vertex excepted). There are an infinity of admissible solutions (x,y,z,t) and the research for an optimal one presupposes the choice of an objective function (optimality criterion).

2. An allocation program when there are strata:

Even for our experiment on 39 farms, the problem was more difficult. With 4 "benchmark" variates, the primal space R^{39} has a R^4 subdual space. Now m would be 806 + 990 + 305 = 2101. The $p_1, p_2, \ldots, p_{39}$ are substituted for x, y, z, t. Suppose we know the optimal division of 2101 among the 39 points of sub-dual space: we cannot realize, in fact, this division because there are also the more restrictive constraints:

$$p_1+\ldots+p_4 = 806;\ p_5+\ldots+p_{24} = 990;\ p_{25}+\ldots+p_{39} = 305;$$

as well as $$p_1+\ldots+p_{39} = 2101.$$

Proposition: When an allocation program includes strata the center of gravity G remains the same. But the constraints are equivalent to the definition of a new convex set Γ^*. G can be: out of Γ^*, a vertex of Γ^*, on the boundary of Γ^*, in Γ^*. The existence of feasible solutions presupposes $G \in \Gamma^*$. Unicity presupposes G to be a vertex of Γ^*.

First of all, let us define a mixture of convex hulls.

<u>1st example</u>: Two points M and M' remain in two circles C and C'. What is the set defined by the middle point of MM'?

<u>Answer</u>: The set is a circle, the center is the middle point of the centers of C and C', the radius is the mean of radii.

<u>2nd example</u>: Substitute triangles for circles. Suppose C' obtained by a translation of C. Then any mixture of C and C' may be obtained in the same way. Suppose C' symmetric of C: it became possible to obtain hexagons.

<u>More generally</u>: Let C and C' be convex sets, M and M' being elements of those sets; a, b real positive numbers, a+b = 1. All elements aM + bM' define a set C*, we can name this a <u>mixture</u> of C and C'. C* also is convex. If the boundary of C and C' is composed of vertices, faces, etc... so is the boundary of C*. If S and S' are vertices for C and C', any vertex S* of C* is an element of $\{aS + bS'\}$. The reciprocal is not true.

<u>Mixtures of three or more convex sets</u>: Immediate extension.

3. <u>Pournin's problem</u>:

We consider a 4-dimension space with points (h,b,q,s). In this space (subdual space) every farm is a point. We distinguish 3 <u>clouds</u>, having respectively 4, 20, 15 points. (M. Divisia has coined the expression: <u>nuage de points</u>, cloud of points, for any subset of points.)

Every "cloud" defines a <u>convex hull</u>: the smallest polyhedron for which any point (of the subset) stands inside or on the boundary (a very classic notion).

Now the weights are masses attached to 39 points. Every "cloud" has a center of gravity, and this center is in the corresponding convex hull. We consider the mixture of the 3 convex hulls or sets, respectively weighted by 806/2101, 990/2101, 305/2101. The center of gravity of the 39 points is also the center of the 3 partial sets, with these same weights. There is a feasible program P if (and only if) the "bench-mark" point ($\bar{H}$, $\bar{B}$, $\bar{Q}$, $\bar{S}$) is a point of the mixed convex hull Γ^*.

Practical aspect:

1st method: Consider every point for mixture (4 x 20 x 15 = 1520 points) then find their convex hull (vertices, in fact).

2nd method: First find vertices of convex hull for the 3 clouds then proceed to the mixture of vertices; finally find the convex hull (in fact the same Γ^*).

It is doubtful whether the second way is any shorter than the first. In both cases, the problem is to find vertices of the convex hull defined by a set (cloud) of points. In R^2 , graphical methods are very efficient. In R^3 we have to study different projections of the set, together with computation of determinants in dubious cases. In R^4 we gave up.

APPENDIX 3

Unicity of the optimal program

The following point is rather trivial:

On a convex set, the objective function W'D W is minimal on only one point (but it is not possible to say the same for the maximum).

Proof: We suppose W'D W minimal on two points, W_1 and W_2 and consequently W'D W cannot be less than this minimal value r, say anywhere on the convex set.

But (definition of a convex set) W can be chosen on the segment W_1W_2, i.e., $W=aW_1 + bW_2$, $a+b = 1$.

In such a case we have:

$$W'D\,W = (a^2 + b^2)\ r + ab\ (W_1'\ D\ W_2 + W_2'\ D\ W_1)$$

or

$$W'D\,W = (a^2 + b^2)\ r + ab\ x$$

On the other hand we have:

$$z = (W_1' - W_2')\ D(W_1 - W_2) = 2\ r - x$$

so

$$x = 2\ r - z$$

and

$$\begin{aligned} W'D\,W &= (a^2 + b^2)\ r + ab\ (2\ r - z) \\ &= (a + b)^2\ r - ab\ z \\ &= r - ab\ z \end{aligned}$$

Suppose z positive definite (practically here D is a diagonal matrix, and not 0; then suppose the whole diagonal to be positive); if a or b is not 0, we obtain $W'D\,W < r$.

So it is proven that W_2 cannot be different from W_1.

APPENDIX 4

Program without inequalities - then with inequalities

Let N_h and n_h be respectively population and sample sizes in the section h; without other constraints, the weight given to sample unit (h,i) would be N_h/n_h.

Now if a condition such as $\Sigma_h S_i w_{hi} a_{hi} = A$ relating to the bench-mark variate (a) is imposed, it is necessary to correct the weight in the following way:

$$w_{hi} = (N_h/n_h) + \alpha(a_{hi} - \bar{a}_h)$$

where α is determined by $\alpha\Sigma_h S_i(a_{hi} - \bar{a}_h) = A - \Sigma_h N_h \bar{a}_h$.

In the same way, using several "bench-mark" variates a, b, c, we have

$$w_{hi} = (N_h/n_h) + \alpha(a_{hi} - \bar{a}_h) + \beta(b_{hi} - \bar{b}_h) + \gamma(c_{hi} - \bar{c}_h)$$

where α, β, γ,... are solutions of a system of linear equations.

If we decide to have $\underset{h}{\Sigma} S_i w_{hi}^2$ minimum under constraints A,B,C,... it appears that α, β, γ,... are Lagrange multipliers. It is hardly more difficult to minimize $\Sigma_h \gamma_h S_i w_{hi}^2$.

(Remark: It is possible we had some reminiscences of Bowley's 1925 paper in the Bulletin de l'Institut International de Statistique XXII, 1926, I, 51, partly reproduced in our book. Etude Théorique No. 5, I.N.S.E.E. 1953, 209-215 (out of print). However, for Bowley, the linear constraints are only approximate and data are supposed random (not the case here). This paper of Bowley's is mentioned in the Journal of the Royal Statistical Society in the Discussion following (1934) J. Neyman's well-known paper.)

Now let us consider the point W having w_{hi} as its coordinates. The constraints define for W an admissible set.

Equations: Define a linear multiplicity (hyperplane) L;

Inequalities: Define a convex cone (positive orthant) Ω.

Our purpose is to find in $L \cap \Omega$ a point $\hat{W}$, with

$$\hat{W}'D\,\hat{W} < W'D\,W, \quad W \in L \cap \Omega.$$

We have just obtained W^o with:

$$W^{o\prime} D\, W^o < W' D\, W, \quad W \in L \quad .$$

When $W^o \in \Omega$, then $W^o = \hat{W}$. Now suppose this is not the case. We have

$$W^{o\prime} D\, W^o < \hat{W}' D\, \hat{W} \quad (\hat{W} \text{ being unknown}).$$

Suppose the set $L \cap \Omega$ be non-void, so that some W exist. We have

$$W' D\, W - W^{o\prime} D\, W^o = (W-W^o)' D\, (W-W^o) + W^{o\prime} D (W^o-W) + (W^o-W)' D W^o$$

Now it is useful to choose D (diagonal matrix) in such a way that the last two terms are 0; in fact it is sufficient to put

$$W^{o\prime} D\, (W^o - W) = 0$$

Put $\qquad (11 \ldots 1)\, D = A' \iff A'(W^o-W) = 0$

But W^o and W both are in L and, consequently $M\,(W^o-W)=0$. So A is defined by linear combinations of columns of M'; in particular all terms in A can be 1 (i.e., D may be the matrix I.

Having

$$W' D\, W = W^{o\prime} D\, W^o + (W-W^o)' D\, (W-W^o)$$

we have also

$$(\hat{W}-W^o)' D\, (\hat{W}-W^o) < (W-W^o)' D (W-W^o)$$

So $\hat{W}$ can only be on the <u>boundary</u> of the convex set $L \cap \Omega$.

So $\hat{W}$ is an element of a subset (0 V), where 0 represents some null components (on the boundary of Ω).

Practically it looks possible to try first $w_{hi}=1$ (or p = 0) for every negative component of W^o, then every pair of these negative components, etc...(supposing these are rather few).

APPENDIX 5

Sampling theory and mathematical programming

This is by no means the first irruption of Mathematical Programming in Sample Survey Theory. Let us recall some other cases.

Stratification:

Optimal stratification (Dalenius) is defined from an infinite parametric population point of view; another' possibility is to deal with a finite moderately-sized population and to sort every sampling unit so as to constitute H strata minimizing the variance for an auxiliary variate. This operation - named pooling or grouping - was studied by Fisher (Walter D.), Cox, Ward, Lazar (in France) in another context. Of course, this method of stratification would be too costly with populations of great size.

Optimum Allocation of sample in strata:

In Neyman's optimum allocation (as in our Belgrade paper), we have to minimize an objective non-linear function subject to a linear equation (the total size or cost being imposed). Here the sample size in each stratum is unknown, a difference having some importance for the high number of unknown variates constitutes a real practical difficulty in the present problem. Lagrange's multipliers are used in both cases. Now it is possible for Neyman's agrument to be inapplicable. In some stratum, the "optimal" sample would be greater than the whole stratum. Of course, this is no practical difficulty; for it is very easy to obtain a second optimal allocation, these strata being excluded.

In recent papers, Kokan, Jagannathan, Folks and Antle, Khan, have solved various problems on optimal stratification and we should not forget earlier work such as allocation between first and second-stage operations.

It is convenient to take the inverses of unknown sizes as new variates; limitations on variances become linear constraints, research for minimal cost gives a convex objective function. The solution components have to be positive. This is a convex program problem.

Goodman and Kish problem:

We refer to our own paper, reproduced in *Estadistica* (1961). It is possible (but not very practicable) to reduce the variance of a stratified sampling (estimation remaining unbiased) by non-independent selections in strata, so as to add negative covariances (between strata) to positive variances (within strata). Real assay was made toward 1950 at the I.N.S.E.E. But only several years after did we realize that optimal procedure would be given by convex programming.

Here the objective function has terms of both signs; it is the variance of the stratified estimator of the mean of a leading variate. The population is distributed into cells of a rectangular table, whose rows are strata and columns false strata. We aim to obtain a stratified sample which is balanced in regard to the false strata. Thus it is necessary to attribute a certain probability of choice to every cell. Unbiasedness of estimators (whatever the variate be) implies a set of linear constraints relating these probabilities which, of course, have also to be positive.

REFERENCES

Cox, D. R. (1957). Note on grouping, Jour. Amer. Statist. Assoc. 52, 543-547.

Fisher, W. D. (1953). On a pooling problem from the statistical view point, Econometrica 21, 567-585.

Fisher, W. D. (1958). On grouping for maximum homogeneity, Jour. Amer. Statist. Assoc. 53, 789-798.

Folks, J. L. and Antle, C. E. (1965). Optimum allocation of sampling units to strata, when there are R responses of interest, Jour. Amer. Statist. Assoc. 60, 225-233.

Jagannathan, R. (1965). A method for solving a nonlinear programming problem in sample surveys, Econometrica 33, 841-846.

Kokan, A. R. (1963). Optimum allocation in multivariate surveys, Jour. Royal Statist. Soc. A 126, 557-565.

Kokan, A. R. and Khan, Sanaullah (1967). Optimum allocation in multivariate surveys: An analytical solution, Jour. Royal Statist. Soc. b 29, 115-125.

Lazar, Ph. (1966). Partition d'un groupe hétérogène en sous-groupes homogènes, Revue de Statist, Appli. 14, 39-43.

Pournin, M. (1967). Utilisation des échantillons des centres de gestion pour l'établissement des revenus régionaux en agriculture, Jour. Soc. Statist. de Paris 108, 245-289 (with discussion).

Pournin, M. (unpublished). L'utilisation d'un "mauvais" échantillon pour déterminer le revenu d'une population d'exploitations agricoles, -mémoire présenté le 7-2-1967 pour l'obtention du Diplôme d'Aptitude de l'I.S.U.P.

Thionet, P. (1953). La théorie des Sondages, p. 177, in Etudes Théoriques de l'I.N.S.E.E., No. 5, Paris.

Thionet, P. (1960). Quelques aspects de la théorie des Sondages, _Jour. Soc. Statist. de Paris_ 101, 99-111. Reprint in _Estadistica_ (1961).

Thionet, P. A. (1965). Une façon d'exploiter un mauvais échantillon, _Bull. Intern. Statist. Institute_, 35th Session, Belgrade 41, 871-874.

Ward, J. H. (1963). Hierarchical grouping to optimize an objective function, _Jour. Amer. Statist. Assoc._ 58, 236-244.

THE VALUE OF PRIOR INFORMATION

Norman R. Draper and Irwin Guttman

University of Wisconsin

Summary

Suppose we have a sample $y_1, y_2, \ldots, y_n$ of observations from some specified parent distribution, together with some prior information on one or more parameters of interest of the distribution. They by applying Bayes theorem we can obtain the overall distribution (or the marginal distribution if nuisance parameters are involved) of parameters of interest, and can make inferences about these parameters. A natural question to ask is how inferences might be affected by a change in the prior distribution, i.e. by the availability of more, or less, information on the parameters. It would be convenient to have a quantity which could be used to indicate the value of more or less prior information.

We suggest, in this paper, for the one parameter case, the use of the ratio of the expected values of the posterior variance of the parameter, expectation being taken over the future distribution of the sampling statistics which occur in the posterior variance. We illustrate the use of this ratio for a number of well-known distributions.

Cases involving more than one parameter will be treated by taking the ratio of the determinants of the expected value of the posterior variance-covariance matrix, again expectations being taken over the future distribution of the sampling statistics involved. This procedure is illustrated in Section 2.

1. Introduction

The prior information in a Bayesian analysis may arise in a number of ways. For example, it might arise from the first phase of a two-phase sampling scheme (discussed by Draper and Guttman, 1968), might be formulated as a result of expert opinion (Winkler, 1967), or might be contained in plant records which are on file. When we apply Bayes' theorem to combine prior and sample information on a parameter, we are often interested in assessing the value of the prior information and in answering questions such as

(a) How useful is the prior information and is it in fact of any value at all, i.e. does it contribute at all to the precision of the estimator obtained via the posterior?

(b) If no more sample information is obtainable at present, is it worthwhile to obtain extra "prior" information, for example by examining additional plant records, so that some desired precision of estimation is attained?

Examining past records would, of course, entail additional expense which would have to be balanced against the value of any increase in precision.

Our preliminary analysis in this paper will take the following form. Suppose we are interested in a vector parameter $\underline{\theta}$; we can then

1. Obtain a posterior distribution for $\underline{\theta}$ by combining prior and sample information.

2. Find the posterior variance-covariance matrix of $\underline{\theta}$, say, $\underline{V}(\underline{\theta})$.

3. Find det $E\underline{V}(\underline{\theta})$, expectation being taken over the future (preposterior) distribution of the sample observations.

4. See how det $E\underline{V}(\underline{\theta})$ is affected by changing the prior distribution in a specified manner.

In Section 2, we shall consider the general linear regression situation with homogeneous variance. This will enable the Normal distribution ($N(\mu,\sigma^2)$) situations to be derived as special cases. We shall also consider some other distributional situations in subsequent sections.

2. Linear Regression Model

General Case.

Let $\underline{y} = (y_1, y_2, \ldots, y_n)'$ be an $n \times 1$ vector of observations subject to the model $\underline{y} = \underline{X}\underline{\beta} + \underline{\epsilon}$ where $\underline{X}$ is $n \times p$ and $\underline{\beta}$ is $p \times 1$ and where $\underline{\epsilon} \sim N(o, \underline{I}\sigma^2)$. We shall assume that prior information on the parameters $\underline{\beta}$ and σ^2 is abailable in the form of a conjugate prior (Raiffa and Schlaiffer, 1961),

$$p(\underline{\beta},\sigma^2) = \frac{|\underline{C}_o|^{\frac{1}{2}} \omega_o^{\frac{n_o - p}{2}}}{\pi^{p/2} 2^{\frac{n_o}{2}} \Gamma(\frac{n_o - p}{2})} \quad \frac{\exp\{-\frac{1}{2\sigma^2}[\omega_o + (\underline{\beta}-\underline{b}_o)'\underline{C}_o(\underline{\beta}-\underline{b}_o)]\}}{(\sigma^2)^{\frac{n_o}{2}+1}}, \tag{2.1}$$

where $\underline{C}_o$, $_o$, and n_o are constants assumed given. (The rationale behind this representation of prior information is that it has a reasonable form which would arise a posteriori if n_o observations subject to the model $\underline{y}_o = \underline{X}_o\underline{\beta} + \underline{\epsilon}$ were taken, $\underline{\epsilon} \sim N(o, \underline{I}\sigma^2)$ and if the prior information available were represented by independent locally uniform priors in $\underline{\beta}$ and $\log \sigma^2$, where $\underline{b}_o = (\underline{X}_o'\underline{X}_o)^{-1}\underline{X}_o'\underline{y}_o$, $\underline{C} = \underline{X}_o'\underline{X}_o$ and $\omega_o = (\underline{y}_o - \underline{X}_o\underline{b}_o)'(\underline{y}_o - \underline{X}_o\underline{b}_o)$. Thus, in a very real sense, the conjugate prior (2.1) represents "n_o units of information" on the parameters β and σ^2. Alternatively, $\underline{C}_o$, ω_o and n_o can be simply regarded as constants appearing in the prior distribution (2.1).)

The likelihood can be written as

$$\ell(\underline{\beta},\sigma^2 | \underline{y}) = \frac{1}{(2\pi)^{n/2}} \frac{\exp\{-\frac{1}{2\sigma^2}[\omega + (\underline{\beta}-\underline{b})'\underline{C}(\underline{\beta}-\underline{b})]\}}{(\sigma^2)^{n/2}} \tag{2.2}$$

where $\underline{C} = \underline{X}'\underline{X}$, $\underline{b} = \underline{C}^{-1}\underline{X}'\underline{y}$ and $= (\underline{y}-\underline{X}\underline{b})'(\underline{y}-\underline{X}\underline{b})$. Combining the prior and likelihood we obtain the posterior

$$p(\underline{\beta},\sigma^2|\underline{y}) = k(\sigma^2)^{-(\frac{n+n_o}{2}+1)} \exp\{-(Q_1+Q_2)/(2\sigma^2)\}$$

where

$$k = \frac{|\underline{C}+\underline{C}_o|^{\frac{1}{2}} Q_2^{\frac{n+n_o-p}{2}}}{2^{\frac{1}{2}(n+n_o)}\Gamma(\frac{n+n_o-p}{2})\pi^{p/2}} \tag{2.3}$$

and where, completing the square in $\underline{\beta}$ in the exponent, we can write

$$Q_1 = (\underline{\beta}-\underline{a})'(\underline{C}+\underline{C}_o)(\underline{\beta}-\underline{a}), \quad \underline{a} = (\underline{C}+\underline{C}_o)^{-1}(\underline{C}\underline{b}+\underline{C}_o\underline{b}_o),$$

$$Q_2 = \omega_o + +\underline{b}'\underline{C}\underline{b}+\underline{b}'\underline{C}_o\underline{b}_o-(\underline{b}'\underline{C}+\underline{b}'\underline{C}_o)(\underline{C}+\underline{C}_o)^{-1}(\underline{C}\underline{b}+\underline{C}_o\underline{b}_o). \tag{2.4}$$

By integrating out σ^2 we obtain the marginal posterior of $\underline{\beta}$ as proportional to

$$\left\{1 + \frac{t'Rt}{n+n_o-p}\right\}^{-\frac{1}{2}\{p+(n+n_o-p)\}}, \tag{2.5}$$

where $\underline{t} = \underline{\beta} - \underline{a}$ is a p-dimensional multivariate -t variable with variance-covariance matrix

$$R^{-1} = (\underline{C}+\underline{C}_o)^{-1} Q_2/(n+n_o-p-2). \tag{2.6}$$

Similarly, we can obtain the marginal distribution of σ^2 by integrating out $\underline{\beta}$. Since

$$\int \exp\{-Q_1/(2\sigma^2)\}d\underline{\beta} = |\underline{C}+\underline{C}_o|^{-\frac{1}{2}}(2\pi\sigma^2)^{p/2}, \qquad (2.7)$$

$$p(\sigma^2|\underline{y}) = \frac{Q_2^{\frac{n+n_o-p}{2}}}{2^{\frac{1}{2}(n+n_o-p)}\Gamma(\frac{n+n_o-p}{2})}(\sigma^2)^{-(\frac{n+n_o-p}{2}+1)}\exp\left\{\frac{Q_2}{2\sigma^2}\right\}. \qquad (2.8)$$

It follows that

$$E(\sigma^2)^{2r} = \frac{Q_2^r}{2^r}\frac{\Gamma(\frac{n+n_o-p-2r}{2})}{\Gamma(\frac{n+n_o-p}{2})} \qquad (2.9)$$

from which

$$V(\sigma^2) = \frac{2Q_2^2}{(n+n_o-p-2)^2(n+n_o-p-4)} . \qquad (2.10)$$

Furthermore it is easy to verify that $E\{\sigma^2(\beta_j-a_j)\} = 0$. It follows that the variance-covariance matrix of the vector $(\underline{\beta},\sigma^2)'$ is

$$\underline{V} = \begin{bmatrix} \underline{R}^{-1} & \underline{0} \\ \underline{0} & V(\sigma^2) \end{bmatrix} , \qquad (2.11)$$

where $\underline{R}$ is given by (2.6) and $V(\sigma^2)$ by (2.10). The terms of this matrix contain the sample quantities $\underline{b}$ and ω through Q_2. We now take the expectation of (2.11) with respect to the future distribution of $\underline{b}$ and ω, $h(\underline{b},\omega)$. We can write

$$h(\underline{b},\omega) = \int\int p(b,\omega|\underline{\beta},\sigma^2)\,p(\beta,\sigma^2). \qquad (2.12)$$

It is well known that $\underline{b} \sim N(\underline{\beta}, \underline{C}^{-1}\sigma^2)$ and that $\omega \sim \sigma^2\chi^2_{n-p}$ and that $\underline{b}$ and ω are independent. Combining this with (2.1) we obtain

$$h(\underline{b},\omega) = \iint \frac{|C|^{\frac{1}{2}}}{(2\pi)^{p/2}(\sigma^2)^{p/2}} \exp\{-\frac{1}{2\sigma^2}(\underline{b}-\underline{\beta})'\underline{C}(\underline{b}-\underline{\beta})\} \times$$

$$\frac{\omega^{\frac{n-p}{2}} e^{-\frac{\omega}{2\sigma^2}}}{2^{\frac{n-p}{2}}\Gamma(\frac{n-p}{2})(\sigma^2)^{\frac{n-p}{2}}} \times$$

$$\frac{|C_o|^{\frac{1}{2}}\omega_o^{\frac{n_o-p}{2}}}{\pi^{\frac{1}{2}p}2^{\frac{1}{2}n_o}\Gamma(\frac{n_o-p}{2})(\sigma^2)^{\frac{n_o}{2}+1}} \exp\{-\frac{1}{2\sigma^2}[\omega_o+(\underline{\beta}-\underline{b}_o)'\underline{C}_o(\underline{\beta}-\underline{b}_o)]\}d\underline{\beta}d\sigma^2$$

$$= k'\iint(\sigma^2)^{-(\frac{n+n_o}{2}+1)} \exp\{-\frac{1}{2\sigma^2}(Q_1+Q_2)\}d\underline{\beta}\ d\sigma^2 \tag{2.13}$$

where

$$k' = \frac{|C|^{\frac{1}{2}}|C_o|^{\frac{1}{2}}\omega_o^{\frac{n_o-p}{2}}\omega^{\frac{n-p}{2}-1}}{2^{\frac{n+n_o}{2}}\pi^p\Gamma(\frac{n-p}{2})\Gamma(\frac{n_o-p}{2})}. \tag{2.14}$$

Integrating with respect to $\underline{\beta}$ we have, from multivariate Normal properties,

$$h(\underline{b},\omega) = k' \frac{(2\pi)^{p/2}}{|\underline{C}+C_o|^{\frac{1}{2}}} \int (\sigma^2)^{-(\frac{n+n_o-p}{2}+1)} \exp\left\{-\frac{Q_2}{2\sigma^2}\right\} d\sigma^2$$

$$= k' \frac{(2\pi)^{p/2}}{|C+C_o|^{\frac{1}{2}}} \frac{\Gamma(\frac{n+n_o-p}{2})\, 2^{\frac{1}{2}(n+n_o-p)}}{Q_2^{\frac{n+n_o-p}{2}}} . \qquad (2.15)$$

Since this density must integrate to 1 it follows that

$$\iint \frac{\omega^{\frac{n-p}{2}-1}}{Q_2^{\frac{n-p+n_o}{2}}} d\underline{b}\; dw = \frac{1}{k''} \frac{|\underline{C}+\underline{C}_o|^{\frac{1}{2}} \pi^{p/2} \Gamma(\frac{n-p}{2}) \Gamma(\frac{n_o-p}{2})}{|\underline{C}|^{\frac{1}{2}} |C_o|^{\frac{1}{2}} \omega_o^{\frac{n-p}{2}} \Gamma(\frac{n-p+n_o}{2})} \qquad (2.16)$$

for $n+n_o-p-1 > 0$ and $n_o-p-2 > 0$. Moments of Q_2 are thus easily obtained since $E(Q_2^r)$ is k" times (the right hand si e of (2.16) with (n_o-2r) replacing n_o). In fact

$$E(Q_2^s) = \frac{(n-p+n_o-2)(n-p+n_o-4)\ldots(n-p+n_o-2s)}{(n_o-p-2)(n_o-p-4)\ldots(n_o-p-2s)} \omega_o^s, \qquad (2.17)$$

provided that $n_o-p-2s > 0$. We propose, and will use as an overall measure of variation, the quantity

$$\det E(\underline{V}) = \{\det E(\underline{R}^{-1})\}\{EV(\sigma^2)\}$$

$$= \frac{1}{|\underline{C}+\underline{C}_o|} \left\{\frac{E(Q_2)}{n+n_o-p-2}\right\}^p \times \frac{2E(Q_2^2)}{(n+n_o-p-2)^2(n+n_o-p-4)}$$

$$= \frac{1}{|\underline{C}+\underline{C}_o|} (\sigma_o^2)^p \times \frac{2(\sigma_o^2)^2(n_o-p-2)}{(n+n_o-p-2)(n_o-p-4)}, \tag{2.18}$$

where $\sigma_o^2 = \omega_o/(n_o-p-2)=(n_o-p)\, s_o^2/(n_o-p-2)$ is the prior mean of σ^2,

$$= \frac{1}{|\underline{C}+\underline{C}_o|} \frac{2(n_o-p-2)(\sigma_o^2)^{p+2}}{(n+n_o-p-2)(n_o-p-4)}. \tag{2.19}$$

(See also Kiefer, 1968, p. 681.)

Thus by taking the ratio of "(2.19) with n_o inserted" and "(2.19) with kn_o inserted for n_o" we can obtain a measure of the value of kn_o "units of prior information" compared with "n_o" such units, and thus see how the precision of the parameters <u>a posteriori</u> is affected by changes in the quality of the prior information. We call this ratio an information ratio and we will denote it by R. Note that since $\underline{\beta}$ and σ^2 are uncorrelated, results for $\underline{\beta}$ (with σ^2 unknown) and for σ^2 (with $\underline{\beta}$ unknown) are obtained, respectively, from the left and right portions of (2.18).

<u>Special case, $N(\mu,\sigma^2)$.</u>

It is easy to derive results for the $N(\mu,\sigma^2)$ distribution by setting p=1, $\underline{C}_o=n_o$, $\underline{C}=n$, $\underline{X}=\underline{1}$, $\beta=\mu$ in (2.18). We obtain

$$EV(\mu) = \frac{\sigma_o^2}{n+n_o}, \text{ if } \sigma^2 \text{ is unknown,} \qquad (2.20)$$

$$EV(\sigma^2) = \frac{2(n_o-3)\sigma_o^4}{(n_o-5)(n+n_o-3)}, \text{ if } \mu \text{ is unknown,} \qquad (2.21)$$

since $\underline{\beta}$ and σ^2 are uncorrelated as we have said. Thus if σ^2 is unknown and the prior mean for σ^2, namely σ_o^2, is the same for both the prior distributions involved, "information ratios" of

$$R_\mu = \frac{n+kn_o}{n+n_o} \qquad (2.22)$$

and

$$R_{\sigma^2} = \frac{(n_o-3)(kn_o-5)(n+kn_o-3)}{(kn_o-3)(n_o-5)(n+n_o-3)} \qquad (2.23)$$

for μ, if σ^2 is unknown, and for σ^2, if μ is unknown, respectively are obtained. For both μ and σ^2 considered together, results using the products of (2.20) and (2.21) and of (2.22) and (2.23) apply.

What can we infer from results such as these? As an example consider the case n_o=20. Then we can write

$$R = \frac{17}{15(n+17)} \{20k+(n-3)\}\{1-\frac{2}{20k-3}\} .$$

If n is specified this equation is almost linear in k so that it will be, almost, a straight line with slope 340/{15(n+17)} through the point (R,k)=(1,1). The smaller the value of n, the greater will be the slope, i.e., the more valuable will be the addition of extra prior information. Specifically, for example, when n=3, the use of kn_o=20 units, allows estimation of σ^2 to be 5.55 times as precise; or, the use of kn_o=2(20)=40, units instead of 20, allows the precision of σ^2 to be increased by a factor of 2.14. As n increases, smaller increases in precision are achieved as can be easily calculated. Table 1 provides some figures for the increase in precision for $n_o=20$ and for various values of n and k. Similar calculations can be performed for other values of n_o.

Table 1. Increase in precision when kn_o, instead of n_o, units of prior information are available, for $n_o=20$.

n \ k	2	3	4	5
3	2.14	3.28	4.42	5.55
5	2.05	3.08	4.12	5.15
6	2.00	3.00	3.98	4.97
10	1.87	2.71	3.56	4.40
15	1.74	2.46	3.17	3.89
20	1.65	2.28	2.90	3.51
100	1.26	1.47	1.67	1.87

Another way of using the posterior variance (see, for example, Raiffa and Schlaifer, 1961, p. 125) is to plot v(n), $v=EV(\sigma^2)/\sigma_o^4$ against n_o for various values of n. Suppose such a plot is made. Then for any n_o we can read off the comparative precisions of taking, say, n=10 observations versus n=5. For example, when $n_o=10$, v(10)=0.16 and v(5)=0.23 to two places of decimals. Some idea of the sort of improvements which can occur by changing n, for various values of n_o, can be obtained by referring to Table 2. As one would expect, the more prior information is available, the less worth while it is comparatively to take additional observations to increase the precision.

This concludes our illustrative numerical work. In the remainder of the paper we evaluate the appropriate variance function for the parameters of some standard distributions. Numerical calculations similar to those above can also be carried out, of course.

Table 2. Values of v(n) for various n_o and n.

n \ n_o	6	10	15	20
5	.75	.23	.14	.10
10	.46	.16	.11	.08
15	.33	.13	.09	.07
20	.26	.10	.08	.06
100	.06	.03	.02	.02

Cases where some parameters are known

Reverting to the general regression situation we can consider the special cases (i) σ^2 known; we can set $\sigma^2=1$ without loss of generality (ii) $\underline{\beta}$ known; we can set $\underline{\beta}=0$ without loss of generality. When σ^2 is known the appropriate conjugate prior for $\underline{\beta}$ is

$$p(\underline{\beta}) = \frac{|C_o|^{\frac{1}{2}}}{(2\pi)^{p/2}} \frac{\exp\{-\frac{1}{2\sigma^2}[(\underline{\beta}-\underline{b}_o)'\underline{C}_o(\underline{\beta}-\underline{b}_o)]\}}{(\sigma_o^2)^{p/2}}, \tag{2.24}$$

while the likelihood is

$$\ell(\underline{\beta}|\underline{y},\sigma^2) = \frac{1}{(2\pi)^{n/2}} \frac{\exp\{-\frac{1}{2\sigma^2}[\omega+(\underline{\beta}-\underline{b})'\underline{C}(\underline{\beta}-\underline{b})]\}}{(\sigma^2)^{n/2}}. \tag{2.25}$$

"Why use σ_o^2 in (2.24) when we know σ^2 in (2.25)?", it may be asked. This course of action would be appropriate in a situation such as the following. An experimenter has n_o "units of information" obtained by method A which has a standard deviation of σ_o and wishes to perform n further experiments employing method B, which has a standard deviation of σ. He now would like to know how useful it might be to examine past records further to obtain additional prior information, i.e., to increase n_o to $kn_o (k>1)$. Thus the posterior is proportional to

$$\exp\left\{-\tfrac{1}{2}(\underline{\beta}-\underline{c})'\left(\frac{\underline{C}}{\sigma^2}+\frac{\underline{C}_o}{\sigma_o^2}\right)(\underline{\beta}-\underline{c})\right\} \tag{2.26}$$

where

$$\underline{c} = \left(\frac{\underline{C}}{\sigma^2}+\frac{\underline{C}_o}{\sigma_o^2}\right)^{-1}\left(\frac{\underline{C}}{\sigma^2}\,\underline{b}+\frac{\underline{C}_o}{\sigma_o^2}\,b_o\right) \tag{2.27}$$

i.e., a multivariate Normal distribution with variance-covariance matrix

$$\left(\frac{\underline{C}}{\sigma^2}+\frac{\underline{C}_o}{\sigma_o^2}\right)^{-1}.$$

Thus for the $N(\mu,\sigma^2)$ case we have

$$E(V(\mu)) = \left(\frac{n}{\sigma^2}+\frac{n_o}{\sigma_o^2}\right)^{-1} = \frac{\sigma_o^2}{n\frac{\sigma_o^2}{\sigma^2}+n_o} \tag{2.28}$$

when σ^2 is known. We can compare this with (2.20), namely $\sigma_o^2/(n+n_o)$ when σ^2 is unknown. Note that, when $\sigma^2 < \sigma_o^2$, (2.28) < (2.20) i.e. knowledge of σ^2 (but knowledge that it is less than the σ_o^2 of the prior information) leads to a smaller expected variance. However, if $\sigma^2 > \sigma_o^2$, (2.28) > (2.20); i.e. knowledge of σ^2 (but knowledge that it exceeds σ_o^2 of the prior information) results in less precise knowledge of μ than if we did not know σ^2 at all. This seemingly paradoxical result is reminiscent of that mentioned by Stuart (1955), but is somewhat different. When $\sigma^2 = \sigma_o^2 = 1$, (2.28) reduces to $(n+n_o)^{-1}$ and in this case we have, for the ratio of variances of μ when n_o and kn_o "units of prior information" are available,

$$R'_\mu = \frac{n+kn_o}{n+n_o} . \tag{2.29}$$

When $\underline{\beta}=0$, the appropriate conjugate prior for σ^2 is the inverted χ^2 distribution

$$p(\sigma^2|\omega_o) = \frac{\omega_o^{n_o/2} \exp\{-\omega_o/2\sigma^2\}}{2^{n_o/2}\,\Gamma(\frac{n_o}{2})\,(\sigma^2)^{\frac{n_o}{2}+1}} \tag{2.30}$$

with mean $\sigma_o^2 = \omega_o/(n_o-2)$. The likelihood is

$$\frac{1}{(2\pi)^{n/2}} \; \frac{\exp\{-\frac{1}{2\sigma^2}(\Sigma y_i^2)\}}{(\sigma^2)^{n/2}} \tag{2.31}$$

so that the posterior is, with $\omega = \Sigma y_i^2$,

$$p(\sigma^2|\underline{y}) = \frac{(\omega+\omega_o)^{\frac{n+n_o}{2}}}{2^{\frac{n+n_o}{2}}\,\Gamma(\frac{n+n_o}{2})} \cdot \frac{\exp\{-(\omega+\omega_o)/2\sigma^2\}}{(\sigma^2)^{\frac{n+n_o}{2}+1}} . \tag{2.32}$$

Clearly,

$$E(\sigma^2)^r = \frac{\Gamma(\frac{n+n_o-2r}{2})\,(\omega+\omega_o)^r}{2^r\,\Gamma(\frac{n+n_o}{2})} \tag{2.33}$$

whereupon

$$V(\sigma^2) = \frac{2(\omega+\omega_o)^2}{(n+n_o-2)^2(n+n_o-4)} . \tag{2.34}$$

For the future distribution of ω, we have

$$h(\omega|\omega_o) = \int f(\omega|\sigma^2)\, p(\sigma^2|\omega_o)$$

$$= \frac{\omega_o^{n_o/2}\, n^{n/2}}{2^{\frac{n+n_o}{2}}\,\Gamma(\frac{n}{2})\Gamma(\frac{n_o}{2})} \int_o^\infty \frac{1}{(\sigma^2)^{\frac{n+n_o}{2}+1}} \exp\{-\frac{\omega+\omega_o}{2\sigma^2}\}\, d\sigma^2$$

$$= \frac{\Gamma(\frac{n+n_o}{2})\,\omega_o^{n_o/2}\,\omega^{\frac{n}{2}-1}}{\Gamma(\frac{n}{2})\Gamma(\frac{n_o}{2})\{\omega+\omega_o\}^{\frac{n+n_o}{2}}} . \tag{2.35}$$

Since this integrates to 1, it is clear that

$$E(\omega+\omega_o)^2 = \frac{(n+n_o-2)(n+n_o-4)}{(n_o-2)(n_o-4)}\,\omega_o^2 = \frac{(n_o-2)(n+n_o-2)(n+n_o-4)\sigma_o^4}{n_o-4} \tag{2.36}$$

Thus

$$EV(\sigma^2) = \frac{2(n_o-2)\sigma_o^4}{(n_o-4)(n+n_o-2)} . \tag{2.37}$$

Note that this is like (2.21) but with n_o replacing n_o-1. Thus knowledge of the mean in this situation is worth one prior observation. It follows that, if the prior mean σ_o^2 is the same for both prior distributions involved when there are "no units of prior information" or "kn_o units of prior information", the "information ratio" is

$$R'_{\sigma^2} = \frac{n_o-2}{kn_o-2} \cdot \frac{(kn_o-4)(n+kn_o-2)}{(n_o-4)(n+n_o-2)} . \tag{2.38}$$

Two normal distributions, regression case, $r=\sigma_1^2/\sigma_2^2$.

Suppose we have two normal populations denoted by subscripts 1 and 2 and are interested in the ratio $r = \sigma_1^2/\sigma_2^2$. Using (2.8), and with an obvious notation, the joint posterior distribution of σ_1^2 and σ_2^2 is

$$\frac{Q_{21}^{\frac{n_1+n_{10}-p}{2}} Q_{22}^{\frac{n_2+n_{20}-p}{2}}}{2^{(n_1+n_2+n_{10}+n_{20}-2p)/2}\,\Gamma(\frac{n_1+n_{10}-p}{2})\Gamma(\frac{n_2+n_{20}-p}{2})}$$

$$(\sigma_1^2)^{-\left\{\frac{n_1+n_{10}-p}{2}+1\right\}}(\sigma_2^2)^{-\left\{\frac{n_2+n_{20}-p}{2}+1\right\}}\exp\left\{-\frac{Q_{21}}{2\sigma_1^2}\right\}\exp\left\{-\frac{Q_{22}}{2\sigma_2^2}\right\}. \tag{2.39}$$

Set $r = \sigma_1^2/\sigma_2^2$ and $s = \sigma^2$ so that $\sigma_1^2 = rs$ and $\sigma_2^2 = s$, and the Jacobian $J\{(\sigma_1^2\sigma_2^2)/(r,s)\}=s$. Then the joint distribution of r and s is proportional to

$$r^{-\left\{\frac{n_1+n_{10}-p}{2}+1\right\}}s^{-\left\{\frac{n_1+n_2+n_{10}+n_{20}-2p}{2}+1\right\}}\exp\left\{-\frac{1}{2s}\left[\frac{Q_{21}}{r}+Q_{22}\right]\right\}. \tag{2.40}$$

Integrating out s using a gamma integral provides the marginal posterior distribution of r proportional to

$$r^{\frac{n_2+n_{20}-p}{2}-1} \Big/ \left\{1+\frac{Q_{22}}{Q_{21}}r\right\}^{\left[\frac{n_2+n_{20}-p}{2}\right]+\left[\frac{n_1+n_{10}-p}{2}\right]} \tag{2.41}$$

whereupon it follows that

$$\frac{Q_{22}}{Q_{21}}r = \frac{n_2+n_{20}-p}{n_1+n_{10}-p}F \tag{2.42}$$

where F is an F-variable with $(n_2+n_{20}-p)$ and $(n_1+n_{10}-p)$ degrees of freedom. Using known results for the mean and variance of the F-distribution we obtain

$$E(r) = \frac{Q_{21}}{Q_{22}}\,\frac{n_2+n_{20}-p}{n_1+n_{10}-p-2} \tag{2.43}$$

$$V(r) = \left(\frac{Q_{21}}{Q_{22}}\right)^2 \frac{2(n_2+n_{20}-p)(n_1+n_2+n_{10}+n_{20}-2p-2)}{(n_1+n_{10}-p-2)(n_1+n_{10}-p-4)}\,. \tag{2.44}$$

We now wish to take the expectation of V(r) with respect to the distributions of Q_{21} and Q_{22}. Since Q_{21} and Q_{22} are independent, the numerator and denominator can be handled separately, using (2.16). In fact $E(Q_{21}^2)$ follows from (2.17) with s=2, while

$$E(Q_{22}^{-2}) = \frac{(n_{20}-p+2)(n_{20}-p)}{(n_2+n_{20}-p+2)(n_2+n_{20}-p)}\,. \tag{2.45}$$

Combining these results we find

$$EV(r)=\frac{2(n_1+n_2+n_{10}+n_{20}-2p-2)(n_{20}-p+2)(n_{10}-p-2)(n_{20}-p)}{(n_1+n_{10}-p-2)(n_2+n_{20}-p+2)(n_{20}-p-2)^2(n_{10}-p-4)}\cdot\frac{\sigma_{10}^4}{\sigma_{20}^4} \tag{2.46}$$

where $\sigma_{io}^2=\omega_{io}/(n_{io}-p-2)$ is the prior mean of σ_i^2. The "information ratio" can be calculated in the usual way.

When $\underline{\beta}$ is known, we can take $\underline{\beta}=0$ without loss of generality and set p=0 in (2.46).

Two Normal distributions, difference between means $\mu_1-\mu_2$.

When we are sampling from two Normal distributions $N(\mu_1,\sigma_1^2)$ and $N(\mu_2,\sigma_2^2)$ we can derive results for the difference between means very simply. For example, for the difference between two means we have, since μ_1,μ_2 are independent,

$$EV(\mu_1-\mu_2) = \frac{\sigma_{10}^2}{n_1+n_{10}} + \frac{\sigma_{20}^2}{n_2+n_{20}}$$

with an obvious notation-see(2.20)-when σ_1^2 and σ_2^2 are unknown, or, when σ_1^2 and σ_2^2 are known,

$$EV(\mu_1-\mu_2) = \left(\frac{n_1}{\sigma_1^2}+\frac{n_{10}}{\sigma_{10}^2}\right)^{-1}+\left(\frac{n_2}{\sigma_2^2}+\frac{n_{20}}{\sigma_{20}^2}\right)^{-1}.$$

"Information ratios" can be calculated as usual.

3. The Binomial Distribution

<u>One binomial, θ</u>.

Suppose the variable r is binomially distributed with parameter θ. Then if a sample of n observations is taken, the likelihood is

$$\ell(r|\theta) = \frac{\Gamma(n+1)}{\Gamma(r+1)\Gamma(n-r+1)} \theta^{r}(1-\theta)^{n-r}. \qquad (3.1)$$

We can take, for the form of the prior information on θ, the conjugate prior beta distribution

$$p(\theta|n_o,r_o) = \frac{\Gamma(n_o+2)}{\Gamma(r_o+1)\Gamma(n_o-r_o+1)} \theta^{r_o}(1-\theta)^{n_o-r_o} \qquad (3.2)$$

where n_o,r_o are constants assumed known. Thus (3.2) represents "n_o units of prior information containing r_o successes". Combining (3.1) and (3.2) and normalizing, we obtain the posterior distribution

$$p(\theta|n,r,n_o,r_o) = \frac{\Gamma(n+n_o+2)}{\Gamma(r+r_o+1)\Gamma(n+n_o-r-r_o+1)} \theta^{r+r_o}(1-\theta)^{n+n_o-r-r_o}$$

(3.3)

from which it easily follows that $E(\theta) = (r+r_o+1)/(n+n_o+2)$ and

$$V(\theta) = \frac{(r+r_o+1)(n+n_o-r-r_o+1)}{(n+n_o+2)^2(n+n_o+3)}. \qquad (3.4)$$

We now wish to take the expectation of $V(\theta)$ with respect to the future distribution of r. Now

$$p(r|r_o,n_o,n) = \int_o^1 p(r|\theta,n)p(\theta|r_o,n_o)d\theta$$

$$= C\int_o^1 \theta^{r+r_o}(1-\theta)^{n+n_o-r-r_o}d\theta$$

$$= \frac{\Gamma(n+1)}{\Gamma(r+1)\Gamma(n-r+1)} \cdot \frac{\Gamma(n_o+2)}{\Gamma(r_o+1)\Gamma(n_o-r_o+1)} \cdot \frac{\Gamma(r+r_o+1)\Gamma(n+n_o-r-r_o+1)}{\Gamma(n+n_o+2)}$$

$$= \frac{\Gamma(n+1)\Gamma(n_o+2)}{\Gamma(r_o+1)\Gamma(n_o-r_o+1)\Gamma(n+n_o+2)} \; \frac{\Gamma(r+r_o+1)\Gamma(n+n_o-r-r_o+1)}{\Gamma(r+1)\Gamma(n-r+1)} .$$

Since the total probability equals 1, it follows that

$$\sum_{r=o}^{n} \frac{\Gamma(r+r_o+1)\Gamma(n+n_o-r-r_o+1)}{\Gamma(r+1)\Gamma(n-r+1)} = \frac{\Gamma(r_o+1)\Gamma(n_o-r_o+1)\Gamma(n+n_o+2)}{\Gamma(n+1)\Gamma(n_o+2)} .$$

We can thus find $E(r+r_o+1)(n+n_o-r-r_o+1)$ by using the same equality with (r_o+1) and (n_o+2) replacing r_o and n_o respectively, from which it follows directly that

$$EV(\theta) = \frac{(r_o+1)(n_o-r_o+1)}{(n_o+2)(n_o+3)(n+n_o+2)} = \frac{(n_o+2)(1-\theta_o)\theta_o}{(n_o+3)(n+n_o+2)}$$

where $\theta_o = (r_o+1)/(n_o+2)$ is the mean of the prior distribution (3.2). Thus, if the prior mean θ_o is the same for both distributions involved when there are "n_o units of prior information" or "kn_o units of prior information", the "information ratio" is

$$R_\theta = \frac{(kn_o+3)(n+kn_o+2)}{(n_o+3)(n+n_o+2)} \cdot \frac{n_o+2}{kn_o+2} .$$

Two binomial $\theta_1 - \theta_2$.

If we have two binomial distributions with parameters θ_1 and θ_2 and are interested in $\theta = \theta_1 - \theta_2$ we can obtain, since θ_1 and θ_2 are independent *a posteriori*,

$$EV(\theta) = \frac{\theta_{10}(1-\theta_{10})(n_{10}+2)}{(n_{10}+3)(n_1+n_{10}+2)} + \frac{\theta_{20}(1-\theta_{20})(n_{20}+2)}{(n_{20}+3)(n_2+n_{20}+2)}$$

with an obvious notation. The "information ratio" can be derived in the usual way.

4. THE NEGATIVE BINOMIAL DISTRIBUTION

5. THE POISSON DISTRIBUTION

6. THE EXPONENTIAL DISTRIBUTION

The details of these sections have been omitted in response to the editors' request for a reduction in the length of the original paper. A manuscript version which contains these details is available from the authors.

7. Acknowledgments

The authors would like to thank Mrs. Julia Grey of the Mathematics Research Center, University of Wisconsin, and Mrs. Betty Chambers, Imperial College, University of London, for computing assistance. This research was sponsored in part by the Mathematics Research Center, United States Army, Madison, Wisconsin, under Contract No: DA-31-124-ARO-D-462, and in part by the Wisconsin Alumni Research Foundation.

References

Draper, N. R. and Guttman, I. (1968) Some Bayesian stratified two-phase sampling results, _Biometrika_ _55_, 131-139, April issue.

Kiefer, J. (1958) On the randomized optiomality and randomized nonoptimality of symmetrical designs, _Ann. Math. Statist._ _29_, 675-699.

Raiffa, H., and Schlaifer, R. (1961) _Applied Statistical Decision Theory_, Harvard University.

Stuart, A. (1955) A paradox in statistical estimation, _Biometrika_ _42_, 725-729.

Winkler, R. L. (1967) The assessment of prior distributions in Bayesian analysis, _J. Amer. Statist. Assoc._ _62_, 776-800.

SUBJECTIVE BAYESIAN MODELS IN SAMPLING FINITE POPULATIONS: STRATIFICATION

W. A. Ericson[1]

University of Michigan

Abstract

In the present paper we first review some of the major aspects of a subjective Bayesian view of inference to finite populations which we put forth earlier (Ericson (1967)). The main emphasis of that paper was on subjectivistic equivalents to simple random sampling. As a straightforward generalization of these earlier results a general class of prior distributions is put forth and illustrated under which a stratified sample generally constitutes the optimal design. Under this model the design question regarding the optimal within stratum sample size or the allocation of sampling effort among stratum is considered. For a wide variety of specific prior distributions this problem may be formulated as one specific convex programming problem. A new special algorithm is obtained for its solution.

1. Review of Preliminaries

We take as a basic model of a finite population that which was initially put forth and subsequently used by Godambe (1955, 1965, 1966) and others. The author, Ericson (1967), has presented a general Bayesian view of inference regarding simple characteristics of a finite population based on this model. Essentials of that earlier work are first reviewed as a basis for the new results which are to be given in the present paper. Further details may be found in our earlier paper.

---

[1]This research was partially sponsored by the National Science Foundation under Grant No. NSF-GP-7350.

1.1. The Basic Model:

We define a finite population of N elements or units by letting $\underset{\sim}{N} = \{1,2,\ldots,n\}$, the label set, and $\underset{\sim}{X} = (X_1,X_2,\ldots,X_N)$, where X_i is the unknown value of some characteristic possessed by the ith population element. Inference concerns the N-dimensional real vector parameter $\underset{\sim}{X}$ or, more realistically, some simple function $g(\underset{\sim}{X})$ of $\underset{\sim}{X}$ say.

A sample of size m, s^*, is defined quite generally to be an ordered sequence of m of the population elements $i_1^*,i_2^*,\ldots,i_m^*$ ($i_j^* \varepsilon \underset{\sim}{N}$, $j=1,\ldots,m$, repetitions allowed) together with the sequence of their associated observed characteristic values

$$\underset{\sim}{x}^* = (x_{i_1^*},x_{i_2^*},\ldots,x_{i_m^*}),$$

i.e., we observe for each i_j^* that $X_{i_j^*} = x_{i_j^*}$, $j=1,2,\ldots,m$.

A sample design is then defined by some countable set S^* of ordered sequences, s^*, together with a probability measure assigned by choosing a function $p(s^*) \geq 0$, $\sum_{s^* \varepsilon S^*} p(s^*) = 1$, where $p(s^*)$ is the probability of choosing the sample s^*.

For any such sample $(s^*,\underset{\sim}{x}^*)$ we define the statistic $(s;x_i,i\varepsilon s)$ to be the set of distinct population elements $s = \{i_1,\ldots,i_n\} \subseteq \underset{\sim}{N}$ included in the observed sequence s^* together with the observed values x_j of X_j, $j\varepsilon s$. For notational convenience, given any sample s^* containing the n distinct units $s = \{i_1,\ldots,i_n\}$ we define the (matrix) operator S such that $S(\underset{\sim}{X}) = (X_{i_1},\ldots,X_{i_n})$, for definiteness assuming $i_1<i_2\leq\ldots<i_n$; the complementary operator $\bar{S}$ such that $\bar{S}(\underset{\sim}{X}) = (X_{j_1},X_{j_2},\ldots,X_{j_{N-n}})$ for all $j_i\varepsilon\underset{\sim}{N}-s$ $(j_1<j_2<\ldots<j_{N-n})$; and the vector $\underset{\sim}{x} = (x_{i_1},\ldots,x_{i_n})$ of observed values of $S(\underset{\sim}{X})$.

It is obvious under this model that for any joint prior probability distribution on the vector parameter

$\underset{\sim}{X}$ given by the general density $p'(X_1,\ldots,X_N)$ the posterior probability distribution of $\underset{\sim}{X}$ given the sample $(s^*,\underset{\sim}{x}^*)$ is precisely the same as that given only the statistic $(s,\underset{\sim}{x})$. Thus, by the Bayesian definition of sufficiency[2] $(s,\underset{\sim}{x})$ is a sufficient statistic, a fact previously demonstrated or noted using equivalent definitions by Basu (1958), Hajek (1959), and others. It is further clear that given $(s^*,\underset{\sim}{x}^*)$ the likelihood function of $\underset{\sim}{X}$ is given by

$$\ell(\underset{\sim}{X};\ (s^*,\underset{\sim}{x}^*)) = \ell(\underset{\sim}{X};\ (s,\underset{\sim}{x})) = \begin{cases} kp(s^*) & \text{for } \underset{\sim}{X}|S(\underset{\sim}{X}) = \underset{\sim}{x} \\ 0 & \text{otherwise,} \end{cases} \tag{1}$$

where $k > 0$ is an arbitrary constant (it, of course, being assumed that $p(s^*)$ is independent of $\underset{\sim}{X}$).

Finally, it follows that given any joint N-dimensional prior distribution on $\underset{\sim}{X}$, $p'(X_1,\ldots,X_N)$, the posterior distribution on $\underset{\sim}{X}$ given a sample of the sort described above is, using (1), given by the general density.

$$p(\underset{\sim}{X}|(s,\underset{\sim}{x})) = \begin{cases} p'(\underset{\sim}{X})/p'_{S(\underset{\sim}{X})}(\underset{\sim}{x}) & \text{for } \underset{\sim}{X}|S(\underset{\sim}{X}) = \underset{\sim}{x} \\ 0 & \text{otherwise,} \end{cases} \tag{2}$$

where $p_{S(\underset{\sim}{X})}(\underset{\sim}{x}) \neq 0$ is just the marginal prior density of $S(\underset{\sim}{X})$.

Several writers, notably Godambe (1966), have viewed the likelihood function (1) as being almost, if not completely, uninformative and have investigated principles of inference eschewing the likelihood principle. Our view here, quite to the contrary, is that when reasonable prior distributions, $p'(\underset{\sim}{X})$, are introduced, their revision by sample data can lead to meaningful and useful inferences on the functions of $\underset{\sim}{X}$ which are typically of interest.

The problem then is the choice of the prior distribution, $p'(\underset{\sim}{X})$, for once it is chosen, terminal personal

---

[2] Essentially that a statistic is sufficient if for any p prior the posterior given that statistic is the same as the posterior conditional on the whole sample.

inference regarding any $g(\underset{\sim}{X})$ is obtainable from the joint posterior given in (2). Also questions of optimal sample design and size may then be resolved by application of decision-theoretic concepts. These may include full specification of a loss function and preposterior analysis, in the sense of Raiffa and Schlaifer (1961), or adoption of less formal criteria such as choosing a design so that the posterior distribution of $g(\underset{\sim}{X})$ has smallest expected variance for a given total cost. Such approaches take formal cognizance of the incontrovertible fact that practical choice of sample or experimental designs and sizes is dictated by prior information, tempered by economic considerations.

1.2. Exchangeable Prior Knowledge:

In an earlier paper, Ericson (1967), we have presented what seems to be the simplest useful general class of prior distributions, namely those representing exchangeable prior knowledge regarding the X_i's. The property of exchangeability is one of symmetry;[3] random variables $X_1,\ldots,X_N$ are said to be exchangeable if the joint probability distribution of each of the N! permutations of the variables is the same. Exchangeability thus expresses the prior knowledge that while the units of the finite population are identifiable by their labels (here the integers $1,2,\ldots,N$) there is no information carried by these labels regarding the associated X_i's; that is, under exchangeability given $1 \leq r \leq N$ one's initial betting behavior regarding events defined by the unknown quantities $X_{i_1},\ldots,X_{i_r}$ is the same for every ordered set of indices $i_1,\ldots,i_r$.

There are close ties between an objectivistic sampling distribution under simple random sampling and corresponding properties of subjectivistic exchangeable prior distributions.

Also truly exchangeable prior distributions reflect the knowledge or belief that the population is, in effect, already randomized. This manifests itself by the fact

---

[3]For further discussion see B. deFinetti (1964).

that the symmetry of such a prior on $\underset{\sim}{X}$ implies that the posterior distribution of any symmetric function of $\underset{\sim}{X}$ depends only on the set of observed variate values and neither on the particular subset of units included in the sample nor on the identity of the units assuming the observed values. This under exchangeability and for inference on symmetric functions of $\underset{\sim}{X}$ all samples of size n, or designs involving a fixed sample size, n, are, other things being equal, viewed as indifferent. The only relevant design question is that of determining the economic sample size. Under such priors a good case may be made for random selection of the n units rather than the first or most conveniently obtained n units since the cost of randomization may be more than offset by the increased utility of the resulting sample to other of its potential users.

Prior knowledge which may be roughly approximated by exchangeability seems to characterize many of those situations where thoughtful objectivistic practitioners deem simple random sampling to be the appropriate design. Also there appears to be good reason for the subjectivist having exchangeable or roughly exchangeable prior knowledge to select a simple random sample.

2. Stratification--Partitioning for Exchangeability

In our earlier paper (Ericson, 1967) we examined the situation where one's prior knowledge was such that he views the X_i's, i=1,... N as exchangeable. It will be assumed that, on the basis of his prior information regarding the population units, one is able to partition those units in such fashion that the X_i's within each element of the partition are viewed as exchangeable or approximately so. This, in terms of the specification of a joint prior on $\underset{\sim}{X}$, is directly akin to the well-established principle of stratification aimed at within stratum homogeneity. For obvious reasons the elements of the partition will then be termed strata.

For notational convenience the units will now be labelled or identified by double subscript. Thus, the label (i,j) will denote or identify the ith population unit in the jth stratum and X_{ij} will denote the variate value possessed by the unit identified by the label or name (i,j). We also suppose that there are N_j units in the jth stratum $\Sigma_1^k N_j = N$.

Letting

$$\underset{\sim}{X}_j = (X_{1_j},\ldots,X_{N_j j}), \quad j=1,\ldots,k, \tag{3}$$

and

$$\underset{\sim}{X} = (\underset{\sim}{X}_1,\ldots,\underset{\sim}{X}_k) \tag{4}$$

respectively denote the vectors of the variate values attached to the units in the jth stratum and in the whole finite population, we shall consider the class of prior distributions on $\underset{\sim}{X}$ given by the general density

$$p'(\underset{\sim}{X}) = \prod_{j=1}^{k} p_j'(\underset{\sim}{X}_j), \tag{5}$$

where each $p_j'(\underset{\sim}{X}_j)$ reflects exchangeability regarding the X_{ij}'s $i=1,\ldots,N_j$. Under this class of priors on $\underset{\sim}{X}$ the $\underset{\sim}{X}_j$'s are viewed as <u>a priori</u> independent. While such a model may often roughly capture one's real prior knowledge there certainly are many practical cases, requiring other classes of prior distributions, where knowledge of the variate values in one stratum yields some information regarding the values assumed by the X_{ij}'s in other strata.

For any particular sample (s,x) obtained by any sampling design the likelihood function certainly remains as given in (1) and the posterior density, using the prior (5), is exactly given in (2). However, in dealing with the present class of priors some additional notation will be useful. Given any sample, s^*, containing the n distinct elements whose labels, the ordered pairs (i,j), are specified by the set s, we will let n_j denote the number of those observed units which fall in stratum j, $\Sigma_{j=1}^{k} n_j = n$. Also we will denote by s_j the set of identifying labels of those units falling in the jth stratum; that is, we will define $s_j = \{i|(i,j)\varepsilon s\}$ $j=1,\ldots,k$ so that

$$\bigcup_{j=1}^{k} (s_j \times \{j\}) = s.$$

We will designate an observed value of X_{ij} by x_{ij} for $(i,j)\varepsilon s$, will let $\underset{\sim}{x}_j = (x_{i_1 j},\ldots,x_{i_{n_j} j})$, $i_1<i_2<\ldots<i_{n_j}$. be the vector of observed values of X_{ij} for $i\varepsilon s_j$ and will let $\underset{\approx}{x} = (\underset{\sim}{x}_1,\ldots,\underset{\sim}{x}_k)$. Finally for any sample containing the distinct units having labels in s, we define the matrix operators S_j such that

$$S_j(\underset{\sim}{X}_j) = (X_{i_1 j},\ldots,X_{i_{n_j} j}), \quad j=1,\ldots,k. \tag{6}$$

Using this notation it follows immediately, using (2) and the prior (5), that the posterior density of $\underset{\approx}{X}$ is given by

$$p(\underset{\approx}{X}|(s,\underset{\approx}{x}))=\begin{cases}\prod_{j=1}^{k} [p_j'(\underset{\sim}{X}_j)/p_{s_j}'(\underset{\sim}{x}_j)] & \text{for } \underset{\sim}{X}_j|S_j(\underset{\sim}{X}_j)=\underset{\sim}{x}_j, j=1,\ldots k \\ 0 & \text{otherwise,}\end{cases} \tag{7}$$

where $p_{s_j}'(\underset{\sim}{x}_j) > 0$ is the marginal prior density of the X_{ij}'s for $i\varepsilon s_j$, evaluated at the observed $\underset{\sim}{x}_j$.

2.1. A Useful Sub-Class of Priors:

A wide and useful subclass of such exchangeable prior densities, $p_j'(X_j)$, may be generated by viewing the components of X_j as independent and identically distributed conditional on some real or hypothetical parameter $\underset{\sim}{\theta}_j = (\theta_{1j},\theta_{2j},\ldots,\theta_{m_j j})$, having general density given by $p_j(\underset{\sim}{X}_{ij}|\underset{\sim}{\theta}_j)$, and where $\underset{\sim}{\theta}_j$ is assigned the probability distribution function $F_j(\underset{\sim}{\theta}_j)$. The joint prior on $\underset{\sim}{X}_j$ is then taken as the marginal density given by the mixture

$$p_j'(\underset{\sim}{X}_j) = \int_{\underset{\sim}{\theta}_j} \prod_{i=1}^{N_j} p_j(X_{ij}|\underset{\sim}{\theta}_j)dF_j(\underset{\sim}{\theta}_j). \tag{8}$$

The generation of exchangeable joint priors on $\underset{\sim}{X}_j$ by this approach results in some economy of thought and effort and is equivalent, barring differences in probability interpretation, to the notion of viewing the finite population as a sample from a super-population defined by the density $p_j(\cdot|\underset{\sim}{\theta}_j)$ and having unknown parameter $\underset{\sim}{\theta}_j$. Within the general context of the present model, a wide spectrum of prior uncertainty regarding $\underset{\approx}{X}$ may be given formal expression by the assessment of $p_j(X_{ij}|\underset{\sim}{\theta}_j)$ and $F_j(\underset{\sim}{\theta}_j)$, $j=1,\ldots,k$.

With such priors on $\underset{\sim}{X}_j$ the posterior on $\underset{\approx}{X}$ is, easily obtained as

$$p(\underset{\approx}{X}|(s,\underset{\approx}{x}))=\begin{cases}\prod_{j=1}^{k}\int_{\underset{\sim}{\theta}_j}\prod_{i\notin s_j}p_j(X_{ij}|\underset{\sim}{\theta}_j)\,dF_j(\underset{\sim}{\theta}_j|(s_j,\underset{\sim}{x}_j)),\ \underset{\approx}{X}|S_j(\underset{\sim}{X}_j)=\underset{\sim}{x}_j \\ \qquad\qquad\qquad\qquad\qquad j=1,\ldots,k \\ 0 \qquad \text{otherwise,}\end{cases} \tag{9}$$

where $F_j(\underset{\sim}{\theta}_j|(s_j,\underset{\sim}{x}_j))$ is merely the posterior c.d.f. on $\underset{\sim}{\theta}_j$ given the sample observations.

2.2. Discussion:

Several points should be noted explicitly. First, for the type of exchangeable prior opinion regarding the X_{ij}'s as formulated here the choice of the prior distribution $p_j'(X_j)$ may be conveniently thought of as involving (a) the choice of a parametric model $p_j(X_{ij}|\underset{\sim}{\theta}_j)$ and (b) the assessment of a prior distribution $F_j(\underset{\sim}{\theta}_j)$ on the unknown parameter $\underset{\sim}{\theta}_j$. These distributions, each representing assessments of personal probability, are precisely those required for most inference problems from the subjectivistic Bayesian viewpoint, no more and no less. However, unlike the usual Bayesian posterior inference problem where the primary focus is on the posterior distribution of $\underset{\sim}{\theta}_j$, under the present finite population model it is the predictive distribution on the unobserved components of $\underset{\sim}{X}_j$ (or a future sample of N_j-n_j units) which is primarily of interest. The posterior on the real or hypothetical $\underset{\sim}{\theta}_j$ is but a useful and convenient intermediate step in obtaining this predictive distribution. (See Roberts (1965) for further discussion of the notion of predictive distributions).

There appears great flexibility afforded within this model regarding one's prior uncertainty about the within stratum finite population distribution. Even though N_j may be large, by the choice of $p_j(\cdot|\underset{\sim}{\theta}_j)$, including the dimensionality of $\underset{\sim}{\theta}_j$, and $F_j(\underset{\sim}{\theta}_j)$ a wide spectrum of prior uncertainty may be subsumed under this formulation. There is certainly no necessary relation among the distributions assumed or assessed within each of the k strata.

It seems to us that subjectivistic Bayesian views of inference to finite populations involves no new essential difficulties; that the choice of a prior on $\underset{\sim}{X}$ may often be satisfactorily viewed as equivalent to the choice of a parametric model and a prior distribution on the parameter. In such assessments the sampler would often appear to us to possess prior information at least as good as that underlying choices of models and priors in other areas of statistics.

Finally, it should be noted that all of our previous discussion in Section 1 relating randomization, exchangeability, and approximate exchangeability applies equally here with respect to the establishment of the prior within each of the k strata.

2.3. Examples:

To illustrate some of the previous ideas we briefly present two examples. The first involves rather strong prior knowledge that the unknown within stratum population frequency distribution of X_{ij} is roughly normal (especially if the stratum sizes N_j, are large). The second example is one appropriate in cases where there is extreme vagueness regarding the shape and other properties of the true distribution of the X_{ij}'s within strata. The results given are simple extensions of those presented and derived in our earlier paper (Ericson, 1967). Further details and discussion may be found in that paper

2.3.1. A Normal Model:

As a first example we suppose that within the jthk stratum the prior distribution on $\underset{\sim}{X}_j$ is reasonably approximated by viewing the X_{ij}'s as a sample from a normal superpopulation with unknown mean θ_j and unknown variance $1/h_j$, and thus will take $\underset{\sim}{\theta}_j = (\theta_j, h_j)$. Let

$$\mu_j = \frac{1}{N_j}\sum_{i=1}^{N_j} X_{ij} \text{ and } \sigma_j^2 = \frac{1}{N_j}\sum_{i=1}^{N_j}(X_{ij}-\mu_j)^2,\ j=1,\ldots,k$$

are the (unknown) means and variances of the variate values associated with the population units in the jth stratum.

We will further assume that $F_j(\underset{\sim}{\theta}_j)$ can reasonably be approximated by a normal-gamma distribution with density

$$f_j(\underset{\sim}{\theta}_j) \propto e^{-\frac{1}{2}h_j n_j'(\theta_j - m_j')^2}\, h_j^{\frac{1}{2}\delta(n_j')}\, e^{-\frac{1}{2}h_j \nu_j v_j'}\, h_j^{\frac{1}{2}\nu_j' - 1},\ j=1,\ldots,k, \tag{10}$$

for $0 < h_j < \infty$, $-\infty < \theta_j < \infty$, n_j', ν_j', $v_j' \geq 0$, and where $\delta(n_j') = 0$, if $n_j' = 0$ and is unity otherwise. We will also briefly consider the degenerate prior obtained by setting $n_j' = \nu_j' = 0$ in (10), i.e., the prior

$$f_j(\underset{\sim}{\theta}_j) \propto h_j^{-1}, \quad -\infty < \theta_j < \infty, \quad 0 < h_j < \infty. \tag{11}$$

Prior and posterior inference on any one particular μ_j under this model has been discussed by us in the earlier paper. We here will only summarize some of those results without proof and indicate their extensions to inference on

$$\mu = \frac{1}{N}\sum_{j=1}^{k} N_j \mu_j,$$

the overall population mean.

The exchangeable joint prior on $\underset{\sim}{X}_j$ in the present example is an N_j dimensional symmetric Student distribution. The prior on μ_j can then be deduced as a one-dimensional Student distribution, that is, a priori μ_j is distributed like the quantity

$$m_j' + t_{\nu_j'} \left[\frac{(N_j + n_j') v_j'}{n_j' N_j} \right]^{\frac{1}{2}} , \tag{12}$$

where $t_{\nu_j'}$ is a Student's "t" random variable on ν_j' d.f.

It then follows that the prior mean and variance of μ_j are given by $E(\mu_j) = m_j'$ and

$$V(\mu_j) = \frac{\nu_j' v_j'}{\nu_j' - 2} \left[\frac{N_j + n_j'}{n_j' \; N_j} \right] .$$

Thus under the present stratified model the prior on μ may be characterized as having mean,

$$E(\mu) = \frac{1}{N} \sum_{j=1}^{k} N_j m_j' , \tag{13}$$

variance,

$$V(\mu) = \frac{1}{N^2} \sum_{j=1}^{k} N_j^2 \frac{\nu_j' v_j'}{\nu_j' - 2} \left[\frac{N_j + n_j'}{n_j' \; N_j} \right] , \tag{14}$$

and distributed as a linear function of independent "t" random variables.

Turning to the posterior distribution, the following results have been shown. Let $\bar{x}_j$ denote the mean of the n_j observed x_{ij}'s falling in the jth stratum and let

$$s_j^2 = \sum_{i=1}^{n_j} (x_{ij} - \bar{x}_j)^2 / (n_j - 1)$$

be the observed within stratum variance. The posterior on μ_j given $(s, \underset{\sim}{x})$ is exactly that of the random variable

$$\frac{n_j \bar{x}_j + (N_j - n_j) m_j''}{N_j} + t_{\nu_j''} \left[\frac{(N_j - n_j + n_j'')(N_j - n_j) v_j''}{n_j'' N_j} \right]^{\frac{1}{2}} , \tag{15}$$

where $t_{\nu''_j}$ is a standard "t" random variable on ν''_j d.f. and

$$m''_j = (n'_j m'_j + n_j \bar{x}_j)/(n'_j + n_j), \tag{16}$$

$$n''_j = n'_j + n_j, \tag{17}$$

$$\nu''_j v''_j = \nu'_j v'_j + (n_j - 1)s_j^2 + \frac{n_j n'_j}{n_j + n'_j}(\bar{x}_j - m'_j)^2, \tag{18}$$

and

$$\nu''_j = \nu'_j + \delta(n'_j) + n_j - 1. \tag{19}$$

These results may be recast in a more informative and intuitive form. It can be shown that given the sample $(s,\underset{\sim}{x})$, μ_j is distributed like the quantity

$$\frac{V(\mu_j)x_j + V(X_{n_j}|\mu_j)E(\mu_j)}{V(\mu_j) + V(\bar{X}_{n_j}|\mu_j)} + t_{\nu''_j}\left[\frac{N_j - n_j}{N_j}\,\frac{V(\theta_j|(s,x))}{V(\theta_j)}V(\mu_j)\frac{\nu''_j - 2}{\nu''_j}\right]^{\frac{1}{2}} \tag{20}$$

where $t_{\nu''_j}$ is as in (15) and where $V(\theta_j|(s,\underset{\sim}{x})) = \nu''_j v''_j / n''_j(\nu''_j - 2)$ and $V(\theta_j) = \nu'_j v'_j / n'_j(\nu'_j - 2)$ are the posterior and prior variances of θ_j, the super-population mean, and $V(\bar{X}_{n_j}|\mu_j)$ is merely the prior variance of the mean of any n_j of the X_{ij}'s conditional on knowing the finite population mean, μ_j.

From (20) it follows that the posterior distribution of the overall population mean,

$$\mu = \frac{1}{N}\sum_{j=1}^{k} N_j \mu_j,$$

is given by that of a linear combination of independent t random variables and has mean and variance given by

$$E(\mu|(s,\underset{\sim}{x})) = \frac{1}{N}\sum_{j=1}^{k}\left[\frac{V(\mu_j)\bar{x}_j+V(\bar{X}_{nj}|\mu_j)E(\mu_j)}{V(\mu_j)+V(\bar{X}_{n_j}|\mu_j)}\right] \tag{21}$$

and

$$V(\mu|s,\underset{\sim}{x})) = \frac{1}{N^2}\sum_{j=1}^{k} N_j^2 \frac{N_j-n_j}{N_j}\frac{V(\theta_j|(s,x))}{V(\theta_j)} V(\mu_j). \tag{22}$$

In the case of the diffuse prior, (11), on the superpopulation mean and variance the posterior on μ is that of the random variable

$$\frac{1}{N}\sum_{j=1}^{k} N_j\bar{x}_j + \frac{1}{N}\sum_{j=1}^{k} N_j \left[\frac{N_j-n_j}{N_j}\frac{s_j^2}{n_j}\right]^{\frac{1}{2}} t_{n_j-1}, \tag{23}$$

where the t_{n_j-1}'s are independent "t" random variables on n_j-1 d.f.

2.3.2. Extreme Prior Vagueness:

Analyses of the sort given in the preceding subsection can, of course, be carried through under various alternative assumptions regarding $p_j(X_{ij}|\underset{\sim}{\theta}_j)$. While such examples may often adequately approximate one's uncertainty regarding the unknown $\underset{\sim}{X}_j$, nevertheless since $\underset{\sim}{\theta}_j$ is of low dimensionality, they do assume strong prior knowledge regarding the shape of the finite population distribution within the jth stratum by taking the superpopulation form as known.

Another example, based on the multinomial distribution and which encorporates extreme vagueness regarding the shape of the finite population, has been put forth by the author (Ericson, 1967). The essentials of this model are summarized briefly below in the present context of within stratum exchangeability and for inference on μ. Many other details are given in the earlier paper.

We suppose that each X_{ij} may only assume one of the finite set of numerical values $Y_j = \{y_{1j}, y_{2j}, \ldots, y_{m_j j}\}$ where m_j may be an extremely large integer having no relation whatever to N_j, the total stratum size, and where it is assumed that $y_{1j} \leq y_{2j} \leq \cdots \leq y_{m_j j}$. This assumption clearly recognizes the inherent discreteness of almost all observations owning to the limitations of measuring instruments, etc. We further supoose that the probability that $X_{ij} = y_{\ell j}$ is $p_{\ell j}$, i.e.,

$$p_j(X_{ij}=y_{\ell j}|p_j) = p_{\ell j}, \ \ell=1,\ldots,m_j, \ \sum_{\ell=1}^{m_j} p_{\ell j} = 1, \quad (24)$$

where $\underset{\sim}{p}_j = (p_{1j},\ldots,p_{m_j j})$. Here $\underset{\sim}{p}_j$ is assumed unknown and plays the role $\underset{\sim}{\theta}_j$ in (8). For given j the X_{ij}'s are assumed independent and identically distributed with the density function (24).

We will let $M_{\ell j}$ denote the unknown number of the N_j population elements for which X_{ij} assumes the value $y_{\ell j}$, thus

$$\sum_{\ell=1}^{m_j} M_{\ell j} = N_j.$$

It is clear that under the present formulation and for inference regarding symmetric functions of $\underset{\sim}{X}_j$ (our only concern here) one may equivalently deal with either the distributions (prior and posterior) of $\underset{\sim}{X}_j$ or those of the $M_{\ell j}$'s, $\ell=1,\ldots,m_j$. We will here find it convenient to work with the distributions of $M_{\ell j}$'s. It is immediate that given $\underset{\sim}{p}_j$, $\underset{\sim}{M}_j = (M_{1j},\ldots,M_{m_j j})$ is multinomially distributed. The real finite population is defined by the unknown frequencies, $\underset{\sim}{M}_j$, while the superpopulation is defined by the unknown $\underset{\sim}{p}_j$.

To this point the setting above seems quite straightforward, realistic, and flexible. However, there are difficulties involved in the choice of appropriate and useful priors $f_j(\underset{\sim}{p}_j)$ on $\underset{\sim}{p}_j$. A mathematically convenient choice of such a prior is the m_j-1 dimensional Dirichlet distribution (see, for example, Wilks, 1962) with density

$$f_j(\underset{\sim}{p}_j) = \frac{\Gamma(\varepsilon_j) \times \prod_{\ell=1}^{m_j-1} p_{\ell j}^{\varepsilon_{\ell j}-1}\left(1-\sum_{\ell=1}^{m_j-1} p_{\ell j}\right)^{\varepsilon_j - \sum_{\ell=1}^{m_j-1} \varepsilon_{\ell j}-1}}{\prod_{\ell=1}^{m_j-1} \Gamma(\varepsilon_{\ell j})\Gamma(\varepsilon_j - \sum_{\ell=1}^{m_j-1} \varepsilon_{\ell j})} \tag{25}$$

for $0 \leq p_{\ell j} \leq 1$, $\sum_{\ell=1}^{m_j-1} p_{\ell j} \leq 1$ and for parameter values $\varepsilon_{\ell j} > 0$, $\ell=1,2,\ldots,m_j$ and $\varepsilon_j = \sum_{\ell=1}^{m_j} \varepsilon_{\ell j}$. While a mathematically convenient choice, the difficulty with this class of priors on $\underset{\sim}{p}_j$ and the resulting prior on $\underset{\sim}{M}_j$ is that it is incapable of expressing the realistic prior knowledge that the $p_{\ell j}$'s and the $M_{\ell j}$'s (at least when adjacent values are grouped) are quite likely to be "smooth". We proceed using this distribution with the stipulation that the parameters $\varepsilon_{\ell j}$ and their sum ε_j are all very small (ε_j perhaps less than unity); for, as discussed in more detail in our earlier paper, such a prior seems to represent an extreme position of initial vagueness not even incorporating the typical belief regarding "smoothness". We have shown that even though, given a typical sample from the jth stratum, almost all cells will have an observed frequency of zero, seemingly realistic and useful posterior inferences are available regarding those simple properties of the units in the jth stratum which are typically of interest in sampling. Also such inferences are little dependent on the actual choice of the $\varepsilon_{\ell j}$'s, provided they and their sum are all small.

Under this model of extreme prior vagueness the following results relating to inference on the overall finite population mean μ, here given by

$$\mu = \frac{1}{N} \sum_{j=1}^{k} \sum_{\ell=1}^{m_j} y_{\ell j} M_{\ell j} = \frac{1}{N} \sum_{j=1}^{k} N_j \mu_j, \tag{26}$$

may be shown using results given in Ericson (1967).

The joint prior on M_j is the Dirichlet-multinomial distribution having density

$$p_j(M_j)=\frac{\Gamma(N_j+1)\prod_{\ell=1}^{m_j-1}\Gamma(M_{\ell j}+\varepsilon_{\ell j})\Gamma(N_j+\varepsilon_j-\sum_{\ell=1}^{m_j-1}(M_{\ell j}+\varepsilon_{\ell j}))\Gamma(\varepsilon_j)}{\prod_{\ell=1}^{m_j-1}[\Gamma(M_{\ell j}+1)\Gamma(\varepsilon_{\ell j})]\Gamma(N_j-\sum_{\ell=1}^{m_j-1}M_{\ell j}+1)\Gamma(\varepsilon_j-\sum_{\ell=1}^{m_j-1}\varepsilon_{\ell j})\Gamma(N_j+\varepsilon_j)}$$

$$M_{\ell j}=0,1,\ldots,N_j,\ \sum_{\ell=1}^{m_j-1}M_{\ell j}\le N_j,\quad j=1,\ldots,k, \tag{27}$$

The joint prior on $\underset{\sim}{M}=(\underset{\sim}{M}_1,\ldots,\underset{\sim}{M}_k)$ is then simply $p'(\underset{\sim}{M}) = \prod_{j=1}^{k} p_j(\underset{\sim}{M}_j)$. It then follows, most easily by first conditioning on $\underset{\sim}{p}_j$, that the prior means and variances of μ_j and μ are given by:

$$E(\mu) = \frac{1}{N}\sum_{j=1}^{k} N_j \sum_{\ell=1}^{m_j} y_{\ell j}\varepsilon_{\ell j}/\varepsilon_j, \tag{28}$$

and

$$V(\mu) = \frac{1}{N^2}\sum_{j=1}^{k} N_j^2\,\frac{N_j+\varepsilon_j}{N_j(\varepsilon_j+1)}\left[\sum_{\ell=1}^{m_j} y_{\ell_j}^2\,\varepsilon_{\ell j}/\varepsilon_j-\left(\sum_{\ell=1}^{m_j} y_{\ell j}\varepsilon_{\ell j}/\varepsilon_j\right)^2\right]. \tag{29}$$

Turning to properties of the posterior distribution conditional on any sample $(s,\underset{\sim}{x})$ comprised of n distinct elements, we shall utilize the following definitions. We shall again let n_j be the number of those elements falling in the jth stratum and let $n_{\ell j}$ be the number of those n_j which have observed variate values equal to $y_{\ell j}$

$\sum_{\ell=1}^{m_j} n_{\ell j} = n_j$. The observed sample mean within the jth stratum is then given by

$$\bar{x}_j = \frac{1}{n_j} \sum_{\ell=1}^{m_j} y_{\ell j} n_{\ell j}$$

and the observed sample variance within the jth stratum is given by

$$s_j^2 = \frac{1}{n_j - 1} \left(\sum_{\ell=1}^{m_j} y_{\ell j}^2 n_{\ell j} - n_j \bar{x}_j^2 \right).$$

Finally we let

$$U_{\ell j} = M_{\ell j} - n_{\ell j} \tag{30}$$

be the (unknown) number of unobserved X_{ij}'s in the jth stratum which equal $y_{\ell j}$, $\underset{\sim}{U}_j = (U_{1j}, \ldots, U_{m_j j})$ and $\underset{\sim}{U} = (\underset{\sim}{U}_1, \ldots, \underset{\sim}{U}_k)$.

The posterior distribution on U (and, by (30), that of M) is also of the form

$$p(\underset{\sim}{U} | (s, \underset{\sim}{x})) = \prod_{j=1}^{k} p_j(\underset{\sim}{U}_j | (s, \underset{\sim}{x})) \tag{31}$$

where each term $p_j(\underset{\sim}{U}_j | (s, \underset{\sim}{x}))$ is of the Dirichlet-multinomial form (27), where the $\varepsilon_{\ell j}$'s and ε_j are replaced respectively by $\varepsilon'_{\ell j} = \varepsilon_{\ell j} + n_{\ell j}$ and $\varepsilon'_j = \varepsilon_j + n_j$. The posterior expectation of the finite population mean μ is given equivalently by

$$E(\mu | (s, \underset{\sim}{x})) = \frac{1}{N} \sum_{j=1}^{k} N_j \frac{n_j (N_j + \varepsilon_j) \bar{x}_j + (N_j - n_j) \varepsilon_j E(\mu_j)}{N_j (\varepsilon_j + n_j)} \tag{32}$$

where $E(\mu_j)$ is the prior mean, (28) of μ_j, or by

$$E(\mu | (s, \underset{\sim}{x})) = \frac{1}{N} \sum_{j=1}^{k} N_j \left[\frac{\bar{x}_j V(\mu_j) + E(\mu_j) E[V(\bar{X}_{nj} | \mu_j)]}{V(\mu_j) + E[V(\bar{X}_{nj} | \mu_j)]} \right], \tag{33}$$

where $V(\mu_j)$ is the prior variance of μ_j as given in (29) and $E[V(\bar{X}_{n_j}|\mu_j)]$ is the prior expectation of the conditional variance of the mean of any n_j observations, $\bar{X}_{n_j}$, given the overall mean, μ_j, of that stratum. This last form, (33), has the intuitive appeal, as in the first example, of being an easily interpretable weighted average of $\bar{x}_j$ and $E(\mu_j)$. Also it follows easily from (32) that since the ε_j's are assumed small the posterior mean of μ may often be well approximated by

$$E(\mu|(s,\underset{\sim}{x})) \doteq \frac{1}{N}\sum_{j=1}^{k} N_j\bar{x}_j .$$

The variance of the posterior distribution of μ may be shown to be given by

$$V(\mu|(s,\underset{\sim}{x})) = \frac{1}{N^2}\sum_{j=1}^{k}\frac{(N_j-n_j)(N_j+\varepsilon_j)}{(n_j+\varepsilon_j)^2(\varepsilon_j+n_j+1)}\{\sum_{\ell=1}^{m_j} y_{\ell j}^2(\varepsilon_{\ell j}+n_{\ell j})(\varepsilon_j+n_j-\varepsilon_{\ell j}-n_{\ell j})$$

$$-\sum_{\ell\neq q} y_{\ell j}y_{qj}(\varepsilon_{\ell j}+n_{\ell j})(\varepsilon_{qj}+n_{qj})\}; \tag{34}$$

or letting $\mu_j(\underset{\sim}{p}_j) = \sum_{\ell=1}^{m_j} y_{\ell j}p_{\ell j}$, $j=1,\ldots,k$, be the unknown "superpopulation" means by

$$V(\mu|(s,\underset{\sim}{x})) = \frac{1}{N^2}\sum_{j=1}^{k} N_j^2\,\frac{N_j-n_j}{N_j}\left[\frac{V(\mu_j(\underset{\sim}{p}_j)|(s,\underset{\sim}{x}))}{V(\mu_j(\underset{\sim}{p}_j))}\right]V(\mu_j), \tag{35}$$

where the term in square brackets is just the ratio of the posterior to the prior variance of $\mu_j(\underset{\sim}{p}_j)$. Finally if the ε_j's are very small (and the n_j's are moderate) then this posterior variance may well be approximated by

$$V(\mu|(s,\underset{\sim}{x})) \doteq \frac{1}{N^2}\sum_{j=1}^{k} N_j^2\,\frac{N_j-n_j}{N_j}\,\frac{n_j-1}{n_j+1}\,\frac{s_j^2}{n_j}\ \left(\doteq \frac{1}{N^2}\sum_{j=1}^{k} N_j^2\,\frac{N_j-n_j}{N_j}\,\frac{s_j^2}{n_j}\right), \tag{36}$$

where s_j^2 is again the within-stratum sample variance.

2.3.3. Generalizations of Examples:

While these two examples are included mainly to illustrate the possible range of exchangeable prior knowledge which may be subsumed under the model (8), they also serve to illustrate certain common properties of the posterior distribution of μ which hold under a variety of other priors of this general form. It has been shown (Ericson, 1967) that for priors $p_j'(\underset{\sim}{X}_j)$ of the form (8) obtained by taking $p_j(X_{ij}|\underset{\sim}{\theta}_j)$ in a restricted exponential family (including at least the Poisson, binomial, gamma, exponential, normal with known variance, and the two examples above, multinomial and normal with unknown mean and variance) and for natural conjugate distributions $f_j(\underset{\sim}{\theta}_j)$ on $\underset{\sim}{\theta}_j$, the posterior mean and variance of μ_j, the finite population jth stratum mean, may be put in the intuitively appealing forms:

$$E(\mu_j|(s,\underset{\sim}{x})) = \frac{\bar{x}_j V(\mu_j)+E(\mu_j)E[V(\bar{X}_{nj}|\mu_j)]}{V(\mu_j)+E[V(\bar{X}_{nj}|\mu_j)]}, \tag{37}$$

and

$$V(\mu_j|(s,\underset{\sim}{x})) = \frac{N_j-n_j}{N_j}\left[\frac{V(\mu(\underset{\sim}{\theta}_j)|s,\underset{\sim}{x}))}{V(\mu(\underset{\sim}{\theta}_j))}\right]V(\mu_j). \tag{38}$$

Here $E(\mu_j)$ and $V(\mu_j)$ are the prior means and variances of μ_j, respectively, $\mu(\underset{\sim}{\theta}_j)$ is the unknown "superpopulation" mean (i.e., $\mu(\underset{\sim}{\theta}_j) = E(X_{ij}|\underset{\sim}{\theta}_j)$), and $E[V(\bar{X}_{n_j}|\mu_j)]$ is the prior expectation of the conditional "sampling" variance of the mean of <u>any</u> n_j of the X_{ij}'s given μ_j. For this same restricted exponential class it can also be shown that the prior expectation of $V(\mu_j|(s,\underset{\sim}{x}))$, (38), is always of the form

$$E_{(s,\underset{\sim}{x})}[V(\mu_j|(s,\underset{\sim}{x}))] = \frac{N_j-n_j}{N_j^2}\,\frac{N_j+n_j'}{n_j+n_j'}\,E_{\underset{\sim}{\theta}_j}[V(X_{ij}|\underset{\sim}{\theta}_j)], \tag{39}$$

where $E_{\underset{\sim}{\theta}_j}[V(X_{ij}|\underset{\sim}{\theta}_j)]$ is the prior expectation (with respect to $f_j(\underset{\sim}{\theta}_j)$) of the variance of X_{ij} given $\underset{\sim}{\theta}_j$ and $n_j' > 0$ is the parameter of the natural conjugate prior on $\underset{\sim}{\theta}_j$ which may be interpreted as a fictitious sample size. These results turn out to be most useful, as we shall see, in consideration of optimal design questions.

3. Optimal Design

Under the present general class of prior distributions on $\underset{\sim}{X} = (\underset{\sim}{X}_1,\ldots,\underset{\sim}{X}_k)$, which is characterized by exchangeability within each stratum and independence among strata, the discussion given in Section 1.2 above applies equally within each of these k strata. In particular, the posterior on any symmetric function of $\underset{\sim}{X}$ given $(s,\underset{\sim}{x})$ depends only on having observed n_j units in the jth stratum with variate values given by $\underset{\sim}{x}_j$ and neither on which particular n_j units were observed nor on the particular correspondence of observed values with the units. As a consequence of this all fixed size sample designs $(S^*,p(s^*))$ which invariably lead to n_j distinct units being observed within the jth stratum are viewed indifferently or as being equivalent. The interesting design question then consists of a determination of the sample sizes $n_1,n_2,\ldots,n_k$. Once the decision as to these sample sizes is made, the selection of the n_j units from the jth stratum would typically involve simple random sampling.

3.1. Criteria for Optimal Design

We will consider two criteria for optimal design. The first criterion we shall adopt concerns the variance of the posterior distribution of μ, $V(\mu|(s,\underset{\sim}{x}))$. Specifically, since $V(\mu|(s,\underset{\sim}{x}))$ is most often unknown prior to the sample $(s,\underset{\sim}{x})$ we will adopt the criterion of choosing the design, the vector $\underset{\sim}{n}=(n_1,\ldots,n_k)$ of sample sizes, to minimize the prior expectation of the posterior variance of μ subject to a fixed total budget for sampling. That is, given that observations cost c_j per unit within the jth stratum we will seek that $\underset{\sim}{n}$ which minimizes

$$E_{(s,\underset{\sim}{x})}[V(\mu|(s,\underset{\sim}{x}))] \tag{40}$$

subject to the constraint that

$$\sum_{j=1}^{k} c_j n_j \leq C, \tag{41}$$

where C is the total prespecified sampling budget. Also since here the population is assumed finite we have the added constraint that $0 \leq n_j \leq N_j$, $j=1,\ldots,k$. For obvious reasons this problem will be termed an optimal allocation problem.

The other criterion we will adopt and discuss is obtained via a more formal decision theoretic approach. We will assume that μ is to be estimated with quadratic losses, i.e., the terminal loss is given by $\ell(\mu,\hat{\mu})=\gamma(\mu-\hat{\mu})^2$. It will also be assumed that in loss units the sampling costs or losses are additive so that the total loss of the sequence of choosing the design $\underset{\sim}{n}$, observing $(s,\underset{\sim}{x})$, and choosing the estimate $\hat{\mu}$, when μ is the true mean given by $\ell(\mu,\hat{\mu},n) = \gamma(\mu-\hat{\mu})^2 + \underset{\sim}{c}\underset{\sim}{n}^t$ where $\underset{\sim}{c} = (c_1,\ldots,c_k)$ and the superscript "t" denotes transpose. Under this decision theoretic criterion it is well-known, e.g., Raiffa and Schlaifer (1961), that the optimum estimate given $(s,\underset{\sim}{x})$ is the posterior mean of μ, $E(\mu|(s,\underset{\sim}{x}))$. Also the optimal design is simply that $\underset{\sim}{n}$ which minimizes

$$\gamma E_{(s,\underset{\sim}{x})}[V(\mu|s,\underset{\sim}{x})] + \underset{\sim}{c}\underset{\sim}{n}^t. \tag{42}$$

We will demonstrate that if a general solution to the allocation problem is available then the choice of an optimal design minimizing (42) is very easily obtained.

3.2. The Allocation Problem and its Application:

As mentioned in Section 2.3.3, for a side subclass of prior distributions, $p_j'(\underset{\sim}{X}_j)$, of the form (8) the prior expectation of the posterior variance of μ_j is of the form (39) and thus assuming among strata independence

$$E_{(s,\underset{\sim}{x})}[V(\mu|s,\underset{\sim}{x}))] = \sum_{j=1}^{k} \left(\frac{N_j-n_j'}{n_j+n_j'}\right) \frac{K_j}{N^2}, \tag{43}$$

where $K_j = (N_j+n_j!)E_{\underset{\sim}{\theta}_j}V(X_{ij}|\underset{\sim}{\theta}_j)$ and n_j' are known positive constants. For this class of priors the allocation problem becomes one of choosing $\underset{\sim}{n} = (n_1,\ldots,n_k)$ to minimize (43) subject to the constraints (41).

Before proceeding into the details of the solution to this problem, we indicate beiefly the forms of K_j and n_j' for several specific priors in the class for which (43) holds. Table 1 gives several simple examples of priors

Table 1

$p_j(X\mid\underset{\sim}{\theta}_j)$	$\underset{\sim}{\theta}_j$	$f_j(\underset{\sim}{\theta}_j)$	K_j	n'_j
$e^{-\theta_j}\theta_j^x/x!$	θ_j	$\frac{\alpha_j^{\eta_j}}{\Gamma(\eta_j)}\theta_j^{\eta_j-1}e^{-\alpha_j\theta_j}$	$\frac{(N_j+\alpha_j)\eta_j}{\alpha_j}$	α_j
$\frac{1}{\sqrt{2\pi}\sigma_j}e^{-\frac{1}{2\sigma_j^2}(x-\theta_j)^2}$	θ_j	$\frac{1}{\sqrt{2\pi v'_j}}e^{-\frac{1}{2v'_j}(\theta_j-m'_j)^2}$	$\sigma_j^2(N_j+n'_j)$	σ_j^2/v'_j
$\frac{h_j^{\frac{1}{2}}}{\sqrt{2\pi}}e^{-\frac{1}{2}h_j(X-\theta_j)^2}$	(θ_j,h_j)	Normal-gamma, (10)	$\frac{(N_j+n'_j)\nu'_j v'_j}{\nu'_j-2}$	n'_j
$N_j!\prod_{\ell=1}^{m_j}P_{\ell j}^{M_{\ell j}}/\prod_{\ell=1}^{m_j}M_{\ell j}!$	$\underset{\sim}{p}_j$	Dirichlet, (25)	$\frac{N_j+\epsilon_j}{\epsilon_j(\epsilon_j+1)}W_j^*$	ϵ_j

$$^*W_j=\left[\sum_{\ell=1}^{m_j}y_{\ell j}^2(\epsilon_j-\epsilon_{\ell j})-\sum_{\ell\neq q}y_{\ell j}y_{qj}\epsilon_{\ell j}\epsilon_{qj}\right]$$

of the form (8) (generated by $p_j(X_{ij}|\underset{\sim}{\theta}_j)$ and $f_j(\underset{\sim}{\theta}_j)$) where the prior expectation of the posterior variance of μ_j is of the form $(N_j-n_j^!/n_j+n_j^!)(K_j/N_j^2)$ so that if the prior in each stratum is of this form then (43) holds. The table gives the specific forms of $p_j(X_{ij}|\underset{\sim}{\theta}_j)$, $\underset{\sim}{\theta}_j$, $f_j(\underset{\sim}{\theta}_j)$, K_j, and $n_j^!$.

This table is easily extendable to cover a variety of other distributions.

We now proceed to derive an algorithm for the solution to the allocation problem, which by the results above would seem to have wide applicability within the class of priors discussed here.

3.3. A Solution to the Allocation Problem:

It is easily seen that the objective function, (43) to be minimized by the choice of n_j (subject to the constraints (41) and (44)) is a convex function of $\underset{\sim}{n}$. The well-known theorem of Kuhn and Tucker (1950) is thus applicable to this problem. By a simple adaptation of their theorem to the specific present allocation problem we have the following result.

<u>Theorem 1</u>: The minimum of $\sum_{j=1}^{k} \frac{N_j-n_j}{N^2} \frac{K_j}{n_j+n_j^!}$ subject to the constraints $\sum_{j=1}^{k} c_j n_j - C \leq 0$, $0 \leq n_j \leq N_j$, $j=1,\ldots,k$ is given by $\underset{\sim}{n}^0 = (n_1^0,\ldots,n_k^0)$ if and only if there exist non-negative numbers $\lambda, \lambda_1, \lambda_2,\ldots,\lambda_k$ such that

$$-\frac{K_j(N_j+n_j^!)}{N^2(n_j^0+n_j^!)} + \lambda c_j + \lambda_j \geq 0, \quad j=1,\ldots,k, \tag{45}$$

$$\sum_{j=1}^{k} \{\frac{-K_j(N_j+n_j^!)}{N^2(n_j^0+n_j^!)} + \lambda c_j + \lambda_j\}n_j^0 = 0, \tag{46}$$

and

$$\lambda[\sum_{j=1}^{k} c_j n_j^0 - C] + \sum_{j=1}^{k} \lambda(n_j^0-N_j) = 0. \tag{47}$$

The obvious result that if $C \geq \sum_{j=1}^{k} c_j N_j$ then the optimal allocation is a census, i.e., $n_j^0 = N_j$, $j=1,\ldots,k$, follows as a trivial corollary of this theorem. It is also clear that if $C < \sum_{j=1}^{k} c_j N_j$, then by the convex, strictly decreasing nature of the objective function, the budget constraint will hold with equality.

In many practical applications where n^0 is desired only for some prespecified total sampling budget, C, that optimal allocation may often be quickly and easily obtained by use of the following theorem. Here, and in the sequel, we define

$$w_j = \left[\frac{K_j(N_j+n_j')}{N^2 c_j}\right]^{\frac{1}{2}}, \quad j=1,\ldots,k. \tag{48}$$

Theorem 2: If $C < \sum_{j=1}^{k} c_j N_n$ and if $\min_j w_j/n_j' > \max_j w_j/(N_j+n_j')$ then for all budgets in the interval

$$\frac{\sum_{i=1}^{k} w_i c_i}{\min_j w_j/n_j'} - \sum_{i=1}^{k} n_i' c_i < C < \frac{\sum_{i=1}^{k} w_i c_i}{\max \frac{w_j}{N_j+n_j'}} - \sum_{i=1}^{k} n_i' c_i, \tag{49}$$

the optimum allocation is given by $0 < n_j^0 < N_j$ where

$$n_j^0 = \left[\frac{C+\sum_{i=1}^{k} c_i n_i'}{\sum_{i=1}^{k} c_i w_i}\right] w_j - n_j', \quad j=1,\ldots,k. \tag{50}$$

Proof: By taking $\lambda_j = 0$, $j=1,\ldots,k$, and

$$\lambda = \left[\sum_{j=1}^{k} c_j w_j/(C + \sum_{j=1}^{k} c_j n_j')\right]^2$$

it is easily verified that all the conditions (45)-(47) of Theorem 1 are satisfied with strict equality in (45). Furthermore, the condition (49) guarantees that $0 < n_j^0 < N_j$ for all j.

While the optimal allocation, n^0, is often yielded by Theorem 2, the complete solution for n^0 for all $0 \leq C \leq \sum_{j=1}^{k} c_j N_j$ is given by repeated application of the results contained in the next theorem.

<u>Theorem 3</u>: Let $\alpha_j = w_j/(N_j+n_j')$ and $\beta_j = w_j/n_j'$, $\alpha_j < \beta_j$ for $j \varepsilon K = \{1,...,k\}$. Let $x_1 \leq x_2 \leq ... \leq x_{2k}$ be the ordered values of the 2k numbers α_j, β_j, j=1,...,k. Let x_i and x_{i+1} be any two consecutive values of these ordered α_j's and β_j's such that $x_i < x_{i+1}$. Finally let

$$B_i = \{j | \alpha_j \geq x_{i+1}\} \subseteq K,$$

$$O_i = \{j | \beta_j \leq x_i\} \subseteq K,$$

and

$$P_i = \bar{B}_i \cap \bar{O}_i,$$

where $\bar{B}_i$ and $\bar{O}_i$ are the complements of B_i and O_i with respect to K. For all C in the non-degenerate sub-interval of

$$[0, \sum_{j=1}^{k} N_j c_j]$$

given by

$$\sum_{j\varepsilon P_i} w_j c_j/x_{i+1} - \sum_{j\varepsilon P_i} n_j' c_j + \sum_{j\varepsilon B_i} N_j c_j \leq C \leq \sum_{j\varepsilon P_i} w_j c_j/x_i - \sum_{j\varepsilon P_i} n_j' c_j + \sum_{j\varepsilon B_i} N_j c_j \qquad (51)$$

the solution to the allocation problem is given by

$$n_j^0 = \begin{cases} N_j, & j \varepsilon B_i \quad (52) \\ 0\ , & j \varepsilon O_i \quad (53) \\ \left[\dfrac{C + \sum\limits_{j \varepsilon P_i} n_j' c_j - \sum\limits_{j \varepsilon B_i} c_j N_j}{\sum\limits_{j \varepsilon P_i} w_j c_j} \right] w_j - n_j' \ , & j \varepsilon P_i . \quad (54) \end{cases}$$

Proof: It suffices to show first that (51) is indeed a sub-interval of

$$[0, \sum_{j=1}^{k} c_j N_j],$$

for (51) is clearly a non-degenerate interval since $x_i < x_{i+1}$. And second to show that for all C in that interval there exist non-negative values $\lambda, \lambda_1, \ldots, \lambda_k$ such that for $\underset{\sim}{n}^0$ given by (52)-(54) the conditions of Theorem 1 hold.

To accomplish this we first note that $j \varepsilon P_i$ implies

$$\beta_j = \frac{w_j}{n_j'} \geq x_{i+1}$$

and hence

$$\sum_{j \varepsilon P_i} w_j c_j \geq x_{i+1} \sum_{j \varepsilon P_i} n_j' c_j$$

from which it follows that the leftmost side of (51) is always non-negative. Similarly, $j \varepsilon P_i$ implies

$$\alpha_j = \frac{w_j}{N_j + n_j'} \leq x_i$$

and thus

$$\sum_{j \varepsilon P_i} w_j c_j / x_i \leq \sum_{j \varepsilon P_i} c_j N_j + \sum_{j \varepsilon P_i} c_j n_j'$$

from which it follows that the right side of (51) is less than or equal to

$$\sum_{j\varepsilon P_i} c_j N_j + \sum_{j\varepsilon B_i} c_j N_j \le \sum_{j=1} c_j N_j .$$

To show that for C in the interval (51) the solution is indeed given by (52)-(54) we set

$$\sqrt{\lambda} = \sum_{j\varepsilon P_i} c_j w_j / [C + \sum_{j\varepsilon P_i} c_j n_j' - \sum_{j\varepsilon B_i} c_j N_j]. \tag{55}$$

It then follows that for C in the interval (51)

$$x_i \le \sqrt{\lambda} \le x_{i+1} \tag{56}$$

and for all such C and $j\varepsilon P_i$

$$\alpha_j = \frac{w_j}{N_j + n_j'} \le x_i \le \sqrt{\lambda} \le x_{i+1} \le \frac{w_j}{n_j'} = \beta_j \tag{57}$$

It then follows using (57) and (55) that $\lambda \ge 0$ and n_j^0, given by (52)-(54), is such that for all j $< 0 \le n_j \le N_j$. Also, if we let

$$\lambda_j = \begin{cases} 0 & j \not\in B_i \\ \dfrac{c_j w_j^2}{(N_j + n_j')^2} - c_j \lambda , & j\varepsilon B_i , \end{cases} \tag{58}$$

then, since $j\varepsilon B_i$ implies

$$\frac{w_j}{(N_j + n_j')} = \alpha_j \ge x_{i+1} \ge \sqrt{\lambda},$$

it follows that λ and λ_j as defined above are all nonnegative. We now assert that for $\underset{\sim}{n}^0$ as given in (52)-(54) and for these lambdas all the conditions of Theorem 1 hold for all C in the interval (51). It is easily verified by direct computation that

$$\sum_{j=1}^{k} c_j n_j^0 = C,$$

and since $\lambda_j = 0$ except when $n_j^0 = N_j$ condition (47) is seen to hold. By a straightforward computation it is seen that (45) holds with equality as long as $j \not\in O_i$, and thus (46) holds. Finally for $j\varepsilon O_i$, $\lambda_j = 0$ and

$$\frac{K_j(N_j+n_j')}{N^2 n_j'^2} = \frac{w_j^2 c_j}{n_j'^2} = \beta_j^2 c_j \leq x_i^2 c_j \leq c_j \lambda,$$

and thus (45) holds for all j. This completes the proof.

This theorem may be used as i varies from 1 to 2k-1, ignoring i's for which $x_i = x_{i+1}$, to n_j^0 for all $0 \leq C \leq \sum_{i=1}^{k} c_i N_i$. It may be shown that the sequence of intervals, (51), yielded by this procedure are non-overlapping (except at endpoints) and exhaust the interval $[0, \sum_{i=1}^{k} c_i N_i]$, e.g., for i=1 the right side of (51) equals $\sum_{i=1}^{k} c_i N_i$ and for i=2k-1 the left side of (51) is zero. The procedure or algorithm will be illustrated in the next subsection.

Another result which is of some use concerns the behavior of the minimum value of the objective function as a function of the budget, C. We let

$$v^*(C) = \sum_{j=1}^{k} \frac{N_j - n_j^0}{N^2} \frac{K_j}{n_j^0 + n_j'} \tag{59}$$

and have the following theorem.

Theorem 4: $v^*(C)$ is a convex and decreasing function of C on the interval $[0, \sum_{j=1}^{k} c_j N_j]$ and possesses a continuous first derivative in this interval. In the interval (51) $v^*(C)$ has the explicit form

$$v_i^*(C) = \sum_{j\varepsilon O_i} \frac{N_j K_j}{N^2 n_j'} - \sum_{j\varepsilon P_i} \frac{K_j}{N^2} + \frac{[\sum_{j\varepsilon P_i} w_j c_j]^2}{C + \sum_{j\varepsilon P_i} c_j n_j' - \sum_{j\varepsilon B_i} c_j N_j}. \tag{60}$$

Also in that interval the derivative of $v^*(C)$ is given by

$$\frac{dv^*(C)}{dC} = -\lambda,$$

where λ is as in (55)

3.5. Optimal Design with Quadratic Losses:

Given the results above for the solution of the allocation problem a solution to the problem described in Section 3.1 of optimal design under quadratic losses is immediately obtainable. Recall that this design problem was one of choosing n_j, $0 \leq n_j \leq N_j$, $j=1,\ldots,k$, to minimize (42) or, using (43) to minimize

$$\ell(\underset{\sim}{n}) = \gamma \sum_{j=1}^{k} \frac{N_j - n_j}{N^2} \frac{K_j}{n_j' + n_j} + \sum_{j=1}^{k} c_j n_j . \tag{61}$$

It is intuitively clear, and demonstrated generally in Ericson (1967a), that this minimization may be carried out in two stages, (a) minimize (61) subject to the condition that $\sum_{j=1}^{k} c_j n_j \leq C$, as a function of C, is given by:

$$\ell^*(C) = \min_{\underset{\sim}{n} \mid \underset{\sim}{n}\underset{\sim}{c}^t \leq C} \ell(\underset{\sim}{n}) = \begin{cases} \sum_{j=1}^{k} c_j N_j, & C \geq \sum_{i=1}^{k} c_i N_i \\ \gamma v_i^*(C) + C, & C \text{ in the ith interval in (51)} \end{cases} \tag{62}$$

where v_i^* is as given in (60)

The solution to the design problem is then to determine that C, C^0 say, which minimizes (62), then using the results for optimal allocation to find the optimal allocation $\underset{\sim}{n}^0$ for the budget C^0. This is very easily accomplished for the derivative of $\ell^*(C)$ with respect to C is given by

$$\frac{d\ell^*(C)}{dC} = \begin{cases} 0 & \text{for } C \geq \sum_{i=1}^{k} c_i N_i \\ \gamma \dfrac{dv_i^*(C)}{dC} + 1 = -\gamma\lambda + 1 & \text{for C in the ith interval (51)} \end{cases} \quad (63)$$

where λ, a function of C, is given in the ith interval, (51), by (55). This derivative is zero when $C \geq \sum_{i=1}^{k} c_i N_i$ or when $\lambda = 1/\gamma$ for some C in the interval $[0, \sum_{i=1}^{k} c_i N_i]$. By Theorem 4, λ is a continuous decreasing function of C and for C in the ith interval, given by (51), $x_i \leq \lambda \leq x_{i+1}$. It then follows that if $1 \leq i^* \leq 2k-1$ is such that $x_{i^*} \leq 1/\gamma \leq x_{i^*+1}$ then by solving the equation $\lambda=1/\gamma$, where the definition of λ for the i^*th interval is given by (55), for C one finds that

$$C^0 = \gamma^{\frac{1}{2}} \sum_{j\varepsilon P_{i^*}} c_j w_j - \sum_{j\varepsilon P_{i^*}} c_j n_j' + \sum_{j\varepsilon B_{i^*}} c_j N_j$$

and that then the solution to the optimal design problem is given by (52)-(54) for $C=C^0$ and $i=i^*$. If $1/\gamma < x_1 = \min_j \alpha_j$ the solution is a census $n_j^0=N_j$, $j=1,\ldots,k$, while if $1/\gamma > x_{2k} = \max_j \beta_j$ then it is easily seen that $\ell^*(C)$ is non-decreasing for all $C > 0$ and the optimal design is $\underset{\sim}{n}^0 = \underset{\sim}{0}$, i.e., no sampling is best.

Thus by a simple comparison of $1/\gamma$ with the ordered values x_i of Theorem 3 the solution to the decision problem of optimal design with quadratic losses is immediately obtained.

References

Basu, D. (1958) On sampling with and without replacement, Sankhya 20, 287-294.

Box, G. E. P., and Tiao, G. C. (1962) A further look at robustness via Bayes' theorem, Biometrika 49, 419-432.

DeFinetti, B. (1964) La prevision ses lpis logiques, ses sources subjectives, Annales de l'Institut Henri Poincare, 7, 1-68 (1937). Appearing in English translation with new notes in Studies in Subjective Probability edited by Kyburg, H. E., and Smokler, H. E., John Wiley and Sons, 1964.

Ericson, W. A. (1965) Optimum stratified sampling using prior information, J. Amer. Statist, Assn. 60, 750-771.

Ericson, W. A. (1967a) On the economic choice of experiment sizes for decision regarding certain linear combinations, J. Roy. Statist. Soc. (B), 29, #3, 503-512.

Ericson, W. A. (1967) Subjective Bayesian models in sampling finite populations, I. Manuscript submitted for publication.

Ericson, W. A. (1969) A note on the posterior variance of a finite population mean, unpublished manuscript.

Godambe, V. P. (1955) A unified theory of sampling from finite populations, J. R. Statist. Soc. (B), 17, 267-278.

Godambe, V. P. (1965) A review of the contributions towards a unified theory of sampling from finite populations, Inter. Statist. Inst. Rev. 33, 242-285.

Godambe, V. P. (1966) A new approach to sampling from finite populations I and II, J. R. Statist. Soc. (B) 28, 310-328.

Hajek, J. (1959) Optimum strategy and other problems in probability sampling, Casopis Pest. Mat., 84, 387-423.

Kuhn, H. W., and Tucker, A. W. (1950) Non linear programming, Proc. 2nd Berkeley Symp. Math. Statist. and Prob. (Ed. by J. Neyman), University of California Press, Berkeley.

Raiffa, H., and Schlaifer, R. O. (1961) Applied Statistical Decision Theory. Boston: Graduate School of Business Administration, Harvard University.

Roberts, H. V. (1965) Probabilistic prediction, J. Amer. Statist. Assn., 60, 50-62.

Savage, L. J. (1962) The Foundations of Statistical Inference. London: Methuen and Co., Ltd.

Wilks, S. S. (1962) Mathematical Statistics. New York: John Wiley and Sons, Inc.

APPLICATIONS OF LIKELIHOOD AND FIDUCIAL PROBABILITY TO SAMPLING FINITE POPULATIONS

J. D. Kalbfleisch and D. A. Sprott*

University of Waterloo, Canada

1. Introduction

Statistical Inference has usually concerned itself with the problem of making an informative statement about a hypothetical population based on an observed sample from that population. Usually this implies measuring the plausibility of values of a parameter θ defining the population. Many problems of inference, however, seem to depend on measuring the plausibility of future events whose probabilities of occurrence depend on the hypothetical population from which the original sample was drawn. That is, the inferences depend on measuring the plausibility of an unobserved x, where the actual probability of x is a function of θ. Such problems have been considered by Fisher (1956) under the titles of Fiducial Prediction, Bayesian Prediction, and Inference from Likelihoods. At least two problems which can be made to depend on this approach are discrimination and inferences about a real finite population.

This paper generalizes Fisher's approaches to prediction of future observations and applies the results to the estimation of statistics which are functions of the finite population under consideration. The actual sampling model that this entails is discussed and compared with other approaches to finite populations not involving hypothetical background populations. The problem of randomization (Godambe (1967)) is also discussed.

---

*With a grant from National Research Council of Canada.

2. Estimation of future observations

Suppose a sample $x_1,\ldots,x_n$ from the distribution $f(x;\, \theta)$ is observed. It is required to predict a future observation y from this distribution, or more generally, to predict the function $g(y_1,\ldots,y_m)$ of m future observations.

a) Application of the fiducial agrument.

As Fisher (1935, 1956) pointed out, the fiducial argument leads to statements about the future contingency. Suppose there exists a set of statistics T which (conditioned on the ancillaries A if necessary) is sufficient for the unknown parameter θ. Suppose further necessary conditions for a fiducial distribution are satisfied so that

$$f(T;\, \theta | A)dT$$

may be inverted to yield the fiducial distribution of θ as

$$f(\theta;\, T,A)d\theta. \tag{1}$$

The problem is to estimate a function $g(y_1,\ldots,y_m)$ of m future observations where the distribution of g is

$$f(g | \theta)dg;$$

the notation $f(g|\theta)$ is used to indicate that θ has the logical status of a random variable distributed according to (1). The joint density is

$$f(g,\, \theta;\, T,A) = f(g|\theta)\, f(\theta;\, T,A).$$

To make statements about g in the absence of knowledge of θ, the marginal fiducial distribution can be used:

$$f(g;\, T,A) = \int_\theta f(g,\, \theta;\, T,A)d\theta$$

Example 2.1.

Fiducial prediction for a normal and exponential

distribution has been considered by Fisher (1956). An example which perhaps better illustrates the scope of the statements possible arises in life testing. Suppose the failure times $t_1,\ldots,t_n$ have been observed on n items and it is required to predict the number k of items which fail when m items from the same population are observed over a period of length τ. Assuming an exponential distribution for failure time t

$$f(t;\ \theta) = \frac{1}{\theta}\exp(-\frac{t}{\theta})$$

then if θ were known the distribution of k would be

$$f(k;\ \theta) = \binom{m}{k}[1 - \exp(-\tau/\theta)]^k \exp[-(m-k)\tau/\theta]. \quad (2)$$

Having observed $T = \Sigma t_i$, the fiducial distribution of θ is

$$\frac{T^n \exp(-T/\theta)}{\theta^{n+1}\ (n-1)!}\ . \quad (3)$$

Thus, the joint distribution of θ and k for the observed T is the product of (2) and (3). Integrating over θ gives the distribution of k in the absence of knowledge of θ as

$$f(k) = T^n\binom{m}{k}\sum_{i=0}^{k}(-1)^i\binom{k}{i}\frac{1}{[T+(m-k+i)\tau]^n}\ . \quad (4)$$

The distribution (4) takes account of both the uncertainty about θ given by (3) and the randomness of the process. Thus (4) measures the probability of a future value k.

b) The application of likelihood.

Often the necessary conditions for obtaining the fiducial distribution of θ on the basis of the sample are lacking; then only a likelihood function for θ arising from the distribution of the observations

$$f(x_1,\ldots,x_n;\ \theta) \quad (5)$$

is available.

Suppose we wish to estimate a function g of the future observations $y_1,\ldots,y_m$. Knowing θ, any statements about g would be made from the distribution of g.

$$f(g;\ \theta)dg \tag{6}$$

All the information about θ is contained in the distribution (5) and the problem is then to combine (5) and (6) to make as informative a statement as possible about g. If g had in fact been observed and no other information were available (6) could be inverted to obtain the fiducial distribution of θ for the fixed g

$$f(\theta;\ g)d\theta .$$

Then the joint distribution of the observations and θ for the given g can be integrated to yield the marginal distribution

$$f(x_1,\ldots,x_n;\ g) = \int_\theta f(x_1,\ldots,x_n|\theta)\ f(\theta;\ g)d\theta\ . \tag{7}$$

This is the distribution of the observations $x_1,\ldots,x_n$ based on a hypothesized fixed g. It is thus the likelihood of g (considered as a fixed parameter) in the light of the actual observations $x_1,\ldots,x_n$. Note that (7) could be considered a "fiducial likelihood" since it incorporates in it the fiducial uncertainty about θ. Statements of likelihood are then available for g from (7).

Example 2.2.

In example 2.1 suppose n items are observed over a period τ and k of these fail at times $t_1,\ldots,t_k$ while the n-k others are still operating at the end of the test period. The likelihood of θ is then,

$$\theta^{-k}\exp\{-\frac{\Sigma t_i + (n-k)\tau}{\theta}\}$$

which does not yield a fiducial distribution for θ. Consider a future sample in which all of the observations are observed to failure at times $s_1,\ldots,s_m$. If

$s_1,\ldots,s_m$ were known, then $S = \Sigma s_i$ would be sufficient for θ; given S and no other information; the fiducial distribution of θ would be

$$f(\theta;\ S) = \{\frac{S^m}{\theta^{m+1}} \frac{\exp(-S/\theta)}{(m-1)!}\} .$$

The distribution of $t_1,\ldots,t_k$ and k for the hypothesized S is proportional to

$$\int_\theta S^m \theta^{-k-m-1} \exp[-\frac{S+\Sigma t_i+(n-k)\tau}{\theta}]d\theta/(m-1)!$$

or

$$\frac{S^m}{(S+T)^{m+k}}$$

where $T = \Sigma t_i+(n-k)\tau$. This attains its maximum at the value $\hat{S} = \frac{mT}{k}$. Thus, given T and k (observed) the relative likelihood function for any hypothesized S is

$$R(S) = \frac{(m+k)^{m+k}}{m^m k^k} \frac{S^m T^k}{(S+T)^{m+k}} . \tag{9}$$

It is interesting to note that if one of the items had been chosen in advance to be observed to failure, a fiducial distribution for θ based on $t_1,\ldots,t_k$ could have been obtained (Fisher (1956)).

c) Application of Fisher's likelihood prediction.

The argument in (a) requires the existence of a fiducial distribution based on the observed sample. The argument in (b) requires at least the existence of a fiducial distribution based on the statistic of interest in the future sample. In some cases neither of these fiducial distributions exists. For instance, if the distribution from which the data arise is binomial or Poisson, neither of the above methods would be applicable.

Suppose $x = (x_1,\ldots,x_n)$ is observed when the underlying value of θ is θ_x; plausible values of θ_x are those for which the relative likelihood function $R(\theta_x; x)$ is large. If $y = (y_1,\ldots,y_m)$ independent of x is observed when the underlying θ value is θ_y then plausible values of θ_y are those for which $R(\theta_y; y)$ is large. Since x and y are independent, the plausibility of the pair (θ_x, θ_y) can be measured by $R(\theta_x; x)\ R(\theta_y; y)$. The plausibility of the event $\theta_x = \theta_y = \theta$ is then

$$R(\theta; x)\ R(\theta; y)$$

which is maximized at $\hat{\theta}$, the maximum likelihood estimate for a common θ given x and y. Thus the maximum of the relative likelihood of the event $\theta_x = \theta_y$ is

$$R(\hat{\theta}; x)\ R(\hat{\theta}; y) \tag{10}$$

and this gives the plausibility of the triple $(x, y, \theta_x = \theta_y)$.

Now, suppose that y is a future observation to be drawn from the same population as x. It is then known that $\theta_x = \theta_y$ and x has been observed. The plausibility measured by (10) must then apply to the hypothesized future y. Thus if a postulated y entails a small value of (10), this implies that it is unlikely that $\theta_x = \theta_y$; however this is known to be true, and thus doubt is cast on that particular y. Indeed if any two elements of the triple $(x, y, \theta_x = \theta_y)$ are known, the plausibility (suitably standardized) can be taken to apply to the third element. The principle used here is not unlike that forming the basis of the test of significance, namely our reluctance "to accept rare events". That is, if two elements A and B are in conflict, and if A is known to be true the doubt is cast on the truth of B. The expression (10) may be called the likelihood of the future y given x or equivalently of x given y. This is not a likelihood function in the usual sense since it does not arise directly from a probability distribution. Fisher (1956) has given an example of the use of (10) applied to the binomial distribution. Sprott (1968) also gave an example of the use of (10).

In cases where the likelihood function (7) discussed in (b) above exists, the two likelihood functions (7) and (10) are often the same.

Example 2.3.

Suppose, as in example 2, we have observed the number k of failures and T the total life time over a period of length τ (i.e. $T = \Sigma t_i + (n-k)\tau$). It is required to estimate r the number of failures (at times $s_1, \ldots, s_r$) and $S = \Sigma s_i + (m-r)\eta$ when m further items are observed over a period of length η. Here neither the past nor future observations yield a fiducial distribution of θ. It may be checked that the likelihood (10) is

$$\frac{T^k \; S^r \; (k+r)^{k+r}}{(S+T)^{k+r} \; k^k r^r}$$

which is identical to (9) when $r = m(\eta \to \infty)$.

3. Applications to sampling theory

As has been pointed out by several authors, in a sampling survey the finite population $(X_1, \ldots X_N)$ is a vector in R_N. The units in the finite population are numbered and distinguishable. The finite population may be sampled by picking at random n integers from $1, \ldots, N$ and examining the variate values associated with these. On the basis of the sample values $x_1, \ldots, x_n$, it is required to make an inference about the finite population as a whole or more usually about some function of the finite population such as the mean $\bar{X} = \Sigma X_i / N$ or the variance $S^2 = \frac{1}{N} \Sigma (X_i - X)^2$.

One suggested approach (Cochran (1939, 1946)) is to consider the finite population as having been selected at random from a hypothetical population (often referred to as a super population). The observed sample, being a random sample from the finite population, is also a random sample from the hypothetical population. An inductive inference may first be made to the hypothetical population and then a deductive inference may be made about the finite population from which this sample may have arisen. This is equivalent to predicting a finite set of future observations which together with the observed sample forms the finite population under discussion. Thus the methods of section 2 are applicable.

This approach might be criticized on the ground that the inference concerns a hypothetical reconstruction of the finite population and not the objective finite population itself. As in most cases concerning fiducial probability and repeated sampling from the same hypothetical population, this criticism is not well founded. In fact, the finite population might well have arisen from a random physical process described by the hypothetical distribution. An example of this might be a finite population of heights, weights or intelligence; these might well be considered as large samples from a normal distribution defined by the genetical mechanisms giving rise to the finite population. In any case, the logic and interpretation are the same as those used in making fiducial statements of uncertainty concerning a fixed parameter (Fisher 1956). An empirical verification of the resulting fiducial probability statement would entail sampling a different finite population each time, these being random selections from the same hypothetical distribution.

A second approach which has been developed by Godambe (1955) avoids any reference to a hypothetical population. In this case an empirical verification would entail repeatedly sampling the same objective finite population.

There is some doubt as to whether this would be a meaningful procedure since knowledge would be gained about the finite population by repeatedly sampling it. In verifying statements of probability it is necessary that the state of knowledge remain the same. It would seem therefore that even in this approach one would have to consider a hypothetical population of finite populations (the reference set of Fisher 1959) in all relevant respects like the given one and to which it belongs.

Attempts at inferences using this second model lead to difficulties since the structure of the finite population of which the experimenter has knowledge is usually too weak to allow statements any more informative than an unbiased estimator with no bound on the variance unless at least the variance of the finite population is known. It should be noted that when there is reason to believe the finite population to have arisen from a hypothetical population, this forms an important part of the information. It is as dangerous to neglect this information as it is to assume too much (Fisher, p. 46, 1966). The differences between these two approaches in relation to a problem in sampling for proportions are discussed in section 7.

4. Sampling with replacement

The identification of the units in the finite population is often mentioned in the literature and the question arises as to what role this plays in the estimation problem (Godambe 1967). This is perhaps best illustrated in the case of sampling with replacement where there is a positive probability of drawing the same unit twice. Suppose that n units are selected at random from the finite population $(X_1, X_2, \ldots, X_N)$ which may be considered a random sample from the hypothetical population defined by the density $f(X; \theta)$. Suppose further that s_i of the units appear i times in the sample with variate values $x_{i1}, x_{i2}, \ldots, x_{is_i}$; then

$$\Sigma s_i = s \text{ (say)}$$

and

$$\Sigma i s_i = n.$$

The probability of drawing the i'th unit and observing variate value x is then

$$\Pr(x | \text{i'th unit}) \Pr(\text{drawing the i'th unit}) = f(x; \theta/N$$

if the i'th unit has not been drawn before and is 1/N if the i'th unit has been drawn and had variate value x. The probability of the sample, ignoring only the order of drawing, is

$$n! \prod_{i=1}^{r} \prod_{j=1}^{s_i} f(x_{ij};\ \theta)/[\pi(i!)^{s_i} N^n] .$$

Clearly we are left with a reduced sample space of s variate values consisting of the different units sampled. It should be noted that if the actual number on the units drawn has no influence on inferences, this sample is then equivalent to any other sample with the same variate values, allowing the numbers on the sampled units to range over all possible permutations and combinations. The probability of such a sample is

$$s!\ n!\ \binom{N}{s} \prod_i \prod_j f(x_{ij};\ \theta)/[\pi(s_i!)(i!)^{s_i} N^n] .$$

Since sampling with replacement is equivalent to sampling without replacement if we use the reduced sample, most of the subsequent arguments in this paper will consider only sampling without replacement. The role of the identification of the units will be discussed further in section 7.

5. Examples from a normal population

Example 5.1.

Suppose that a finite population $X_1,\ldots,X_N$ can be considered as a random sample from the normal distribution $N(\mu,\sigma^2)$ and n numbers $i_1, i_2,\ldots,i_n$ are selected at random from 1 to N. Suppose the corresponding $X_{i1}, X_{i2},\ldots,X_{in}$ are observed to have variate values $x_1, x_2,\ldots,x_n$. Then $x_1,\ldots,x_n$ is a random sample from $N(\mu,\sigma^2)$. We shall consider the problem of inference concerning the mean $\bar{X}$ and variance S^2 of the finite population.

a) If σ^2 is known then $\Sigma x_i/n = \bar{x}$ is sufficient for μ; given $\bar{x}$ the fiducial distribution of μ is $N(\bar{x}, \sigma^2/n)$. If $\bar{y}$ is the mean of the remaining (unsampled) N-n variate values the distribution of $\bar{y}$ given μ is

$N[\mu, \sigma^2/(N-n)]$. Using the method of section 2 (a) the fiducial distribution of $\bar{y}$ given $\bar{x}$ is $N(\bar{x}, \frac{\sigma^2}{n} + \frac{\sigma^2}{N-n})$. The mean of the finite population $\bar{X} = [(N-n)\bar{y} + n\bar{x}]/N$ is then distributed as

$$N(\bar{x}, \frac{(N-n)\sigma^2}{Nn}) .$$

The classical point estimate of $\bar{X}$ is $\bar{x}$, the expected value of $\bar{X}$ over the fiducial distribution. The variance of $\bar{X}$ above is also the usual result.

b) If μ and σ^2 are unknown then $\bar{x}$ and $s_x^2 = \Sigma(x_i-\bar{x})^2/(n-1)$ are jointly sufficient for μ, σ^2. The fiducial distribution of μ and σ^2 has been given many times by Fisher (e.g. Fisher 1956, p. 119). If $\bar{y}$ is the mean of the remaining variate values then the distribution of $\bar{y}$ for a given μ, σ^2 is $N[\mu, \sigma^2/N-n)]$. The marginal distribution of $\bar{y}$ depending only on the observed $\bar{x}$, s_x^2 is

$$f(\bar{y}; \bar{x}, s_x^2) = \int_{\mu} \int_{\sigma^2} f(\bar{y}|\mu, \sigma^2)\, f(\mu, \sigma^2; \bar{x}, s_x^2) d\mu\, d\sigma^2 .$$

This gives

$$\frac{(\bar{x}-\bar{y})\sqrt{(N-n)n}}{\sqrt{N}\, s_x} = \frac{(\bar{X}-\bar{x})\sqrt{Nn}}{\sqrt{N-n}\, s_x}$$

as a t variate with n-1 degrees of freedom. From this, exact probability statements about the unknown $\bar{X}$ are calculable. Here again the expected value of $\bar{X}$ and its variance are the same as the classical results (Kendall and Stuart 1962).

c) More generally if s_y^2 is the variance of the remaining variate values in the population, then

$$\frac{(\bar{x}-\bar{y})\ \sqrt{n(N-n)(N-2)}}{\sqrt{N}\ \sqrt{(n-1)s_x^2 + (N-n-1)s_y^2}}\ ,\ \frac{s_x^2}{s_y^2}$$

are a $t_{(N-2)}$ and for $F_{(n-1),N-n-1)}$ variate respectively as given by Fisher (1956). The joint distribution of $\bar{X}$ and S^2 of the finite population can be obtained from this.

d) Stratified Sampling.

In some cases where there is reason to believe that all the population units were not drawn from a common hypothetical distribution it is desirable to divide the finite population into strata. If there are k such strata, one could assume the observations in the strata to have been drawn at random from k different hypothetical populations with one or more unknown parameters.

Thus if $X_{i1},\ldots,X_{iN_i}$ are the variate values in the i'th stratum we might assume that these subpopulations are independent samples from $N(\mu_i, \sigma_i^2)$. If n observations $x_{i1},\ldots,x_{in_i}$ are taken from the i'th stratum and if

$$\bar{x}_i = \sum_{j=1}^{n_i} x_{ij}/n_i\ ,$$

then μ_i is fiducially a $N(\bar{x}_i, \sigma_i^2/n_i)$ variate. In the i'th stratum the mean of the remaining variates is $\bar{y}_i$, so that the fiducial distribution of $\bar{y}_i$ for a given $\bar{x}_i$ is $N[x_i, \sigma_i^2/n_i + \sigma_i^2/(N_i-n_i)]$ where σ_i^2 is assumed known. The population mean is

$$\bar{X} = \sum_i \sum_j X_{ij}/N = \sum_{i=1}^{k} [(N_i-n_i)\bar{y}_i+n_i\bar{x}_i]/N.$$

Hence the fiducial distribution of the mean $\bar{X}$ is

$$N(\Sigma N_i\bar{x}_i/N,\ \Sigma(N_i-n_i)N_i\sigma_i^2/n_iN^2).$$

Here again the mean and variance of $\bar{X}$ are the classical estimates of $\bar{X}$ and the variance of that estimate. Thus if we wish to minimize the variance of $\bar{X}$ for a fixed sample size $\Sigma n_i = n$ the n_i's should be chosen so that

$$n_i = N_i\sigma_i n/\Sigma N_i\sigma_i$$

as given by Kendall and Stuart (1962). If the σ_i are unknown then the population total is then distributed as a linear combination of t variates.

If we may assume the unknown variances to be equal then this becomes a single t variate with $\Sigma n_i - k$ degrees of freedom. In this case the n_i's should be chosen proportional to the stratum sizes.

e) Sampling with a Covariate.

Stratification (like the use of blocks in experimental design) can increase precision when there are qualitative factors affecting the observations. Similarly when there are quantitative factors the information can sometimes be increased by the use of a covariate. Suppose the values of the covariate are known beforehand.

Example 5.2.

(i) Consider a finite population of pairs (Y_i, X_i) such as N fields with unknown yields Y_i and known acreages X_i. Assume that the Y_i have the distribution

$$N(\alpha X_i + \beta X_i^2,\ \sigma^2 X_i^2),$$

that is, the average fertility (yield/acre) is a linear function of the acreage. The total $T = \Sigma Y_i$ is to be estimated from an observed sample of n pairs (y_i, x_i). From the above y_i/x_i is a $N(\alpha' + \beta\ (x_i - \bar{x}),\ \sigma^2)$ variate where $\bar{x}$ is the mean of the sample x_i's. Then α' and β have the fiducial distributions

$$N(\bar{a},\ \sigma^2/n),\ N[n(\bar{y} - \bar{a}\bar{x})/\Sigma(x_i - \bar{x})^2,\ \sigma^2/\Sigma(x_i - \bar{x})^2]$$

independently, where

$$\bar{a} = \Sigma y_i/(nx_i),\ \bar{y} = \Sigma v_i/n\ .$$

If $(y_1', x_1', \ldots, y_{N-n}', x_{N-n}')$ are the remaining variate values in the finite population then $Y' = \Sigma y_i'$ has the distribution

$$N(\alpha'\Sigma x_i' + \beta\Sigma x_i'(x_i'-\bar{x}),\ \sigma^2\Sigma x_i'^2)\ .$$

Thus given the sample, Y' is fiducially distributed

$$N[\bar{a}\Sigma x_i' + \frac{n(\bar{y}-\bar{a}\bar{x})}{\Sigma(x_i-\bar{x})^2}\Sigma x_i'(x_i'-\bar{x}),\ \sigma^2\{\frac{(\Sigma X_i^2-N\bar{X}\bar{x})^2}{\Sigma(x_i-\bar{x})^2} + \frac{N^2\bar{X}^2}{n} - \Sigma X_i^2\}]$$

and the population total T has the fiducial distribution

$$N[\Sigma y_i + \bar{a}\Sigma x_i' + \frac{n(\bar{y}-\bar{a}\bar{x})\ \Sigma x_i'(x_i'-\bar{x})}{\Sigma(x_i-\bar{x})^2},\ \sigma^2\{\frac{(\Sigma X_i^2-N\bar{X}\bar{x})^2}{\Sigma(x_i-\bar{x})^2} + \frac{N^2\bar{X}^2}{n} - \Sigma X_i^2\}]\ .$$

The expected value of T is then

$$\frac{n(\bar{y}-\bar{a}\bar{x})\ (\Sigma X_i^2-N\bar{X}\bar{x})}{\Sigma(x_i-\bar{x})^2} + N\bar{a}\bar{X}\ .$$

If a sample is chosen so that $\bar{x}$ is close to $\bar{X}$ then

$$\frac{\Sigma X_i^2-N\bar{X}\bar{x}}{\Sigma(x_i-\bar{x})^2} \rightarrow \frac{\Sigma(X_i-\bar{X})^2}{\Sigma(x_i-\bar{x})^2} = \frac{(N-1)\sigma_X^2}{(n-1)s_x^2}\ .$$

If the sample x's are representative of the population X's then σ_X^2/s_x^2 should be close to unity and as a limiting case this estimator approaches the Hartley-Ross estimator

$$N\bar{a}\bar{X} + n(\bar{y}-\bar{a}\bar{x})\ (N-1)/(n-1)\ .$$

It can be shown that this is unbiased for simple random sampling without replacement.

(ii) If it is known that β is zero (so that fertility is independent of acreage) we have

$$E(Y_i|X_i) = \alpha X_i, \quad var(Y_i|X_i) = \sigma^2 X_i^2 ,$$

(cf. Hanurav 1967). Suppose then that Y_i has the distribution

$$N(\alpha X_i, \sigma^2 X_i^2) \quad . \tag{11}$$

The fiducial distribution of α is $N(\bar{a}, \sigma^2/n)$ and the distribution of Y' for a given α is $N(\alpha\Sigma x_i', \sigma^2\ \Sigma x_i'^2)$. If σ^2 and α are unknown the quantity

$$(y'-\bar{a}\Sigma x_i')/(s_a\sqrt{\Sigma x_i'^2 + \frac{(\Sigma x_i')^2}{n}}$$

has the t distribution with n-1 degrees of freedom, where $s_a^2 = \Sigma(y_i/x_i-\bar{a})^2/(n-1)$, from which the distribution of T can be found. The estimate of T is

$$E(T) = \Sigma y_i + (\Sigma X_i - n\bar{x})\bar{a}$$

which differs from the usual estimator

$$N\bar{a}\bar{X} = n\bar{a}\bar{x} + a(\Sigma X_i - \bar{n}\bar{x})$$

in replacing the estimated yield of the sample $n\bar{a}\bar{x}$ by the observed yield Σy_i. This would appear to be a definite improvement.

(iii) If the dependence of Y_i on X_i is of the form $Y_i = \alpha+\beta X_i+\epsilon_i$ where ϵ_i is a $N(0, \sigma^2)$, variate, then if σ is known the distribution of $\bar{Y} = T/N$ on the basis of the observed $x_1, y_1, \ldots, x_n, y_n$, is given by the standard normal variate

$$\frac{\bar{Y}-\bar{y}-b(\bar{X}-\bar{x})}{\sigma\sqrt{\frac{N-n}{Nn}+\frac{(\bar{X}-\bar{x})^2}{\Sigma(x_i-\bar{x})^2}}}$$

where

$$\bar{X} = \Sigma X_i/N, \quad \bar{x} = \Sigma x_i/n, \quad \bar{y} = \Sigma y_i/n$$

and b is the usual maximum likelihood estimate of β. The estimate of $\bar{Y}$ is $E(\bar{Y}) = \bar{y}+b(\bar{X}-\bar{x})$ which is the usual estimate. The variance of $\bar{Y}$ differs from the usual asymptotic result (Kendall and Stuart 1962) although if N is much larger than n the two variances are almost identical.

In such cases as the above it would seem that the sampling plan can be designed to increase the amount of information available, just as in experimental design (Fisher 1966). Thus, in (iii) the sample could be chosen so as to give the observed x_i's as wide a spread as possible. In (ii) the sample could be chosen so that only the largest X values are included since the Y's with the largest variances are then known. This is the logical extension of sampling schemes which attach a higher probability of inclusion to the larger X values, and indicates, perhaps, why, if this model is at all appropriate, such schemes are considered superior to simple random sampling in cases like (i) or (ii).

6. Other Examples

Example 6.1.

Consider a finite population of N lifetimes which may be considered a sample from an exponential distribution. When sampling for lifetimes it is usually necessary to observe to failure the item tested. Suppose therefore the inference concerns only the unsampled items in the finite population. Suppose we observe failure times $t_1, \ldots, t_n$ by drawing and testing n items at random from the population. Then $T = \Sigma t_i$ is sufficient for θ and fiducially

$$f(\theta;\ T) = \frac{T^n}{\theta^{n+1}} \frac{\exp(-T/\theta)}{(n-1)!}\ d\theta,$$

as before. If the sum of the failure times for the remaining items is S, then

$$f(S;\ T) = S^{N-n-1}T^n(N-1)!/(S+T)^N(N-n-1)!(n-1)!$$

Thus

$$\frac{N\ S}{(N-n)T}$$

is an F variate on (2(N-n), 2n) degrees of freedom. If $\bar{S}$ is the mean of the unsampled values then

$$E(\bar{S}) = T/(n-1) \text{ and } \operatorname{var}(\bar{S}) = \frac{(N-1)}{(n-1)^2\ (n-2)(N-n)}\ T^2$$

which differ from the corresponding estimates in the normal distribution.

If the data were of the form given in example 2.2 where $T = \Sigma t_i + (n-k)\tau$ gives the total lifetime observed in the sample with k failures, the relative likelihood for S is (from (9))

$$R(S) = \frac{T^k\ S^{N-n}}{(T+S)^{N-n+k}} \frac{(N-n+k)^{N-n+k}}{(N-n)^{N-n}\ k^k} \quad .$$

In the special case where k = n and the fiducial argument is available the maximum likelihood estimate of $\bar{S}$ is

$$\hat{\bar{S}} = T/n$$

which differs from $E(\bar{S})$.

Example 6.2.

In many situations due to the frequent occurrence

of short lifetimes, the items are placed in parallel. The problem is then, on the basis of the sample T and k, to make inferences concerning t, the maximum lifetime of N-n unsampled variate values. The distribution of the observations is proportional to the likelihood function

$$\theta^{-k} \exp(-T/\theta) \ . \tag{12}$$

If θ were known, any statement about t would be made from the distribution of t

$$g(t;\ \theta) = (N-n)[1-\exp(-t/\theta)]^{N-n-1} \exp(-t/\theta)/\theta \ ; \tag{13}$$

but all of the information about θ in the sample is contained in the likelihood (12). The solution then lies in the combining of (12) and (13) as in 2(b). Given t and no other information, t is sufficient for θ and (13) would yield the fiducial distribution

$$f(\theta;\ t) = (N-n)[1-\exp(-t/\theta)]^{N-n-1} \exp(-t/\theta)t/\theta^2 \ .$$

Combining this with (12) and integrating over θ yields the distribution (from (13)) of the observations given t as proportional to

$$\int_\theta t\ \theta^{-k} \exp(-\frac{T+t}{\theta})[(1-\exp(-t/\theta)]^{N-n-1}\ d(1/\theta)$$

or

$$t \sum_{m=0}^{N-n-1} \binom{N-n-1}{m}(-1)^m(T+t+mt)^{-k-1} \ .$$

Observing T and k yields the likelihood function of t proportional to the above.

Suppose T = 308, k = 7, and N-n = 3, 5, 7; Table 1 gives the corresponding likelihoods of t. The relative likelihoods are plotted in Figure 1.

Table 1

RELATIVE LIKELIHOODS OF THE MAXIMUM
LIFETIME OF 3, 5, 7 ITEMS

N-n	t	10	50	90	130	170	230	310	390
3	R(t)	.052	.80	.98	.77	.544	.30	.14	.068
5	R(t)	.005	.60	.997	.90	.69	.39	.19	.094
7	R(t)	.000	.45	.94	.96	.77	.465	.23	.115

<u>Figure 1</u>

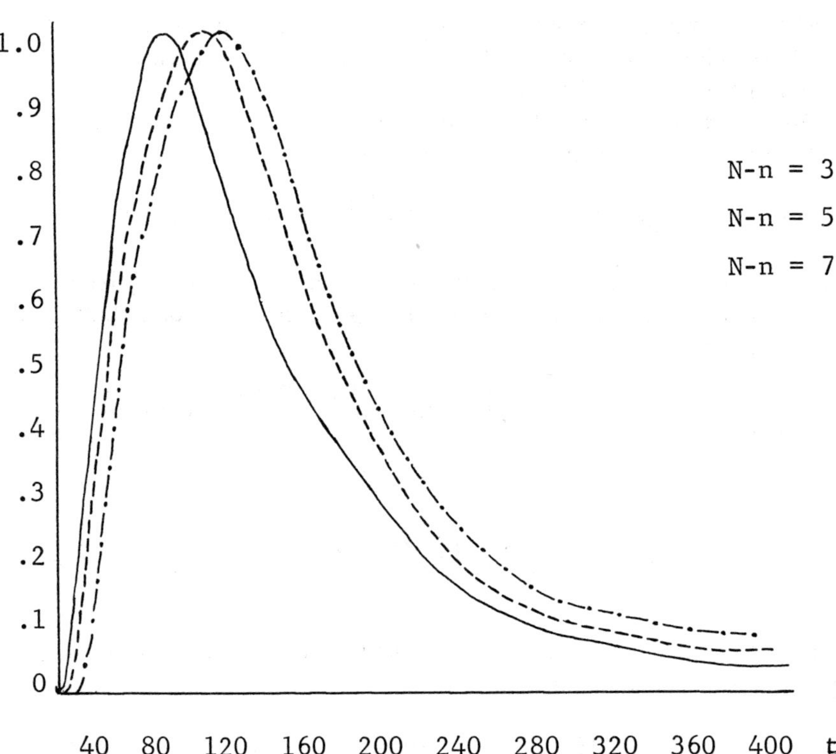

Relative likelihoods for the maximum of 3, 5, 7, items.

7. A problem in sampling for proportions

A problem proposed by Godambe (1965) is of considerable interest since it allows of both of the approaches discussed in Section 3. N red and blue balls in an urn are numbered 1, 2,...,N. Sampling is with replacement to estimate the proportion k/N where k is the unknown number of red balls in the urn of N balls.

Suppose n observations $x_1, x_2, \ldots, x_n$ are taken at random with replacement and n_i balls are drawn i times $(i = 1,2,\ldots,n)$. Of the η (say) different balls in the sample s are red and t are blue where s_i red balls are drawn i times and t_i blue balls are drawn i times. The probability of such a sample (allowing all possible permutations of the unit members) is

$$\frac{n!}{\pi(i!)^{n_i}} \frac{s!}{\pi(s_i)!} \frac{t!}{\pi(t_i)!} \binom{k}{s} \binom{N-k}{t} / N^n \quad . \tag{14}$$

When N is known this leads to a discrete likelihood function for k

$$L(k) = \binom{k}{s}\binom{N-k}{t} \qquad s \le k \le N-t$$

$$= 0 \quad \text{otherwise} \quad . \tag{15}$$

The additional information obtained in marking the balls is easily seen if we consider the case where N is unknown. The balls are marked with distinguishing marks rather than consecutive numbers (i.e. the mark on the ball yields no information about N). In this case the joint likelihood of k and N arising from (14) is

$$L(k,\, N) = \binom{k}{s}\binom{N-k}{t} N^{-n} \tag{16}$$

from which statements about N are available. This may be compared with the binomial situation where the balls are indistinguishable; all the information is centred on the ratio k/N and nothing may be said about either k or N separately, i.e.

$$L(k, N) = (k/N)^r (1-k/N)^b \tag{17}$$

where r and b are the total numbers of red and black balls drawn respectively.

It has been stated that the likelihood of the population is proportional to

$1/N^{b+r}$ if the population is consistent with the observations

0 otherwise

which is independent of k, so that no information is contained in the likelihood function. This likelihood arises from noting that i'th ball is distinguishable and so yields a relevant subset, relative to which statements of probability should be made.

However, assuming the experimenter has no knowledge of the method of marking the balls, although the i'th ball is distinguishable, this does not indicate to him a smaller subset about which he has any different knowledge. Here again, as in section 3, in order to make meaningful probability statements it seems necessary to consider the hypothetical reference set of all urns similar to the given one. The marking on a ball certainly defines a recognizable subset. The lack of knowledge concerning the marking and colour is taken into account in (14) by assuming the fraction of red balls within this subset to be the same as in the whole reference set. The additional information as to the marking of the ball then provides no further knowledge as to its colour and thus specifies no relevant subset. Indeed if such information is not ignored (in the absence of knowledge linking marking to colour) then by accumulating enough information to distinguish an item all statements of probability could be made meaningless. For eventually the recognizable subset to which the item belongs will consist of only the item itself. Similar discussions about disregarding information have been given by Fisher (1959), by Williams (1967), and by Royall (1967).

<u>Example 7.1.</u>

(i) In a situation as described above a random sample of size n = 8 is drawn from an urn and it is found

$$s_1 = 1,\ s_2 = 0,\ s_3 = 1,\ t_1 = 2,\ t_2 = 1\ .$$

The experimenter wishes to know what may be said about

a) k, the number of red balls when N is 10;

b) k and N when there is no information about N.

From the previous discussion the likelihood functions are (15) and (16).

a) $L(k) = \binom{k}{2}\binom{10-k}{3}$;

the relative likelihood is tabulated in Table 2. The plausible values of k are 2 to 6, while 7 is fairly plausible. For (b) the likelihood function is

$$L(k,\ N) = \binom{k}{2}\binom{N-k}{3}N^{-8}\ .$$

The relative likelihood is plotted in Figure 2. For the relative likelihood $R \geq .1$ we have

$$1/6 \leq k/N \leq 3/4$$

$$2 \leq k \leq 9$$

$$\text{and} \quad 5 \leq N \leq 19\ .$$

The maximum likelihood estimates are 2/5, 2 and 5 respectively for k/N, k, and N.

If the same sample were drawn at random and the balls are not distinguishable, we would be restricted to the binomial model (17) where the relative likelihood is

$$R(k) = 256(k/10)^4(1-k/10)^4$$

in (a) and

$$R(k, N) = 256(k/N)^4(1-k/N)^4 \qquad (18)$$

in (b). R(k) is tabulated in Table 2 and R(k, N) is plotted in Figure 3. From Figure 3 and (18) it can be seen there is no information about k or N individually. The binomial model (18) gives an estimate of k/N for $R \geq .1$ or

$$.15 \leq k/N \leq .85$$

compared to

$$.16 \leq 1/N \leq .75$$

if the balls are distinguishable.

Table 2

RELATIVE LIKELIHOODS FOR EXAMPLE 7.1

k	2	3	4	5	6	7
Distinguishable Units (15)	.467	.875	1.0	.833	.50	.175
Binomial	.167	.50	.85	1.0	.85	.50
R(k) (19)	.194	.785	1.0	.818	.422	.064

(ii) (b) In certain cases this problem may be considered from the first approach mentioned in section 3. In many situations we can assume that the finite population was generated by a series of Bernoulli trials with an unknown probability p of a red ball. Suppose further that we have no prior information about p so that all the information about p is contained in the likelihood function arising from the observations

$$p^s(1-p)^t \ .$$

The method of section 2c is applicable to this model. The relative likelihood of p is then

$$R_1(p) = \frac{p^s(1-p)^t(s+t)^{s+t}}{s^s t^t} \ .$$

The remainder of the finite population also constitutes a random sample from the binomial distribution which, if observed, would yield the relative likelihood of p

$$R_2(p) = \frac{p^{k-s}\ (1-p)^{N-k-t}\ (N-s-t)^{N-s-t}}{(k-s)^{k-s}\ (N-k-t)^{N-k-t}} \ .$$

The most likely value of p based on the entire finite population is k/N, so that the relative likelihood of k given s and t is (from (10))

$$R(k) = R_1(k/N)R_2(k/N) = \frac{k^k(N-k)^{N-k}(s+t)^{s+t}(N-s-t)^{N-s-t}}{N^N s^s(k-s)^{k-s}\ t^t(N-k-t)^{N-k-t}} \ . \tag{19}$$

This is maximized at the nearest integer to k = Ns/(s+t).

The likelihoods (15) and (19) indicate the differences in the approaches discussed in section 3. The hypothetical population for which (19) is appropriate is the collection of all urns in which the number k of red balls is a random variable distributed according to the binomial distribution with a constant value of p. The likelihood (15) is appropriate only if there is no

<u>Figure 2</u>

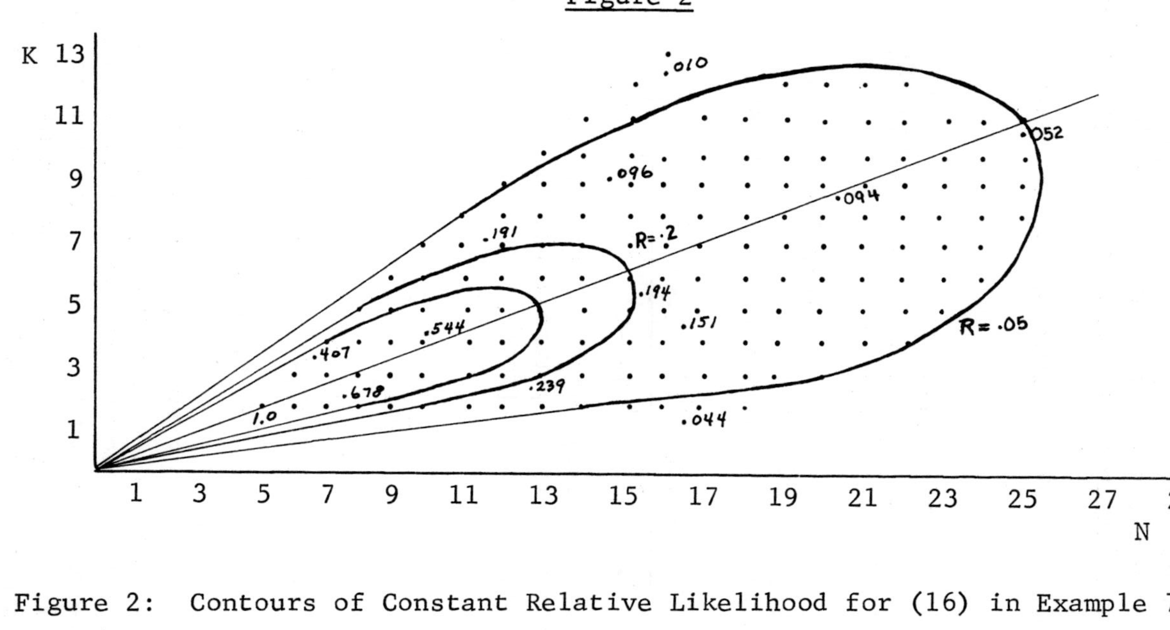

Figure 2: Contours of Constant Relative Likelihood for (16) in Example 7.1.

Figure 3

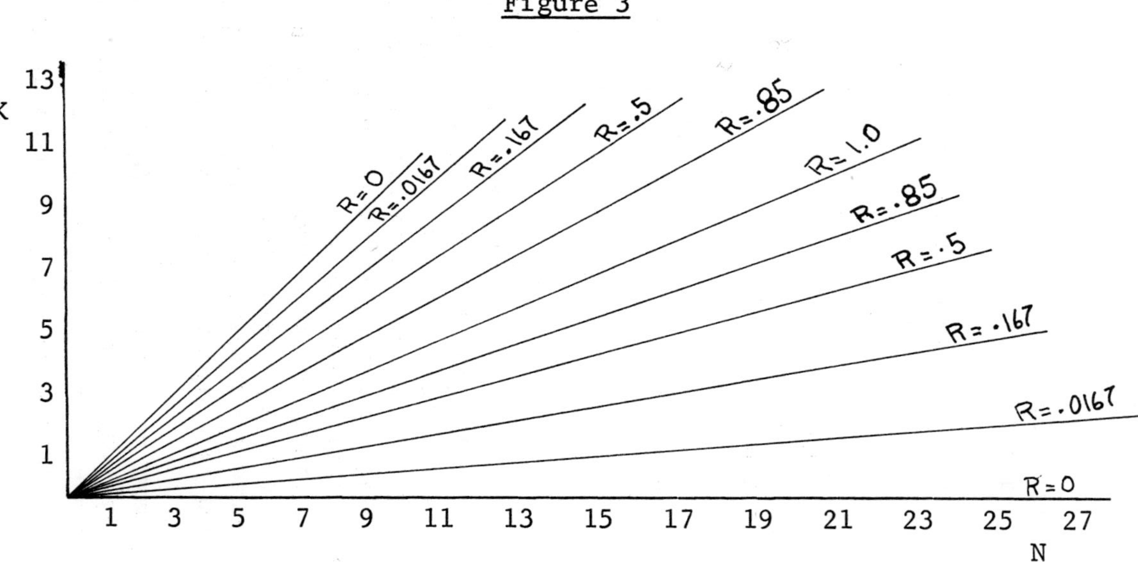

Figure 3: Contours of Constant Relative Likelihood for (18). The Contours are of the form k/N = c.

reason to believe the population is generated by a Bernoulli process. If N and k are large compared to n and s then (15) tends to the binomial likelihood

$$(k/N)^{s}\ (1-k/N)^{t}\ . \tag{20}$$

It can be shown that the likelihood (19) is more informative than (20), as would be expected since more information was assumed in deriving (19). This is illustrated in Table 2 where (19) is tabulated for Example 7.1.

It should be noted for Example 7.1, however, that in dealing with a problem like acceptance sampling neither of these methods is appropriate. In such cases it is often reasonable to assume that the batches (finite populations) are generated by Bernoulli trials with some probability p of a defective article. Past observations however contain information about p and this must be used to determine a prior distribution for p. This prior distribution would then form a part of the specification of the problem of determining the plausibility of k (Sprott 1968).

8. Artificial randomization

In the above discussion the arguments depend on the assumption that the sample on which we base the inference is a random sample from the hypothetical population. For this to be so, the finite population must itself be a random sample from the hypothetical population; after selection, the N items are labelled (they become distinguishable). Two cases must be considered.

(i) The labels are assigned at random to the population members (randomization at stage 1).

(ii) The labels are assigned to suit some purpose and hence may contain information about the variate values.

If randomization has occurred at stage 1 then any sampling scheme will ensure a random sample from the hypothetical population since randomization has already taken place. If, however, randomization is not known to have occurred at stage 1, the sample must be selected at random from the finite population to ensure it is a random sample from the hypothetical population. This will be called randomization at stage 2. This is necessary since there may be an unknown systematic relationship between the labelling and the variate values. This point has been discussed by Godambe (1966).

It can thus be seen the arguments here depend very definitely on randomization. This may not seem to be the case when there is randomization at stage 1 since an additional randomization at stage 2 is then unnecessary. It is essential however that randomization has taken place at some stage.

That randomization at stage 1 and stage 2 are equivalent is intuitive but may be illustrated in a model by Godambe (1967). He obtains a fiducial distribution which depends on simple random sampling (no fiducial distribution exists if the sampling is not simple random, Godambe, 1967). The same distribution arises as a Bayes posterior if the prior attaches equal probability to each variate value appearing on each unit. This prior is equivalent to randomization at stage 1, and as we would expect, the posterior distribution depends in no way on the sampling scheme used at stage 2. It is because of the assumption of such a prior, thus ensuring randomization at stage 1, that the Bayesian posterior is independent of the sampling scheme employed at stage 2.

9. Conclusions

It is frequently realistic and in a sense necessary to assume a hypothetical population from which the finite population is drawn at random. In fact, the large samples considered in many sampling surveys allow a good possibility of actually checking these assumptions with goodness-of-fit tests and other tests of significance. This assumption does not contradict the intuitive concept of selecting a sample at random; rather the methods developed here require very definitely that the finite populations under consideration be sampled using simple random sampling within a stratum or a covariate value.

Under the above assumption, the problem of estimation of the finite population can be considered as the problem of predicting a finite set of future observations from the hypothetical distribution. The predicted observations together with the observed sample then form the finite population under consideration. An advantage of this approach is that the observed part of the finite population is taken to be a fact. The uncertainty is then limited to the unobserved portion of the finite population under discussion. The larger the observed sample is as a fraction of the whole finite population, the less is the uncertainty concerning the remainder. This is illustrated by the ratio estimator of the population total

$$T = \sum_1^N Y_i$$

$$N\bar{a}\bar{X} = n\bar{a}\bar{x} + \left(\sum_1^N X_i - n\bar{x}\right)\bar{a} \tag{21}$$

as compared to the estimate

$$\sum_1^n y_i + \left(\sum_1^N X_i - n\bar{x}\right)\bar{a} \tag{22}$$

given in section (5,e,ii). The estimated sample total $n\bar{a}\bar{x}$ in (21) is replaced by the actual observed sample total Σy_i in (22). Thus, if n = N so that the whole finite population has been observed then the population total T is known exactly. The latter estimate (22) becomes the population total T whereas the former estimate (21) does not.

It is pointed out that stratification and the use of covariates are slight extensions of the above and serve a purpose similar to their corresponding use in experimental design and regression. Sampling schemes (like experiments) can be designed using the concepts of likelihood and information so as to increase or maximize the expected information (cf. Fisher 1966).

Finally, the concepts of likelihood and probability seem to provide a unifying approach to sampling theory (just as they do to statistical inference in general). Many of the classical results can be obtained and explained from this approach. By examining the effects of changing the hypothetical distribution it is possible to gain some insight as to the appropriateness of the various estimates and to note what distributional assumptions give rise to them. This would seem to bring sampling theory more into line with standard statistical inference, treating it as the estimation of future observations to be randomly drawn from the hypothetical population. The concepts of likelihood and probability then play the same role in sampling theory as they do in estimation theory. In this regard it should be noted that the estimates and variances of this paper were calculated in order to compare them with classical results. It should be emphasized that under the model used in this paper considerably more informative statements can be made than merely quoting a point estimate and a variance. In fact the entire probability distribution (or likelihood) can be plotted or tabulated giving plausibilities for all parameter values.

Acknowledgment

We would like to thank Professor V. P. Godambe for many helpful discussions, for reading a previous draft of the manuscript, and for the interest he has shown in writing of this paper. We would also like to thank Professor James G. Kalbfleisch for many helpful criticisms, discussions, and extensive suggestions for revising the manuscript.

REFERENCES

1. Cochran, W. G. (1939), The use of analysis of variance in enumeration by sampling, J. Amer. Statist. Assoc. 34, 492-510.

2. Cochran, W. G. (1953), Sampling Techniques, John Wiley & Sons Inc., London.

3. Cochran, W. G. (1946), Relative accuracy of systematic and stratified random samples for a certain class of populations, Ann. Math. Stat. 17, 164-177.

4. Ericson, W. A. (1967), Subjective Bayesian Models in Sampling Finite Populations, I, Manuscript submitted for publication.

5. Fisher, R. A. (1935), The fiducial argument in statistical inference, Annals of Engenics, Vol VI, Pt. IV, pp. 391-398.

6. Fisher, R. A. (1956), Statistical Methods and Scientific Inference, Oliver and Boyd, London.

7. Fisher, R. A. (1959), Smoking, the Cancer Controversy, Oliver and Boyd, London.

8. Fisher, R. A. (19-6), Design of Experiments, 8th Edition, Oliver and Boyd, London.

9. Godambe, V. P. (1955), A unified theory of sampling from finite populations, J. Roy. Stat. Soc. B, 17, 269-278.

10. Godambe, V. P. (1965), A review of the contributions towards a unified theory of sampling from a finite population, Rev. Int. Statist. Inst. 33, 2, 1965, 242-258.

11. Godambe, V. P. (1966), Bayesian Sufficiency in survey sampling. Technical Report 63, Dept. of Statistics, John Hopkins Univ., Baltimore.

12. Godambe, V. P. (1967), A fiducial argument with applications to survey sampling, 36th Session, I.S.I., Sydney, Australia.

13. Hanurav, T. V. (1967), Optimum utilization of auxiliary information; πps sampling of two units from a stratum, J. Roy. Statist. Soc. B, 29, 374-391.

14. Kendall, M. G. and Stuart, A. (1962), The Advanced Theory of Statistics, Vol. 3, Butler and Tanner Ltd., London.

15. Royall, R. M. (1967), An old approach to finite population sampling theory, Presented at I.M.S. Contributed Papers, Session II, Joint Statistical Meetings, Washington, D. C.

16. Sprott, D. A. (1968), The estimation of prior distributions in problems of decisions, to appear in Sankhyā.

17. Williams, J. S. (1967), The Role of Probability in Fiducial Inference, Sankhyā, Series A, 29, 271-296.

DESIGNING DESCRIPTIVE SAMPLE SURVEYS

Tore Dalenius

University of Stockholm, Stockholm

Introduction

It has been stated by a sampling expert that sampling is both art and science. This statement implies a challenge to those of us who are responsible for training statisticians who are to work in government agencies, market survey agencies and the like:

i) In many countries, most surveys are designed and carried out by statisticians who are not sampling experts.

ii) Thus, there is a need for some set of "rules of thumb" to guide these statisticians in their responsible endeavours.

The prime purpose of this paper is to present a tentative set of such "rules of thumb" which would help these statisticians, not to achieve the very best, but hopefully "the second best".

These rules are presented in Chapters II-IV against the background of a review of statistical studies presented in Chapter I.

I. A REVIEW OF STATISTICAL STUDIES

1. Practical problems and statistical studies.

Statistics is a young science in the midst of a rapid development. In the course of this development, strikingly different points of view have been propounded on the very subject of statistical theory.[1] By the same

---

[1] For a discussion of the need for "a unified theory of random experiments", reference is given to Dalenius and Matérn (1964).

token, the role of statistics in coping with real life problems has been conceived of in different ways.

With the situation just described in mind, it is perhaps not surprising to find in the literature an abundance of divergent versions of the role of 'statistical studies'.

In order to provide a background to the description in par. 2 of 'the descriptive survey', we will present here one possible point of view concerning the role of statistical studies. The presentation follows closely the discussion in Deming (1950), Chapter 8.

According to this point of view, the ultimate aim of statistical studies is to provide a rational basis for <u>action</u>. Some action is called for, because there is a problem, and something is to be done about it. The nature of this problem will, in principle, determine the kind of study to carry out, in order to get the rational basis for action.

It is possible to distinguish two classes of problems, descriptive and analytical problems, respectively. In the descriptive problem, the action is to be taken on (some portion of) a 'population' as it is as of a certain date. The choice of action will depend upon the answers to such questions as "How many --- are there in the population?", or "How much --- is there in the population?".

In the analytic problem, however, the action is to be taken on the cause system producing populations - now and in the future. The choice of action will depend upon the answers to such questions as "Which populations will be produced in the future?", or "What will be the effect, if we change this factor?". Coping with the analytic problem clearly calls for the search for mechanisms capable of explaining the various frequencies observed, and also capable of predicting future frequencies. The ultimate aim in studying the cause system may be to learn to control it, in order to be able to produce desired populations.

The point of view concerning the role of statistical studies presented above will be made the basis of the present paper. The question then arises: What is - in terms of the <u>kind</u> of statistical study - a rational basis for action? We will discuss this question separately for descriptive and analytic problems.

(1) If the problem is descriptive, the study called for is a 'descriptive survey', or 'survey' for short. This term will be used here to denote a study, based on observational data: the statistician records for each unit of a sample or a population the value of some measurable characteristic. These values are thereafter used to compute (an estimate of) some quantity characterizing the population, or some part of it.

(2) If the problem is analytic, a 'survey' or an 'experiment' will be called upon; the choice between these two kinds of studies will reflect the kind of cause system under consideration.

In such cases, where the cause system is beyond our control, the immediate aim of a study may be to predict the future populations that the system will produce. The statistician constructs - exploiting existing subject matter knowledge about the mechanisms of the cause system - a model of the system. This model will contain certain parameters, say α and β. By means of a <u>survey</u>, α and β are estimated by $\hat{\alpha}$ and $\hat{\beta}$, respectively. Granted the realism of the model, it may now be used to predict the future populations that the cause system will produce.

In other cases, the cause system is subject to some control. A realistic example deals with the comparison of two treatments with respect to their effects. Then, a controlled experiment would be called for. It is characteristic of a controlled experiment that the statistician determines by design not only <u>which</u> factors to include in the experiment, but also the <u>levels</u> of these factors.

The previous discussion of the role of statistical studies presents a <u>principal</u> point of view; as a consequence, the picture of this role is rather pure.

In real life, the picture is much more complicated: the study designs used cover the whole spectrum, the endpoints of which are represented by the descriptive survey and the controlled experiment. Two examples, corresponding to intermediate cases, will be given.

i) The terms 'critical survey' and 'analytic survey' are sometimes used to denote studies which involve comparisons between different subgroups of a population "in order to discover whether differences exist among them that may enable us to form or to verify hypotheses about the forces at work in the population", Cochran (1963).

ii) The second example refers to "controlled experiments imbedded in survey", as discussed in Cochran (1963), p. 387-389.

2. Basic concepts of descriptive sample surveys.

In this paragraph, we will present the minimum of concepts that are necessary to make this paper self-contained.

The survey concerns a set of '<u>units</u>' of some kind; these units may be objects, events or the like.

There is a '<u>rule of inclusion</u>' for determining whether or not a given unit belongs to the set of units which is the concern of the survey. Thus, this rule defines the <u>population</u> of the survey.

This population is to be quantitatively characterized with respect to a measurable <u>property</u>. To this end, measurements are to be performed by a '<u>method of measurement</u>': a specification of the equipment to be used, the operations to be performed, the sequence in which they are to be carried out, and the conditions under which they are respectively to be carried out.[2]

The 'measurement process', i.e. the realization of the method of measurement, yields 'measurements'

---

[2] Cf Eisenhart (1963)

('observations'), which, depending upon the method of measurement, may be of different kinds:

i) nominal measurements;

ii) ordinal measurements;

iii) interval measurements; or

iv) ratio-scale measurements.

The quantitative characterization of the population is thereafter performed by applying a 'method of summarizing' the measurements. In a simple case, this means computing the frequency of units having the property under consideration. Or it may mean computing the amount of the property (as represented by ratio-scale measurements) in the population. In the case of a sample survey, the method of quantitative characterization just discussed is referred to as 'estimation'.

3. The main subjects of survey design.

There is as yet no universally accepted 'survey design formula' that provides an explicitly formulated solution to the intricate problem that survey design represents. This state of things may explain why most textbooks on sample survey theory and methods devote comparatively small space to the design of a survey; an exception proving the rule is Stephan and McCarthy (1958), which devotes one of the three parts, which make up the book, to the design of sample surveys.

It may be argued, however, that the problem of design is dealt with in the textbooks implicitly. The following quotation from Hansen et al (1953) is illuminating:

> An effort is made in the following chapters to provide a guide for the efficient design of sample surveys. In applications of sampling to various problems the same fundamental principles of sampling theory appear again and again, and an attempt is made in what follows to introduce illustrative problems in which these common

principles are applied. At the same time, some principles have a great deal more importance than others in certain sampling problems, and the differences in the importance of various principles are emphasized.

Lacking a universally accepted 'survey design formula', the author of a paper on design of a survey has the option - and faces the problem! - of choosing one specific approach that may serve as the main thread of the paper. The approach that we have chosen is as follows.

Following Hansen et al (1967), we will view a survey from three perspectives: requirements, specifications, and operations.

First, there are the requirements imposed by some problem. Corresponding to these requirements, there is a design defined by reference to a relevant population, a relevant property, a method of measurement, and a method of summarizing the measurements. If this design is properly carried out, the survey will yield a set of statistics, the <u>ideal</u> goals (Z) of the survey.

Second, there are the specifications of the actual survey to be carried out. These specifications are arrived at by the choice of the population to be studied, the property to measure, the method of measurement to use, and the method of summarizing the measurements. Thus, the specifications identify the <u>defined</u> goals (X) of the survey.

Third, there are the actual operations of the survey. These operations yield the actual set of statistics, <u>the results</u> (y) of the survey.

For a particular statistic y in the set of statistics (y), we may define the following two errors:

i) The error y - X, i.e. the error relative to the defined goal.

ii) The error y - Z, i.e. the error relative to the ideal goal.

The aim of the design is to provide procedures that will make it possible and feasible to measure and control these errors, in an efficient way.

This aim prerequisites measures of the errors y - X and y - Z, respectively. Such measures are the mean-square errors, with respect to the defined goal:

$$MSE_X(y) = E(y - X)^2$$

and with respect to the ideal goal:

$$MSE_Z(y) = E(y - Z)^2$$

of some 'survey model'.

One example of a survey model that may be used in the present context is developed at the U.S. Bureau of the Census; a comprehensive presentation is given in Hansen et al (1964). According to (a straight-forward generalization of) this model, the mean-square error with respect to the ideal goal is

$$MSE_Z(y) = (X - Z)^2 + 2(Y - X)(X - Z) + E(y - Y)^2 + (Y - X)^2$$

where Y = E(y). Here:

$(X - Z)^2$	reflects the relevance of the survey specifications as related to the requirements;
$E(y - Y)^2$	is the total variance (sampling and nonsampling);
$(Y - X)^2$	is the square of the bias of the survey operations as related to the survey specifications; and
$2(Y - X)(X - Z)$	finally, is an interaction term.

Viewing a survey from the three perspectives just discussed, leads naturally to the following approach to the design of a survey:

i) The specification of the ideal goals (Z).

ii) The analysis of the survey situation.

iii) The construction of a (relatively small) number of alternative designs $D_1, D_2, \ldots, D_s$. Associated with these designs, there are alternative defined goals $(X_1), (X_2), \ldots, (X_s)$.

iv) These designs are evaluated by reference to the associated mean-square errors and costs, and the decision is taken to carry out the survey on the basis of one of these alternative designs (or possibly a modification of one of them), or not to carry it out.

v) The development of the administrative machinery necessary to carry out the survey.

The operations i) - ii) will be referred to as 'the preparatory stages'; they are discussed in Chapter II. The operations iii) - iv) will be referred to as 'the methodological design', to be discussed in Chapter III. The last operation, i.e. v), finally, is discussed in Chapter IV, which is devoted to some administrative considerations.

Throughout Chapters II - IV, the emphasis of the presentation will be on the sample design: this term will be used here in a wide sense which encompasses the estimation procedure. Even so, however, the presentation has a rather narrow scope; especially, there will be no explicit discussion of design problems associated with the measurement procedure (e.g. the construction of a questionnaire).

II. THE PREPARATORY STAGES

4. The specification of the ideal goals.

As the correct solution of any problem depends primarily on a true understanding of what the problem really is, and wherein

> lies its difficulty, we may profitably pause upon the threshold of our subject to consider first, in a more general way, its real nature; the causes which impede sound practice; the conditions on which success or failure depends; the directions in which error is most to be feared.
>
> Wellington

According to the point of view expressed in par. 1, the aim of a survey is to provide a rational basis for action. What constitutes such a basis will clearly depend upon the problem calling for action.[3]

The initial step in the design of a survey must necessarily be a careful analysis of the problem under consideration. This analysis is clearly of decisive importance in the endeavours to specify the ideal basis for action; as sometimes said, it is better to have a reasonably good solution of the proper problem than optimum solutions of the wrong problems.

As a preliminary to that specification, it is often helpful to prepare a <u>small</u> set of 'dummy tables'. These tables may serve two important purposes:

i) The table titles, stub entries, and column heads play an instrumental role for the specification of the population to be enumerated, of the method of measurement to use, etc.

ii) Filling in a few alternative sets of figures in the tables (covering the range of results that may appear) and stating separately for each such set the action to be taken, serves as a powerful guide to the level of accuracy to aim at.

---

[3] We will assume here that there is only one prospective user; this user is identified prior to the design. For a discussion of the much more complicated situation where there are several prospective users, all of which are not identified prior to the design, reference is given to Dalenius (1967).

The final step is the specification of the survey that would yield the ideal basis for action, that is the ideal goals (Z). This specification would identify the following design features:

i) The relevant population

ii) The relevant property to be measured

iii) The method of measurement

iv) The method of summarizing the measurements.

These design features and the associated ideal goals (Z) will serve as a point of aim for the specification of the survey to be carried out.

5. Conditions to be met.

Proper design calls for a careful analysis of the conditions to be met by the survey. The purpose of this analysis is to determine not only if these conditions can be met at all, but also if they can be met effectively.

In most realistic cases, certain conditions are set in terms of the accuracy and cost of the survey.

If X denotes the defined goal, and y denotes the survey result, $MSE_X(y)$ may be used to measure the accuracy. In simple cases, the condition to be met may be written:

$$MSE_X(y) \leq a$$

where a is some constant. In more complicated cases, the <u>direction</u> of the error must as well be taken into account; Blythe (1945) provides an illuminating illustration.

In an analogous way, a cost condition

$$C(y) \leq b$$

is set with respect to the cost C(y) of the survey: if it is to be carried out, the cost must not exceed an amount b.

In addition, other conditions may have to be met. As a realistic illustration, a date D_0 may be set before which y must be available. In some cases, the usefulness of the result y may depend upon the date when it is available in a simple fashion, as illustrated in Törnqvist (1948).

6. Resources that may be used.

Proper design calls, as discussed in par. 5, for a careful analysis of the conditions to be met. By the same token, it calls for a careful analysis of which resources are available, and how they may effectively be used.

We will group the potential resources under three broad headings, and discuss them separately in par. 6.1 - 6.3, respectively.

6.1. Financial resources

The design of a survey and its execution demands "sufficient financial resources".

It is self-evident that we must know, at least within certain limits, the financial resources that are at our disposal. But it is also important to know how these resources are allocated over time. It may make a distinct difference if all money is available from the very beginning, or only is made available by portions. As an example, the statistician may wish to use stratified sampling or ratio estimation as a means of exploiting auxiliary information. If all money is available from the very beginning, he has in fact a choice between these two means; if, however, most of the money will be available only at a very late date, there may be no choice he may have to use ratio estimation, as the expense for using this means comes later than would the expense for using stratified sampling.

6.2. A priori information

We will limit the discussion hère to two important aspects: costs and variances.

If a survey similar to the one being considered has previously been carried out, records of the costs associated with the various operations are often available - or may be made available. Such information may prove extremely useful in the sample design.

Likewise, it is often possible to get useful variance estimates. Such estimates may be based on previous surveys, or be in the nature of computations exploiting information about the distribution of the population. This latter kind of information may in addition prove useful as a means of identifying the sources that contribute most to the overall variance.

6.3. Methodological resources

The goal of a survey implies the use of a certain set of methods: measurement methods, methods for collecting the measurements, and methods for summarizing the measurements.

As a first step, the statistician may clarify the situation with respect to which methods are at all available. In addition, however, he has to evaluate these methods from the point of view of the feasibility of using them.

Suppose that the purpose of the survey is to estimate the audience of a certain periodical. This may be done using one of several approaches in terms of questions to be asked to (a sample of) respondents. One such approach may be to ask the question: "Which periodicals, if any, did you read yesterday?" The feasibility of this question will obviously depend upon the possibility of controlling the field-work with respect to the days of interview (and thus, with respect to what "yesterday" is for any specific respondent).

In the last few years, considerable interest has been devoted to the development of "resource oriented models". For a brief review of this notion, reference is given to Dalenius (1962). We will limit the discussion here to two aspects of special importance.

If the survey is likely to be a sample survey, the inventory of resources must include an inventory of any material that provides "access to the population", that is material that may serve as (part of) a frame.

Another aspect of great importance in this context is the data processing equipment available. Access to a powerful computer may allow the use of estimation procedures - for example methods for composite estimation - which are otherwise out of reach.

III. THE METHODOLOGICAL DESIGN

7. A feasible approach to the methodological design.

In par. 3, we pointed out the design of a survey proceeds through a sequence of operations, one of each is the construction of a small number of alternative designs $D_1, D_2, \ldots, D_s$.

The main part of this chapter will be devoted to the presentation of a specific method for constructing such sample designs. The presentation is based on the assumption that the condition in terms of the accuracy is represented by:

$$MSE_X(y) \leq a$$

and not by

$$MSE_X(y) = 0$$

in which case a sample survey is out of reach.

It is characteristic of our method for constructing sample designs that the construction advances in stages:

i) Construction of 'small variance sample designs'

ii) Choice between classes of sample designs

iii) Choice within a class of sample designs

iv) Optimization

v) Choice of <u>the</u> sample design .

We will discuss these stages in par. 8-12. In par. 13, we will discuss the design for an evaluation of the design.

8. Construction of 'small variance sample designs'.

The first step towards the construction of a sample design in the set of feasible alternative designs $D_1,\ldots,D_s$, is the construction of a small set of "small variance sample designs". Such designs play an instrumental role for the further work.

In order to explain the notion of a 'small variance sample design', we consider the special case of estimating the mean-per-element $\bar{X}$. A sample design will be denoted a 'small variance sample design', if it comprises n elements and has a variance $\sigma^2(\hat{x})$ considerably smaller than the variance of the following standard sample design: simple random sampling of n elements, with the mean-per-element estimator $\bar{x}$ with variance $\sigma^2(\bar{x})$.

Having constructed a set of 'small variance sample designs', they are used as follows. Each such design is carefully analyzed from a variance and cost point of view.

(1) The analysis from the variance point of view serves the purpose of identifying those design features that are especially powerful by way of reducing the variance. If such features can be identified, an effort should be made to 'save' them for the design actually to be used.

(2) The analysis from the cost point of view serves the purpose of identifying those design features that contribute heavily to the survey cost. If such features can be

identified, an effort should be made to eliminate them from the final design, or at least to moderate their impact.

The use of 'small variance sample designs' as just indicated may be illustrated by reference to the design of a survey to estimate the total sales in a population of retail firms, in a geographically large country.

In a realistic case, the 'sales' exhibit a skew distribution:

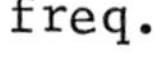

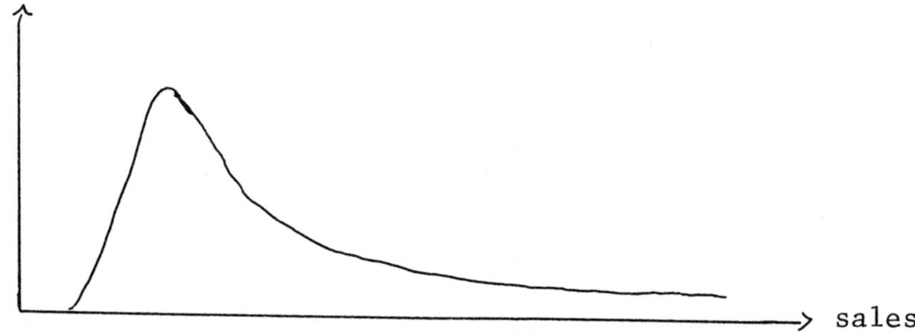

As indicated by this figure, there are many 'small' firms; while each such firm contributes very little to the total sale, they contribute much when taken together. At the same time, however, each one of a few 'large' firms contributes very much to the total sale.

One 'small variance sample design' would be as follows. The population of N firms is divided into 3 strata on the basis of 'size': N_S small, N_M medium-sized and N_L large firms, $N_S + N_M + N_L = N$. A small fraction n_S/N_S of the small firms, a larger fraction n_M/N_m of the medium-sized firms, and a considerable fraction n_L/N_L of the large firms, say $n_L/N_L = 1$, is selected; the total sale is estimated by

$$x' = N_S\bar{x}_S + N_M\bar{x}_M + X_L$$

where $N_S\bar{x}_S$ estimates X_S, the sales volume in the stratum of small firms, $N_M\bar{x}_M$ estimates X_M, the sales volume in the stratum of medium-sized firms, and X_L is the sales volume in the stratum of large firms.

The analysis from the sampling variance point of view will reveal that the variance is relatively small because of the kind of disproportionate allocation used.

The analysis from the cost point of view may show, however, that a design of the type just presented is not efficient: the inefficiency reflects the fact that interviewers may have to be employed to collect the measurements from (especially) the small and medium-sized firms.[4]

A reasonable compromise between the variance and cost considerations may call for a design of the following kind:

i) All 'large' firms are selected, irrespective of geographic location.

ii) All 'medium-sized' firms in a sample of first-stage units (which may be groups of contagious counties or the like) are selected.

iii) A sample of the 'small' firms in the same sample of first-stage units is selected.

A second but important use of a set of 'small variance sample designs' is the following one. If the analysis from the variance point of view is carried out with respect to each one of the population characteristics which are to be estimated, it will throw light on the 'design conflict' which may prevail, and thus guide in the choice of means to cope with this problem.

Mention should finally be made of the fact that the analysis of a set of 'small variance sample designs' may prove useful to elucidate problems caused by measurement errors.

---

[4]This is an example of the implicit way in which in this paper attention is paid to the important design problems associated with the measurement procedure; cf. the discussion at the end of par. 3.

9. Choosing between classes of sample designs.

The approach to the design of a survey discussed in par. 8-12 is based on the possibility of identifying a small number of 'classes' of sample designs. The sample designs that make up such a class should be similar to each other with respect to the design features that have a decisive impact on the variances and the cost.

A reasonable approach to the problem of arriving at a useful classification of sample designs is as follows. We classify the sample designs with respect to three dimensions A - C with subclasses defined for each dimension as follows:

A: the kind of hierarchy of sampling units:

a_1: for one-stage sampling

a_2: for multi-stage sampling, with small first-stage sampling units

a_3: for multi-stage sampling, with large first-stage sampling units

B: the selection scheme (for the first stage in multi-stage sampling):

b_1: selection with equal probabilities

b_2: selection with unequal probabilities

C: the estimation scheme:

c_1: estimation without use of auxiliary information

c_2: estimation with use of auxiliary information .

A sample design may thus be written as D (a_i, b_j, c_k) According to the discussion above, we distinguish twelve classes of sample designs. Thus, D (a_1, b_1, c_1) is one class; D (a_1, b_1, c_2) is a second class, etc.

Given a set of classes - for example the set of 12 classes just presented - a choice is made between these classes; the further work is to be directed towards the choice of a design within the class chosen.

This choice between the classes may be made by a step-wise procedure. First, a choice is made with respect to the A-dimension; the outcome is D $(a_i; -; -)$. Second, a choice is made with respect to the B-dimension, given the previous choice; the outcome is D $(a_i; b_j; -)$. Finally, a choice is made with respect to the C-dimension; the outcome is D $(a_i; b_j; c_k)$. These three choices should, however, not be carried out independently of each other; a dependent choice is called for because the three steps represent choices between various ways of using information that may be available.

If the information available is both quantitative and qualitative, it is often advantageous to use the qualitative information for the selection scheme, and the quantitative information for the estimation scheme.

We will briefly discuss the choice with respect to each dimension.

(1) As to the A-dimension, a satisfactory choice between a_1, a_2, and a_3 may often be made by means of a very simple cost analysis, which compares the following survey costs:

$$C(a_i) = C(f; a_i) + C(m; a_i)$$

for i = 1, 2 and 3, respectively. Here:

$C(f; a_i)$ = the cost of creating a frame, using a_i.

$C(m; a_i)$ = the cost of collecting the measurements, including the cost of 'quality control' of the collection process,[5] using a_i.

---

[5]This is a most important aspect; it means that considerations of 'measurement errors' may exercise a decisive impact on the sample design. For an illuminating illustration, the reader is referred to the literature on the original design of the U. S. Bureau of the Census 'Current Population Survey', and the subsequent revisions of this design.

(2) As to b_1 and b_2, the following may be said. A design based on b_2 may be called for, if the population exhibits a skew distribution; the control of the variability thus achieved may, however, alternatively be exercised by the choice of an estimation scheme of the C_2-type.

(3) As to C_1 and C_2, an analogous analysis may be carried out.

10. Choosing within a class of sample designs.

Next, a specific sample design is chosen within the class of designs chosen in the previous step.

Experience shows that this choice of a specific sample design is of a small importance relative to the choice between the classes of sample designs; the sample designs belonging to a given class are roughly equivalent. As a consequence, it is not only possible but desirable to pay great attention to administrative considerations - for example simplicity - when choosing the specific sample design.

11. Optimization.

The choice of a class of sample designs is a means of narrowing down the scope of choice. The choice of a specific sample design means that the scope of choice is further narrowed down.

The outcome of this second choice leaves, however, still a margin for choice. It is the function of the last step, the optimization, to eliminate this margin by picking out a unique sample design.

The optimization may suitably be carried out in two stages. In the first stage, a 'theoretical' optimum is derived. In the second stage, this optimum is evaluated from the point of view of deriving a 'practical' optimum, a 'super-optimum'. As the distinction between these kinds of optima is important, we will give two examples.

The first example is discussed in some detail in Hansen et al (1953), vol. I, p. 223. This example concerns a survey based on stratified sampling, with L=2 strata. A straightforward application of the theory of optimum allocation called for the following sampling fractions:

$$f_1 = \frac{1}{11} \quad ; \quad f_2 = \frac{1}{19} \quad .$$

This 'theoretical' optimum was, however, replaced by the following 'practical' optimum:

$$f_1' = \frac{1}{10} \quad ; \quad f_2' = \frac{1}{20} \quad .$$

The superiority of these latter sampling fractions was largely due to the fact that the measurements were to be processed by means of punched-card equipment; considerably simpler processing operations were possible with the f'-values than with the f-values.

The second example is as follows. A sample of m first-stage units is selected. In each sample unit, a directory of elements is established, by having interviewers list the unit. Thereafter, $\bar{n}$ of the $\bar{N}$ elements are subsampled. Here $\bar{n}$ is determined by means of a formula for "optimum $\bar{n}$" which takes into account the cost of listing the units. Now the formula may give:

$$\text{optimum } \bar{n} < \bar{N} \quad .$$

This value of $\bar{n}$ is, however, not necessarily equal to the 'practical' optimum. Thus, it may prove preferable to choose:

$$\bar{n} = \bar{N}$$

as this would eliminate the listing operation. The saving thus achieved may more than counterbalance the increase in cost due to taking a larger sample of elements.

12. Choosing the sample design.

The operations discussed in par. 8-11 result in the construction of a small set of sample designs, $D_1, D_2, \ldots, D_s$, which represent alternatives among which a choice of one is to be made, to serve as <u>the</u> sample design to be used.

This choice may be made on the basis of a criterion which takes into account such relevant considerations as the (estimated) survey cost and variances, the time schedule, the administrative feasibility, etc.

13. Design for an evaluation of the survey design.

The design of a survey involves a certain amount of risk-taking; the information on which the design is based is always fragmentary, and sometimes erroneous. What ex ante looks like a reasonable approach may prove, ex post, to be unsatisfactory.

Carrying out a survey with the defined goals (X) offers an opportunity to collect additional data which may prove useful in the design of future surveys, and especially to collect data on costs and variances. The following quotation is instructive, United Nations (1964):

> An important reason for the use of sampling (instead of complete enumeration) is lower cost. Information on costs is therefore of great interest. Cost should be classified so far as possible under such heads as preparation (showing separately the cost of pilot studies), field work, supervision, processing, analysis and overhead costs. In addition, labour costs in manweeks of different grades of staff, and also time required for interview and journey time and transport costs between interviews, should be given. The compilation of such information, although often inconvenient, is usually worth undertaking as it may suggest substantive economies in the planning of future surveys. Efficient design demands a knowledge of the various components of costs, as well as of the components of variance.

It is characteristic of the survey activity of such leading agencies as the U. S. Bureau of the Census and the Indian Statistical Institute that this opportunity is systematically exploited.

If the collection of cost and variance data is to be fully effective, it must be carried out within the framework of a design which is an integrated part of an overall survey design. This means, inter alia, that a portion of the total budget must be taken away for this specific purpose.

IV. SOME ADMINISTRATIVE CONSIDERATIONS

14. The administrative design.

The discussion in Chapter III is oriented towards the development of the methods to be used. In this Chapter, the discussion will be oriented towards the administration of the survey.

As emphasized in Hansen and Steinberg (1956) "it is one thing to specify a good survey design, and quite another to set in motion administrative procedures and controls which insure that the operations are carried through substantially as specified". This calls for the development of an administrative design, to go with the methodological design.

The aim of the administrative design must be to create a 'machinery' by means of which the survey can be carried out "substantially as specified".

It is clearly not sufficient to develop procedures which are satisfactory, granted that everything develops according to expectation. Something more is called for, to cope with deviations from plans, unexpected events, etc.

A useful approach is to develop a '<u>signal system</u>' that records continuously all events taking place, and therefore warns when there are deviations from plans. A simple example of such a signal system is the following. In a mail survey, a daily, cumulative record is kept of the responses. Such a record makes it possible to initiate action, if the response rate is too low.

In addition, there must be an '<u>emergency plan</u>' that makes it feasible to cope with (tendencies towards) deviations from the specification. In the case of the mail survey just discussed, it is not sufficient to keep record of the responses; there must, in addition, be resources set aside which may be exploited to cope with the situation, should the response rate prove to be too low.

Mention should here be made of a class of tools for planning and administration which has a great potentiality in the present context. These tools are referred to as "net work planning"; CPM ("Critical Path Method") and PERT ("Program Evaluation and Review Technique") are but two brand names. For an illustration of the application of such a tool, reference is given to International Business Machines Corporation (undated).

15. The documentation.

The design of a survey is not properly carried out, until there is a self-contained document presenting it; there is indeed need for various kinds of documents, as argued in United Nations (1964) to which reference is given for details.

One kind of document - not mentioned in the reference just quoted - is a 'plan tree'. The main function of a 'plan tree' is to present a comprehensive picture of the main aspects of the design; therefore, the 'plan tree' must not be belaborated with details - they should be discussed in separate attachments.

In the interest of perspicuity, we present a concrete example referring to the "Housing Starts Statistics" produced by the U. S. Bureau of the Census.

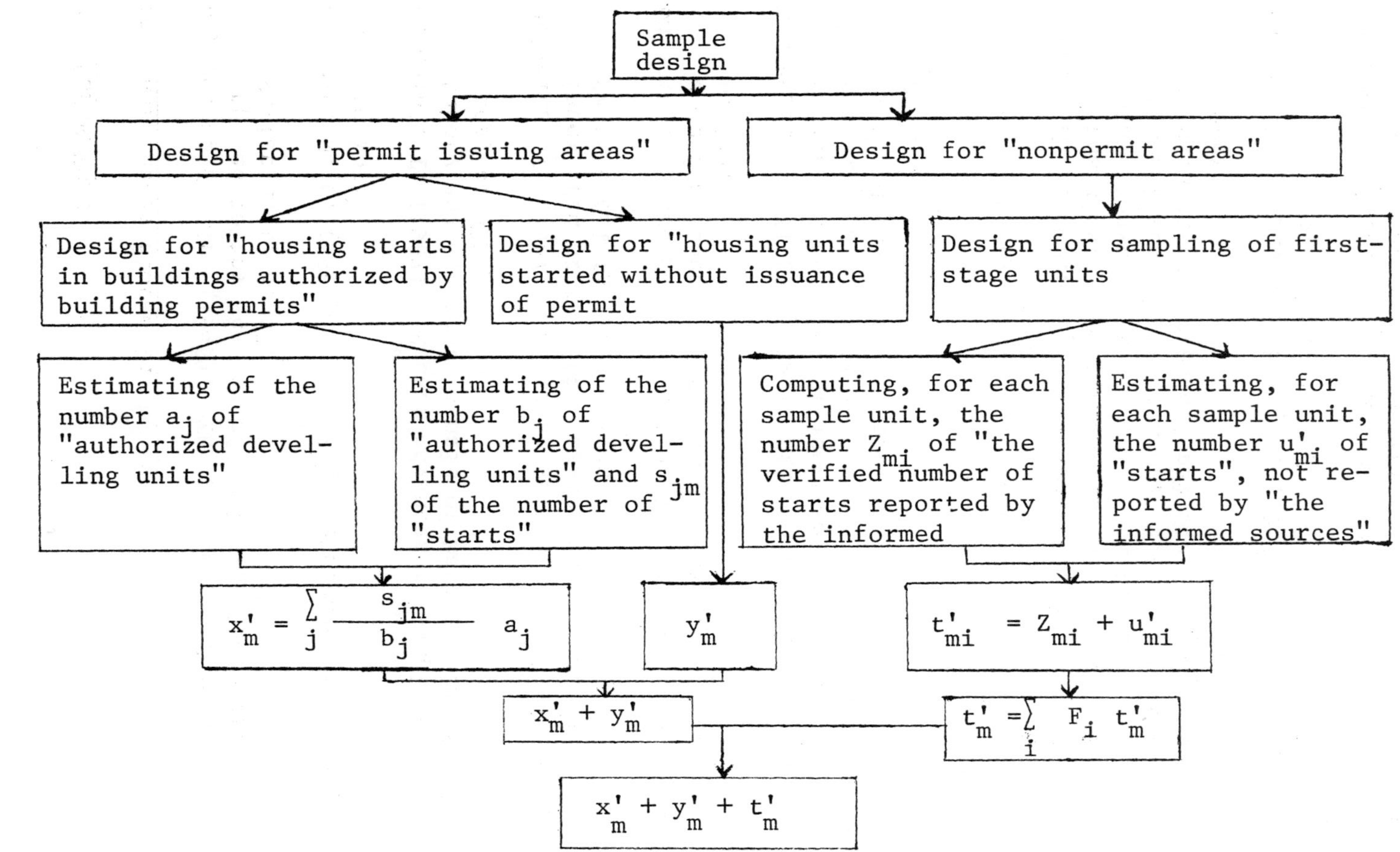

Sample design
Design for "permit issuing areas"
Design for "nonpermit areas"
Design for "housing starts in buildings authorized by building permits"
Design for "housing units started without issuance of permit
Design for sampling of first-stage units
Estimating of the number a_j of "authorized develling units"
Estimating of the number b_j of "authorized develling units" and s_{jm} of the number of "starts"
Computing, for each sample unit, the number Z_{mi} of "the verified number of starts reported by the informed
Estimating, for each sample unit, the number u'_{mi} of "starts", not reported by "the informed sources"
$x'_m = \sum_j \frac{s_{jm}}{b_j} a_j$
y'_m
$t'_{mi} = Z_{mi} + u'_{mi}$
$x'_m + y'_m$
$t'_m = \sum_i F_i t'_m$
$x'_m + y'_m + t'_m$

V. REFERENCES

16. References quoted in the text.

Blythe, R. H., Jr. (1945). The economics of sample size applied to the scaling of sawlogs. Biometrics Bulletin, 1, p. 67.

Cochran, W. G. (1963). Sampling techniques, 2nd. ed., New York.

Dalenius, T. (1962). Recent advances in sample survey theory and methods. Annals of Mathematical Statistics, p. 325.

Dalenius, T. (1967). Official statistics and their uses. Paper read at the 36th session of the International Statistical Institute, Sydney.

Dalenius, T. and Matérn, B. (1964). Is there a need for a unified theory of random experiments? Metrika, p. 235.

Eisenhart, C. (1963). Realistic evaluation of the precision and accuracy of instrument calibration systems. Journal of Research of the National Bureau of Standards - C. Engineering and Instrumentation, p. 161.

Hansen, M. H., Hurwitz, W. N. and Madow, W. G. (1953). Sample survey methods and theory. Vol. I., New York.

Hansen, M. H., Hurwitz, W. N. and Pritzker, L. (1967). Standardization of procedures for the evaluation of data: measurement errors and statistical standards in the Bureau of the Census. Paper read at the 36th session of the International Statistical Institute, Sydney.

Hansen, M. H. and Steinberg, J. (1956). Control of errors in surveys. Biometrics, p. 462.

International Business Machines Corporation (undated). "PERT" in action at the Bureau of Labor Statistics, U. S. Department of Labor, Washington, D. C.

Stephan, F. F., and McCarthy, P. J. (1958). Sampling Opinion. An analysis of survey procedures. New York.

Deming, W. E. (1950). Some theory of sampling. New York.

Törnqvist, L. (1948). An attempt to analyze the problem of an economical production of statistical data. Nordisk Tidskrift för Teknisk Ökonomi, p. 265.

United Nations (1964). Recommendations for the preparation of sample survey reports (provisional issue). Statistical papers Series C, No. 1, Rev. 2, New York.

17. Additional references.

Ackoff, R. L. (1953). The design of social research. Chicago.

Hansen, M. H., Hurtwitz, W. N., and Jabine, T. B. (1964). The use of imperfect lists for probability sampling at the U. S. Bureau of the Census. Bulletin of the International Statistical Institute, Vol. 40, p. 497.

Hansen, M. H., Hurwitz, W. N., Nisselson, H., and Steinberg, J. (1955). The redesign of the Census current population survey. J. Amer. Statist. Assoc., p. 701.

Hyman, H. (1955). Survey design and analysis. Clencoe, Ill. 1955.

Kish, L. (1952). A two-stage sample of a city. American Sociological Review, p. 761.

Monroe, J. and Finkner, A. L. (1959). Handbook of area sampling. Philadelphia, Pa.

Statistisches Bundesamt (Wiesbaden) (1960). Stichproben in der amtlichen Statistik, Wiesbaden.

DESIGN AND ESTIMATION FOR SUBCLASSES, COMPARISONS, AND ANALYTICAL STATISTICS*

Leslie Kish

Survey Research Center, University of Michigan

1. Introduction

The theory and literature of survey sampling have been developed almost exclusively for means and aggregates based on the entire sample. These estimates may provide the focus of some sample censuses of population, agriculture, goods, and services. Historically, these fields were the sources of sampling theory and their problems were more immediately amenable to theoretical elaborations.

Separate consideration of distinct domains, such as regions, requires only simple modifications. However, in most actual surveys we need to estimate characteristics of diverse subclasses of the population, which cannot be designed economically as separate domains. Further, most social surveys aim chiefly at discovering relationships between subclasses; this is now done commonly by comparisons of subclasses, and usually by the differences between the means of subclasses.[1]

---

*Research and writing was supported by Grant GS-777 from the National Science Foundation.

[1]In dealing with the subclass means and their differences our emphasis will be on their variances. The mean of a subclass from a probability sample can be generally justified by considering it a ratio mean. Difference of two subclass means is commonly used; two problems are mentioned in sections 3 and 4. We assume throughout that compared subclasses are mutually exclusive, without common elements; if they overlap, covariance terms must also be considered (Kish 1965, 12.4). That variances give us the information needed for inference is implicit here, and in the literature of probability sampling. We do need generalizations of central limit theorems for probability samples which are not simple random (Tharakan 1968).

Increasingly more sophisticated methods, such as linear regressions and other multivariate techniques are also being applied to sample survey data to uncover relationships among variables. Unfortunately the assumption of independent selection, almost universal in the literature on these analytical statistics, is often grossly violated by correlations that occur between selected elements in practical designs, and especially in clustered samples.

We shall begin with an outline of the problems, the techniques and the designs for subclasses and their differences. Much of this is an organized, brief review of results from scattered literature, although it is still largely absent from most textbooks. This will be followed by results of recently developed and applied techniques for balanced repeated replications, which are widely useful for computing standard errors for analytical statistics, where mathematical techniques are lacking.

Emphasis throughout is on the applied and applicable; on clarity and simplicity to aid the samplers who need these techniques; on approximations easy to compute, rather than on refinements. In the same spirit, instead of variances I often give expressions for their sample estimators. The lack of rigor and precision in places hopefully may serve as stimulus for those whose abilities and tastes run along those lines.

2. Allocations for domains and comparisons

Early in the design process attention should be directed to possible conflicts in allocating the entire sample among various domains. Serious conflict can arise, for example, when equal precision is specified for domains whose sizes differ greatly. Whereas equal sample sizes are optimal for estimating and comparing domain means, for the mean of the entire sample the optimal sizes are proportional to the stratum sizes. How much is lost for one aim when the allocation is optimal for the other?

Denote the diverse element costs c_h and the variances S_h^2. The total cost is $\Sigma c_h n_h$ and the variance of the combined mean is $\mathrm{Var}(\bar{Y}_w) = \Sigma(1-f_h)W_h^2 S_h^2/n_h$. For domain means and their comparisons, I suggest a simple and symmetrical criterion, the mean domain variance $(\Sigma(1-f_h)S_h^2/n_h)/H$. Optimal allocation is proportional to $W_h S_h/\sqrt{c_h}$ for the combined mean, and to $S_h/\sqrt{c_h}$ for the domain means.

Now we want to express the relative increase of the variances in terms of departures of the actual n_h from optimal n_h'. Let any actual allocation be $n_h \propto K_L n_h'$, where the K_h are proportionality factors denoting deviations from the optimal allocation n_h'. Then it can be shown (Kish, 1969) that the ratio of the variance of the actual allocation to the optimal allocation is

$$\frac{\text{actual variance}}{\text{optimal variance}} = (\Sigma W_h'/K_h)(\Sigma W_h'K_h)$$

$$= (\Sigma w)(\Sigma w_h K_h^2) = 1 + CV^2(K_h).$$

For the combined mean use $W_h' = W_h S_h/\sqrt{c_h}/\Sigma W_h S_h \sqrt{c_h}$; for t the mean domain variance use $W_h' = S_h \sqrt{c_h}/\Sigma S_h \sqrt{c_h}$. Increase over the minimum of 1 is seen to depend on the relvariance of the departures K_h from optimal, with the factors W_h'/K_h as weights.

Tables for the function (2.1) are given by Kish [1965, pp. 431-2; also p. 97]. (The discussion there concerns the simple case when S_h and c_h are constant, so that the $W_h' = W_h$ for the combined mean, and $W_h = 1/H$ for the mean domain variance.) Great increase of the variance result from U-shaped distributions of the weights and large departures for the K_h. For rectangular and triangular distribution of the weights, the increases are moderate.

Sedransk (1965, 1967) has recently made intensive investigations of special problems of optimum allocations for analytical studies of the differences of means from survey samples, but these apparently have not received applications. In an excellent, comprehensive article, Cochran (1965) gives motives and techniques for diverse weighting methods for differences of subclass means. Sampling methods for multi-factor comparisons from census data were explored by Keyfitz (1953; also in Yates 1960, 9.6).

3. Subclasses in stratified random selection

The problem of a subclass (m) from a simple random sample (n) need not detain us long. Almost everyone, either on theoretical grounds or as good approximations, would use $\bar{y}_m = y_m/m = \Sigma y_j/m$ as the estimator of the subclass mean $\bar{Y}_m = Y_m/M$, and $\text{var}(\bar{y}_m) = (1-f)s_m^2/m$ for its variance, where $s_m^2 = (\Sigma y_j^2 - \bar{y}^2)/(m-1)$ (Durbin, 1968). (In the rare situation when M is known one could also use $f = m/M$, rather than n/N.)

The above terms are useful for defining analogous terms in formulas for estimators and their variances in stratified element sampling, when simple random selections of n_h from the N_h elements yield random numbers of m_h members from the M_h subclass members in the hth stratum. We need not dwell on simpler situations when the subclasses are domains that merely represent entire strata (N_h and n_h), so that formulas of stratified sampling apply.

The problem may also be relatively simple if the M_h are known. This may allow us to allocate the m_h to be the desired sample sizes if the practical constraints of the situation permit such control. More often we must allow the m_h to remain random variables, but we may still use, with post-stratification, the weights $W_h = M_h/M$, where $M = \Sigma M_h$.

Most often, however, the M_h are unknown, and we must depend entirely on the random sizes m_h yielded by the sample. The variance of the simple unbiased expansion estimate $\tilde{Y}_w = \Sigma N_h y_h/n_h$ of ΣY_h is

$$\text{var}(\tilde{Y}_w) = \Sigma(1-f_h)\frac{N_h^2}{n_h-1}\bar{m}_h[v_h^2 + (1-\bar{m}_h)\bar{y}_h^2]. \quad (3.1)$$

Here $\bar{y}_h = \sum_j y_{hj}/m_h$, $\bar{m}_h = m_h/n_h$ and

$$v_h^2 = (m_h-1)s_h^2/m_h = (\sum_j y_{hj}^2 - m_h\bar{y}_h^2)/m_h.$$

The penalty for not knowing the M_h is represented approximately by the second term within the brackets, which can be large. A Bayesian introduction of distributions of M_h may help; but reasonable knowledge about all the M_h is probably rarer than about the entire M, which can be used with the sample mean.

The sample mean is the ratio mean

$$\bar{y}_w = \frac{\Sigma N_h y_h / n_h}{\Sigma N_h m_n / n_h} = \Sigma w_h \bar{y}_h, \text{ where } w_h = \frac{N_h m_n / n_h}{\Sigma N_h m_h / n_h} \quad (3.2)$$

is used instead of $W_h = M_h/M$, because these are unknown. This is a ratio mean because the m_h are random variables, and its estimated variance is the useful approximation

$$\text{var}(\bar{y}_w) \doteq \sum_h (1-f_h) \frac{w_h^2}{m_h'}[v_h^2 + (1-\bar{m}_h)(\bar{y}_h - \bar{y}_w)^2], \quad (3.3)$$

where $m_h' = m_h(n_h-1)/n_h$. Using w_h instead of W_h results approximately in the quantity in the brackets instead of s_h^2. Furthermore, the effect of small subclasses may be seen even more clearly by taking the deviations around the sample mean. Thus (3.3) becomes

$$\text{var}(\bar{y}_w) \doteq \sum_h (1-f_h)\frac{w_h^2}{m_h'}[t_h^2 - \bar{m}_h(\bar{y}_h - \bar{y}_w)^2], \quad (3.4)$$

where $$t_h^2 = \sum_j (y_{hj} - \bar{y}_w)^2/m_h = v_h^2 + (\bar{y}_h - \bar{y}_w)^2.$$

For proportionate sampling $f_h = n_h/N_h$ is constant (f), $w_h = m_h/\Sigma m_h$, $\bar{y} = \Sigma y_h/\Sigma m_h$, and

$$\text{var}(\bar{y}) \doteq \frac{1-f}{(\Sigma m_h)^2} \sum_h \frac{n_h}{n_h-1}[\sum_j (y_{hj} - \bar{y}_w)^2 - m_h \bar{m}_h (\bar{y}_h - \bar{y}_w)^2]. \quad (3.5)$$

The second terms in (3.4) and (3.5) refer to the between stratum variances with the factors $\bar{m}_h$. Hence for small subclasses the gains of proportionate sampling tend to vanish; for such situations the simple random variance is a good approximation.

These simple and useful approximations hold even better for the difference $(\bar{y}_{wa} - \bar{y}_{wb})$ of the means of subclasses a and b. The variance is well approximated with

$$\text{var}(\bar{y}_{wa} - \bar{y}_{wb}) \doteq \Sigma(1-f_h)\left[\left(\frac{w_h^2 t_h^2}{m_h'}\right)_a + \left(\frac{w_h^2 t_h^2}{m_h'}\right)_b\right]. \quad (3.6)$$

This approximation neglects merely a product of $\bar{m}_{ha}\bar{m}_{hb}$ with the differential of the between stratum components; this should usually be negligible, and was so found in one large body of data (Kish 1961 and 1965). <u>For differences of subclass means from proportionate samples, simple random variances are most often good approximations</u>.

For the approximations above we assume that any inequalities in the sampling fraction f_h are countered with weights proportionate to $1/f_h$ introduced into the w_h. Disproportionate weights may result , of course, in substantial losses of efficiency. However, gains of efficiency that may be obtained with optimum allocation can be shown (Cochran 1963, 5.6 and Kish 1965, 4.6) to be

$$\frac{\text{var(optimum)}}{\text{var(srs)}} \doteq 1 - \frac{\Sigma W_h(\bar{Y}_h - \bar{Y})^2}{S^2} - \frac{\Sigma W_h(S_h - \bar{S})^2}{S^2}, \quad (3.7)$$

where $\bar{S} = \Sigma W_h S_h$.

The second term denotes the gains from proportionate sampling, and they tend to vanish for small subclasses. The third term refers to <u>gains from optimum allocation, and they may tend to persist</u> to the degree that the weights and standard deviations for the subclasses tend to be proportional to those for the entire sample.

For optimum allocation of a subclass, the sampling fractions f_h should be made proportional to

$\sqrt{\bar{M}_h[U_h^2]}$, where $[U_h^2]$ refers to element variances of the kind shown within brackets [ ] in several variance formulas above. If we need to take into account costs per element c_h that differ between strata, the allocation f_h should be proportional to $\sqrt{\bar{M}_h[U_h^2]/c_h}$. If we use $\sqrt{\bar{M}_h[U_h^2]c_h n_h/\Sigma c_h n_h}$ for S_h, then (3.7) also holds. For element cost we should consider not only the cost c_h' for the m_h subclass element, but also d_h for the (n_h-m_h) eliminated elements, so that $c_h=\bar{m}_h c_h' + (1-\bar{m}_h)d_h$.

Yates first published the basic formulas and optimum allocations for subclasses (1953, 9.1-9.5). Derivations, the decline from proportionate gains to srs approximations, and large scale applications were presented in 1954 to the American Statistical Association (Kish 1961, 1965, 4.5). Independently Durbin (1958) and Hartley (1959) also gave derivations; those of the latter are especially complete.

When we consider several subclasses as well as several characteristics (variables), the problems of multipurpose design can soon overwhelm us. Often it is not even practical to control the sizes of subsamples. Fortunately, moderate departures from optimum allocation increase variances only slightly (section 2). I mention, with caution, three approximations that seemed useful in a large empirical study (Kish, 1961).

(1) Let $(K_h)_{cs}$ denote the optimal "loading" factor in the hth stratum for the cth characteristic and the sth subclass. If the stratifying variables have systematic relationships to the survey variables, try to find a set of exponents g_h such that $(K_h)_{cs} = k_{cs}^{g_h}$, the k_{cs} are constant over strata, and the g_h over characteristics and subclasses. In our study we had 3 strata based on economic ratings of dwellings, and the exponents 0:1:2 fit well; the computed optimal loading factors were close to the curve $1:k_{cs}:k_{cs}^2$.

(2) It also seemed that the separate factors k_{cs} could be fairly well represented by the product $k_c k_s$

of 2 constants. Thus we could deal with a set of constants equal to the sum of the numbers of strata, subclasses, and characteristics, rather than to their product.

(3) We formed a small set of approximate "types", each a group of variates that behaved fairly similarly. This approximation further reduced the number of parameters to be considered in the design.

4. Subclasses and comparisons in cluster sampling

When diverse subclasses are completely segregated in separate clusters and domains (as for regional estimates), their problems can be handled as mere extensions of section 2. We concentrate here on the ubiquitous problem of subclasses that cut across clusters; for example, age, sex, and social classes in area samples of counties and blocks.

Dealing with subclasses instead of the entire sample produces two principal effects: reduction in the average cluster size and loss of control over the sub-sample sizes. Imagine a national sample of 5000 persons from 100 counties. A subclass of 200 or 500 members means 2 or 5 members per county on the average. If there are 5 persons per segment, the average subclass content is fractional.

Good sample designs usually maintain moderate control of size over the entire sample; control of size of subsample clusters is sometimes attained by creating clusters of equal sizes, but mostly by selection with probabilities proportional to size (PPS). The coefficient of variation, around the average of 50 persons per county, may be kept between 0.2 and 0.5, let us say. For the entire sample a coefficient of variation of $0.2/\sqrt{100} = 0.02$, or 100 persons, should do, and even 250 may be acceptable. For a subclass that contains 1/25 of the sample the coefficient becomes 0.10 instead of 0.02, if the subclass is randomly distributed in the sample. This is roughly true for most subclasses, although some subclasses will be unevenly clustered: extreme examples may occur with some ethnic groups and a few occupations, such as miners and fishermen. Control of overall size of subclasses may sometimes be achieved through differential selection rates and double sampling. But control of cluster sizes can rarely be attained for subclasses, though some may be improved by stratifying the sampling units for average subclass proportions.

Unequal cluster sizes usually lead to using ratio means of the form $r = y/x = \Sigma y_h/\Sigma x_h$, and this estimator also serves subclasses. But methodological difficulties are exacerbated by the loss of control over cluster sizes; this can become serious for some subclasses. A low coefficient of variation of the sample size, CV(x), assures a low bias for the ratio mean; it is also assumed in the Taylor expansion for deriving its variance (Hanson, Hurwitz and Madow 1953 II, 4.12; Kish 1965, 6.6, 7.1). For small and unevenly distributed subclasses CV(x) may become large; this may undermine our confidence in the ratio mean and its variance for small subclasses.

However, the actual situation is usually much better. The bias ratio is $\text{Bias}(r)/\sqrt{\text{Var}(r)} = -\rho_{rx}\text{CV}(x)$, which is small when either CV(x) or ρ_{rx} (the correlation of the ratio mean and its denominator) is small. It should be most unusual for both of these to be large without the investigator being aware of it. They have been found to be always negligible in large scale empirical investigations (Kish, Namboodiri, Pillai 1962); and simulation experiments on the ratio mean also give encouraging results (Rao, 1967, Tepping, 1968). The situation is considerably more serious for "separate" ratio estimators $(\Sigma y_h/x_h)$; they should probably be avoided for small and uneven subclasses.

Survey results are often produced for so many subclasses that computing and presenting sampling errors for all of them becomes too costly and cumbersome. Hence some methods for summarizing and generalizing become necessary (Kish 1965, 14.1). Curves have been fitted to computed values of relvariances as functions of sample sizes by the U. S. Census Bureau and others (Hansen, Hurwitz and Madow, I, 1953). Here we add a few brief remarks about models relating the variances of subclasses to their sizes m.

The relationships can be expressed as $\underline{S^2 Deff/m}$, where S^2/m is the simple random variance; and $\underline{Deff}$ is the ratio of actual variance to S^2/m, the "design effect" (Kish 1965, 8.2). Now we may often suppose that S^2, the element variance, is roughly the same for subclasses as for the entire sample, when m=n. Values of Deff can differ among characteristics; they are higher for measuring more clustered variables, like income or ethnicity, than for age or sex. For a fixed characteristic Deff may vary between subclasses. We can construct two easy and contrasting models, which may serve as rough empirical upper and lower limits for Deff as the subclass size m is decreased. (U) Suppose that the $Deff_n$, estimated for the entire sample when m=n, remains constant as m decreases; thus $Var(\bar{y}_m) = S_n^2 Deff_n/m$. This may be roughly true for subclasses that are completely segregated, such as geographic subclasses. (L) On the other hand, for subclasses that are evenly or randomly distributed among clusters, we may suppose that Deff can be approximated by the function $[1 + roh(\bar{m}^* - 1)]$, where the intraclass correlation $\underline{roh}$ remains constant as the average cluster size $\bar{m}^*$ declines to 1. Deff is then allowed to decline linearly with m, to a minimum of 1, reached when m equals the number of primary selections. If there are more than two effective selection stages, this problem should become more complex theoretically. We have used these straight lines to fit subclass variances from many surveys; they seem to fit either on or between those lines. But we acknowledge the need here for more research.

The comparison of subclasses, in the form of differences of ratio means (r-r'), raises new issues. The variance and its estimate, var(r) + var(r') - 2 cov(r,r'), are merely extensions of var(r), and its computation is not difficult (Kish and Hess 1959,

Kish 1965, 6.5). The bias ratio of the difference, $\text{Bias}(r-r')/\sqrt{\text{Var}(r-r')}$, causes theoretical problems because it can increase without limit if the denominator should tend to vanish when the correlation between r and r' approaches 1. We can offer a plausible upper limit in the larger of CV(x) and CV(x'); also a large set of empirical results, in all of which the bias ratio was negligible (Kish, Namboodiri, Pillai, 1962).

Concerning the variances of subclass differences, we have obtained valuable empirical generalizations from computations on multitudes of variables from many surveys of our Survey Research Center, since about 1952. We compute them routinely with standard programs written (and rewritten several times) for high speed computers; a greatly improved program is being written in 1968. I regret to state that we seem to have almost a monopoly on such computations, and I hope this statement will be contradicted.

Most frequently, it seems that comparisons are simply ignored when sampling errors are computed and presented for survey results. When not ignored, one of two assumptions is presented as approximations for var(r-r'):

(1) that it is equal to $S^2/n + S'^2/n'$, the sum of the simple random variances; or

(2) that it is equal to var(r) + var(r'), the sum of the two subclass variances.

The second would hold if the covariances were zero, leaving the design effects on the differences the same as for the variances. The first would hold if the effect of covariances would completely eliminate the design effects of clustering.

Usually neither of these assumptions is correct, and empirical reality falls between the two. On the basis of the many results mentioned above we find generally that

$$\frac{S^2}{n} + \frac{S'^2}{n'} \leq \text{Var}(r-r') \leq \text{Var}(r) + \text{Var}(r'). \quad (4.1)$$

Equality seldom holds, and the results generally fall well within these extremes, though more often nearer the lower extreme. This seems to be a dependable and useful empirical law. We have used those as upper and lower limits for tables of standard errors of percentages (Kish, 1965, Table 14.1.III). Among hundreds of computations I have not seen a contradiction reliable beyond sampling fluctuation. On the contrary, seeming contradictions tend to disappear when faced with new computations and more reliability. Similar results also hold for many computations of differences of comparable means from two periodic surveys from the same PSU's (Kish, 1965, Table 14.1.IV).

If estimates of the two extremes are close to each other we may take their mean as a plausible approximation for var(r-r'). That is, if estimates of Deff for r and r' are only a little beyond 1, we may use half their sum as a fair approximation for the Deff of var(r-r'). However, if Deffs for r and r' are much beyond 1, computing var(r-r') is needed to see where it actually falls between those extremes.

It is interesting to speculate jointly about the contrasting effects on small subclasses of proportionate stratification and clustering designed for a large population. In both cases the variances seem to approach those of simple random sampling as the proportion of subclass members decreases; in clustering the approach is usually from above, and in stratification from below. Together we can generalize to the operation of a kind of entropy. The strength of the average correlation induced among all sample elements by the controls of selection decrease, and tend to vanish, as the hold of the controls becomes loosened on smaller subclasses.

5. Linear Combinations, ratios of ratios, and index numbers

Differences of means are not the only way, though the most common, of making comparisons. The most immediate extension of the difference (r-r') is the difference of differences $[(r_a-r_a') - (r_b-r_b')]$; for example, the difference of incomes for young versus old, between white and blue collar workers. This leads to computing forms for linear combinations. If $r = \Sigma y_h/\Sigma x_h$, and its variance is $\text{var}(r) = \sum_h dz_h^2/x^2$, then with W_j constant for the jth variate,

$$\text{var}(\sum_j W_j r_j) = \sum_h [\sum_j \frac{W_j dz_{hj}}{x_j}]^2 \tag{5.1}$$

is a good computing form (Kish 1965, 12.11). We may conjecture that the design effect, generally decreased for the difference of means, would tend to be similarly, and perhaps further, decreased for the difference of differences. We hope to investigate this conjecture with a new computing program, and again invite corroboration. We need to know the behavior of these statistics to design efficient samples for them.

Instead of the difference (r-r') between subclass means, sometimes their ratio (r/r') may appear more useful. Variances for r/r' have been given by J.N.K. Rao (1957) and Keyfitz (Yates 1960, 10.5). Convenient computing forms for this and for its linear combinations, which are index numbers of the form $(\Sigma W_j r_j/r_j')$, are given by Kish (1965, 12.11). An empirical investigation of the behavior of ratios of ratios, and of index numbers for 15 variables in several surveys, is reported by Kish (1968).

The decrease in the difference we noted in the design effect of var(r-r') is also useful when designing for r/r'. Its variance can be computed as

$$\mathrm{var}(r/r') \doteq \bar{r'}^{2}\left[\mathrm{var}(r) + \left(\frac{r}{r'}\right)^{2} \mathrm{var}(r') - 2\left(\frac{r}{r'}\right) \mathrm{cov}(r,r')\right], \qquad (5.2)$$

and the decrease in design effect, due to cov(r,r'), should appear here also. For the difference $(r_1-r_2)/r'$, if it is small (say within $\pm$ 0.10) the variance can be well approximated with the simpler $\mathrm{var}(r_1-r_2)/r'^2$ (Kish 1968).

6. Response variance and bias in subclasses

Each interviewer may be supposed to have an individual bias, averaged over the responses of his workload; we may express the variation between interviewers, in its joint effect on the variance of the sample mean, as an interviewer variance (Hansen, Hurwitz and Madow 1953, II, Ch. 12). Its contribution to the variance of the sample mean resembles other variance components, specifically that of cluster variance; the workloads of individual interviewers may be viewed as clusters of responses subject to similar effects. The effect of the between-interviewer variance is a component proportional directly to the average variance per interviewer, and inversely to the number of interviewers. This effect on the element variance can most usefully be expressed, like the cluster effect, by the approximation $[1 + \mathrm{roh}(\bar{n} - 1)]$; here roh is the intra-interviewer correlation coefficient and $\bar{n}$ is the average workload. What happens to this effect in subclasses?

One answer is given by the investigations of Hansen, Hurwitz, and Bershad (1961) into the large effects of interviewer variance on estimates for small places that are covered by only a few enumerators. The Census enumerator's workload $\bar{n}$ is about 250 households, or about 50 households in the 20 percent sample; even small correlations may produce large effects on census estimates for small places. For example, an average roh of 0.005 will result in adding 1.25 to the unit variance, and 0.25 for the items in the 20 percent sample. These effects are doubled, quadrupled, or larger for difficult census items for which the roh's may be 0.01, 0.02, and even higher (Kish, 1962, and 1965, 13.2).

However, most subclasses cut across interviewer workloads in rather random fashion, and for them the interviewer effects are proportionately and drastically reduced. If roh remains about constant, and if $\bar{m}^*$ is the average workload for a subclass, then we can take $[1 + \text{roh}\ (\bar{m}^*-1)]$ to represent approximately the interviewer effect. Like the clustering effect, the increase due to interviewer variance diminishes as the subclass size approaches the number of interviewers. For example, roh's for attitudinal items averaging 0.01 or 0.02 increase the variance by 0.50 or 1.00 for the entire sample when the workload averages 50 interviews. But these effects may be cut to 0.025 to 0.05 for subclasses which are 1/20th of the sample. Furthermore, for differences of subclass means, interviewer biases tend to be similar for both means, and the effects tend to be further reduced and to vanish. These conjectures of clusterlike behavior were well borne out in an empirical investigation (Kish, 1962); deviations from the model held no significance. Of course, more and larger investigations may modify these first approximations.

The design of interviewer workloads may greatly affect which of the two situations contrasted above exists for the subclasses; similar considerations may apply also to other sources of error, such as coding in the office. For example, some geographic subclasses may be forced by cost considerations into the first type of interviewer effect, because the workload remains at $\bar{n}$ for the subclass. But for other subclasses we may be able to achieve a spread, approaching eveness or randomness among the interviewers and coders.

The relative importance of constant, systematic biases tends to be less for subclasses than for the entire sample. This decrease in importance can be noted in the mean square error (bias^2 + variance), or in the bias ratio ($\text{bias}/\sqrt{\text{variance}}$). If the bias remains relatively constant, its relative importance diminishes as the variance increases, roughly in proportion to the decrease in the subclass size. Furthermore, the variance of comparisons involves the sum of two variances; on the contrary the bias involves the difference of biases, which often tend to be similar; to that degree the effects of constant biases greatly diminish.

7. Analytical statistics from complex samples

Standard statistical literature has been developed on the assumption of independent observations; this greatly facilitates obtaining interesting theoretical results. On the other hand the designs of most survey samples violate this assumption of unrestricted sampling. Practical, economic designs often use clusters which induce strong correlations among sample elements. These correlations can have serious effects on statistics based on complex samples. They also pose formidable theoretical obstacles. It is difficult to unravel the effects of one complex design on the distribution of a specific, analytical statistic (see McCarthy, 1965). It is not reasonable to expect separate derivations for each statistic for all major designs. We may never overcome these obstacles separately and analytically; we need general methods for getting around them.

To clarify this discussion, let us think specifically of the coefficients of multiple linear regression, which typify the broad class of analytical statistics for which we need estimates. The available derivations of distributions imply independence of observations (not of the variables); specific proofs of asymptotic theories, laws of large numbers, and central limit theorems, assume independence. Yet we do conjecture that regression coefficients (and other statistics) based on large complex samples do approach the corresponding population parameters; the approach will be merely slowed by the correlation between elements of the same clusters. We are not alone I think, in acting on such conjectures, and on the belief that proofs will be provided (Kish 1965, 2.8C; Tharakan, 1968).

However, we must not expect that variance formulas, based on assumptions of independence, for the regression coefficients (or other statistics) will hold. We should expect that they will tend to underestimate the variance, as they did in section 4 for the differences of subclass means. This is what we found in a series of investigations.

Sometimes one can avoid obstacles by restricting the design to simple random sampling. Or he may design an interpenetrating sample comprised of a set of independent replications; the set of independent statistics (regression coeffieients) can yield a measure of the standard error. But usually this is not satisfactory because of the conflict it raises between stratification and sufficient degrees of freedom (Jones 1956, Kish 1965, 4.3, 7.3). Many samples are highly stratified, with clustered selections from the strata. A model of two independent primary selections from each stratum is probably the most basic design for many practical designs. Our discussion is based on that model, as were our investigations.

8. Balanced repeated replications

We propose this descriptive name for a method we have used in a series of investigations since 1956, but especially since 1964. For deeper justification and more details see McCarthy (1966, 1968) and Kish and Frankel (1968). These first applications to analytical statistics will, I hope, be followed by many others, probably with modifications in details.

A. A <u>replication</u> is a half-sample created by selecting one of the two primary selections (replicates) from each stratum. We had 47 strata of roughly equal sizes; these were imposed by "collapsing" and "combining" as needed (Kish 1965, 8.6). Next the desired statistics are computed from the data in the replication; for example, the regression $y = b_{1j}x_1 + b_{2j}x_2 + b_{3j}x_3$ from the jth half-sample. Next obtain the difference $(b_{1j}-b_{1*})$, where $y = b_{1*}x_1 + b_{2*}x_2 + b_{3*}x_3$ was previously computed for the entire sample. Then $(b_{1j}-b_{1*})^2$ is an estimate of the variance of b_{1*}; similarly for the variances of b_{2*} and b_{3*}.

Instead of $(b_{1j}-b_{1*})$ one could use the difference $(b_{1j}-b'_{1j})/2$ between the two complimentary half-samples. Present theory does not indicate a clear choice. Our investigations show only negligible differences between results with the two methods; but the estimate $(b_{1j}-b_{1*})$ is cheaper to compute.

B. <u>Repeated replications</u> are needed because estimates from (A), based on one replication and a single degree of freedom, are extremely variable; almost uselessly so in most cases. But we can repeat the process (A) by drawing new replications (half-samples), obtaining new estimates b_{1i} and their variances $(b_{1i}-b_{1*})^2$. With I repetitions the average estimate of the variance is their mean

$$\sum_i (b_{1i} - b_{1*})^2/I.$$

The precision of the estimate of the variance increases with the number of repetitions. With H strata, full precision can be obtained from the 2^{H-1} possible ways of forming half-samples.

C. Balanced repeated replications reduce the number of repetitions needed. Most of the precision from 47 strata, for example, can be obtained from 48 balanced repetitions. This, carefully managed, does not place an impossible burden on a large modern computer. The precision it yields is only moderately adequate for the usual need: the coefficient of variation of the standard error is about $\sqrt{1/2(47)}$, or about 0.10.

D. However, a variability of 10 percent is somewhat high when we want to detect design effects as low as 1.05 or 1.10. Hence we prefer to rely on averages based on groups of variates. To obtain average results we computed design effects by dividing actual variances by simple random variances. But instead of that ratio deff, we have been averaging values of $\sqrt{\text{deff}}$, the ratios of the standard errors. The theoretical grounds for preferring one or the other are not clear; $\sqrt{\text{deff}}$ is less subject to extreme values, hence may be preferred as a square-root transformation even if biased; the difference between the two averages has not been great.

The simple random variances of most statistics are computed cheaply, with analytical formulas and the entire sample; they are usually based on many degrees of freedom, and are far more precise than the computed actual variances based on replications. When analytical formulas are not available, simple random variances can be computed from simple random splits of the sample.

We have three aims in computing and accumulating design effects. First, they should help us to design better samples. Second, they can help us make better guesses about standard errors when we have only their simple random estimates. Third, they may lead to more accurate estimates of the standard errors of specific statistics than their own computed standard errors, when these are highly variable.

We are aware of course that especially here in D, but also in this whole section, our needs, dictated by the demands of empirical research, have run beyond rigorous theoretical foundations. We hope this will be seen by some as a challenge.

9. Summary of results from five investigations

These results are given in detail elsewhere (Kish and Frankel, 1968). Each investigation involved multivariate analyses based on results from a complex, clustered sample for a large scale survey.

A. Based on a national household sample, 7 social and economic predictors were used to "explain" 3 predictands in 20 regression equations. These linear regressions used from 2 to 5 variables at a time. The mean effect on the standard error was $\sqrt{deff} = 1.06$.

B. For a study of geographic mobility 6 regression equations used 14 "dummy variables" to represent nonmetric, multinomial predictors. The average effect on the standard errors was $\sqrt{deff} = 1.10$.

C. "Multiple classification analysis" is a technique of multivariate analysis for nonmetric data, also used to circumvent nonlinearity of the variables. An equation of 6 social predictors of a "receptivity index" was computed. The effect on the standard errors averaged $\sqrt{deff} = 1.10$ for the unadjusted deviations and 1.02 for the adjusted deviations. The simple random variance had to be devised here; it was based on a reasonable model and checked with simple random splits.

D. From a national survey of political behavior, a regression based on 4 predictors was computed, and the mean effect on the standard errors was $\sqrt{deff} = 1.03$. Similar results were obtained in our first computations with balanced repeated replications (Stokes, 1958).

E. For a national health examination survey, age, height and weight were used as predictors for 28 physiological variables, each in a separate equation (McCarthy, 1966, 1968). The average design effects on the variances of regression coefficients and of multiple R's was deff = 1.81, or $\sqrt{\text{deff}}$ = 1.35. This large effect does not contradict our other results, we believe; on the contrary, it is in line with our expectations and conjectures. It is not due to the nature of the variables, we believe, but to the large clusters of the sample. The sample of 3,091 males came from clusters of 4 per segment, clustered in 42 primary areas. For the means of the characteristics the variances were increased by an effect of deff = 3.22 ($\sqrt{\text{deff}}$ = 1.80), and this is in line with the smaller effects for the regression coefficients.

REFERENCES

Cochran, W. G. (1963), Sampling Techniques, 2nd ed., New York: John Wiley and Sons.

Cochran, W. G. (1965), "The planning of observational studies of human population," JRSS-A (128), 234-265

Cochran, W. G.(1968), "The effectiveness of adjustment of subclassification in removing bias in observa-' tional studies," Biometrics (24), 295-313.

Deming, E. W. 91956), "On simplification of sampling design through replication with equal probabilities and without stages," JASA (51), 24-53.

Durbin, J. (1958), "Sampling theory for estimates based on fewer individuals than the number selected," BISI (36,3), 113-119.

Durbin, J. (1968), "Inferential consequences of randomness of sample size in survey sampling," for Symposium on Foundations of Survey Sampling, University of North Carolina, Chapel Hill.

Hansen, M. H., Hurwitz, W. N., and Bershad, M. A. (1961), "Measurement of error in census and survey," <u>BISI</u> (38), 359-374.

Hansen, M.H., Hurwitz, W. N., and Madow, W. G. (1953), <u>Sample Survey Methods and Theory</u>, Vols. I and II, New York: John Wiley and Sons.

Hartley, H. O. (1959), "Analytic studies of survey data," Instituto di Statistica, Rome, Volulme in onora di Corrado Gini.

Jones, H. L. (1956), "Investigating the properties of a sample mean by employing random subsample means," <u>JASA</u> (51), 54-83.

Keyfitz, N., (1953), "A factorial arrangement of comparisons of family size," <u>Am. Jour. Soc.</u> (58), 470-480.

Kish, L. and Hess, I. (1959), "On variances of ratios and their differences in multistage samples," <u>JASA</u> (54), 416-446.

Kish, L. (1961), "Efficient allocation of a multipurpose sample," <u>Econometrica</u> (29), 363-385.

Kish, L., Namboodiri, N. K., and Pillai, R. K. (1962), "The ratio bias in surveys," <u>JASA</u> (57), 863-876.

Kish, L. (1968), "Standard Errors for indexes from complex samples," <u>JASA</u> (63), 512-529.

Kish, L. (1969), "Optima and Proxima in Linear Designs", manuscript in preparation.

Kish, L. and Frankel, M. (1968), "Balanced repeated replications for analytical statistics," <u>Proceeding of the Social Statistics Section of ASA</u> and submitted to JASA.

McCarthy, P. J. (1965), "Stratified sampling and distribution-free confidence intervals for median," JASA (60) 772-783,

McCarthy, P. J. (1966), Replication: An Approach to the Analysis of Data from Complex Surveys, Washington: Public Health Service, Series 2, No. 14.

McCarthy, P. J. (1968), "Pseudo-replication, half samples," for Symposium on Foundations of Survey Sampling, University of North Carolina Chapel Hill,

Rao, J.N.K. (1957), "Double ratio estimate in forest surveys," Jour. Indian Soc. of Agr. Stat. (9), 191-204.

Sedransk, J. (1957), "Designing some multi-factor studies," JASA (62), 1121-1139.

Sedransk, J. (1965), "Analytical surveys with cluster sampling," JRSS-B (27), 264-278.

Stokes, D. E. (1958), Partisan Attitudes and Electoral Decisions, Ph.D. dissertation, Yale University.

Tharakan, T. C. (1968), "Inference based on complex samples from finite population,I," manuscript, Institute for Social Research, University of Michigan.

U.S. Bureau of the Census (1963), The Current Population Survey: A Report on Methodology, Techincal Paper No. 7, Washington: Superintendent of Documents.

Yates, F. (1960), Sampling Methods for Censuses and Surveys, 3rd ed., London: Chas. Griffin and Company.

THE ROLE OF ACCURACY AND PRECISION OF RESPONSE IN SAMPLE SURVEYS

Lester R. Frankel

Audits and Surveys Company, New York

With the increasing utilization of the sample survey in the fields of government, industry, social studies, public health and other endeavors it is natural that great attention is constantly being directed to improve the accuracy and reliability of this instrument.

In the early days of sampling the main attention was focused upon the methods of selecting representative samples so that the principles of the mathematical probability could be applied. Later on and continuing to the present emphasis began being placed upon the problem of sample design and problems of sample estimation so as to achieve maximum reliability.

As the sample survey became more widely used in areas where human populations were sampled a set of problems related to the execution of the sample design appeared. While it became possible to specify the particular set of human individuals to be included in the sample it often became impossible from the practical point of view to obtain information from each individual. The question of non-response received attention and methods were developed to minimize its effect.

Another area of investigation arose when it was recognized that in sample surveys when information from a respondent was elicited, either in a personal face-to-face interview or by means of a telephone interview, the phenomenon of interviewer variation came into play. That is, when different interviewers were assigned and interviewed different random samples from the same population, the variation among interviewers was larger than what could be explained by sampling alone. This problem is currently receiving attention.

In the post-World War II period the sample survey became a widely used tool in marketing studies. In this application many sample estimates were subject to check from independent sources. On occasion it was discovered that discrepancies occurred that could not be explained by sample variance, sample bias, non-response bias, poor interviewing or processing error. This forced analysts to make a critical examination of the questions that were used to elicit information from respondents. It was discovered that often the questions themselves were at fault. Even today, some questions do notcommunicate what is in the mind of the person who framed the questions. Words are sometimes used that have different meanings to people in different parts of the country. Technical terms are sometimes employed, and these terms are not understood by the respondent. Often questions do not specify a frame of reference to the respondent. Other questions invite the respondent to give answers just to impress the interviewer. Sometimes questions are so complicated that the response, when it does come, contains no more information than a random number. The art of questionnaire writing has now developed to the stage where a competent survey analyst can avoid the above pitfalls. In 1966 William A. Belson took the first step to convert this art to a science. By means of a series of formal experiments he was able to examine the communicability of questions and develop hypotheses concerning the workings of survey questions.[(1)]

There is still another source of error and variation that is present in the sample survey instrument. This relates to the respondent's answers to the questions that have been posed. Assume that a precise question has been asked; where there is no problem of communication from the mind of the one who framed the question to the respondent. Is the response accurate? And how much useable information is contained in the response? These two concepts are similar to terms "accuracy" and "precision" encountered in the early days of sampling.

Today, I would like to discuss this aspect of the question-response interplay and its implications on the sample survey design. We shall assume that there is no problem of semantics present in the question, that there is perfect communication between interviewer and respondent answers the question to the best of his or her ability. The response that is forthcoming may be biased and/or subject to sampling error.

RESPONSE BIAS

Let us first examine response bias. When a respondent does attempt to respond truthfully to a precise and clear question of a factual nature there are situations where the response does not report the truth. Fortunately, it is often possible to make a direct comparison, on an individual level, between the verbal response and the true situation. We shall consider two cases.

In preparation for a national sample study of the ownership of household appliances and the brands owned it was planned to select a large representative sample of households and conduct a personal interview with the housewife. In the process of preparing the questionnaire we made a small pretest with 100 housewives where by means of a check list to aid the respondent, the interviewer asked each housewife to indicate what appliances she had in the home and then to give the brand name of each appliance. The interviewer then asked to examine each appliance that was mentioned in order to determine its condition. When examining the appliance the interviewer recorded the actual brand name. It turned out that in those 100 homes there were 931 appliances of which 906 brands were reported correctly and 25 were reported incorrectly. The gross response error was 2.7%.

In a more recent case, we in our company, had occasion to test the validity of the telephone coincidental method for determining the in-home radio station rating. In this procedure telephone calls are made at random times during each one-quarter hour period of the day to a random sample of individuals in listed telephone households. The selected person is asked if the radio is on and if it is to identify the station that was tuned in either by reporting the call-letters, the dial position or the program. In a test which we conducted in the New York Metropolitan area we followed this standard procedure. However, each interviewer who did the calling had at her disposal an electronic device whereby she could transmit over the telephone she was using, the broadcast currently coming from any one of the leading twenty A.M. stations in the New York area merely by pressing any one of twenty buttons. In the interview, when a respondent indicated to the interviewer the station that was on, the respondent was asked if he could hear the radio playing while he was at the phone and if not, whether he could tune the radio up a little louder. When the respondent could hear the radio while he was talking on the phone, the interviewer pressed the test button which corresponded to the station that was reported. If the respondent reported that what was coming over the telephone was the same as what was on the radio the original response was considered to be correct. A total of 854 responses was validated. Of these 91.3% of the stations mentioned were correct and 8.7% were incorrect.

In these two examples it was possible to determine the existence and magnitude of inaccuracies by comparing individual responses with certain checks that were built into the interview. Since it was possible to obtain accurate measurements by a more time-consuming and hence more expensive process the question arises as to when we should accept such biased measurements. If the bias is small relative to the sampling error we could better expend our resources by accepting the bias and increasing the sample size. If the bias is large relative to the sampling error then we should take steps to reduce the bias. Perhaps a combination of biased and unbiased measurements should be employed so as to obtain a minimum mean square error for a given expenditure. Let us examine these alternatives.

Suppose a random sample of n individuals is selected and measurements of a given factor are made upon all of these individuals. Let the measurement on the ith individual equal m_i.

now, $m_i = x_i + b_i$

where x_i = true value of the ith individual

b_i = bias of response of the ith individual

Suppose that we make our observations in such a manner that we allow a randomly designated fraction, f of the observations to be biased and make the remaining fraction, (1-f), unbiased. Then

$$\bar{m} = \frac{1}{n}\sum_{i=1}^{n} m_i = \frac{1}{n}\left(\sum^{n} x_i + \sum^{fn} b_i\right)$$

$$= \bar{x} + f\bar{b}$$

where $\bar{b}$ = average bias for the fn observations

then, $\text{MSE} = \frac{\sigma^2}{n} + f^2 b^2$

where b = bias in the population

In setting up a cost function let C_1 equal the direct cost of a measurement and C_2 be additional cost to eliminate the bias inherent in the measurement. Then

$$C = C_1 n + C_2(1-f)n$$

With the above error and cost function the optimum allocation is achieved by setting

$$\hat{f} = \frac{1}{2}\frac{C_2}{C}\frac{\sigma^2}{b^2}$$

It follows that if C_2, the extra cost per observation to eliminate the bias, is small then the fraction of the uncorrected biased observations should be made small.

The value C, the cost of making all of the observations, appears in the denominator. If C relative to C_2 is small it means that relatively few total observations can be made. All other things being equal it means that in dealing with few observations it would be better to allocate the resources to make more observations albeit with biases than to make fewer observations without biases.

On the other hand if C is large relative to C_2 we are generally dealing with large samples. We reach a point where by increasing the sample size the sampling variance tends to be smaller than the bias. In these situations it pays to allocate some of the resources towards removing the bias.

In actual practice it turns out that in more marketing and attitude surveys the cost structure is such that the optimum value of f comes out to be very close to zero or greater than unity. In this situation the choice of design is determined. If optimum f is between .25 and .75 an estimate of the mean square error under this optimum condition is made and it is compared to the mean square errors when f=0 and when f=1. Since there are certain advantages in utilizing a uniform procedure an attempt is made if it is at all possible, to go one way or the other depending upon differences in mean square errors.

RELIABILITY OF RESPONSE

We shall now examine the matter of the reliability of response. It sometimes happens that the designer of a sample survey may have various procedures available to measure response. While each of the methods may yield unbiased responses some procedures will yield more precise information than others. Let us consider an actual case where either two alternative procedures can be used.

With the first alternative we obtain an exact measurement of the variate on the individual level, while with the other procedure we obtain a measurement of the individual that is subject to a random error. In order to dissociate the problem of response bias we shall first consider a situation where the response does not involve questioning human beings.

Consider a sample survey study designed to estimate the sales of specific products and brands of merchandise at the retailer level during a fixed period of time.

This type of study is known as a store audit and the period of time for which sales are determined is usually two months. Manufacturers of high turnover consumer goods make use of the information obtained for marketing purposes.

In obtaining the estimates of sales, a representative sample of retail outlets that carry the products is selected. At the beginning of the time period interviewers, or auditors as they generally are called, visit these outlets and count the "opening" inventory of each brand in that outlet.

At the end of the period--two months later-- the same stores are revisited and a closing inventory of the same item is taken. At that time the auditor examines all of the stores' purchase invoices and records the number of the units of the item that was purchased. The sales made by the ith retail outlet is then determined by

$$S_i = O_i + P_i - C_i$$

where S_i = sales to consumers

O_i = opening inventory

P_i = purchases

C_i = closing inventory

Now total sales is estimated from a sample of n outlets

$$S = \frac{N}{n} \sum_{i=1}^{n} S_i$$

The variance of the average sales per outlet is

$$\sigma_{\bar{s}}^2 = \frac{\sigma_s^2}{n}$$

Where σ_s^2 is the design variance.

The relative error squared may be expressed as

$$(RE_s)^2 = \frac{C_s^2}{n}$$ where C_s^2 is the design relative variance.

Let us consider an alternative procedure where opening and closing inventory measurements are not made, and sales at the ith retail outlet is assumed to be equal to the purchases in this outlet. The direct interviewing costs for this method is considerably less than by the original method because inventory is not counted.

For this procedure

$$S_i' = P_i$$

and

$$S' = \frac{N}{n} \sum^{n} P_i$$

The mean square error of the estimate of the average sales per store, is made up of two components.

$$\sigma^2_{\bar{s}'} = \frac{1}{n} \{\sigma^2_s + \sigma^2_m\}$$

Where σ^2_m is the variance introduced by ignoring inventory change in the outlets. In other words

$$\sigma^2_m = E(S_i - P_i)^2$$

$$= E(O_i - C_i)^2$$

Now let us examine some of the practices engaged in by retailers. Each retailer in business maintains some form of inventory control so with a minimum in stock he is able to maintain a sufficient inventory so as to satisfy his customers. The amount of space is limited and as his inventory approaches zero he re-orders.

The distribution of inventory at random points in time, that is when the interviewer calls, has a rectangular distribution. The distribution of $O_i - C_i$ is the distribution of the difference between two rectangular distributions each having the same range.

Therefore

$$E(O_i - C_i) = 0$$

and

$$E(O_i - C_i)^2 = \frac{R_i^2}{6}$$

where R_i is the maximum range of inventory in the i<u>th</u> outlet

Now, for a random sample of n outlets

$$E(\overline{O}-\overline{C})^2 = \frac{\sum_{i=1}^{n} R_i^2}{6n^2}$$

but

$$\frac{\sum_{i=1}^{n} R_i^2}{n} = \overline{R}^2(1+C_R^2)$$

where C_R^2 = coefficient of variation squared of the R_i's

A concept used in retail management is that of inventory turnover (t).

$$t_i = \frac{S_i}{\frac{R_i}{2}}$$

It happens that t_i tends to be a function of the particular product that is sold rather than the retailer. For example in a two-month period t for milk, bread, and other high perishable products will be about 50. This is because the dealer will tend not to stock up on more than he can sell in a day. Because of limited storage space other products will have high turnover rates. Thus, for gasoline t in a two-month period will be about 20 to 25.

For some products such as for example jewelry and and diamonds t is less than one.

Since the turnover rate is more a function of the product than for the outlet we can let

$$t = \frac{2\bar{S}}{\bar{R}}$$

Then

$$(RE_{s'})^2 = \frac{1}{n}\{C_S^2 + \frac{2}{3t^2}(1+C_R^2)\}$$

This is obviously greater than

$$(RE_s)^2 = \frac{C_S^2}{n}$$

In designing the data collection procedure at the individual outlet level the question arises as to what procedure should be employed. Should funds be allocated to collect inventory for a given size sample or should this money be saved and the sample of stores be expanded, or should a combination be used?

If the cost function is similar to what we were dealing with previously, namely

$$C = C_1 n + C_2(1-f)n$$

where

C_1 = cost of measuring dealer purchases in a single outlet

C_2 = cost of measuring inventory in a single outlet

f = the proportion of outlets where retail inventory is not measured.

We have a peculiar situation. One of the alternatives: either (1) measuring inventory in all outlets or (2) not measuring inventory at all yields a smaller relative error than by taking inventory in some fraction of the outlets.

That is, the relative error reaches a <u>maximum</u> when $o<f<1$. The choice is simply between one of two alternatives.

- -

Another example where one can control response reliability occurs in the measurement of attitude in sample surveys. An attitude of a person toward an object is his personal disposition, either positive or negative, towards the object. Each person's characteristic position on the attitude continuum is the average of the values of the opinions towards the object that he endorses. Psychologists have developed large numbers of opinion questions that can be used to define particular attitudes. In a sample survey designed to obtain an attitude measurement the survey designer will generally ask many opinion questions of each respondent so as to control the within-individual variation. The more opinion questions that are asked, the more expensive the interview. By using the principles of optimum design as encountered in cluster sampling, the statistician can, for a given expenditure, balance the number of opinion questions to be used in the questionnaire with the size of sample.

RESPONSE VARIATION

In the preceding examples and in similar situations the degree of response error is controllable by the measuring instrument. The value of the variate attached to the individual can be obtained with different degrees of precision, but for reasons of economy or feasability we have the option of obtaining measurements that of themselves are subject to pre-determined degrees of error.

Now let us consider another aspect of response. I call this "response variation" in contrast with response reliability and response bias. Upon examination of many marketing and opinion studies, in particular longitudinal studies, it becomes apparent that each human being behaves in a probabilistic manner. Thus, if 20 percent of the people watch television on an average day it does not mean that the same 20 percent watch television every day. There are some people who watch every day. Others may watch every other day, some may watch only occasionally, etc. When we conduct interviews we generally ask our respondents if they had viewed television the preceding day. A particular respondent may have answered in the affirmative. However, if we had interviewed the day before or the day later he may not have watched. The same type of phenomenon occurs for practically all human activity that is of a reoccurring nature. The use of a particular brand of shaving cream; the brands of soft drinks consumed on an average day; the brand of soap powder obtained when the last purchase was made; the length of time travelling to work "yesterday"; the number of cigarettes smoked daily are just a few of the examples that come to mind.

There are usually many questioning approaches that can elicit measures of behavior from a respondent. As a first step those procedures which, on the basis of past experience, lead to biased responses are eliminated. The remaining approaches are then examined from the point of view of the amount of information obtained and a choice can be made as to which approach to use. I would like to present an example where two different questioning approaches and their variations can be used to measure the same phenomenon.

AMOUNT OF INFORMATION ELICITED BY A QUESTION

There are many sample surveys conducted in marketing research where it is desired to determine the number of people in the United States who read an average issue of a particular magazine. The number of readers of an issue of a magazine is not the same as the circulation of that magazine. This is because a single issue of a magazine may generate anywhere from 2 to 7 readers.

Not only do people who buy or subscribe to the magazine read it but other members of the household and people outside the household have an opportunity to read it. It is important for a manufacturer who advertises in a magazine to have an estimate of the size of the audience of an average issue.

The sample survey is the only method possible to obtain estimates of the audience, or the percent of the population reading, a magazine. However, several questioning procedures may be applied to a representative sample of the population. We shall examine some of these procedures from the view-point of the amount of information obtained. For purposes of exposition let us assume that we are dealing with a weekly magazine.

The standard method used in the United States is known as the "Recognition" technique. The respondent in an interview is presented with a five-week old issue of the weekly magazine under study; he is shown practically all of the editorial matter; and then he is asked if he read that particular issue. If the interviewing phase of the study takes several weeks more recent issues are shown so that the respondent is always shown an issue that is more than five weeks old or less than six weeks. The reason for using a five-week-old issue is that for all practical purposes an issue will generate its total audience before it is five weeks old. Issues are changed so that more than one issue can be measured. The percent of respondents reading the test issue is an estimate of the percent of the population reading an average issue.

For an unrestricted random sample the variance of this estimate is simply

$$\sigma_p^2 = \frac{\pi(1-\pi)}{n}$$

Where π is the proportion of the population who read the issue.

Another questioning approach that is sometimes used is known as the "Recall" method. The questioning procedure is much simpler. The respondent is asked if he had read any issue of the magazine "yesterday" and, if he had, whether or not yesterday was the first day he had read the issue. This method yields an estimate of the number of readers being added to the audience of a magazine on an average day. If the magazine is published weekly then the daily average is multiplied by seven. This is an estimate of the number of readers accumulated during the publication cycle of the magazine. Since it does not make any difference which issue of the magazine is read for the first time the total audience of an average issue is obtained. Needless to say interviewing has to be conducted equally over all days of the week.

Let us examine the variance of this type of estimate. If, as before π is the proportion of the population reading an average issue then $\pi/7$ is the proportion of the population that have become readers of any issue during an average day.

Then

$$P' = 7(P/7)$$

and

$$\sigma^2_{P'} = \frac{7^2(\pi/7)(1-\pi/7)}{n} = \frac{(7-\pi)}{n}$$

The relative amount of information inherent in the one day recall method as compared with the recognition technique described previously is

$$RI = \frac{1-\pi}{7-\pi}$$

In spite of the relative statistical inefficiency of the method it is sometimes used because the interview time is much less per magazine than that required by the recognition technique. It makes possible the inclusion of a large number of magazines in the study and also makes it possible to collect a great deal of ancilliary data.

Another variation of the recall technique is a seven-day recall. The relative statistical efficiency of this design for a weekly magazine is the same as that of the recognition technique.

Going back to the recognition technique it is possible in the interview situation to have the respondent questioned on more than one issue of each magazine being studied. The interviewer would show the respondent two or more issues of the same magazine--the youngest issue being five weeks old. In this case some additional information is obtained and the gain in information can be determined once we know the intra-class correlation of reading.

If we consider the behavior of individuals with respect to reading a magazine and assume a static population we can postulate that each magazine generates a probability distribution f(p)[2]. This is sometimes known as the Latent Behavior Function. The average issue audience expressed as a proportion of the population is simply the mean of the distribution, i.e., $\bar{p} = \int p f(p) dp$

Schreiber[3] and Stock and Green[4] have postulated that f(p) is a Beta function so that

$$f(p) = \frac{1}{B(u,v)} p^{u-1}(1-p)^{v-1} \qquad o<p<1$$

On the basis of empirical data the latter two authors have evaluated u and v for many different magazines. For example, u = .052 and v = .433 for <u>Newsweek</u> magazine. In the case of <u>Reader's Digest</u> u = .096 and v = .220.

The average issue audience assuming the Beta distribution is

$$\bar{p} = \frac{1}{B(u,v)} \int_o^1 p^u (1-p)^{v-1} dp$$

$$= \frac{u}{u+v}$$

If a single issue is used in the course of an interview then, assuming unrestricted random sampling,

$$\sigma^2_{\overline{p}} = \frac{1}{n} \frac{uv}{(u+v)^2}$$

The intra-class correlation of reading behavior is simply

$$\rho = \frac{1}{u+v+1}$$

If k issues of a magazine are used during the course of an interview the variance of the estimate of the average issue audience is

$$\sigma^2_{\overline{p}(k)} = \frac{1}{nk} \frac{uv}{(u+v)^2} \frac{u+v+k}{u+v+1}$$

The relative amount of information obtained by using k issues in the interview as compared with a single issue is

$$RI = k \frac{u+v+1}{u+v+k}$$

Thus, if two issues of Newsweek are used in the interview instead of a single issue RI = 1.20. For Reader's Digest RI = 1.13.

Thus, it would appear that if the purpose of the study is to obtain the average issue audience the use of multiple issues in the interview would be inefficient. However, in order to obtain estimates of the parameters of Beta Function Distribution for a particular magazine the use of at least two issues is necessary.

In concluding we see:

First, that in addition to the components of sample design and estimation, interview variation and bias, non-response bias, and processing error; factors relating to response should be taken into account in examining the mean square error of a sample survey.

Second, factors relating to response fall into two categories response bias and response precision.

The latter, response precision, can be analyzed into two parts. The first, response error, is a function of the measuring instrument--sometimes referred to in a sample survey as the interviewing procedure. The other part, response variation, is present in certain measurements of behavior because individuals behave in a probabilistic manner.

Third, procedures are available that tend to minimize response bias and response error. Since the costs of making use of these procedures are easily determinable and their effects can be ascertained optimumization of the survey design can be achieved.

Fourth, in order to cope with response variation alternative forms of questioning should be examined and a choice can be made on the basis of the relative amount of information obtainable per dollar of expenditure.

As a final remark, I should like to point out that in our discussion today examples were selected so as illustrate the effects of the three response factors taken in isolation. In actual practice it is not unusual to find two or three of them operating jointly.

REFERENCES

(1) Respondent Understanding of Questions in the Survey Interview, A report of the Survey Research Center; The London School of Economics and Political Science, 1966, London.

(2) Lester R. Frankel, "Mass Media: The Process of Communication", paper presented at International Marketing Federation Conference, October 22, 1963, Hamburg, Germany.

(3) Robert J. Schreiber, "A Method for Calculating Reach and Frequency for Multiple Insertions of Multiple Media", memorandum, Corporate Research Department, Time, Inc., January 20, 1966, New York.

(4) J. Steven Stock and Jerome D. Greene, Advertising Reach and Frequency in Magazines, Vol. 2, Reader's Digest Association and Marketmath, Inc., 1967, N.Y.

from independent sources. On occasion it was discovered that discrepancies occurred that could not be explained by sample varinace, sample bias, non-response bias, poor interviewing or processing error. This forced analysts to make a critical examination of the questions that were used to elicit information from respondents. It was discovered that often the questions themselves were at fault. Even today, some questions do not communicate waht is in the mind of the person who framed the quest tions. Words are sometimes used that have different meanings to people in different parts of the country. Technical temms are someiimes employed, and these terms are not understood by the respondent. Often questions do not specify a frame ofreference to the respondent. Thert

SOME "MASTER" SAMPLING FRAMES FOR SOCIAL AND STATISTICAL SURVEYS IN CALIFORNIA*

R. J. Jessen

University of California, Los Angeles

0. Summary

A statewide "master" frame of area units for sampling people in California is described and its actual and potential uses are indicated. The frame consists of four component frames:

1) 58 counties
2) 602 places (of which 58 are "quasi")
3) 3,396 tracts (of which 511 are "quasi")
4) 92,461 blocks (of which 3,396 are "quasi")

which are wholly or partially interdependent.

A feature of particular interest is a derived mean household income value imputed for each of the 92,000 blocks covering the state. Having a quantitative value of income level for each block may be a useful stratification or an estimation device for socio-economic surveys where income is an important correlate. Moreover, it provides a convenient device for differential sampling of households in different income groups.

The use of the frames in the design of a state-wide survey of households for an immunization study is briefly described.

---

* Based on Technical Paper No. 3 of the Survey Research Center, UCLA. Supported in part by a grant from UCLA Academic Senate.

1. Introduction

When a number of surveys dealing with the same universe, for example, of people or households, are required from time to time, there may be advantages in having a basic or master frame from which various kinds of samples can be drawn as needed. So it seems to be the case of California, and perhaps other states, where there is an increasing demand for more and speedier surveys at statewide, local, and special group levels. Suitable samples for such surveys can in most cases be designed quite independent of one another; but there are times when a coordinate scheme can greatly increase the speed, effectiveness and economy of a set of interlocking surveys. A master sampling frame properly laid out and constructed could aid the sampler in carrying out his tasks.

Some advantages to be expected from a master sampling frame would include:

a) More accurate frames. An established frame permits adequate time and effort for assembling the relevant information, as well as checking that for its accuracy and adequacy.

b) A frame properly constructed and kept in readiness should reduce the time required to designate a sample for field use.

c) Costs for a given quality level will be reduced where frame costs can be amortized over the several surveys using it.

d) Specialized surveys can sometimes be easily developed as subsamples or supplements to a larger and more general survey taken either currently or at some time in the past.

e) The precision of surveys can sometimes be increased by better sample design (either structure or estimation), using information available from previous surveys through a master frame. It is presumed that certain key information from previous surveys would be collected and stored as an operation adjunct to that of the maintenance of the Master Frame.

f) Some problems of conducting surveys through time may be substantially reduced through the appropriate use of some form of a master frame.

Moreover, it was hoped at the beginning of this undertaking that it might be of technical interest to utilize some of the potential power of modern computing facilities to experiment with more complex sampling designs which appear to be interesting and perhaps useful.

Background

The beginnings of the project leading to a master frame for California took place during the late summer of 1963. A rather comprehensive study of immunization of childhood diseases in California was being considered at the time and Mr. Mannheimer of the State Department of Public Health, who was planning the study, asked me to assist him by designing the sample. It appeared to both of us that although more or less conventional sampling schemes would probably be adequate for this particular study, there was a need for some exploration into the possible different approaches of selecting representative samples of the households in a state. The sponsor of the study, the U. S. Public Health Service, agreed to support the special request for such methodological exploration as well as the overall proposal for the immunization study.

This report is a presentation of one of the results of these explorations in sample design - the development of a basic sampling frame for California which should be useful for generating a variety of household samples, either statewide or local, either of households or of persons, for inquiries of a general nature or for those on certain ethnic, color or income groups.

A Sampling Frame

Suppose we are interested in studying the data comprising the ages and immunization status of all the people of some area, for example, Los Angeles County. Let i denote a specific person, Y_i whether that person has received a particular immunization or not and X_i, his age, where i = 1,2,3,...,N. The people of the county constitute our universe of interest (the physical set of N objects or elements) and the two sets of observations under study, immunization status on one hand and ages on the other, constitute the populations of measurements or observations (the y- and x- populations, respectively).

When a sample (that is, any subset of this universe) is to be selected it is convenient, if we are concerned with its statistical properties, to have some sort of procedure in mind. In the classic case where simple random sampling is desired, and the universe consists of some balls in an urn, the "random" selection is carried out by "mixing" the balls (presumed to be "identical" physically) in the urn and withdrawing the desired number "without looking." Playing cards or poker chips are also used to illustrate the drawing random samples (with or without replacement) by a physical procedure. Moreover, the cards or chips are also used to show that they can be used as convenient stand-ins for the actual universe of interest, which may be people, widgets, and so forth.

A procedure only slightly less simple in selecting random samples from a finite universe is that of numbering the elements serially from 1 to N and selecting random numbers from 1 to N until the desired sample size is reached. Or, more formally, we have a listing of the elements in a form such that we can decide on the selection of those elements for a sample by performing certain operations on the listing. Used in this capacity, we may regard the listing as a simple sampling frame where each unit in the frame and each element in the universe has a one to one correspondence. Many practical frames do not have this simplicity. They may contain frame units (FUs) which are not mutually exclusive, or FUs which are aggregates or clusters of elements. For example, a typical frame for selecting households in cities consists of unlisted households within blocks, a list of blocks within each tract and a list of tracts within the city. This is a multi-stage frame where the FUs are tracts, blocks, and households at stages one, two and three respectively. Each FU then, is a simple aggregation of the FUs at a lower stage. Stages one and two consist of area FUs, at stage three it is a household, but the number of such units for any block may not be known prior to a survey and then will be determined only at the time of the survey for those blocks falling into the sample. Hence, a sample of the frame itself is completed as an integral part of some survey and the total number of the third stage FUs (households) in the frame and their distribution (by block) is available only as estimates from surveys.

A complete list of persons or households in Los Angeles County does not exist. There are some partial lists of householders, such as telephone directories (limited to those having telephones and also to those agreeing to be listed) and city directories (usually rather incomplete and out-of-date). These frames are quite inadequate for many surveys. An area-list frame consisting of tracts, blocks and households-to-be-listed is a suitable and widely used alternative. The job of constructing such a frame is simply to assemble the relevant information (mostly from published reports of the Bureau of the Census) and put it in a form convenient for performing the sampling operations. In some cases, particularly in non-metropolitan counties, it is desirable to extend the basic plan to portions of the State not covered in the usual urban manner, e.g. non-tracted and non-blocked areas.

Since some properties of samples, such as biases, variances and costs, for example, may be strongly affected by the kind and nature of the frame used, the frame becomes not only a practical means for carrying out the mechanics of selection but also an important device for achieving efficient and valid samples. The design to be presented here being somewhat more ambitious in scope and purpose than conventional schemes, it may be convenient therefore to designate it as a master frame.*

Auxiliary Data

The data to be associated with each FU (frame unit) depend on their availability and probable utility for the particular survey that is to be dealt with. Data may be available directly or indirectly from publications - or in some cases, derivable from accessible data.

Following are the characteristics for which data were obtained for the FUs in the frame:

1) Total occupied housing units (OHUs) or its equivalent, total households (HHs).

2) Total population.

3) Negro population.

4) Other non-white population.

---

* The idea of a "master" sampling frame to facilitate the drawing of samples goes back at least as far as the Master Sample of Agriculture, described by King (1945) and Jessen (1945). This frame for agriculture was national in scope and, at first, aimed at providing a base to select samples of relatively small areas which then could be used to designate farms, households, tracts of land or other such which may be of interest in an investigation of the rural zones of the nation as a whole or any portion of it. It was designed to possess compatibility in case frames dealing with non-agricultural matters were to be coupled with it.

5) Number of households whose heads are white and have a Spanish surname.

6) Number of OHUs who own their housing unit (O-OHUs).

7) Number of OHUs who rent their housing unit (R-OHUs).

8) Median value of home for O-OHUs.

9) Median monthly contract rent paid for R-OHUs.

10) Median income (annual) for household (i.e. for family and unrelated family members).

Item 1), OHUs, is useful as a general measure of size for sampling when households or housing units are being investigated. Items 3), 4), and 5) are useful for studies involving ethnic groups. Item 10) may be a useful variable for statistical control in sampling and items 6) through 9) are to be considered later in this paper.

2. Design and Construction

California presents most of the problems with which the sampler may be confronted, such as rapid population growth, large and varied ethnic groups, some of which are highly concentrated, others widely scattered. California has 10% of the entire U. S. population and much of that (40%) is confined to one county(Los Angeles). Its 58 counties range in population (1960 census) from 397 (Alpine) to 6,038,771 (Los Angeles). Population changes from the 1950 to 1960 censuses range from a decrease of 26.4% (Lassen County) to an increase of 225.6% (Orange County). Los Angeles County alone increased by 500 persons each day during the 1950s.

For the State as a whole, about 56% of the entire population was covered by block statistics in the 1960 census. Many of those not covered live in what would normally be regarded as urban territory. For example, nearly 2,000,000 people in Los Angeles County live in such urban areas, but were not covered by block statistics.

Moreover, in 1960, about 460,000 (or 7.6%) of Los Angeles County's population were negroes, and another 570,000 (or 9.6%) had a Spanish surname of which about 70% are Mexican in origin. San Francisco and Alameda Counties have sizeable negro populations; and in these and some other counties there are large Mexican American populations. Moreover there are rather large numbers of Japanese Americans and Chinese Americans within the State. Thus, California has a number of special population characteristics and unique features which make problem of sampling somewhat different and complex.

Because of the availability of the data and the desire for flexibility, it appeared that four frames would be fitting: (i) the 58 counties, (ii) a simple frame of 602 places, (iii) a more complex frame of 3,396 tracts and (iv) a frame of 92,461 blocks. Each frame completely covers the state and although places and tracts are only partially related, the block frame is simply a further segmentation of the tract frame.

The work involved in constructing the frame will be described in two stages; first, the determination and establishment of the basic units and, secondly, the kind of information assembled for each frame unit.

A County Frame

The elements of this simple frame are the 58 counties of the state. It will be convenient to distinguish three types of counties: (i) metro, (ii) submetro, (iii) other non-metro. Metro counties are those comprising Standard Metropolitan Statistical Areas (SMSAs). The non-metro counties which include one or more cities for which block statistics are published (by the Bureau of the Census), are designated as submetro counties.

This frame is quite limited in use. It will serve here as a base for the frames to be described below. A summary is provided in Table 2.1.

Table 2.1 THE COUNTY FRAME

	Metro Counties		Non-Metro Counties		All Counties	
Type of County	No. of Counties	No. of OHU s	No. of Counties	No. of OHU s	No. of Counties	No. of OHU s
Metro	17	4,343,092	-	-	17	4,343,093
Submetro	-	-	6	242,745	6	242,745
Other Non-Metro	-	-	35	396,271	35	396,271
Total, State	17	4,343,092	41	639,016	58	4,982,108

A Place Frame

The basic unit of the place frame consisted of two types; the census place (C-place) and the quasi-place (Q-place). A census place is a city, town or village, or community, either incorporated or unincorporated, designated as a place in the 1960 census of population and which at that time had a population of 1,000 or more persons. A Q-place is simply the portion of each county not included in census places of 1,000 population or more. Hence each county has one Q-place. A list of C-places is easily compiled from published tables of the Bureau of the Census.*

In addition to census versus quasi, a number of other classes of places will be distinguished.

Blocked versus Unblocked:	A blocked place is a census place for which statistics by block were published (4).
Metro versus Non-Metro:	A metro place is a census place for which statistics by tract were published (5).
Sub-metro versus Other Non-Metro:	A blocked non-metro place will be called a submetro place.
Incorporated versus Unincorporated:	An incorporated place is census place having corporate status under California law.
Urban versus Rural:	An urban place is a census place having a population in 1960 of 2,500 or more.

---

* Reference 8, table 7.

Place Boundaries: Boundaries of blocked places are quite easily determined from the maps supplied by the blocked cities reports (4). Those of incorporated places, blocked or unblocked, are usually available from commercial maps, although it may be difficult to find those which existed at the time of the 1960 census if annexations or other changes have been taking place. Boundaries of metro places can be approximately determined from maps supplied by the census tract reports (5). Unincorporated non-metro places require special maps from the Bureau of the Census for boundary determination. With these, boundaries of all places can be accurately determined. Those portions of quasi-place boundaries following Census County Divisions can be determined from the special census report (7).

The data compiled for places parallels those for counties above. The completed place frame consists of 602 units of which 544 are census places and 58 are quasi-places, 73 are blocked and 339 are metro. Table 2.2 presents a distribution of FUs (frame units) and OHUs (occupied housing units - identical to households) by size of place and type of county.

A Tract Frame

The basic unit of this frame consists of two types: The census tract (C-tract) and the quasi-tract (Q-tract). Census tracts are areas, established by local authorities, in collaboration with the Bureau of the Census, into which all metro counties are divided to provide a convenient basis, to show the geographic distribution of demographic and other data in the cities and outlying areas of metro counties. The boundaries of tracts frequently follow corporate limits of incorporated cities but they do not always do so - particularly where corporate lines are complex and/or undergoing changes. These conditions are most likely to arise where incorporation law is liberal and where population growth is rapid - both being present in California.

Table 2.2 THE PLACE FRAME

		Metro Counties		Non-Metro Counties		All Counties	
Size Class (Population)	Blocked or not	No. of Places	No. of OHU s	No. of Places	No. of OHU s	No. of Places	No. of OHU s
500,000 and	B	3	1,344,098	-	-	3	1,344,098
over	N.B.	-	-	-	-	-	-
Total		3	1,344,098	-	-	3	1,344,098
100,000 to	B	11	648,438	-	-	11	648,438
500,000	N.B.	1	29,146	-	-	1	29,146
Total		12	677,584	-	-	12	677,584
50,000 to	B	26	584,105	-	-	26	584,105
100,000	N.B.	2	34,309	-	-	2	34,309
Total		28	618,414	-	-	28	618,414
25,000 to	B	9	100,343	5	52,354	14	152,697
50,000	N.B.	39	419,662	2	19,321	41	438,983
Total		48	520,005	7	71,675	55	591,680
10,000 to	B	9	45,079	5	23,949	14	69,028
25,000	N.B.	67	337,245	17	81,165	84	418,410
Total		76	382,324	22	105,114	98	487,438

Table 2.2 (Continued)

Size Class (Population)	Blocked or not	Metro Counties No. of Places	Metro Counties No. of OHU s	Non-Metro Counties No. of Places	Non-Metro Counties No. of OHU s	All Counties No. of Places	All Counties No. of OHU s
2,500 to	B	4	12,212	1	2,242	5	14,454
10,000	N.B.	88	141,138	99	136,352	187	277,490
Total		92	153,350	100	138,594	192	291,944
Total	B	62	2,734,275	11	78,545	73	2,812,820
Urban	N.B.	197	961,500	118	236,838	315	1,198,338
Totals		259	3,695,775	129	315,383	388	4,011.158
1,000 to	B	-	-	-	-	-	-
2,500	N.B.	63	31,224	93	51,955	156	83,179
Total		63	31,224	93	51,955	156	83,179
Q-Places, all sizes		17	616,092	41	270,595	58	886,687
Total Rural		80	647,316	134	322,550	214	969,866
All sizes:	B	62	2,734,275	11	78,545	73	2,812,820
	N.B.	260	992,724	211	288,793	471	1,281,517
Total, C-Places		322	3,726,999	222	367,338	544	4,094,337
Total, Q-Places		17	616,092	41	270,595	58	886,687
Totals, State		339	4,343,091	263	637,933	602	4,981,024*

* This figure agrees with the one found in the Population Census of 1960 but it is 1,084 OHU s less than that of the Housing Census.

In the non-metro counties, quasi-tracts were established to simulate, in a reasonable manner, the census tracts in metro counties. These consisted of two types: (i) places and (ii) portions of each Census County Division (CCD) lying outside such places. All census places having a 1960 population of 1,000 or more were adopted as quasi-tracts. When completed, a total of 511 Q-tracts were established to cover the non-metro counties. These combined with the 2,885 C-tracts* provided a complete frame of 3,396 tracts.

Both C- and Q-tracts can be put into three classes according to whether they are "split" or not and the type of split. The three classes are:

Census-Split: These occur only in metro counties. They are designated by the Bureau of the Census in the Tract report (5), and are listed separately for each SMSA. They are tracts extending beyond a single Metro-place.

Quasi-Split: Census tracts, not included above which extend beyond a single blocked place. Also, quasi-tracts which are places but are split by a CCD boundary, or quasi-tracts which are CCD residuals but include places of less than 1,000 population, will be regarded as quasi-split tracts.

Non-Split: All tracts, whether census or quasi, not included above.

All tracts, whether census or quasi, were also classified into three type groups as follows:

Completely Blocked: Those entirely covered by a single blocked place (4). Hence completely covered by census blocks. Includes both census and quasi-tracts.

---

* Census tracts with a "cv" suffix were excluded.

<u>Partially Blocked</u>: Those covered partly or completely by two or more blocked places. Included are all "split" tracts of (5) in which at least one of the fragments is a blocked place and those "unsplit" tracts which juncture with sub-metro places.

<u>Unblocked</u>: Those completely disjoint with any blocked place. This group will include "split" tracts where none of the portions include a blocked place. Also included are Q-tracts except those that are blocked places.

<u>Tract Boundaries</u>: The boundaries of all census tracts are conveniently indicated by tract outline maps provided in tract reports (5) or as separates. The boundaries of quasi-tracts are established from (7) for Census County Division boundaries and by special census maps for place boundaries if commercial maps are not available.

Data compiled by tract, both census and quasi, follows that for counties and places. A summary of the tract frame is presented in Table 2.3. It will be noted that 3,325,375 or about 66% of the HHs in the state are in completely blocked tracts. The remaining 34% will require further work to obtain the advantages of small area statistics.

A Block Frame

The block frame comprises <u>census blocks</u> and <u>quasi-blocks</u>. The former are the blocks of blocked places and the latter are devices to complete the frame. In the case of partially blocked tracts, the non-blocked portion may be divided into one or more <u>quasi-blocks</u> depending on circumstances. If the

Table 2.3 THE TRACT FRAME

Type and Class of Tract		Metro Counties		Non-Metro Counties		All Counties	
		No. of Tracts	No. of OHU s	No. of Tracts	No. of OHU s	No. of Tracts	No. of OHU s
Completely Blocked:	C-Tracts	1,550	2,415,750	-	-	1,550	2,415,750
	Q-Tracts	-	-	29	87,078	29	87,078
Partially Blocked:	C-Tracts	363	567,684	-	-	363	567,684
	Q-Tracts	-	-	-	-	-	-
Unblocked:	C-Tracts	972	1,360,044	-	-	972	1,360,044
	Q-Tracts	-	-	482	550,852	482	550,852
All Classes:	C-Tracts	2,885	4,343,478	-	-	2,885	4,343,478
	Q-Tracts	-	-	511	637,930	511	637,930
Totals, State		2,885	4,343,478	511	637,930	3,396	4,981,408

unblocked portion is disjoint with places, or is joint with one or more places, then it constitutes a quasi-block. Each portion of a "split"-tract provides data for a quasi-block. Hence a two way split yields two quasi-blocks, a three-way three, et cetera. Moreover, a quasi-tract, if a place, is "split" by a CCD boundary will yield two quasi-blocks.

Normally, in two-way splits involving a metro (or blocked) and a non-blocked place (or non-place) the values for the quasi-block for each variable was obtained as a residual of the tract after removing the blocked or metro place portion. Otherwise the residual may be allocated proportionally to the non-blocked portions (if the split is three-way or more).

Block Boundaries: These boundaries established by the census were accepted where available. In the case of quasi-blocks in quasi-tracts, place or CCD boundaries were accepted depending on the case.
Again, the same kind of data was compiled for blocks as for the previous frames. In addition, a derived mean household income value was calculated for each. The method is described in the next section.

<u>Zero-Blocks</u>: Those blocks which contain no OHUs in the 1960 census reports. If they contain no population, they will not be listed in the block reports. In order to assure that these blocks are included in the frame, an inspection of the block city mpas was made to determine their existence, after which the appropriate block numbers were added to the listing.

A summary of this frame is presented in Table 2.4.

Table 2.4 THE BLOCK FRAME

Type and Class of Block			Metro Counties		Non-Metro Counties		All Counties	
Type of Tract: Block Status	Census or Quasi	Type of Block	No. of Blocks	No. of OHU s	No. of Blocks	No. of OHU s	No. of Blocks	No. of OHU s
CB:	C-Tract	C-Block	70,154	2,384,463	0	0	70,514	2,384,463
		Q-Block	1,549	30,462	0	0	1,549	30,462
	Q-Tract	C-Block	0	0	4,206	77,782	4,206	77,782
		Q-Block	0	0	29	746	29	746
	Total,	CB-Tracts	72,063	2,414,925	4,234	78,528	76,298	2,493,453
PB:	C-Tract	C-Block	14,345	347,585	0	0	14,345	347,585
		Q-Block	363	218,014	0	0	363	218,014
	Q-Block	C-Block	0	0	0	0	0	0
		Q-Block	0	0	0	0	0	0
	Total,	PB-Tracts	14,708	565,599	0	0	14,708	565,599
UB:	C-Tract	C-Block	0	0	0	0	0	0
		Q-Block	973	1,362,447	0	0	973	1,362,447
	Q-Tract	C-Block	0	0	0	0	0	0
		Q-Block	0	0	482	559,405	482	559,405
	Total,	UB-Tracts	973	1,362,447	482	559,405	973	1,921,852
All Tracts:		C-Blocks	84,859	2,732,048	4,206	77,782	89,065	2,809,830
		Q-Blocks	2,885	1,610,923	511	560,151	3,396	2,171,074
Totals, State			87,744	4,342,971	4,717	637,933	92,461	4,980,904

3. A Mean Household Income Value for Each Block

Since many social and economic characteristics of human populations appear to be related to household income, some measure of this characteristic could be a useful variable in sample design, either for control such as in stratification, or for detecting subuniverses (or "domains") of special interest such as "poverty pockets", or for correlate use such as in improving estimates from samples. Several income characteristics for tracts are given by the census, of which "median income for family and unrelated individuals" is perhaps the most useful and convenient for sample design. No comparable income characteristic is available for blocks.

Because of the presumed importance of a measure of the average income of households by block, a simple scheme was devised to provide an estimate based on "average value of home" and "average contract rent paid" which are available in census published reports. The estimated mean value of housing units for a block was obtained by the estimator,

$$\hat{H} = (1-W_R)\ V + kW_R R \qquad (3.1)$$

where:

$\hat{H}$ = estimated mean value of housing units on the block

V = average value of the owned housing units on the block

R = average monthly contract rent paid for the rented housing units on the block

W_R = the fraction of housing units on the block that are rented

k = a constant

And the estimated mean income per household for a block was obtained by the estimator,

$$\hat{I} = a + b\ \hat{H} \tag{3.2}$$

where a and b are constants.

Using tracts as units of observation, since data on owners, renters and median household income are available, it was found that a reasonably good fit of these functions was obtained when:

$$k = 100$$

$$a = 0$$

$$b = 0.5$$

and satisfactory estimates of average household income might be got by the function.

$$\hat{I} = 0.5\ \{(1-W_R)V + 100\ W_R R\} \tag{3.3}$$

Using Eq. (3.3), a derived average household income value was computed for each block in the frame, including all blocks, which according to the census had no occupied housing units in 1960, by simply extending to all "zero" blocks the same income value given the previous "occupied" block.

4. An Application: The Statewide Immunization Survey

As already mentioned, a proposed statewide survey of the extent and nature of immunization practices provided much of the incentive and all of the financial support for the planning and construction of the frame for which this survey was to be the first application (3). Portions of the frame have been used for several surveys since that time. Some of these will be described here to illustrate some features of the frame and perhaps to show its potential in new approaches in sampling design.

The sample requested for this survey was required to include about 12,000 households generally representative of all households in the state as of 1964. Since primary interest was in those households containing children under 5 years of age (estimated to be about 20% of all households), the sample should contain about 2,400 of these YC-HHs (young children households). Moreover, the research workers in charge of the study decided that relatively more of their attention should be directed to the poorer families where special problems were believed to exist. To accommodate this request the sample was set to contain about 1,600 "low" income households (those below median income) and 800 "high" income households (those above median income). Hence, it was required to oversample low income households at about twice the basic rate.

The cluster structure of the sample design was as follows:

1000	primary units, consisting of blocks, where blocks exist, otherwise tracts or quasi-blocks.
7 or 14	secondary units, consisting usually of individual housing units selected at random within the primaries. In terms of occupied housing units given in the 1960 census, 7 were selected in the high income primaries and 14 in the low income primaries. In terms of HHs and YC-HHs in 1964, the OHUs of 1960 were expected to yield 8 HHs and 1.6 YC-HHs; the 14 OHUs, 16 HHs and 3.2 YC-HHs.

The sample was designed to yield a total of 12,000 HHs in 1964. In the survey these were screened to determine those households having children under 5 years old (a yield of 2,400 YC-HHs was expected), on which the mean inquiry was conducted. In order to obtain a generally representative sample (whether children under 5 were present or not) a sample of 800 HHs, or 1 in 15, was selected from the 12,000 to obtain characteristics of households in general.

In an attempt to exploit some of the special features of the master frame a design was adopted which required that sampling rates be balanced, in a probability sense, with the margins in a three-way classification of the universe (that is, frame) data. The three-way classification consisted of

164	geo-political areas (counties, groups of counties or subdivisions of counties)
10	income classes
3	zones (rural, suburban and urban)

The 164 geo-political areas were formed in a fairly conventional manner. Groups of FUs (tracts or blocks) were placed in the same G-P class with the objectives of achieving some degree of equality in sizes (in terms of 1960 households), homogeneity of general "conditions" in the area and obtaining identifiable boundaries. The choice of 10 rather than some other number of income classes was quite arbitrary. The cut-off points were set at levels such that roughly 10% of the households in 1960 would fall into each of the 10 classes - hence they were in approximation to deciles. The limits were set after some experimentation with work done on the data available in the cities of San Diego and Los Angeles and were adopted for the entire state.

The three zones were also quite arbitrarily determined. All block FUs were classed as "urban", all quasi-tracts in the non-tracted counties which excluded all cities and towns of 2,500 population or more were classed as "rural". All remaining tracts and quasi-tracts were classed as "suburban".

The three factors at their several levels contain a total of 4,912 cells. The 63,000 FUs when classified, produced some 1,582 occupied cells, that is, cells containing one or more FUs. The number of households (in the 1960 census) contained within the FUs so classified was compiled for each cell, for use as a measure of size for sampling. Since 500 "points" were to be selected for the sample and the total number of households in the state in 1960 was 4,982,108, a "point" would be selected for approximately every 10,000 households. Hence, cells having 10,000 or more households would be treated as conventional strata. However, special problems of allocation arise if we wish to choose samples where the number of households in the sample is proportional to the number in each of the 164 geo-political areas, as well as in each of the 10 income classes and each of the three zones. Additional ideas for doing this will be presented in another paper.

5. Remarks

The present block frame, or earlier forms of it, has been used to select samples for audit surveys in marketing research in San Diego and Los Angeles. Samples of Los Angeles County, in general, and its poverty areas, in particular, were used to establish some benchmarks for the early anti-poverty program. Also a sample of HHs was selected to determine the incidence of mental retardation in the city of Riverside and a special sample was selected in Los Angeles County for a rather comprehensive study of Mexican Americans. In each of these cases, and for the state-wide immunization survey, the availability of the HH income data on blocks proved useful.

Computerization

The frame data are on magnetic tape and machine programs are available assembling them into structures useful for various sample designs. Plans are under way to computerize the selection procedures - particularly in the more complex designs.

Extension in Rural Areas

The present frame does not have much facility for dealing with rural and agricultural surveys. It is hoped that further segmentation of the rural tracts might be carried out - possibly along the lines of the old Master Sample of Agriculture with some additional features.

6. Acknowledgments

The bulk of the initial task of assembling the rather large amount of data involved, computerizing the operation of deriving the mean household income and printing out relevant tables for sample allocation and selection was carried out by Mr. James D. Forbes. The same operation for the recent revision and extension of the project was carried out by Mr. Marcel Alter. The author is heavily indebted to both for their diligence and good workmanship.

REFERENCES

(1) Jessen, R. J. (1945). The Master Sample of Agriculture II, Design, *Jour. Amer. Stat. Assn.* 40: 46-56.

(2) King, A. J. (1945). The Master Sample of Agriculture I, Development and Use, *Jour. Amer. Stat. Assn.* 40: 38-45.

(3) Mannheimer, D., Mellinger, G. and Kleman, M. T. (1967). *Deterrents to adequate immunization of pre-school children.* (74 pp. multilithed), Los Angeles County Health Dept.

(4) U. S. Bureau of the Census (1961). *U. S. Census of Housing 1960, Vol. 3, City Blocks*, Series HC (3) U. S. Government Printing Office, Washington, D. C.

(5) U. S. Bureau of the Census (1962). *U. S. Censuses of Population and Housing 1960, Census Tracts*, Final Report PHC (1) U. S. Government Printing Office, Washington, D. C.

(6) U. S. Bureau of the Census (1962). *U. S. Census of Housing, 1960, Vol. 1, States and Small Areas*, Part 6, California, U. S. Government Printing Office, Washington, D. C.

(7) U. S. Bureau of the Census (1962). *U. S. Censuses of Population and Housing, 1960, Census County Division Boundary Descriptions*, California PHC (3)-3, U. S. Government Printing Office, Washington, D. C.

(8) U. S. Bureau of the Census (1963). *U. S. Census of Population, 1960, Vol. 1, Characteristics of the Population*, Part 6, California, U. S. Government Printing Office, Washington, D. C.

MEASUREMENT ERRORS IN ANTICIPATED CONSUMER EXPENDITURES

John Neter

University of Minnesota

1. Introduction

Efficient survey design requires knowledge about nonsampling errors and methods of controlling them as well as about sampling errors and their control. In particular, information is needed about optimum methods for controlling nonsampling errors at various magnitudes and their costs, including the relation between these costs and different sample design variables (such as the size of the sample, extent of clustering, etc.). Empirical information on nonsampling errors has been growing rapidly in the last two decades, but it is still far from adequate to permit sophisticated modeling to obtain optimal survey designs.

In this paper, measurement errors in reports of anticipated consumer purchases and the special problems that arise in the study of these measurement errors will be discussed.

2. Measurement Errors

Measurement errors are one type of nonsampling error. A measurement error may be defined as the difference between the response obtained in the survey and what may be called the "desired response". The desired response is the one that would have been obtained with the best measurement procedure available. Deming has called this the "preferred survey technique" [3, p. 63]. The actual response obtained may refer either to the response as given by the respondent or as recorded by the interviewer or as coded by a coder at survey headquarters. If we let a_i denote the actual response obtained for the ith household, and d_i the desired response, then the measurement error e_i for the ith household is:

$$e_i = a_i - d_i \tag{1}$$

and the average measurement error $\bar{e}$ for the population of N households is:

$$\bar{e} = \sum_i^N e_i/N \ . \tag{2}$$

Typically, the actual response obtained and the desired response with the preferred survey technique should be treated as random variables. Thus, one can consider the measurement error E_1 for the ith household as being made up of a bias term and a random error, as follows:

$$E_i = [E(A_i) - E(D_i)] + [A_i - E(A_i)] \ . \tag{3}$$

The average measurement bias $\bar{B}$ for the population is then:

$$\bar{B} = \sum_i^N [E(A_i) - E(D_i)]/N \ . \tag{4}$$

3. Anticipated Consumer Purchases

In the past two decades, much research has been undertaken on predicting expenditures by firms and individuals on the basis of expressed anticipations of various types. Attention is restricted here to consumer purchases, since the problems connected with anticipations for business expenditures differ to a marked degree from those for consumer purchases.

The Survey Research Center pioneered in developing consumer attitudes and anticipations data and in studying the relation of these to actual purchases.[1] An

---

1 A review of the buying intentions and consumer attitudes approaches is given by Maynes [12].

example of the general attitudinal questions on expectations concerning business conditions, personal prospects, and the like that have been used is: "We are interested in how people are getting along financially these days. Would you say that you and your family are better off or worse off financially than you were a year ago?". Answers to these questions are coded into three categories (e.g., better now, worse now, same or uncertain) for the Survey Research Center's Index of Consumer Sentiment.

An example of "buying intentions" questions which seek to ascertain whether or not the consumer intends to buy a specific item is: "Do you expect to buy a car during the next 12 months or so?" (If "yes"): "How much do you expect to pay for your car?" Answers to these questions are coded into three classes (intentions with some certainty, intentions with indecision, no buying intentions).

More recently, Juster [10] experimented with subjective probability questions, and this approach is now being used by the U. S. Census Bureau's Consumer Buying Expectations Survey. The respondent is given a card containing a scale from 0 to 100 and is asked, for instance: "What are the chances that you or any member of your family living here will buy either a new or used car sometime during the next six months?". Answers may be 0, 10, 20, and so on up to 100.

Data on anticipated consumer purchases are useful in a variety of ways. They serve as indicators of consumer expectations, and their behavior over time is studied for clues about ensuing changes in consumer purchases. At a more formal level, anticipated consumer purchases (the Survey Research Center's Index of Consumer Sentiment) are incorporated into the Brookings and the Wharton School econometric models of the U. S. economy.

4. Studies on Measurement Errors in Anticipated Consumer Purchases Data

Most studies in the area of anticipated consumer

expenditures have been concerned with the predictive ability of the anticipations data. Some of the studies have compared the predictive ability of different anticipations data; others have compared anticipations data with "objective" economic variables. In general, however, these studies have not been concerned with measurement errors in anticipations data per se. Almost all of these studies have utilized anticipated purchases data (will buy, will not buy), rather than anticipated expenditures data.[2]

Only a few studies on measurement errors in anticipations data have been reported, and these all involved anticipated purchases data. Furthermore, almost all of these studies on measurement errors have been of the comparative type, where alternative measurement procedures are compared with each other.

a. Conditioning effect. Ferber [5] studied conditioning effects associated with length of panel membership on planned purchases. He found that the average number of planned purchases for families interviewed for the third to seventh time was only about half as great as for families interviewed for the second time. However, the average number of planned purchases increased for families interviewed for the eighth to the tenth time almost to the same level as for families interviewed the second time. These results suggest that the responses to questions on anticipated purchases may be influenced by the length of time a respondent is on the panel.

b. Respondent effect. In another study, Ferber [7] examined the respondent effect in anticipated purchases data. In this study, all adult family members completed a questionnaire simultaneously and independently. The proportions of total reported purchase plans omitted with different respondent procedures are shown in Table 1 for a number of product classes. It would not be reasonable to assume that the union of all purchase plans reported in a family is the total number of family

---

[2] The study by Tobin [19], in which anticipated expenditures data were used, is an exception.

purchase plans. Nevertheless, the data in Table 1 indicate that the choice of respondent may have great effect on the number of purchase plans reported, and that this effect may differ by type of durable good. Indeed, the data in Table 1 raise serious questions as to the reality of family purchase plans.

Table 1

Per Cent of Total Reported Purchase Plans Omitted With Different Respondent Procedures: Ferber Study

Respondent Procedure	Product		
	Cars	Major Household Appliances	Household Furnishings
Any adult male	27	30	57
Any adult female	73	57	38
Any adult member	48	44	49
Head of family	41	33	58
Wife of head (or principal female)	73	59	37

Source: [17, p. 805]

In this same study, Ferber [6] also analyzed the influence which husbands and wives assign to the wife in the making of purchase decisions. Ferber asked husbands and wives to assign independent ratings from 0 to 100 on the wife's influence in making the purchase decision for a variety of products. The coefficient of determination between the two ratings was low, generally less than .21 for any product. Nevertheless, the average ratings assigned by husband and wife were quite close for each product, though with a consistent small underrating by the wife of her importance as compared with the husband's rating of her importance. Thus, major differences within families averaged out to a large extent. This is in contrast to purchase plans, where substantial differences between the head and wife persisted in the averages. Wolgast [22] also found in another study that heads and wives do not differ, on the average, in responses as to who makes the purchase decisions.

c. Questionnaire effect. The dependence of reported "plans" on question wording and context is suggested by "a" small-scale study by Maynes [13] involving air passengers enroute to the Pacific. In one questionnaire, respondents were informed of the then impending jet travel and possible economy fare and asked whether these factors would or would not increase the number of trips to be made by the respondent to the Orient. In the other questionnaire, respondents were first asked about the number of trans-Pacific trips made in the past and expected to be made in the future, and then asked about the effect of jets and economy fares. The assignment of questionnaires to passengers was made at random. With the first questionnaire, 59% of passengers indicated that jets and economy fares would increase the number of trips, but only 18% so indicated with the second questionnaire (a difference far beyond sampling error).

d. Response variance. Prior to the Survey of Consumer Buying Expectations, the Census Bureau conducted the Quarterly Survey of Buying Intentions. Flechsig [8] has reported on some reinterview results for this latter survey. The reinterviews were conducted by supervisory personnel as part of a regular quality control program. Differences between the regular survey results and those from the reinterview may be due to a variety of factors, including random response variation, differences between household respondents, time effect (reinterviews were conducted several days after the regular interview), and variation within and between enumerators (enumerators coded the responses into likelihood classes during the interview). The average planned purchase rate for automobiles for the time period under consideration was 10%, and the average difference between the regular survey and reinterview results was .7% points. Because the reported data are so sketchy, it is difficult to draw any meaningful conclusions from these reinterview studies.

e. Study through model. Byrnes [1] has formulated a simple model for reported subjective probability. It is (in somewhat different notation and terminology:

$$p_i - y_i = r_i + B_i + e_i \quad (5)$$

where:

p_i = reported subjective probability for ith person

y_i = purchase behavior of ith person (y_i=0,1)

P_i = true subjective probability for ith person at time of interview

$r_i = p_i - P_i$ response error (random variable, with $Er_1 = 0$)

A_i = true subjective probability governing behavior for ith person

$B_i = P_i - A_i$ constant for ith person reflecting systematic unexpected events

$e_1 = A_i - y_i$ random variable ($Ee_1 = 0$) reflecting random unexpected events

In a small-scale study, respondents were asked to give the probability of purchase in the next 6 months, 12 months, and 24 months, for a number of items. Byrnes then wished to measure the response errors. To do this, he made a number of tenuous assumptions. All responses of zero probability of purchase (about 40% of the cases for 24-month anticipations for household equipment) were effectively excluded, since the Poisson distribution utilized to fit the purchase behavior did not work for these cases. Byrnes also assumed that the probability of purchase in the next 12 months is half as great as that for the next 24 months, and that similar relations hold for the 6-month purchase probabilities. Such an assumption goes counter to the influence of cyclical business conditions on purchase behavior. Finally, he assumed that the response error is zero for the 24-month probability reports. On the basis of these assumptions, he found smaller response errors for anticipated purchases in the subsequent six months derived from the 12-month probabilities than from the 6-month probabilities. He concluded that much of the difficulty may be due to the scale (e.g., a 6-month probability of .3 would not have a counterpart on the 24-month scale).

In view of the small size of Byrnes' study and the questionable nature of his assumptions, not too much credence should be placed in the results obtained, although his questioning of the scale is proper, as will be discussed shortly. Nevertheless, the approach of the study through formulation of a model would appear to be the most fruitful one for measuring response errors in anticipated expenditures data.

5. Difficulties in Studying Measurement Errors in Anticipations Data

The major difficulty in studying measurement errors in anticipations data is, of course, the absence of an observable "true value". Consider, for instance, the reported "subjective probability" of purchases. Repetition of the survey procedure, if it could be carried on independently, would furnish information about the response variance but not about the response bias.

Direct comparisons of responses with subsequent purchase behavior will frequently not be appropriate because intervening events will have occurred which were not reflected in the reported purchase probabilities and which affect the probabilities of purchase. Examples of such unexpected events might be illness in the family, improvement in economic conditions, and developing threat of inflation. It should be noted in this connection that not all intervening events invalidate a direct comparison between reported probability and actual purchase behavior, since uncertainty arising from some of these are reflected in the reported probability. It would not be unreasonable to expect that the types of events, the uncertainties of which are reflected in the reported probability, differ between persons. Hence it would be necessary to ascertain for each person which intervening events were not recognized in the reported probability of purchase in order to make proper allowance for measuring response bias.

Despite the lack of direct evidence on measurement biases in anticipated purchases data at present, one can surmise that these data are subject to substantial response biases. For instance, in the Census Bureau's Survey of Buying Expectations, among 1,068 persons who stated that the probability was 100% (absolutely certain) that they would buy a car in the six-month period April 1967 to October 1967, 441, or 41%, did not purchase a car.[3] It would appear that intervening events within this relatively short and stable period could not account for this substantial difference between the stated certainty of purchase and the actual purchase behavior. Furthermore, it is likely that response biases behave differentially for different response groups. For instance, in the same study just cited, only 10% of those reporting a zero probability (absolutely no chance) of purchase of a car within the next six months did purchase a car. It might be noted incidentally that these 10% accounted for 57% of total car purchases.

---

[3] Unpublished data kindly furnished by John McNeil.

To be sure, this discussion of the magnitude of the response bias and its possible differential impacts is only speculative, since the effects of intervening events were not isolated. The discussion does serve to highlight, however, important differences between measurement errors in expenditures data and in anticipated expenditures data. These differences are only ones of degree, but nevertheless are most important. The concept of an actual expenditure has some fuzziness, for instance, as to when the expenditure is incurred or whether installment charges are part of the expenditure. This fuzziness is small, however, compared to that in anticipated expenditures. Anticipated expenditures are much more complex a phenomenon than actual expenditures. They may depend on whether a household does advance planning, on the scope of factors considered in the planning, and on a host of other factors. Indeed, the importance of the different factors determining anticipated expenditures data very much resemble attitudinal data. If this characterization is correct, it would follow that there is no basic difference in kind between anticipations questions (which ask whether a person is likely to buy an item) and subjective probability questions (which ask for a number between 0 and 100). Differences between these two approaches would be ones of degree only. In both cases, therefore, it is necessary to investigate the properties of the scale for the responses. This need may be much more apparent when the response categories are verbal; for instance, a shift from "no likelihood of buying" to "some likelihood of buying" is readily perceived as not necessarily being the equivalent of a shift from "some likelihood" to "likely to buy". Nevertheless, the need is just as great for subjective probability responses since it does not necessarily follow that a shift from a subjective probability of .10 to .30 is the equivalent of a shoft from .75 to .95.

6. Scaling Problems

Psychologists have been concerned for some time with scaling problems for attitudinal data,[4] and their work

---

[4] References [2], [4] and [20] contain summary discussions of scaling problems.

can shed some light on scaling problems for anticipations data. Two important questions, among others, that have been considered by them are whether the scales are the same for different individuals and whether the scales have equal intervals.

The first of these, whether the scales are the same for different individuals, is concerned with the possible presence of respondent effects. The limited research on this subject cited earlier suggests that there may be strong respondent effects for purchase anticipations.

The second question, whether the scale has equal intervals, is particularly relevant for subjective probability data since appearance may suggest that this property is automatically possessed by the subjective probability scale. If the subjective probabilities were not to form an equal-interval scale (e.g., if the significance of a change from .05 to .10 were to exceed that of a change from .50 to .55), then changes over time in an index based on average subjective probabilities could be subject to serious difficulties of interpretation. Stevens [17] distinguishes between the method of magnitude estimation and equal-appearing interval estimation (among others) for deriving a scale. Magnitude estimation for apparent length of a line would proceed as follows: First the subject is shown a line of given length and then presented a line of different length. He is told: "If the first line was 10, what would you call the second line? Use any number that seems appropriate...but try to make the number proportional to the apparent length as _you_ see it." [17, p.38] Then lines of other lengths would be presented. A key characteristic of magnitude estimation is that the subject is not limited to responses in an arbitrary range.

Equal-appearing interval estimation, on the other hand, proceeds by showing the subject the shortest line and attaching a number to it, say 1, then the longest line and attaching another number to it, say 7, and then asking the subject to give numbers for the apparent lengths of the other lines on a scale of equally spaced units between the two extremes (e.g., 1,2,3,4,5,6,7). Thus, a line that appears to be halfway between the two extremes would be called 4. It is apparent that the Census subjective probability questions, where 0 and 100 are the two extremes and possible answers are 0, 10, 20, etc. are a form of equal-appearing interval estimation.

Stevens believes, on the basis of a variety of evidence [16,17] that the technique of magnitude estimation produces valid scales for a wide variety of physical (e. g., loudness, brightness) and social (e.g., attitude toward religion, preference for wrist watches, pleasantness of odors) phenomena. Further, he found for many phenomena (e.g., brightness, duration of noise, pleasantness of odor, seriousness of criminal offenses) that the measures derived from equal-appearing interval estimation plot as concave downward functions of the magnitude estimation measures. From the premise that magnitude estimation produces valid scales, Stevens therefore concludes that equal-appearing interval estimation in many instances does not, since the two are not linearly related. The concave-downward function implies that, if Stevens is correct, equal-appearing interval estimation in these cases leads to a scale where a given shift at the lower end of the scale is of lesser significance than the same amount of shift in the middle or at the upper end.

7. A Study of Scale Properties for Subjective Probabilities

To study whether equal-appearing interval estimation and magnitude estimation of subjective probabilities have the type of relationship found by Stevens, Rossow (a graduate student at the University of Minnesota) conducted an experiment using 39 male undergraduate students. For each of two products (a particular brand of after-shave lotion and a specified movie), he asked the subject for an indication of the likelihood of purchase at a given price. With the equal-appearing interval estimation procedure, the measurement was a subjective probability between 0 and 100. With the magnitude estimation procedure, the subject was told to consider the likelihood of purchase at a stated price to be 1,000 and then to give the likelihood of purchase at other prices. Five prices were used, at irregular intervals. The likelihood of purchase at different prices was asked for by one of of the measurement methods for each of the two products, then a distraction experiment occurred, and finally the other measurement method was used. Both the order of the measurement methods and the order of prices were randomized.

Figures 1 and 2 show the relationships between the mean responses obtained with the two measurement procedures for given prices. The first figure is for the movie, and the second for the after-shave lotion. The 39 subjects were grouped into three categories of equal size according to a mean subjective probability measure I - group most likely to buy; II - middle group; III - group least likely to buy. The purpose of the grouping was to study whether the scaling depends on the predisposition of the subject.

Figure 1

Magnitude Estimation and Equal-Appearing Interval Estimation: Movie

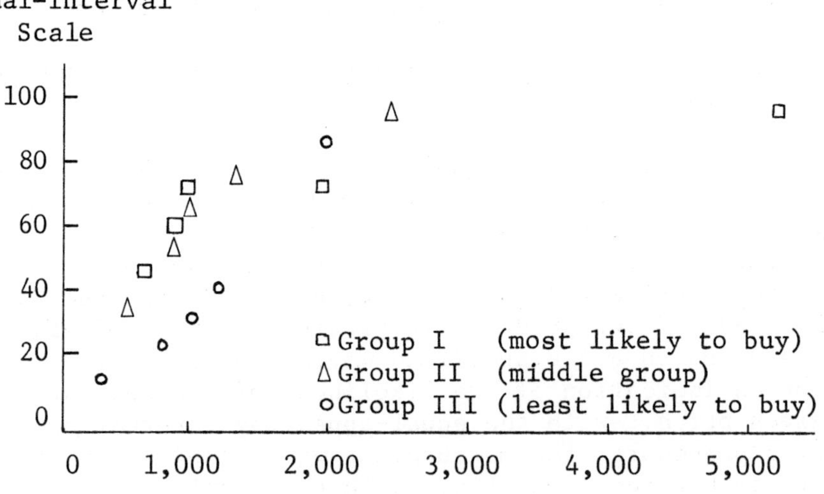

Figure 2

Magnitude Estimation and Equal-Appearing Interval Estimation: After-Shave

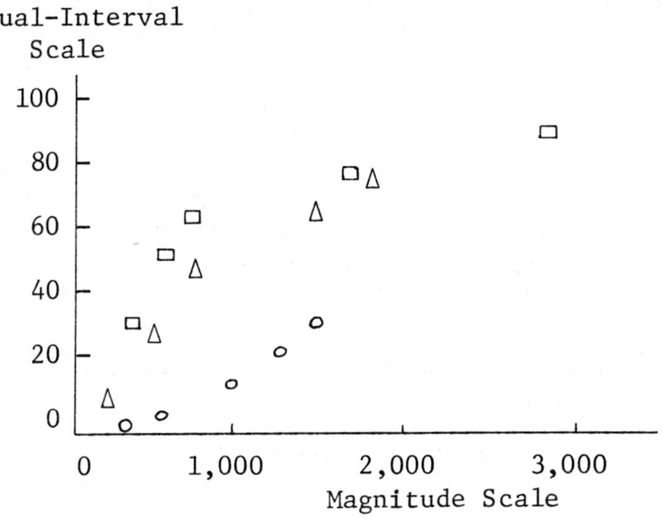

□ Group I (most likely to buy)
Δ Group II (middle group)
○ Group III (least likely to buy)

The data plotted are the mean responses for each of the three groups.

Figure 1 for the movie shows that the relations for groups I and II may be concave downward, and similarly Figure 2 for the after-shave lotion suggests that the relations for groups I and II may be concave downward. Both figures suggest that the shapes of the functions may differ for these groups. More rigorous analysis of the experimental results is not carried out because of certain problems that were discovered in the course of the experiment, involving the choice of products and certain other matters. Nevertheless, this pilot study does suggest that magnitude estimation and equal-appearing interval estimation of subjective probabilities may yield different scales. If this is correct, further research is needed to determine which of the two is the more valid scale. This experiment may also be useful by demonstrating a method for studying the scale properties of subjective probability, namely through the manipulation of factors (e.g., price) which affect the subjective probability.

8. Comparisons of Measurement Instruments in Terms of Predictive Ability

Two or more measurement instruments for anticipations data can always be compared in terms of their effectiveness for predicting actual behavior. For instance, Juster [11] compared different ways of presenting the subjective probability scale in this manner. The presumption with this approach is that the measurement with the best predictive ability has the smallest measurement errors. With this approach, the measurement errors in the anticipations data are confounded with the effects of intervening events. With appropriate randomization, however, the intervening events should be averaged out between the instruments. Since the effects due to given intervening events may not be the same for each of the measurement instruments under study and since different types of intervening events may occur at different times, a number of repetitions over time need to be undertaken with this approach in order to assess the predictive ability of the measurement instruments. A major limitation of this comparative approach is that it cannot isolate the measurement error component. Thus, one cannot ascertain how great are the potential improvements in measurement procedures that are yet possible. After a thorough study of the predictive effectiveness of anticipations data with the responses classified into two categories (will buy, will not buy), Juster concluded that "even complete elimination of the influence of unforeseen events still leaves a good deal of residual error in predictions based on dichotomies. It follows that the potential improvement from better survey design is quantitatively important." [9, p. 11] While this statement was made prior to the development of subjective probability questions, the importance of potential improvements from better survey design undoubtedly is still great.

Anticipated purchase data have been used in two principal ways, each of which reflects the view that these data cannot be taken literally because of the presence of large measurement biases. The first method expresses anticipated purchases as an index, to reveal relative changes over time. Implicit in this approach is the assumption that, although the absolute level and absolute changes in anticipated purchases may not be meaningful, relative changes in anticipated purchases do predict actual relative changes. Comparisons of alternative measurement procedures when the anticipations data are intended for this use clearly need to be made over time in order to be able to ascertain which measurement procedure predicts best.

A second method of using anticipations data employs the relationship between anticipated and actual purchases observed in the past to calibrate current anticipations data into predictions of future purchases (this method is currently used by the Census Bureau). This approach assumes that the relationship between anticipated and actual purchases is stable over time, an assumption that has not been definitively tested, particularly for different cyclical business conditions. For this use of anticipations data, cross-section studies as well as studies over time need to be made for assessing alternative measurement procedures.

a. Prediction of individual behavior. For cross-section studies, the chief problem in comparing measurement procedures as to their effectiveness in predicting actual behavior is the choice of an appropriate criterion. The conventional product-moment correlation coefficient is often used for this purpose, but it may not be appropriate for purchase anticipations where the dependent variable is a 0,1 type.[5] Suppose the purpose of the anticipations data were to predict the purchase behavior for a given individual, using a known relation between subjective probability and actual purchase rate. Since the dependent variable takes on the values 0 and 1 only, the prediction for any individual would normally be either 0 or 1, not some interval. It might be reasonable to use the rule of predicting 0 if the actual probability of purchase P_i is less than .5 for the ith individual, and predicting 1 if $P_i > .5$. (If $P_i = .5$, one may toss a fair coin.) With this rule, the probability of making an error for the ith individual is $\min(P_i, 1 - P_i)$, and hence the average error probability for an individual selected at random in a population of N persons is $\Sigma\min(P_i, 1-P_i)/N$. This quantity does not, however, have a monotonic relation to the correlation coefficient, and hence the correlation coefficient may supply misleading information about the comparative effectiveness of alternative measurement procedures according to the criterion here suggested.

Juster, in one of his studies on subjective probability [10], compared the relation between subjective probability responses using the 11-point Census scale and actual behavior in the enusing 6 months, when the subjective probability questions pertained to anticipations in the next 6, 12, and 24 months. On the basis of the coefficient of determination, he concluded that the 12-month question was most effective. If, however, the criterion above for predicting the behavior of an individual were the appropriate one, employing the actual purchase probability for each subjective probability class, the mean error probability criterion would indicate the 6-month question to be most effective (ignoring

---

[5] A fuller discussion of this topic may be found in a paper by Neter and Maynes [15].

sampling error), as follows:[6]

Questions	r^2	Mean error probability
6-months	.112	.152
12-months	.178	.157
24-months	.151	.167

This tabulation, besides indicating that the coefficient of determination may provide misleading information, is also instructive in another way. The closeness of the mean error probabilities for the 6-month, 12-month and 24-month probability questions for predicting behavior in the next six months supports the view of researchers in this area that these different purchase probability questions may be viewed as different measurement instruments for the same phenomenon, rather than solely as measurement instruments for different phenomenon.

Another possible approach for comparing alternative measurement procedures is through information theory. Theil and Kosobud [18] have analyzed data from the Juster study just discussed by means of a measure of entropy. They found the intentions question with the response coded into 2 categories (no intention, some intention) only slightly inferior to the subjective probability question (11-point scale) in terms of information content for predicting purchases of automobiles.

b. Prediction of aggregate behavior. If the purpose is to predict total purchases, the correlation coefficient again may not be the proper measure for comparing alternative measurement instruments. An appropriate model needs to be formulated so that one can determine the measure to be used for deciding which of

---

[6] Mean error probabilities are based on detailed data kindly furnished by F. Thomas Juster. The published report [10] contains only data grouped into five probability classes.

a number of measurement instruments is most effective for predicting total purchases. An example of such a model, although probably too simple a one, is as follows:

Suppose a population consists of N persons, and that the ith person will purchase, say, a car with probability P_i. Let Y_i be the variable for the ith person, indicating whether he purchases a car ($Y_i = 1$) or does not ($Y_i = 0$). Assume no changes take place just before and during the period, so that information about P_i can be obtained prior to the beginning of the period.

The actual purchases for the perid, $\sum_i^N Y_i$, are then a random variable whose expected value is

$$E\left[\sum_i^N Y_i\right] = \sum_i^N P_i = N\bar{P} \quad . \tag{6}$$

Assuming that the purchase actions are independent, we have:

$$\mathrm{Var}\left[\sum_i^N Y_i\right] = \sum^N P_i(1-P_i) \quad . \tag{7}$$

Suppose now that a survey is conducted in advance, in which each person is to indicate on a scale the likelihood he will buy a car. Suppose further that

$\sum^N Y_i = A$ autos were actually purchased, and that the purchase rate in the jth scale class was B_j. We assume now that the population is so large that all actual purchase rates B_j can be treated as if they contain no random error. Hence, one might treat all persons in the jth scale class as having probability B_j of making a purchase, and calculate the variance of the total purchases (still assuming independence) as:

$$\mathrm{Var}_1[\Sigma Y_i] = \Sigma N_j B_n (1-B_j) \tag{8}$$

where N_j is the number of persons in the jth scale class.

Suppose that a second measurement instrument was used simultaneously and independently, and that for it the purchase rate in the jth scale class was C_j. Since the same persons are involved, we have:

$$\Sigma N_jB_j = \Sigma M_jC_j = A$$

where M_j is the number of persons in the jth scale class according to the second measurement instrument. If one treats all persons in the jth scale class as having probability C_j of making a purchase, one would calculate the variance of total purchases with the second measurement instrument as follows:

$$Var_2[\Sigma Y_i] = \Sigma M_jC_j(1-C_j) \qquad (9)$$

Clearly, Var_1 and Var_2 will generally not be equal to the true variance. It might well be, for instance, that measurement instrument 1 includes, say, 2,000 persons in the jth scale class, 1,000 of which have $P_i = .4$, and 1,000 have $P_i = .5$. Yet in the variance computation above, each of the 2,000 persons would be assigned the probability $B_i = .45$. The effect of this mixing of persons with different probabilities into the same scale-class is to make the calculated variance larger than it should be. Hence, we may say that the smaller the calculated variance for any measurement instrument, given the applicability of the model, the closer it is to the true variance. To put this another way, the measurement instrument with the smallest calculated variance does the best job, according to this model, of sorting persons by probability of purchase for purposes of predicting total purchases.

This model, which assumes that a person makes a purchase decision as if he were using a probability mechanism, shows clearly the unity between verbal and quantitative responses to anticipations questions. The categories of response are perfectly general, and may be verbal or numerical since the model converts the responses into probabilities of purchase. The essence of the problem, according to this model, is to find a measurement instrument which places persons with similar purchase probabilities into the same response class.

9. Needs for Future Research

This review of research on measurement errors in anticipations data clearly indicates that this research is at a very early state of development. Research on all aspects of measurement errors in anticipations data is badly needed. To make progress in this area, it would appear to be crucial that models be developed which will permit the disentangling of intervening effects so that measurement errors can be isolated, at least to a substantial extent. Such models will need to specify the factors taken into account by the respondent in formulating his probability response, as well as the method by which the integration of the various factors takes place. Investigations which obtain (1) information on expectations with respect to future income, family responsibilities, economic conditions and the like at the time the anticipations response is given, and (2) information on actual developments for these factors at the time of ascertaining the purchase behavior, should be most useful for formulating models that will permit at least partial disentangling of intervening effects.

Since anticipations data are used frequently to study changes over time, it would seem particularly important to examine those aspects of measurement errors bearing on the measurement of changes. In this connection, the study of respondent effects looms particularly important. If there are substantial respondent effects in anticipations data, a survey procedure which permits different household respondents to be interviewed on successive occasions may increase the variance of the measure of change to a significant extent. The variance of change may also be affected to a substantial extent by random variation for any given respondent. The use of feedback, by which a respondent is told his previous anticipated purchase measure prior to being asked for his current anticipation, should be investigated to see if it is of help in reducing measurement errors for predicting changes. Feedback here might play a similar role as in "bounding" of actual expenditures. In the later case, feedback of past purchase reports serves to eliminate duplicate expenditures reporting. With anticipations data, feedback of the previously reported anticipated purchase measure may serve to "anchor" the current report relative to the previous report so that the measure of change is subject to smaller random errors.

REFERENCES

[1] Byrnes, James C., "An experiment in the measurement of consumer intentions to purchase", in Proceedings of the Business and Economic Statistics Section, 1964. Washington: American Statistical Association, pp. 265-279.

[2] Coombs, Clyde H., A Theory of Data. New York: John Wiley and Sons, Inc., 1964.

[3] Deming, W. Edwards, Sample Design in Business Research. New York: John Wiley and Sons, Inc., 1960.

[4] Edwards, Allen L., Techniques of Attitude Scale Construction. New York: Appleton-Century-Crofts, Inc., 1957.

[5] Ferber, Robert, "Observations on a consumer panel operation", Journal of Marketing, January 1953, pp. 246-259.

[6] Ferber, Robert, "On the reliability of purchase influence studies", Journal of Marketing, January 1955, pp. 225-232.

[7] Ferber, Robert, "On the reliability of responses secured in sample surveys", Journal of the American Statistical Association, September 1955, pp. 788-810.

[8] Flechsig, Theodore G., "Anticipating consumer purchases of durable goods", in Proceedings of the Business and Economic Statistics Section, 1962. Washington: American Statistical Association, pp. 152-157.

[9] Juster, F. Thomas, Anticipations and Purchases, An Analysis of Consumer Behavior. Princeton, N. J.: Princeton University Press, 1964.

[10] Juster, F. Thomas, "Consumer buying intentions and purchase probability: an experiment in survey design", Journal of the American Statistical Association, September 1966, pp. 658-696.

[11] Juster, F. Thomas, "Some experimental results in measuring purchase probability for durables, and some tentative explanations of time-series demand models", in Proceedings on the Business and Economic Statistics Section, American Statistical Association, 1967, pp. 87-96.

[12] Maynes, E. Scott, "An appraisal of consumer anticipations approaches to forecasting", in Proceedings of the Business and Economic Statistics Section, American Statistical Association, 1967, pp. 114-123.

[13] Maynes, E. Scott, "Sample surveys and prediction: an example of pitfalls to avoid", Business News Notes, No. 47, School of Business Administration, University of Minnesota, 1960.

[14] McNeil, John M. and Stoterau, Thomas L., "The Census Bureau's new survey of consumer buying expectations," in Proceedings of the Business and Economic Statistics Section, American Statistical Association, 1967, pp. 97-113.

[15] Neter, John and Maynes, E. Scott, "On the appropriateness of the correlation coefficient with a 0,1 dependent variable", in Proceedings of the Business and Economic Statistics Section, American Statistical Association, 1967, pp. 447-453.

[16] Stevens, S. S., "A metric for the social consensus", Science, 4 February, 1966, pp. 530-541.

[17] Stevens, S. S., Measurement, psychophysics, and utility", in Measurement: Definitions and Theories (ed. by C. West Churchman and Philburn Ratoosh). New York: John Wiley and Sons, Inc., 1962, pp. 18-63.

[18] Theil, Henri and Kosobud, Richard F., "How informative are consumer buying intentions surveys?", Report 6702, Center for Mathematical Studies in Business and Economics, University of Chicago, 1967.

[19] Tobin, James, "On the predictive value of consumer intentions and attitudes", The Review of Economics and Statistics, February 1959, pp. 1-11.

[20] Torgerson, Warren S., Theory and Methods of Scaling. New York: John Wiley and Sons, Inc., 1958.

[21] U. S. Bureau of the Census, Consumer Buying Indicators, Series P-65, No. 19, September 29, 1967.

[22] Wolgast, Elizabeth H., "Do husbands or wives make the purchasing decisions?" Journal of Marketing, October 1958, pp. 151-158.

SOME SAMPLING PROBLEMS IN TECHNOLOGY

D. R. Cox

Imperial College

SUMMARY

Some general aspects of technological sampling are reviewed. Then some special procedures used in sampling textile fibres and in work study are outlined. Finally some theoretical problems raised by the last two topics are analyzed. In particular, the estimation of frequency distributions from samples based in a known way and from samples of 'recurrence-times' is developed.

1. INTRODUCTION

Most published statistical work on industrial sampling problems is concerned with schemes for acceptance sampling. There the emphasis is not on the actual selection of the sample but rather on the intensity of sampling that is appropriate and on the decision to be taken on the basis of the sample. The relative neglect of sample selection in books and journals on the industrial applications of statistics is probably explained by sampling techniques being largely specific to particular applications. Much practical industrial sampling may be rather haphazard; nevertheless to obtain reliable and reproducible test results, either for research or control purposes, or in connexion with legal agreements to supply material meeting a specification, carefully defined sampling procedures are essential. Thus some British Standards, and comparable U.S. and Japanese documents, contain careful discussion of sampling techniques; see, especially, the British Standards on sampling of coal and coke and of ores (British Standards Institution, 1960, 1967), the former containing an excellent discussion of principles that should be widely applicable. A bibliography on methods for sampling raw materials is given by Bicking (1964); see, especially, Duncan (1962) for some general discussion.

The plan of the present paper is as follows. In Section 2 some points that seem to arise quite commonly in industrial sampling are mentioned briefly. Then in Sections 3 and 4 the special problems of two fields are outlined, the sampling of textile fibres and work sampling. Finally, in Section 5, some theoretical problems raised by the earlier discussion are analyzed.

2. SOME GENERAL POINTS

In this section some brief comments are given on a number of miscellaneous points that seem to arise fairly often in industrial sampling.

(i) Absence of Frame

A quite common difficulty is the initial absence of a good frame, i.e. of a clear specification of the position of each element of the population. Thus, suppose that material is stored say in sacks. It may be straight-forward to draw a random sample of sacks, but a procedure for sampling material from within a sack may be hard to specify unambiguously. One conclusion is that when practicable sampling should be done when material is, for example, moving on a conveyor belt, freely falling, etc., when each element of the population is identified by the time at which it passes a reference point. Then, provided that all material passing the sampling point at suitably chosen sampling points is taken, an unbiased method of sampling is obtained. Another possibility is to obtain a frame in terms of spatial coordinates by spreading the material uniformly over a circular or rectangular sampling area. If these procedures are not practicable, it will be necessary to specify that sample increments are taken from suitably spaced positions from within, say, the sack, but this will often leave open the possibility of appreciable biases.

(ii) Inaccessibility

In some situations, parts of the population may be completely or relatively inaccessible. It may be possible to stratify so that the inaccessible portions from a stratum sampled at low intensity. Another possibility, when sampling is to be repeated over time, is that occasional calibration observations can be used to establish a regression relation between values in the inaccessible and accessible portions of the population , or that there is a theoretical connexion. An example of the latter might arise in studying temperature distribution within a chemical reactor, when a theoretical temperature distribution may be available; measured temperatures at the sampling points can be used, in effect, to estimate an empirical correction to the theoretical temperature at an inaccessible point.

(iii) Distributional Aspects

Sometimes it may be required to estimate simply a mean for the population. In other cases, components of variance may also be required measuring say, important sources of variability in a complex process. In yet other situations, interest may lie in the extremes of the distribution over the population. Thus, a safety requirement may be formulated in terms of the maximum temperature achieved in a reactor, and in other situations the proportion of the population outside specification limits may be of special interest.

(iv) Adjustments

The possible availability of supplementary information and the need to apply adjustments or corrections to the sample observations raise problems on the whole more connected with analysis than with design. Two different aspects are involved. In one, allowance has to be made for differences between the observation obtained on a sampled unit and the "true" value. This may involve theoretical or empirical corrections, the calibration of a "quick" sampling procedure by occasional use of a definitive sampling technique, or, at its simplest, the reduction of the observed variance to allow for the variance explained by errors of measurement. The other possibility is that supplementary information is available for the whole population; see(ii). The statistical analysis is straight-forward and well-known if only means and variances are of interest or if all distributions are normal, but estimation of extremes in non-normal populations may be more difficult.

(v) Components of Variance

The validation of sampling and testing methods for example by inter-laboratory trials, will often raise statistical problems of design and analysis connected with components of variance.

(vi) Formation and Division of Composite Samples

Sometimes the amount of material required for final analysis, e.g. for chemical analysis, is small, much smaller than can be handled without bias in the initial sampling. If separate analyses are not required for the individual sample portions (increments), it will then be common to form a composite sample by combining the increments. The composite sample is then reduced to the required size by thorough mixing, division into two and rejection of a randomly chosen half, the mixing, division and rejection being repeated until a final sample of the required size is obtained. If a direct measure of measurement error is required the process must be repeated independently on duplicate increments drawn at the same time. If estimates are required say of the variation between parallel machines, then the separate machines would normally be regarded as separate populations. More complex schemes of compositing may be useful when components of variance are to be estimated.

(vii) Time Series Aspects.

Quite often, there is a sequence of p pulations to be sampled and the sequence can be considered as a time series. For example, in a batch process, each batch may be regarded as a separate population. There are then typical times series problems in that, in effect, information about a particular population may be contributed by observations on neighbouring populations.

3. Sampling of Textile Fibres

This section is concerned with the problems of a very special field, namely, the sampling of textile fibres, for example for the estimation of fibre length distribution. Palmer and Daniels (1947) have given a general account of the problems and the following remarks are meant to explain the statistical points involved rather than to set out the experimental details.

Consider first an assembly of fibres all parallel to an axis, the position of each fibre along the axis being therefore defined by the coordinate of say the left-hand end and the fibre length. The numbers of fibres in a typical population is usually extremely large; the number crossing a particular cross-section may vary from about 20 for a fine yarn to a thousand or more at earlier stages of processing. The fibres are well mixed in the early stages of processing and their left-ends lie approximately in a Poisson process along the axis.

First there is a very important distribution between unbiased methods of sampling, in which each fibre has an equal chance of selection, and length-biased sampling in which the chance of selecting a particular fibre is proportional to its length.

An unbiased sample is obtained in principle by choosing very short sampling intervals along the axis and taking all those fibres whose left-ends lie in the sam ling intervals. The experimental technique for doing this, cut-squaring, was developed by Daniels (1942). Each fibre left-end is a point, and if the sampling intervals are randomly chosen, each fibre has an equal chance of selection. In practice, because of the near randomness of the fibre arrangement, the crucial step is the isolation of the relevant fibres rather than the positioning of the sampling intervals. Denote by $f(x)$ the density function of the fibre length, X, in the population and $f(x,y)$ the joint density of fibre length, X, and another property Y, for example fibre diameter.

The second method of sampling is in principle as follows. The assembly is gripped at a sampling point and all fibres not gripped, i.e. not crossing the sampling point, gently combed out. The fibres remaining constitute the sample, or an increment when a composite sample is formed. Each fibre has a chance of selection proportional to its length and so the density of length in sampling by this method is

$$g(x) = \frac{x\, f(x)}{\mu}, \qquad (3.1)$$

where μ is the mean of the unweighted density $f(x)$. The joint density of X and another variable Y is

$$g(x, y) = \frac{x\, f(x, y)}{\mu_X}, \qquad (3.2)$$

where μ_X is the mean of X in the unweighted density. The densities $g(x)$ and $g(x, y)$ are called length-biased. The marginal density of Y derived from (3.2) is the length-biased density of Y, etc.

Simple relations hold between the moments of $g(x)$ and those of $f(x)$ and, in particular,

$$E_g(X) = \mu\left(1 + \frac{\sigma^2}{\mu^2}\right), \qquad (3.3)$$

where σ^2, μ are the variance and mean of the unweighted distribution and E_g denotes an expectation with respect to the density $g(x)$. In general

$$E_g(X^r) = \frac{\mu_{r+1}}{\mu}, \qquad (3.4)$$

where μ_s is the s^{u} moment *about the origin* of the unweighted distribution.

Practical difficulties are first that there may be breakage in combing out the unwanted fibres and, more interestingly from a statistical point of view, that non-parallelism of the fibres will mean that the sample is really extent-biased rather than length-biased, where extent is the length of the projection on to the axis of sampling.

In a variant of the method only the fibres to the right of the sampling point are removed, leaving a tuft of projecting fibres, called a *combed tuft*. Electrical or optical scanning, giving the thickness of this as a function of the distance from the sampling point, provides a very quick method of describing fibre length distribution. Imagine first that the length of each projecting fibre is measured. This would give observations of a random variable Z whose distribution is found as follows. Given that a fibre selected is of length x, the sampling point is equally likely to lie anywhere along its length. Thus the conditional density of Z is rectangular over (0, x). Since (3.1) gives the density of the lengths of selected fibres, it follows that the density of Z is

$$k(z) = \int_z^\infty \frac{1}{x} \frac{x\,f(x)}{\mu}\,dx = \frac{1 - F(z)}{\mu}, \quad (3.5)$$

say, where F(z) is the cumulative distribution function of the unweighted distribution. This density is familiar in the theory of stochastic processes as the *recurrence-time* distribution.

It is easily shown, for example by evaluating the moment generating function, that

$$E_k(Z^r) = \frac{\mu_{r+1}}{(r+1)\mu}. \quad (3.6)$$

In particular, $E_k(Z)$ is one-half the length-biased mean fibre length (3.3), as is physically obvious.

Of a tuft of N fibres, the expected number longer than z is

$$N \int_z^\infty k(u)\,du ,$$

so that the expected combed tuft curve is

$$c(z) = N \int_z^\infty k(u)\,du = \frac{N}{\mu} \int_z^\infty \{1 - F(u)\}\,du. \quad (3.7)$$

In practice there are complications, because the scanning methods do not count fibres but measure approximately mass density and therefore depend on the correlation between cross-sectional area and length.

A useful property of (3.7) is that $c'(0)/c(0) = -1/\mu$, implying that the tangent to the combed tuft curve at z = 0 intersects the axis at μ. In the theoretical discussion of Section 5, it will be assumed that the measuring device in effect determines the individual 'recurrence-times'.

Palmer (1948) introduced an ingenious sampling procedure, dye sampling, for use with a plane web of non-parallel fibres. Imagine first that each fibre were to have one end, arbitrarily chosen, named as its left-end.

An unbiased method of sampling would take all fibres with left-ends in randomly selected sampling areas. The procedure which in effect achieves this is to stamp a dyed square on the web; fibres not having an end marked are rejected, those with one end marked receive weight $\frac{1}{2}$, and those with both ends marked receive weight 1. The theoretical variance of the sample mean can be found, assuming a 'Poisson web'.

Size-biased samples arise in other contexts, for example, in work with assemblies of small particles, where surface-area, volume- and mass-biased samples may arise (Herdan, 1960). A rather different application is mentioned by Kriens (1963), in connexion with auditing. Items, of which a running total of values is available, are to be selected for checking. If sampling points are distributed randomly over the total value of the items, a value-biased sample will result.

4. WORK SAMPLING

As a second special field, consider some applications to work study. Suppose that one or more individuals are considered over a certain time period and that at any instant each individual is in one of a number of possible states. For example, an individual may be a machine which at any instant is either running or stopped, or may be an operative who may be busy or idle. More realistically, several types of stop may be distinguished or several types of work by the operative, but for a general description it is enough to consider just two states.

One way of investigating such a system is by detailed recording of its behaviour, for example by a continuous recording device, or by observation with a stop watch. Tippett (1935) discussed the estimation of the proportions of times in the various states from observations merely of the state occupied at discrete sampling points; under suitable assumptions the observed proportion of observations say in the stopped state has a binomial sampling variance. The method is widely applicable to systems for which at any time point one of a number of clearly defined and clearly identified states is occupied. In many situations the absence of quantitative observations is a major advantage.

Practical aspects of the method, which is known

variously as the snap-reading method, the ratio-delay method and as activity sampling, are described for example by Barnes (1957).

In selecting the sampling instants, the usual principles of sample survey design will be relevant. Hill and Hill (1968) have given tables of random clock times for use when some form of random sampling is involved. In some applications systematic sampling will be in order and Davis (1955) and Cox (1961, pp. 86-90) have discussed its theoretical efficiency for particular types of stochastic process. Cox (1966) and Gaver and Mazunder (1967) have made a theoretical comparison of the snap-reading method with alternative schemes of sampling.

In practice, there are two rather different situations to be considered. In one there is a single individual, or at most a few individuals, and sampling is an occasional operation. In other applications, notably those originally considered by Tippett, there are a large number of individuals and a sampler is almost continuously patrolling the whole system. In many applications it will be important that individuals cannot predict when they will be sampled and this will preclude systematic sampling. Another widely important point is that a rule defining the sampling instants should be very precisely formulated; this is particularly important when there are stops that are very frequent but individually of short duration.

Note that the stopped times captured in the snap reading method form a length-biased sample and that the time measured forward from the sampling point to the completion of a stop has the recurrence-time distribution; the arguments of Section 3 are directly applicable.

5. SOME THEORETICAL RESULTS

5.1 Introduction

The techniques outlined in the previous two solutions raise theoretical problems, some of them common to the two situations. In connexion with sampling fibres there is a need to derive information about the unweighted distribution from either the length-biased distribution or from the recurrence-time distribution. Considerations of the relative statistical efficiency of different methods of sampling are likely to be secondary in choosing between methods. Nevertheless it is of interest to know these efficiencies and they are obtained in Sections 5.2 and 5.3. Blumenthal (1967) has given an interesting discussion of these problems, although from a rather different viewpoint; in particular, most of his results assume that the coefficient of variation is known.

5.2 Estimation from Length-Biased Samples

In some contexts in which length-biased sampling is used it may be quite reasonable to regard the length-biased distribution as the object of study and in particular to consider the length-biased mean, variance, etc., as parameters. In other situations, however, it will be required to derive, from length-biased data, estimates referring to the original distribution.

Suppose, therefore, that $X_1, \ldots, X_n$ form a random sample of positive random variables having probability density function (p.d.f.)

$$g(x) = \frac{x\, f(x)}{\mu} \quad (x > 0), \tag{5.1}$$

where is the mean of the p.d.f. $f(x)$. For the reasons explained in Section 3, call $f(x)$ the unweighted p.d.f. and $g(x)$ the weighted or length-biased p.d.f.

We examine methods that apply to general continuous distributions and also those that defend on a paramative form for $f(x)$. First consider the estimation of the unweighted mean μ, assumed finite. Now

$$E_g\left(\frac{1}{X}\right) = \int_0^\infty \frac{1}{x} \frac{x\, f(x)}{\mu}\, dx = \frac{1}{\mu}, \tag{5.2}$$

where E_g denotes expectation with respect to the weighted distribution. Thus

$$\frac{1}{\tilde{\mu}_g} = \frac{1}{n} \sum \frac{1}{X_i}$$

is an unbiased estimate of $1/\mu$. Also

$$E_g\left(\frac{1}{X^2}\right) = \frac{\mu_{-1}}{\mu},$$

where μ_r is the rth moment about the origin of the unweighted distribution; we continue to write μ for μ_1. Thus

$$\operatorname{var}_g\left(\frac{1}{\tilde{\mu}_g}\right) = \frac{1}{n}\operatorname{var}_g\left(\frac{1}{X}\right) = \frac{\mu\mu_{-1}-1}{\mu^2}. \qquad (5.3)$$

Provided that (5.3) is finite, $\tilde{\mu}_g$ is for large n asymptotically normally distributed with mean μ and variance

$$\frac{\mu^2(\mu\mu_{-1}-1)}{n}. \qquad (5.4)$$

On the other hand, the mean of an unweighted sample of size n estimates μ with variance $(\mu_2' - \mu^2)/n$. Therefore for estimating μ by these methods, the asymptotic efficiency of ordinary sampling relative to length-biased sampling is

$$\frac{\mu^2(\mu\mu_{-1}-1)}{(\mu_2 - \mu^2)}. \qquad (5.5)$$

Expression (5.5) can take any non-negative value. At one extreme there are unweighted distributions, for instance those with non-zero oridnate near the origin, for which μ_{-1} is infinite and μ, μ_2 finite. On the other hand, distributions with a suitably long upper tail will have μ_2 infinite and μ, μ_{-1} finite. Special unweighted distributions of interest are the log normal, for which (5.5) is unity, and the gamma distribution of index β, for which (5.5) is $\beta/(\beta - 1)$. Note especially that (5.5) refers to the estimation of a special parameter by methods that will not in general be efficient if the distribution has a given parametric form.

Next consider the estimation of the unweighted cumulative distribution function (c.d.f.), say for the purpose of graphical analysis of goodness of fit to a particular parametric form. To estimate the c.d.f. at say z, let

$$U_i(z) = \begin{cases} 1/X_i & \text{if } X_i \leq z, \\ 0 & \text{if } X_i > z, \end{cases}$$

and

$$V_i = 1/X_i .$$

Then define

$$\tilde{F}_g(z) = \Sigma\, U_i(z)/\Sigma V_i . \qquad (5.6)$$

We now show that for any fixed z, $\tilde{F}_g(z)$ is an estimate of F(z), the unweighted c.d.f.

The asymptotic distribution of (5.6) is found by first examining the joint distribution of numerator and denominator. Write

$$\mu_r(z) = \int_0^z x^r f(x)\,dx$$

for the incomplete moments of the unweighted distribution; note that in particular $\mu_o(z) = F(z)$. Then it is easily shown that

$$E_g\{U_i(z)\} = \mu_o(z)/\mu, \; E_g(V_i) = 1/\mu, \qquad (5.7)$$

and that

$$\text{var}_g\{U_i(z)\} = \frac{\mu\mu_{-1}(z) - \{\mu_o(z)\}^2}{\mu^2}, \quad \text{var}_g(V_i) = \frac{\mu\mu_{-1}-1}{\mu^2},$$

$$\text{cov}_g\{U_i(z),\, V_i\} = \frac{\mu\mu_{-1}(z) - \mu_0(z)}{\mu^2}$$

Provided that all terms are finite, the numerator and denominator of (5.6) have asymptotically a bivariate normal distribution and hence $\tilde{F}_g(z)$ is asymptotically normally distributed with mean

$$\frac{E_g\{U_i(z)\}}{E_g(V_i)} = \mu_0(z) = F(z)$$

and variance

$$\frac{\mu\mu_{-1}(z) - 2\mu\mu_{-1}(z)\mu_0(z) + \mu\mu_{-1}\{\mu_0(z)\}^2}{n} . \qquad (5.8)$$

Now, for an unweighted sample, the proportion of observations below z estimates F(z) with the binomial variance

$$\frac{F(z)\{1 - F(z)\}}{n} . \qquad (5.9)$$

Hence the efficiency of unweighted relative to weighted sampling for the estimation of F(z) by these methods is the ratio of (5.8) to (5.9).

Table 5.1 gives a few values for this when the unweighted distribution is log normal,

$$f(x) = \frac{1}{\sqrt{(2\pi)}\tau x} \exp\left\{-\frac{(\log x-\lambda)^2}{2\tau^2}\right\}; \qquad (5.10)$$

it is easy to express the incomplete moments of (5.10) in terms of the standardized normal integral. The tabulated values of F(z) correspond to simple points on a normal probability scale. The general conclusions are clear on general grounds. The upper tail is better estimated by length-biased sampling, the lower trial by unweighted sampling. More specifically, the distribution in the neighbourhood of the median is always better estimated by unbiased sampling, the relative efficiency there being $e^{\frac{1}{2}\tau^2}$. When τ is large and the distribution extremely skew, the region in which length-biased sampling is better is concentrated in the extreme upper tail; if we consider instead of a log normal distribution a long tailed distribution bounded away from zero the same effect would probably not occur.

Table 5.1. Efficiency of unweighted relative to weighted sampling for estimating F(z); log normal distribution, dispersion parameter τ

τ	0.2	0.5	1	2
F(z)				
.023	1.59	3.27	11.2	158
.159	1.30	2.00	4.53	33.2
.500	1.02	1.13	1.65	7.39
.841	0.793	0.601	0.503	1.44
.977	0.631	0.327	0.130	0.172

Suppose next that it is reasonable to base an analysis on a postulated functional form for the p.d.f. f(x). Then the likelihood of a length-biased sample can be written down and routine methods of estimation applied. Note that sufficient statistics are the same for sampling g(x) as for sampling f(x). In particular, if f(x) is a gamma density of mean μ and index β, then

$$g(x) = \frac{x\ f(x)}{\mu} = \frac{x\beta}{\mu^2}\left(\frac{\beta x}{\mu}\right)^{\beta-1} \frac{\exp(-\beta x/\mu)}{\Gamma(\beta)},$$

which is a gamma density of mean $\nu = \mu(1 + 1/\beta)$ and index $\gamma = \beta + 1$. It follows from the properties of maximum likelihood estimates that for a length-biased sample the maximum likelihood estimates $\hat{\nu}_g$ and $\hat{\gamma}_g$ are asymptotically independent with

$$\text{var}_g(\hat{\nu}_g) = \frac{\nu^2}{n\gamma}, \quad \text{var}_g(\hat{\gamma}_g) = \frac{1}{n\{\psi'(\gamma) - 1/\gamma\}}, \qquad (5.11)$$

where $\psi(u) = d \log \Gamma(u)/du$. For large γ

$$\text{var}_g(\hat{\gamma}_g) \sim 2\gamma^2/n.$$

Thus, for estimating β length-biased sampling gives lower precision, corresponding to the change in argument in (5.11) from β to $\gamma = \beta + 1$. For estimating the mean μ, we have from the length-biased sample

$$\hat{\mu}_g = \hat{\nu}_g(1 - 1/\hat{\gamma}_g),$$

so that the asymptotic variance of $\hat{\mu}_g$ is

$$\frac{\mu^2}{n(\beta+1)}\left[1 + \frac{1}{\beta^2(\beta+1)\{\psi'(\beta+1) - 1/(\beta+1)}\right]. \qquad (5.12)$$

This is to be compared with a variance of $\mu^2/(n\beta)$ for unweighted sampling. Some numerical values for the efficiency are given in Table 5.2.

Table 5.2. Efficiency of unweighted relative to weighted sampling for estimating by maximum likelihood the mean of a gamma distribution of index β

β	Rel. eff.	β	Rel.eff.
0.2	8.17	2	1.57
0.5	3.65	5	1.21
1	2.22	10	1.10

The distinction between this discussion and that at the beginning of the section is that in the earlier work an estimate $\tilde{\mu}_g$ of μ that is consistent for most distributions was used, whereas now an estimate based on the sufficient statistics for the gamma distribution is used. For small β the two estimates differ very appreciably.

Similar results can be obtained for the log normal distribution. A further case that is simple mathematically, even though rather unrealistic, arises when $f(x)$ is a normal density of mean μ and standard deviation σ, it being assumed that σ/μ is sufficiently small for negative values to arise with negligible probability. It can then be shown that in length-biased sampling

$$\operatorname{var}_g(\hat{\mu}_g) = \frac{\sigma^2}{n(1 - 3\sigma^2/\mu^2)}.$$

The results of this subsection can be generalized in various ways. For example, the bias can be of a more complex form, or the estimation of other parameters can be considered. A particularly important extension arises when the biased sampling of X affects the distribution of a second variable Y correlated with X; see (3.2). A simple extension of (5.6) leads to estimates of the unweighted properties of Y. For example, in textile applications one may be concerned with length-biased distribution of fibre diameter, fibre strength, etc. There is a general connexion also with importance sampling in Monte Carlo work.

5.3. Estimation from Recurrence Distributions

In the previous subsection we considered the estimation of quantities associated with the unweighted distribution from samples of the weighted or length-biased distribution. We now consider the corresponding problems when the recurrence-time distribution is sampled. That is, we consider a random sample $Z_1, \ldots, Z_n$ of positive random variables having the p.d.f.

$$k(z) = \frac{1 - F(z)}{\mu}, \tag{5.13}$$

where F(z) is the c.d.f. of the unweighted distribution and it is assumed that F(0) = 0.

The first problem to be considered is the estimation of μ for arbitrary densities f(x). Since $k(0) = 1/\mu$, it is in effect required to estimate the density at the origin, a terminal of the distribution of the observations. Previous work on the estimation of probability densities seems to concentrate on estimation at points interior to the range of variation. Here we sketch two methods for the present problem.

The simplest method for large n is to choose a suitable grouping interval Δ and to consider histogram heights near the origin. Let p(a, b) be the proportion of $Z_1, \ldots, Z_n$ in the interval (a, b). Then

$$E_k\left\{\frac{p(0,\Delta)}{\Delta}\right\} = \frac{1}{\Delta}\int_0^{\Delta} k(x)\,dx = k_o + \tfrac{1}{2}k_o'\Delta + O(\Delta^2),$$

$$(5.14)$$

$$E_k\left\{\frac{p(\Delta,2\Delta)}{\Delta}\right\} = \frac{1}{\Delta}\int_{\Delta}^{2\Delta} k(x)\,dx = k_0 + \frac{3}{2}k_o'\Delta + O(\Delta^2),$$

where $k_o = k(0)$ and $k'_o = k'(0)$.

Thus

$$\tilde{k}_o = \tfrac{1}{2}\{3p(0,\Delta) - p(\Delta,2\Delta)\}/\Delta \qquad (5.15)$$

is an approximately unbiased estimate of k_o. Its variance can be obtained using the multinomial distribution of $p(0,\Delta)$ and $p(\Delta,2\Delta)$; if both p's are small,

$$\frac{\text{var}_k(\tilde{k}_o)}{k_o^2} \sim \frac{5}{2n\, k_o \Delta}. \qquad (5.16)$$

In this $n\, k_o\, \Delta$ is approximately the expected number of observations in $[0,\Delta)$.

It is very easy to show that the somewhat more biased estimate $p(0,\Delta)/\Delta$ has smaller mean square error than $\tilde{k}_0$ if

$$\frac{k_o'^2}{k_o} < \frac{6}{n\Delta^3}$$

and to obtain from a higher order expansion a formula for the choice of Δ. These formulae are not directly helpful. If, however, it is thought that $k'_o = -f(0)/\mu = 0$, the use of the simpler estimate

$$p(0,\Delta)/\Delta \qquad (5.17)$$

would be sensible.

The major difficulty in estimating k_o is the choice of the amount of smoothing that is appropriate, i.e. the choice of Δ, or in more complex forms of estimate the selection of a weight function. In practice, several values of Δ will be tried, chosen after inspection of the data.

A closely related way of approaching the problem is by the use of the order statistics $Z_{(1)} \le Z_{(2)} \le \ldots \le Z_{(n)}$. Roughly, $Z_{(1)}$ will estimate $(n\, k_o)^{-1}$ and so too will $Z_{(r)} - Z_{(r-1)}$, for r/n small. More precisely, expand the standard formula for the expected value of the order statistics about the origin. Then, for fixed r and large n, we have that

$$n\, E_k\{Z_{(r)}\} = \frac{r}{k_o}\left(1-\frac{1}{n}\right) - \frac{r(r+1)}{2n}\,\frac{k_o'}{k_o^3} + 0\left(\frac{1}{n^2}\right).$$

This suggests writing

$$U_r = \frac{n^2}{n-1}\{Z_{(r)} - Z_{(r-1)}\}, \qquad (5.18)$$

with $Z_{(0)} = 0$, when

$$E_k(U_r) = \frac{1}{k_o} - \frac{rk_o'}{nk_o^3} + 0\left(\frac{1}{n^2}\right). \qquad (5.19)$$

The U_r's are approximately independently exponentially distribution. It is sensible to plot the U_r's against r and to estimate $1/k_o$ by extrapolating back a suitable block of observations to r = 0. The simple averaging of a block of observations, appropriate if $k_o' = 0$, is almost equivalent to the use of (5.17).

Now, having estimated $\mu = 1/k_o$ by say $\tilde{\mu}_k$, we can estimate the unweighted c.d.f. F(x) as follows. First, for a grouping interval Δ', define

$$\tilde{F}_k(z) = 1 - \frac{\tilde{\mu}_k\, p(z - \frac{1}{2}\Delta', z+\frac{1}{2}\Delta')}{\Delta'}. \qquad (5.20)$$

This gives for $z = \frac{1}{2}\Delta', \frac{3}{2}\Delta', \ldots$ estimates based on nonoverlapping intervals. These can then be amalgamated by a suitable algorithm (Brunk, 1955) to exploit the non-decreasing nature of F(z).

If Δ, Δ' are both small, if $z > 2\Delta + \frac{1}{2}\Delta'$ and if we ignore gains in precision from amalgamation, we have that asymptotically

$$\mathrm{var}_k\{\tilde{F}_k(z)\} = \frac{k(z)\{\Delta k_o + c\Delta' k(z)\}}{n\,\Delta\Delta'\, k_o^3}, \qquad (5.21)$$

where $c = \frac{5}{2}$ if the estimate (5.15) is used for k_o and $c = 1$ if (5.17) is used. Now $q(z) \leq q_o$ and it would often happen that Δ >> Δ'. Then (5.21) becomes

$$\operatorname{var}_k\{\tilde{F}_k(z)\} \simeq \frac{k(z)}{n\Delta' k_o^2} = \frac{\mu\{1-F(z)\}}{n\Delta'} . \qquad (5.22)$$

Thus by comparison with (5.9) for unweighted sampling, it follows that the efficiency of unweighted ampling relative to sampling k(x) is at least

$$\frac{\mu}{\Delta' F(z)} . \qquad (5.23)$$

Since $F(z) \leq 1$ and usually $\Delta' \ll \mu$, it follows that sampling the recurrence distribution will be very inefficient, at least when analyzed by the present method.

Suppose next that a parametric form can reasonably be taken for f(x). The likelihood of $Z_1, \ldots, Z_n$ can be obtained; sufficient statistics for sampling k(z) will not in general be the same as those for sampling f(x). The alternative procedure of postulating a convenient non-increasing form for k(z) and then deriving f(x) or F(x) will not be examined here.

If f(x) is exponential on (0,∞), then k(x) = f(x) and the two methods of sampling are equivalent. If f(x) is normal with mean μ and standard deviation $\sigma = \eta^{-1}$, with σ/μ small, it can be shown that the expected second derivatives of the log likelihood L of the sample are given by

$$\frac{1}{n} E_k\left(-\frac{\partial^2 L}{\partial \mu^2}\right) = -\frac{1}{\mu^2} + \frac{\eta}{\mu} \int_{-\infty}^{\infty} \frac{\phi^2(v)}{\Phi(v)}\, dv = -\frac{1}{\mu^2} + \frac{0.903}{\mu\sigma} ,$$

$$\frac{1}{n} E_k\left(-\frac{\partial^2 L}{\partial \mu \partial \eta}\right) = \frac{1}{\eta\mu} \int_{-\infty}^{\infty} v \frac{\phi^2(v)}{\Phi(v)}\, dv = -\frac{\sigma}{\mu} 0.595, \qquad (5.24)$$

$$\frac{1}{n} E_k\left(-\frac{\partial^2 L}{\partial \eta^2}\right) = \frac{1}{\eta^3\mu} \int_{-\infty}^{\infty} v \frac{\phi^2(v)}{\Phi(v)}\, dv = \frac{\sigma^3}{\mu} 1.101$$

Thus for the maximum likelihood estimate of the mean μ, we have that asymptotically

$$\text{var}_k\,(\hat{\mu}_k) = \frac{\sigma^2}{n} \times \frac{1.72\,(\sigma/\mu)^{-1}}{1 - 1.72\,\sigma/\mu}\,. \qquad (5.25)$$

A more realistic case is when f(x) has the log normal form (5.10), where it is convenient to write $\tau = 1/\kappa$. Then

$$\frac{1}{n} E_k \left(- \frac{\partial^2 L}{\partial \lambda^2} \right) = -\tau^{-1} e^{-\frac{1}{2}\tau^2} \int_{-\infty}^{\infty} \alpha'(u)\,\Phi(u) e^{-u\tau}\,du,$$

$$\frac{1}{n} E_k \left(- \frac{\partial^2 L}{\partial \lambda \partial} \right) = -\tau - \tau e^{-\frac{1}{2}\tau^2} \int_{-\infty}^{\infty} u\alpha'(u)\,\Phi(u) e^{-u\tau} du, \quad (5.26)$$

$$\frac{1}{n} E_k \left(- \frac{\partial^2 L}{\partial \kappa^2} \right) = 3\tau^4 - \tau^3 e^{-\frac{1}{2}\tau^2} \int_{-\infty}^{\infty} u^2\alpha'(u)\,\Phi(u) e^{-u\tau}\,du,$$

where $\alpha(u) = \phi(u)/\Phi(u)$, and $\alpha'(u) = -u\alpha(u) - \alpha^2(u)$. Now

$$\log \mu = \lambda + \tfrac{1}{2}\tau^2 = \lambda + \tfrac{1}{2}\kappa^{-2}, \text{ so that}$$

$$\frac{\text{var}_k(\hat{\mu}_k)}{\mu^2} = \text{var}_k(\hat{\lambda}) - \frac{2}{\kappa^3}\,\text{cov}_k(\hat{\lambda}, \hat{\kappa}) + \frac{1}{\kappa^6}\,\text{var}_k(\hat{\kappa}) \quad (5.27)$$

can be obtained from (5.26). Table 5.3 gives some numerical values for n var $(\tilde{\mu})/\mu^2$, for various estimates $\tilde{\mu}$.

Table 5.3. Estimates of mean of log normal distribution. Squared coefficient of variation of estimate x n.

τ	Max.lik. from recurrence dist.	Mean of unweighted sample	Efficient estimat from unweighted sample
0.2	0.410	0.0408	0.0408
0.5	1.30	0.284	0.281
1	4.00	1.72	1.50
2	19.2	53.6	12.0

ACKNOWLEDGMENTS

I am grateful to Messrs. S. L. Anderson, T.W. Anderson, W.R. Buckland, R.D. Elston, W.A. Pridmore and V.J. Small for helpful comments and references, and to Mr. B.G.F. Springer, whose work was supported by the Science Research Council, for the numerical evaluation of some integrals.

REFERENCES

BARNES, R.M. (1957). Work sampling. New York: Wiley.

BICKING, C.A. (1964). Bibliography on sampling of raw materials and products in bulk. Tech. Assoc. of Pulp and Paper Industry, 47, 147-170.

BLUMENTHAL, S. (1967). Proportional sampling in life length studies, Technometrics, 9, 205-218.

BRITISH STANDARDS INSTITUTION (1960). The sampling of coal and coke, BS1017, Pt. 1 Coal, Pt. 2 Coke.

BRITISH STANDARDS INSTITUTION (1967). Sampling of iron and manganese ores, BS4103, Pt. I, Manual sampling.

BRUNK, H.D. (1955). Maximum likelihood estimates of monotone parameters, Ann. Math. Statist., 26, 607-616.

COX, D.R. (1955). Some statistical methods connected with series of events, J.R. Statist. Soc., B 17, 129-164.

COX, D.R. (1961). Renewal theory. London: Methuen.

COX, D.R. (1966). A note on the analysis of a type of reliability trial, J. Siam. Appl. Math., 14, 1133-1142.

DANIELS, H.E. (1942). A new technique for the analysis of fibre length distribution in wool, J. Text. Inst., 33, T 137- T 150.

DAVIS, H. (1955). A mathematical evaluation of a work sampling technique, Naval Res. Log. Q., 2, 111-117.

DUNCAN, A.J. (1962). Bulk sampling: problems and lines of attack. Technometrics 4, 319-344.

GAVER, D.P. and MAZUMDER, M. (1967). Statistical estimation in a problem of system reliability, Naval Res. Log. Q., 14, 473-488.

HERDAN, G. (1960). *Small particle statistics*. 2nd ed. London: Butterworth.

HILL, I.D. and HILL, P.A. (1968). *Random times for activity sampling*. Enfield, Middx: Inst. of Work Study Practitioners.

KRIENS, J. (1963). The procedures suggested by de Wolff and van Heerden for random sampling in auditing, *Statist. Neerlandica*, 17, 215-231.

PALMER, R.C. (1948). The dye sampling method of measuring fibre length destribution, *J. Text. Inst.*, 39, T8-T22.

PALMER, R.C. and DANIELS, H.E. (1947). The sampling problem in single fibre testing, *J. Text. Inst.*, 38, T94-T100.

TIPPETT, L.H.C. (1935). Statistical methods in textile research; uses of the binomial and Poisson distributions. A snap-reading method of making time studies of machines and operators in factory surveys, *J. Text. Inst.*, 26, T51-75.

ON SOME METHODOLOGICAL ASPECTS OF SAMPLE SURVEYS OF AGRICULTURE IN DEVELOPING COUNTRIES

P. V. Sukhatme and B. V. Sukhatme

Rome, Italy, and Iowa State University

1. INTRODUCTION

Agricultural statistics can broadly be classified under two heads (a) those commonly covered by censuses of agriculture to provide data on the structure and resources of agriculture to be tabulated by size and other characteristics of the farm, and (b) those covered by current surveys to provide estimates of total values for the population without classification by size of farm. The farm* is the reporting unit in the census; it need not be so for current statistics.

There is hardly any country which has not initiated some sample survey program or other to collect data covered by agricultural censuses and current surveys. This is a great advance in that it has made it possible for countries to collect data, however few, of known precision where only guesses existed before. At the same time the use of sampling method without ensuring control of biases such as those arising from defective frames, and due to response errors has given rise to a large number of queries concerning the reliability of the data obtained.

In agricultural surveys with farms spread all over the rural area and with majority of people uneducated and illiterate, biases can assume large magnitude. Consistency checks, such as in the form of food supply available for consumption based on results of censuses and surveys, have shown that for many countries these

---

* In this paper "farm" and "agriculture holding" are synonymously used.

results are not too convincing. This paper reviews some of the difficulties encountered and methodological problems involved in the use of sampling in agricultural censuses and surveys under conditions of the developing countries.

2. CENSUSES

Owing to the difficulty of preparing a nation-wide list of farms, it is common to use a two-stage sampling frame with area segment as the primary unit and the farm as the secondary unit. Since in an agricultural census the population consists of the totality of farms, it is clearly necessary to include in each selected area segment only those farms which are associated with it, irrespective of whether part of any farm is outside its boundary. Such a segment is often called an open segment.

2.1. Frame: Their availability and need for updating

Two types of material are usually available for constructing a frame of area segments: (i) the enumeration areas constructed for the population census and, (ii) administrative units, e.g. villages. The use of enumeration areas requires that they should be kept up to date, as otherwise with the passage of time it becomes increasingly difficult to delimit their boundaries. They also get out of date in respect of the information recorded at the time of the population census. Jamaica (Lele, 1968) provides a good example of these difficulties. A pilot survey was carried out in the country during 1967 in eight enumeration areas using sketch maps prepared for the population census of 1960. The sketch maps were found inadequate in descriptions. There was no means of ensuring that the population census of 1960 and the pilot agricultural survey of 1967 related to the same boundaries. The FAO statistician, who was in charge of the survey, reported that there was no unambiguous way of identifying local roads, paths and tracks, old and new constructions, rivers and tributaries, and estates which changed hands. There was also insufficient description of natural contours around the boundary. Even after consulting local people he reported that there was ample room for differing interpretations. Differences in households from different interpretations ranged from 10 to 30% in the different enumeration districts.[1] Liberia provides another example of the biases introduced by the use of a frame prepared for the population census. There were 1200 enumeration districts mapped for the population census conducted during 1962. When an agricultural survey was carried out in 1966 in part of Liberia using this frame, a post enumeration check showed that physical features had changed a great deal and delimitation on the ground of the area segments corresponding to those in 1962 required a great deal of field work. The check showed 17% of the localities, 12% of the structures and 18% of the population were missed from enumeration.

Villages present a different type of difficulty. Where they constitute the smallest geographic subdivisions for government administration, they cover the country completely and without overlapping. Large parts of Pakistan, UAR, India, Ceylon, etc. provide examples of such sub-divisions. In many countries, however, and even in parts of India and Pakistan, a village is an agglomeration of dwellings living as a social and economic group with surrounding land used for agricultural purposes but with precise geographic boundaries generally not mapped. A village may remain compact or its boundaries may continuously change with time with new hamlets springing up in the vicinity. Under certain situations the dwellings may even be scattered over considerable distances. Consequently, there is no assurance that the existing list of villages may cover the whole of the country. There is also the possibility of new villages springing up in far away places which do not belong to any administrative villages already existing. The system of shifting cultivation and the semi-nomadic character of some of the communities further complicates the problem. In Gabon, for example, it is not unusual for villages to be abandoned altogether and be shifted elsewhere.

For these and other reasons the use of villages as area units may result in leaving out hamlets and associated dwellings and consequently lead to underestimation of the characters under study. The underestimation may be particularly serious where the administrative structures are not sufficiently well developed. Isolated dwellings may also be duplicated but this is rare. A frame of villages does not also cover the farmers living in urban areas which may account for the omission of an appreciable number of large farms, particularly orchard operators. It is important to ensure their adequate representation in a census through list sampling or similar means. The task of compiling a reasonably complete list of villages is much more time-consuming and laborious than is usually thought of owing to the immense difficulties of transportation and communication in the developing countries, at least in Africa.

A farm being the reporting unit of censuses of agriculture, it becomes necessary to list all farms for each of the selected area segments. This, in turn, requires that a list of households should be prepared for each selected area segment. At first sight, it may appear easy to compile such lists but this is not so in all countries. In Algeria, for example, it has been observed that lists of households prepared with the help of the chiefs of villages were seriously incomplete. The use of taxation lists, such as are available in the Congo (Simaika, 1962) where every adult male has to pay a tax to the village chief, have also proved of limited value. Assembly of local population and interviewing them in order to compile the list was tried in Tunisia (Olivera, 1963) and Senegal (Farah, 1966) but again lists were found to be incomplete through omission of far away households whose heads were not always available for interview simultaneously with other population. Listing of households by dwelling to dwelling enumeration can provide a satisfactory approach but it is an arduous and costly operation since in many countries dwellings are clustered together in a haphazard manner not amenable to easy listing and in some they are scattered over long distances.

The main object of listing households in selected primary units is to identify farm households. To make this possible a number of questions must be asked when visiting households. The formulation of these questions will depend upon the concept of the farm and on the criteria for associating farms with selected area segments. These criteria have to be unambiguous and must be formulated in advance of the census. They are needed for dealing with situations such as whether the farm cuts across two area segments or whether the farmer lives in one segment but his land lies in another, or whether the different parcels constituting the farm are spread over morethan one segment, or whether the holder has two residences in two segments (not a rare case in the polygamous society). In practice, as we already stated, detailed maps of area segments for the rural area do not exist in many of the developing countries. The criteria of associating farms with selected area segment are also in general not easy to formulate and apply. Consequently, the list of households qualifying as farm households may be inaccurate through omissions and also sometimes through duplication. Of course, the frame of farm households is by its very nature an incomplete one as it does not cover the farms operated by states, cooperatives, etc. There is however little difficulty in preparing a list of such farms.

An alternative method of preparing the list of farms is to visit the fields in the selected area segment one by one and ascertain the names of the farmers operating them but this method has not proved practical. The method was tried in Tunisia (Fojtl, 1966) where the enumerators, accompanied by agents of the local governor, visited each individual field and recorded the names of the respective farmers. To avoid duplication the cultivator was asked the first time his name was recorded, whether he operated another field elsewhere. The method proved operationally difficult and time consuming, largely because of ambiguity of names and also because names of owners were confused with those of the operators. Moreover, as was expected, the method failed to list livestock holdings.

Experience of listing farmers through the cadastral list of fields in India, Pakistan and Ceylon is similar.

Frames may be complete and accurate at the time of their construction but they become rapidly out of date. Although this fact is well known, obsolete frames are generally used, largely for reasons of economy. There is no assurance therefore that the results obtained are satisfactory. The cases of Jamaica and Liberia have already been mentioned. Yet another example of the need to up-date the frame is provided by the rural economy survey carried out in Nigeria (Lele, 1965). Aware of the serious limitations of the results by the use of an inaccurate frame, the results were published as averages per farm household and not in the form of global estimates. However, the probability structure of the sample being unknown, the possibility of the averages themselves being seriously biased certainly cannot be ruled out.

If an agricultural census is taken shortly after the population census, or better still if it can be combined with the population census itself, the frame prepared for the population census should prove satisfactory, not only in respect of enumeration districts but in respect of the list of households as well. However, experience shows that even for rural areas a frame older than about a year might not be usable for delimiting area segments and is much less so for the selection of farm households. Even if dwellings as such may not be demolished (new dwellings of course continue to be added) the list may nevertheless become unusable because the dwellings are largely recognized or identified in terms of the names of their occupants. Consequently, if occupants leave there are no adequate descriptions available to identify dwellings. On the other hand, any project of combining the agricultural census with the population census must be looked at from the viewpoint of the limited resources in men and transport available for the purpose. A statistical office may be inclined to spread the statistical survey program over a decade in order that it may use its limited resources to the best advantage but such approach implies constant updating of the frame. Such updating must also include collection of useful ancillary information which can be used for improving the efficiency of the sampling design. We can give a number of examples from the experience of FAO where we have found it necessary to advise the countries to discard the old frame altogether and prepare a new one. In Basutoland (Lesotho), (Morojela, 1963), for example, the original intention was to use a list of households from the 1950 Agricultural Census as a frame for the 1960 Agricultural Census. A pilot test showed that the lists were so much out of date that a special preliminary survey had to be undertaken in order to prepare a new list.

Frames of villages can be checked by sending out questionnaires to administrative officers and/or sending teams of special investigators who could tour the countries systematically, division by division, and ascertain with the help of the local population and the use of available information, the completeness of the list. They can also collect ancillary information in the course of their visits. Such checking of the old frame is a laborious task and is made even more difficult by the difficulty of terrain and lack of efficient means of transportation and communications. Old maps have to be updated showing the newly introduced man-made features, like roads, buildings, newly created hamlets, settlements, etc., and old frames of farm households have to be revised by house to house and hut to hut enumeration. As already mentioned, the mobility of the population is so high that such updating needs to be carried out immediately prior to the census itself. Little can be done to salvage the results of a census based on an incomplete and obsolete frame. Post enumeration survey, supplemented with list sampling, may help to provide correction factors but this is an expensive remedy of limited value.

An alternative means of reducing bias due to incomplete frame is to supplement the area segment sample by sampling from lists of large farms and plantations such as are available for modern sectors of agriculture in many countries. Such lists are known to be available for the old European farms in Africa. As an example, in Algeria there exists a "secteur autogèré" composed of 2000 operational units with cadastral maps accounting for nearly 20% of the total agricultural land in the country. The use of these lists along with area segment sampling undoubtedly helps to improve the accuracy of the results. Even where these lists are not available, listing of large modern farms such as state farms, cooperative etc., and enumerating them completely may not be a difficult or even costly undertaking for most countries. Nevertheless, such list sampling is unlikely to go too far in improving the accuracy of the results for the reason that their contribution is limited to major commercial crops and their total contribution is not very large although they do contribute substantially to the total agricultural exports of the countries. Therefore, we cannot see multiple frame sampling as a means of covering adequately the traditional sector of agriculture under present day conditions in Africa.

Even where area segment sampling is combined with list sampling of large farms and all the relevant ancillary information is fully made use of, we cannot escape the need for updating the list of households in the selected area segments. In fact, in many surveys such lists need to be prepared immediately before the commencement of the survey. While admittedly, this is an arduous task it also helps to serve the needs not only of agricultural censuses but also of other surveys, such as those of rural industries and other related surveys. It appears imperative that as long as administrative needs of the countries have not developed to a point of preparing lists of small farms or detailed maps of the rural areas, expenditure needed for maintaining the frame for agricultural and related censuses must be incurred if we are to have any worthwhile program for the development of statistics. To allow considerations of economy to weigh at this stage and to use obsolete frames is to ignore the very basis of the foundations of survey sampling. It is true that in most of the censuses conducted in developing countries care has been exercised in ensuring the availability of a good frame in the selection of the sample but these are small scale pilot censuses limited in scope. Considerations of cost begin to come in with nationwide census surveys, such as with agricultural census and it is here that the temptation to use obsolete frames is the greatest. Some idea of the cost of listing can be had from the census of agriculture conducted in Ecuador (Tang, 1960). The census included visits to all villages for purposes of listing large farms (i.e. farms above a specified size) and subsequent enumeration thereof. Besides, the census included in the second stage a visit to a sample of villages for listing of other farms and enumeration of 10% of these farms. Out of the total of 21,000 man days on the operation, Tang estimated that listing accounted for over half of the total man days. The need to visit every village to list large farms arose because large farms were known to account for the major production of agriculture in the country. Other examples are provided by surveys conducted in Panama (Tang, 1960), Dahomey, Nigeria (Lele, 1965), Liberia (Thawani, 1963), Ceylon (Koshal, 1954) and many other countries which indicate that the cost of preparing an area frame and of updating households therein may amount to as much as one fifth of the total cost of the census. Revision of available lists of villages and estates alone may well amount to one tenth of the total cost of a multi-purpose agricultural survey.

2.2. Open segment concept: practical difficulties

The second source of bias in census data arises principally from the difficulty of applying the open segment concept in practice. An open segment, by definition, constitutes all farms associated with it. In terms of households, it can be defined to represent a unit comprising all the farm households located within its boundaries. In terms of land however, it need not be identical with the total area of fields (parcels) within the boundaries. Since a farm, by definition, constitutes all land used for agricultural production irrespective of location so long as it is operated as a single technical unit, the different parcels of the farm may not all be located adjacent to each other; they may not also be within the segment itself. In practice, however, it is inevitably the experience in most developing countries that the farmer under-reports the number of parcels constituting his farm. At best he may report only the part located in the village of his residence. This naturally results in understating the size of the farm. Uneducated and frequently illiterate that he is, he probably is not able to interpret the questions concerning the meaning of farm put to him by the enumerator. It is also possible that he may not even like to cooperate for reasons of suspicion of the enquiry and also of fear that information given by him may be used for increasing taxation or compulsory procurement of produce from his farm. It is also not uncommon that he may suffer from a superstitious belief that any precise information about his farm may attract the evil eye and bring him ill luck. Whatever be the explanation, the fact remains that the farmer does usually understate the number of parcels of land operated by him. In an experiment conducted in Greece it was found that farmers under-reported the number of parcels operated by 36%. It is only when the enumerator has visited him frequently and acquired an understanding of his farm business that the farmer may come out with more accurate information concerning the number of parcels constituting his farm, and even then he may not report parcels outside the village of his residence. Experience also shows that the farther the location of a farm the larger is the degree of understatement in the number of parcels constituting the farm.

Post enumeration surveys using physical measurement have shown that even for parcels of land declared by the farmer, he consciously or otherwise is found to understate the area. The size of farm is thus understated even when all parcels of the farm are included. This is due in part to the fact that the farmer has no precise quantitative knowledge of the size of his farm and is partly intentional. The consequence of all this is that the size of farm is seriously understated in censuses. By way of example, we may mention the survey carried out in Algeria during 1965/66. The results showed that the farm size in the different provinces was grossly underestimated from 20 to 50%. A report that land reform law was under way was partly responsible for this gross understatement. Costa Rica provides another example. The 1963 Census of Agriculture was found to have underestimated the average size of farm by 9.5%. Similar results were reported from India collected in the course of a cost of production study (Panse, 1966). Three surveys were conducted. In the first survey data on the size of the holding was obtained by investigators in the course of a single visit. In the other two surveys the investigators were in the field for a longer period but, whereas in the second survey, information on the size of the holding was collected as one of the items forming the background of the main enquiry, in the third survey the investigation was centred round the cultivator and his holding and the investigators presumably had a closer and more critical view of both. Although there were slight differences in definitions and some variation in coverage, an explanation of the greater part of the differences in the results of the three surveys was to be found in the method of approach and the period for which the enumerator was in personal contact with the farmer and his holdings. The serious underestimation of the size of the holding in the first survey in which the information was obtained by interview in the course of a single visit, as is usually done in an agricultural census, will be seen from Table 1 which we have reproduced in the appendices. It will be seen that in every region of India the farm size was underestimated by some 30 to 50%. Similar, though smaller, biases are reported even from advanced countries. In Japan, for example, the overall underestimation of the size of the farm in the census of agriculture conducted during 1955 was found to be about 10%. In the experiment in Greece mentioned above the area of the fields selected for study was under-reported by 12%. In the United States 1959 Census of Agriculture the total number of farms were undercounted by 8.4% and the area of farms in lands by 6%, the percentage sampling errors of the estimated totals being of the order of 1%.

2.3. Response errors

These errors occur whether the censuses are conducted by complete enumeration or by the use of sample survey techniques. They arise in the course of obtaining information from the farmer and are non-sampling errors. As the examples show, their magnitude can be quite large. In extreme cases they may even vitiate the numerical results of the estimates to such an extent as to distort the picture of the agricultural situation altogether. Along with the size of holding understated, all other characteristics also get under-reported — partly from understatement of size of farm and partly from understatement of the items themselves. Thus to continue the example of Japan, paddy land was under-reported by 7% and the area under other grains like that of upland rice, wheat and paddy got under-reported from 20 to 30%. For fruits and vegetables under-reporting was serious being of the order of 40 to 50%. Another example of under-reporting is provided by the 1960 Census of Agriculture of Malaya (Tsuki bayashi, 1961). A careful comparison of the census results in Malaya for area under paddy, rubber and coconut with material available from other sources, such as the official record of land alienated to small holders during past year for growing these crops and the data from the Rubber Research Bureau on rubber production, new and replanted rubber etc. revealed large underestimation biases in the census results. Area under wet paddy was under-reported by 26%, rubber by 49% and coconut by 59%. Urban areas were omitted from the census but this omission cannot account for more than 10% of this underestimation bias of paddy and coconut areas, although for rubber it might explain about 70% of the bias because a considerable proportion of owner/operators live in urban areas which were excluded from the census. The remaining biases can only be explained as a result of under-reporting by the farmers.

In Indonesia (Sattar, 1966) the 1963 census estimate of area under wet paddy for West Java based on reports from farmers was found to be less by 33% as compared with the area compiled by the land tax office. In Thailand, while no serious bias is observed in farm areas reported by farmers in the plains, under-reporting to the extent of 20 to 30% was observed in the hilly areas. In China (Taiwan) under-reporting of cultivated area to the extent of 10 to 20% was observed during the 1960 Census in different townships. Under-reporting of area and livestock was also noticed in the Philippines. In that country the correlation coefficient between the number of farms per village as reported during the last census and observed in the annual crop survey was only 0.58. In several African countries estimates of production based on local reports of crop yields were so low that it was considered necessary to measure the yield of principal crops by crop-cutting surveys. These examples show that it is of the greatest importance to devise ways and means of ascertaining accurate information on the number of parcels constituting the holding and their location and size as well as of using objective methods in obtaining information on the parcels declared by the farmers. Any effort that may have to be incurred for this purpose will be more than repaid in the form of increased reliability of the census results and their acceptance by the users.

One method of controlling such errors is of course to give intensive field training to the enumerator and to provide him with adequate incentives for his work. Alert and well directed supervision can also help in improving the quality of field work by checking up regularly for consistency, completeness and veracity. However, with the majority of the farmers uneducated and illiterate, there is a limit to which the accuracy of the information can be improved in this way unless the interview is supplemented by physical count and measurement in a sub-sample of farms. Indeed, it is in appreciation of this background that agricultural censuses are coming to be organized in the form of surveys spread over the census year. Thus in many countries surveys of land use and associated agronomic practices are carried out during the appropriate season as part of the program of census of farms when the crops are standing. An essential ingredient of such a survey is a physical check of area on a sub-sample of farms for calibrating the area reported by the farmers and under certain conditions for estimating that area. Similarly, for physical determination of yield, crop-cutting surveys on sample of plots growing important crops, staple as well as commercial, are arranged at appropriate harvesting time in the hope that the results may serve to calibrate the production of crops reported by farmers. Again, certain surveys require more frequent rounds of field enumeration. An example is provided by the survey of agricultural labour for employment which is known to vary considerably with the season and which is best estimated by contacting a sample of holdings with reference to information for the preceding week and spreading the enquiry all round the year. Other examples are provided by surveys of numbers of livestock and their production. The field work is done in several rounds as the production is spread over the entire year. A physical check is exercised on a sub-sample by a count of animals and weighing the produce, like milk. The need for physical measurement is forcefully brought out under conditions of shifting cultivation.

2.4. Sampling Errors

Altogether, a review of the results of the 1960 World Census of Agriculture shows that mapping materials in the developing countries are seriously inadequate, the concept of open segment is difficult to apply and understatement of items reported by the farmers is large. By comparison, the sampling errors reported for census results are small, of the order of 1% for totals of major items compared with biases several times as large. This can be illustrated by comparing results of the post enumeration checks with those of the census itself. The Census of 1960 conducted in East Pakistan provides a good illustration. A post enumeration survey was carried out on a sub-sample of villages selected at random out of villages sampled for the census. The method of obtaining information was made as objective as possible, including a reference to Khasra registers where feasible. The comparison with census results for the selected villages shows that the census understated the total number of holdings by 8% mainly due to the fact that the houses located near the boundaries of the villages had been left out in enumeration. This naturally resulted in under-enumeration of other characteristics as well, e.g. total area by 6.5%, cultivated area by 5.2% and crop area by 6.6%. One cannot rule out the possibility that the post enumeration check over-enumerated the number of holdings but this possibility is rather remote in East Pakistan since villages have precise maps with adequate descriptions to delimit boundaries unequivocally on the ground. In these circumstances consistent underenumeration of houses on the boundaries in the villages selected for the checks must largely be ascribed to the difficulty of applying the open segment concept in practice. What appears probable is that the missed houses on the boundary, although properly resident houses in the selected villages, have their holdings or a major part of their holdings in adjacent villages and were therefore left out in enumeration. A post enumeration check carried out in West Pakistan similarly established that the census enumerators had omitted a good many farms which should have been enumerated and that in consequence the related items, such as those of total area of farms, total area under crops, total livestock and other totals, were also under-reported. Greece provides another example of the same experience. A post enumeration check of the 1960 Census results revealed under-enumeration of parcels by 24% and the resulting under-enumeration of land area by 14%. In many countries where villages are little more than an agglomeration of dwellings without any sketch maps for the surrounding area to delimit boundaries on the ground, the biases are larger, partly because of inadequate coverage and partly from difficulty of applying the open segment concept. In these villages much also depends upon the assistance and knowledge of the village chief in identifying the boundary of area under his charge. In many countries the available external evidence alone is sufficient to show that census results are not reliable and that the sampling errors mentioned in the census reports have little more than academic interest for judging the accuracy of the results. Results of post enumeration checks from many other countries, e.g. Libya, Brunei, Guatamala (Lomeli, 1965), etc., confirm that biases much larger than could be accounted for by sampling errors were present in the results of the 1960 Census of Agriculture.

Clearly, to reduce sampling errors while allowing a bias several times as large to enter into the census results is questionable from the viewpoint of the efficient use of resources. On the other hand, without aiming at sampling errors of the order of 1% or even smaller for totals, one could scarcely hope to estimate sub-totals for principal regions and for size classes with precision which can be considered reasonable in practice. Quite apart from considerations of efficiency of sampling, the available evidence shows the importance of evaluating in advance the magnitude of non-sampling errors and of ensuring that the total mean square error for given resources is minimized to an acceptable figure. Considered from this angle it is important that the technique used for collection of data should be objective as far as possible while simultaneously trying to involve the farmer closer into the census effort. As economy develops with attendant progress in educational and social fields, and as the farmer begins to express his interest in knowing the state of demand and supply for his products, he is likely to show increasing appreciation of the purposes of the census and of the advantages accruing to him from its results. It is only then that he will feel it his duty to cooperate in census taking, thus contributing to the reduction of biases in the census results. The decennial census of agriculture sponsored by FAO looks forward to this date. In the meantime, even while conscious of the difficulties of census taking and of the limitations of census results, FAO continues to promote the census program in the hope that it can provide a broad picture, however rough, of the agricultural structure, agricultural resources and their state of use for planning improvement of the agricultural economy of the countries.

2.5. Nomadic holdings

The above approach of dividing the country into area segments and selecting a sample of farm households may not be suitable for livestock holdings under nomadic conditions. The nomads move to areas where adequate water and grazing can be found during drier seasons of the year and return to normal grazing areas after the rains start. They congregate their animals around watering points, drinking wells and boreholes, all of which may not necessarily be fixed, mapped or known to the administration. The frame for location of the nomads and determining the time and place of enumeration raises a number of difficulties. Broadly, two methods have been used for their enumeration as part of the agricultural census - those based on tribal groups as primary sampling unit, as in Iran (Tuyuda, 1963) and those based on water points, as in Ethiopia (Church, 1958), but their discussion is beyond the scope of this paper.

3. CURRENT STATISTICS

3.1. Sampling design

In surveys for collecting current statistics the aim is to estimate totals for the population as a whole. Consequently, when the population is divided into area segments it is necessary to account for all the area, livestock, etc., as the case may be within the selected area segment. The concept is often referred to as that of closed segment; active cooperation of the farmer may not be necessary in collecting information from closed segments.

The design used for surveys is usually single or multi-state sampling with village (area segment) as the primary sampling unit, fields or cluster of fields as the secondary unit with plot within a field as third stage unit in the case of yield surveys. Geographic divisions usually form the principal strata. Auxiliary information such as on measure of size of village is also often used in sub-stratification and in determining selection probabilities and estimation procedures. This helps to improve the sampling efficiency, particularly when the size of village is related to the type of farming. The field staff is usually drawn from local administration. The efficiency of the design can be continually improved when working with closed segment as more and more mapping materials and auxiliary information become available (Trelogan and Houseman, 1967). Space does not permit us to give examples here.

3.2. Nature of biases in current surveys

Experience of surveys has shown that the magnitude of bias can be better controlled when working with closed rather than open segment. This is so for several reasons. Firstly, the use of closed segment can ensure greater completeness of enumeration since all the reporting units are located within the closed segment and can therefore be accounted for without omission or duplication provided the segment is not unduly large and the boundaries are clearly demarcated on the ground. This is particularly easy to ensure where cadastral maps exist. Even in unsurveyed areas, given adequate resources for preparing sketch maps of villages and tracts within them, it is possible to ensure complete coverage. However, from the viewpoint of sampling efficiency, it may not ensure optimum accuracy for given expenditure since the segment size is arbitrary. Secondly, the value reported for any unit of a closed segment under the practical situations obtaining in the developing countries can be represented as follows:

$$y_{ij} = x_i + \alpha_j + \epsilon_{ij}$$

where y_{ij} represents the value reported by the jth enumerator on the ith unit

x_i represents the true unknown value of the ith unit of population

α_j represents the bias of the jth enumerator

and ϵ_{ij} is simply the deviation of $x_i + \alpha_j$ from the reported value

In this model, unlike the one appropriate for the agricultural census with farm as the reporting unit, there is no component representing the bias due to interaction between the enumerator and the interviewee (farmer), the reason being that the value for the selected field is reported, measured or observed by the enumerator himself uninfluenced by the farmer's judgement. The source of bias resulting from understatement of size of farm and the consequential underreporting of other items in an agricultural census is removed in the case of surveys with closed segments. Thirdly, there is nothing in the model which suggests that bias α_j need be in the same direction; it can assume both positive and negative values depending upon the enumerator himself, his training and attitude to the selected unit at the time of reporting. Experience indicates that given suitable measurement technique and intensive training to the enumerator, the positive and negative values in reporting may be expected to cancel each other on average over all enumerators and that in consequence the bias may become negligible. This cancelling of positive and negative errors may hold even for individual segments, as, for example, in enumerating items of land use field by field since the total area for a closed segment is fixed and often known or can be ascertained independently. In what follows we shall give examples to illustrate the advantage of working with closed segment in controlling non-sampling errors.

Example 1: The village in India is a well defined area with dwellings clustering together and surrounded by the agricultural land belonging to the village. This land is divided into a number of fields which in many parts of the country have been surveyed and mapped and given officially recognized numbers called survey numbers*. It is one of the duties of the village accountant, called Patwari**, to make a field to field inspection of all the fields in his jurisdiction, generally twice a year, and to record the name or names of crops grown thereon along with the areas recorded in the Khasra register. Where more than one crop is grown on the field he is required in addition to indicate the proportion of the area under each. This method of annual census of villages, which constitute closed segments, should provide acreage data for each crop with a high degree of accuracy. This is so because there is a detailed and accurate frame available in the form of cadastral village maps, simple objective methods of enumeration are employed, the data are collected in the course of normal administration which ensures their completeness, and there is provision for adequate supervision. All the same, since the collection of these statistics is carried out by agents of the Revenue Agency the question is raised whether the resulting statistics may not be biased.

A direct sampling check was made in Lucknow district of U.P. in 1949/50 (Sukhatme and Kishen, 1951). The plan of sampling consisted of selecting one village at random from each Patwari's circle with probability proportional to number of fields in the village and further selecting two clusters of four consecutive fields with equal probability for inspection. Altogether 281 villages were selected. The field work

---

* The term survey number is in fact used to deisgnate a field.

** A village accountant is known as Patwari. He may have a charge of one or more villages, depending upon the size. The area under his charge is called a Patwari circle.

consisted of visiting the selected clusters of fields during the harvest period and recording the names of crops grown and also the area under each crop. Where only one crop was grown over the whole field the area was noted from the cadastral records but where more than one crop was grown in the same field the area was apportioned between the two crops by actual measurement. In the case of mixtures the proportions were based on eye appraisement. Corresponding entries made by the Patwari in the village register were noted simultaneously. The work of inspection was entrusted to two agencies; (i) the normal supervisory staff of the land records department and, independently (ii), the statistical staff of the Indian Council of Agricultural Research (ICAR).

Table 2 sets out the distribution of the discrepancies in Patwaris' records expressed as a percentage of the acreage noted by the supervisory staff. It will be seen that in 87% of the cases the names of the crops reported by the supervisory staff agreed with the entries made by the Patwaris. Of the remaining 13% of the cases, a substantial number of discrepancies were found to be ascribable to the fact that the Patwaris are required to record acreage under different crop mixtures according to rules prescribed for the purpose. For example, oilseeds form a minor component of mixture containing cereals like wheat and barley and are required under rules to be omitted when recording crop areas. Thus, a field containing a mixture of wheat and linseed, or wheat, barley and linseed will be entered by the Patwaris as containing only wheat or wheat and barley as the case may be. Eliminating such discrepancies, genuine mistakes in recording crop names on the part of the Patwari owing to his failure to make field to field inspections, amounted to only 4% of the total. However, these discrepancies occurred both in the positive and negative direction and largely cancelled one another in the aggregate for the district. A similar check in another district showed that the total discrepancy from the two sources, namely misreporting of the name of the crop and wrong judgement of area, did not exceed 10% for any of the crops examined; for the two principle crops it was 2.7% and 1.7% only.

The conclusions from these sample checks was that, taken over large areas, Patwaris' records present an essentially reliable picture of crop acreages. This has been confirmed by several other checks carried out in different states all over the country.

The main charge levelled against the Patwari is that he tends to neglect his work of making a vigorous inspection of the fields and relies for his data on reports received from the farmers themselves. The result quoted above shows that even if this were true it does not distort the resulting statistics. This is not surprising, considering that the actual acreages of fields have already been measured and entered in the village register. All that the Patwari does is to enter the name of the crop grown in a particular season against each field as also the proportion of area under each crop in the case of a mixture. It does not seem plausible that the farmers and their neighbours will give wrong crop names to the Patwari even if he were to rely on <u>their</u> information, as one does when farm is the reporting unit. The situation might have been different if farmers were to report not only the name of the crop but also the area of each field. But he is not required to report on the latter since it is already cadastrally measured and recorded in the village register. It is this fact, together with complete enumeration of all units within the closed segment, which ensures freedom from any serious bias in the crop acreage statistics of the Patwari. The only remaining error is that arising from using wrong proportions of areas under different crops grown in separate patches in the same field. Such fields occur infrequently and after compensation the net contribution to error in the acreage data from this source is found to be negligible.

How is it then, one may well ask, that evidence which contradicts these findings has also been produced? Mahalanobis, for example, conducted an enquiry for checking the work of the Patwaris through the National Sample Survey organization in four States of India (1949/50). The results of this check showed that the acreage under wheat was found to have been overestimated by the Patwaris to the extent of 10-26% in the different states with an average overestimate of 15%. On the other hand, cash crops like sugar cane, linseed and other oil seeds were stated to have been considerably underestimated, usually more than 50%. Critical examination of the report which remains unpublished, showed that the differences were almost entirely due to differences in the concepts used by the NSS and those laid down in the land records manual to be followed by the Patwaris. For example, the land records manual prescribes that sugarcane, although planted in the months of January to March, should be entered in the register only at the time of Kharif inspection during September/October. The NSS investigator on the other hand, noticing the crop in the field during his inspection in April and finding no entry by the Patwaris, counted this as the latter's mistake. Again, linseed is largely grown as a minor mixture with crops like wheat. The land records rules prescribe that areas under such mixtures should be recorded only under the major constituents and that an allowance be subsequently made for the minor component when computing acreage figures for the different crops. When allowance was made for these discrepancies in concepts it was found that there was no difference in the picture of acreage statistics presented by the NSS and the Patwari agency.

While it is not suggested that land use statistics in India are 100% accurate or that no revision of the concepts and definitions in recording acreage statistics is necessary, the evidence presented is conclusive in showing that where land is cadastrally surveyed, as in India and many other countries, the bias in estimating items of land use statistics is likely to be small even when the enumerators are drawn from local administration, since positive and negative errors in enumeration largely cancel each other and non-sampling errors are held under control.

Example 2: Plot size in yield surveys

The fact that the area segment is closed and the method of enumeration is objective does not imply that non-sampling errors can always be held under control. For the magnitude of the acreage bias, α, and the mean square error among enumerator biases, S_{α}^{2}, may change with the measurement technique itself, depending upon the item under observation and the agency enumerating it. Thus, it is not unlikely that in measuring areas the method of rectangulation may give larger biases than the method of triangulation, although both can be described as objective in character. To take another example, in yield surveys very small plots of the order of 20 sq. ft., such as are used in the developed countries and whose produce can be collected by the experimenter himself, are known to give biased estimates of the yield per acre. As the size of the plot increases the bias diminishes and so also does the component of variability due to plots. The magnitude of the bias also depends upon the enumerator. A small plot but not too small may give unbiased results when handled by a trained statistician. On the other hand, experience has shown that when crop-cutting surveys are carried out by the local field staff, who are normally entrusted with this work, even plots of size 100 sq. ft. lead to biased results. It is clearly advisable to use sufficiently large plots in order to safeguard against this bias. In India, for example, crop-cutting surveys are normally carried out by the local staff of the Department of Agriculture and Revenue and the plot size used is of the order of 500 sq. ft. The crop is harvested, threshed, winnowed and weighed according to the practice usually followed on the farm thereby providing estimates of the produce as at farmgate level. The size at which the bias becomes negligible will certainly be smaller than 500 sq. ft. but not so small that its produce could be collected by the enumerator himself without the use of hired labour. In fact, under conditions in the developing countries labour charges have to be paid on a daily basis. Consequently, variation in size of plot may not make a difference to the cost of a survey.

The advantage of small plot lies, in general, not only in the fact that it enables the field staff to collect the crop themselves, but also in moving rapidly from one village to another. However, this can often lead to other and more serious biases arising from the differences in the distribution of sample harvesting over time from that of actual harvesting by farmers. Thus, if the enumerator reaches too early he may find that he has sampled more of early maturing crop; if he is late he may have the preponderance of later maturing crop. Since the late maturing crop is generally a vigorous high-yielding crop, this will introduce bias in the estimate of yield. Mahalanobis gives a good example of the bias arising from this source. In one area, according to him, on average eight fields under the crop has to be examined to secure one field for sample harvesting. The difficulty is best summarised in his words, "After struggling with the problem for many years it is becoming clear that crop-cutting work to be done properly must be carried out by a comparatively larger number of investigators who could watch the crop as it grows and collect sample cuts at the right time from the fields situated in the neighbourhood of their normal place of residence". (Mahalanobis, 1946).

This difficulty can be avoided when harvesting is entrusted to the local departmental staff, as is done in the ICAR method (Sukhatme and Panse, 1951), for it is possible in this case to arrange harvesting of sample plot on the date the farmer would normally harvest his field and thereby ensure correspondence between the distribution of sample harvesting over time with that of actual harvesting of fields by farmers. On the other hand, large plots may present other difficulties. One such difficulty arises from over-sampling the centre of the field relatively more than in the case of small plots. The location of a plot at a random pair of coordinates (r,s) in a field implies that the central portion will have relatively higher probability of inclusion in the plot as compared with the areas near the border. The nearer the length and breadth of the plot to half the length and breadth respectively of the field, the more unequal will be the probabilities of selection of the central portion of the field compared to the border areas. In actual practice, the method of estimation followed is based on the assumption that all areas of the field are selected with the same probability. Thus, if L and B units represent the length and breadth of a field with a total number of L x B cells and a and b units represent the dimensions of the plot and if, further, it is assumed that the plot is located along the length and breadth of a field at a random pair of coordinates (r,s), then the method of selection assumes that the field is divided into (L - a + 1) x (B - b + 1) plots of size ab each giving equal probability of selection of 1/(L - a + 1) (B - b + 1) to each of the several overlapping plots. Since the frequency with which a given cell is included in a plot is different, depending upon the position of the cell and the dimensions of the plot and field, this results in assigning unequal probabilities for different cells of the field of being included in the plot. The average of these unequal probabilities for the cells included in the plot can be approximated by

$$P_u = \frac{f(r,s)}{a\,b\,(L - a + 1)(B - b + 1)}$$

where f(r,s) stands for the sum of frequencies for all the cells in the plot located with the pair of random numbers (r,s) and is given by

$$f(r,s) = S_a\,S_b$$

where $S_a = S(L, a, r)$

and $S_b = S(B, b, s)$

and the function S is defined in the tabular form below for positive integers L and a with $L \geq a$ and non-negative integer r:

Values of L and a	Values of r in the range $o \le r \le L - a$	Values of S(L, a, r)
$L > 3(a-1)$	$o \le r \le a - 2$	$a^2 - \frac{1}{4}(a-r-1)(a-r)$
	$a - 1 \le r \le L - 2a + 1$	a^2
	$L - 2a + 2 \le r \le L - a$	$a^2 - \frac{1}{2}(r-L+2a-1)(r-L+2a)$
$2(a-1) < L \le 3(a-1)$	$o \le r \le L - 2a + 1$	$a^2 - \frac{1}{2}(a-r-1)(a-r)$
	$L - 2a + 2 \le r \le a - 2$	$a^2 - \frac{1}{2}(a-r-1)(a-r)$ $-\frac{1}{2}(2a+r-L-1)(2a+r-L)$
	$a - 1 \le r \le L - a$	$a^2 - \frac{1}{2}(r-L+2a-1)(r-L+2a)$
$a \le L \le 2(a-1)$	$o \le r \le L - a$	$a(L-a+1) -\frac{1}{2}(L-a-r)(L-a-r+1)$ $-\frac{1}{2} r(r+1)$

In order to obtain an unbiased estimate of the yield rate, Aggarwal suggested dividing the yield of the plot by this average probability P_u and multiplying it by $\frac{ab}{LB}$ which is the probability of a cell being included in the plot on the assumption that equal chance was given to each one of the L x B cells of the field to be included in the plot.

In actual practice this adjustment was never used since it was found that the bias resulting from this source was negligible. We examined the distribution of yield according to the relative position of the plot in the field and found no evidence of any association between the average yield and the relative position of the plot in the field which can produce biases comparable with sampling errors. The point was examined in a large number of surveys with similar results (Suhatme and Panse 1951).

The main reason for mentioning this adjustment here is that it can be related to the Horvitz-Thompson estimator which is the only unbiased estimator in the class T_2 of linear estimators for estimating the total T of the population. The correction factor, which Aggarwal developed in 1944 long before Horvitz-Thompson estimator was known, remained unpublished, largely because he left for the USA and one of us left for Rome. In view of its importance, however, the original note developing the correction factor, and examining it in the light of the theory of sampling from finite populations with unequal probability as known today, is being reproduced elsewhere (I.J.A.S. 1968).

4. CONCLUDING REMARKS

To summarize, a review of the 1960 Censuses of Agriculture in the developing countries suggests that the quality of census results leaves a lot to be desired and that there would be little point in attempting to improve efficiency in sampling without ensuring greater control of non-sampling errors. In particular greater attention needs to be paid to

1) updating of the sampling frames, 2) involvement of farmers in the census efforts, intensive training of enumerators and provision of supervision and 3) use of objective methods of enumeration whenever feasible. All this implies much greater resources in men and money than are generally made available for censuses. In surveys with closed segments for collecting current statistics, biases appear relatively easy to control and sampling efficiency can be continually improved with relatively modest costs. However, planners need data of the census type which can acquaint them in quantitative terms with the structure and resources of agriculture, their state of use and potential. These data are needed to assist in reaching decisions on questions such as land reform, ceiling on farm-size, scope for supplementing employment in agriculture, etc. While a difficult task, a periodical census is the only means to provide the data they need. Even a broad picture with all its deficiencies may suffice in the early stages of economic development. Their needs for census data are so urgent that if censuses were not taken, they would put together all available information, however scrappy, subject it to the scrutiny of subject field experts and try to build a consistent picture of agriculture economy themselves. This poses a great challenge to the sampling statisticians. Sound planning depends as much on statistics as statistical development in its turn can benefit from planning. It is important therefore that planners and sampling statisticians come together in an effort to appreciate each others' needs and difficulties so that probability sampling may receive the support it needs from Governments.

5. ACKNOWLEDGMENT

The authors have drawn extensively on the reports of censuses and surveys made to Governments by FAO statisticians whose assistance is gratefully acknowledged. They also wish to thank their colleagues, particularly Drs. Simaika, Aggarwal, Alonzo and Zarkovic, who read through parts of the draft and made valuable comments.

REFERENCES

Lele, D. Y. (1968). "Pilot Survey for Agricultural Census — Jamaica", FAO.

Markovic, B. (1964). "Centre de Formation en Statistiques Agricoles et le Développement des Enquêtes Agricoles — Algeria", FAO.

Simaika, J. B. (1962). "Rapport au Gouvernement du Congo (Léopoldville) sur les Statistiques Agricoles" (unpublished).

Oliveira, A. J. (1963). "Le Recensement de l'Agriculture de 1961-1962 — Tunisia", FAO.

Farah, A. (1966). "Développement des Statistiques Agricoles Courantes — Senegal", FAO.

Fojtl, J. (1966). "Le Développement des Statistiques Agricoles — Tunisia", FAO,

Lele, D. Y. (1965). "Agricultural Census Project: Western Nigeria", FAO.

Morojele, C.M.H. (1963). "1960 Agricultural Census — Basutoland", Part 1, Census Methodology, Agricultural Department, Basutoland

Tang, P. C. (1960). "Censo Agropecuario y Estadisticas Agricolas Continuas Ecuador", FAO.

Tang, P. C. (1960). "Los Servicios de Estadisticas — Panama", FAO.

Koshal, R. S. (1954). "The Development of Sample Surveys for the Estimation of Agricultural Production — U.A.R.", FAO.

Thawani, V. D. (1963). "Pilot Census of Agriculture — Nepal", FAO.

Office National de Statistiques (1966). "Résultats du Recensement de l'Agriculture-élévage effectué le mars 1961", Vol. 1, Greece.

Dirección General de Estadistica y Censos (1965). "Censo Agropecuario de 1963 — Costa Rica."

Panse, V. G. (1966). "Some Problems of Agriculture Census Taking", FAO.

Ministry of Agriculture and Forestry. "The 1960 Census of Agriculture and Forestry in Japan", Japan.

U. S. Bureau of the Census (1962). "U.S. Census of Agriculture: 1959", General Report, Vol. II, Statistics by Subjects - Introduction.

Tsukibayashi, S. (1961). "The 1960 Census of Agriculture — Malaya", FAO.

Sattar, A. (1966). "The First Census of Agriculture - Indonesia", FAO.

National Statistical Office. "Census of Agriculture 1963", Thailand.

Committee on Census of Agriculture (1963). "General Report on the 1961 Census of Agriculture, Taiwan", China (Taiwan).

Bureau of the Census and Statistics (1965). "Census of the Philippines 1960, Agriculture", Philippines.

Agricultural Census Organization (1962-63). "1960 Pakistan Census of Agriculture", Vol. I, East Pakistan, Vol. II, West Pakistan, Pakistan.

Ministry of Agriculture (1962). "1960 Census of Agriculture - Report and Tables", Lybia.

State of Brunei (1966). "Report on the 1964 Census of Agriculture, Brunei.

Lomeli, M. (1965). "Censo Agropecuario y Desarrollo de las Estadisticas Agropecuarias continuas - Guatemala", FAO.

Toyoda, H. (1963). "Planning of 1960 Census of Agriculture - Syria", FAO.

Church, M. (1958). "Development of Agricultural Statistics - Ethiopia", FAO.

Trelogan, H. C. and Houseman, E. E. (1967). "Progress towards optimising agricultural area sampling", paper presented to 36th Session of the International Statistical Institute.

Sukhatme, P. V. and Kishen, K. (1951). "Assessment of the accuracy of Patwari's area records", Agriculture and Animal Husbandry, Vol. 1, No. 9.

Mahalanobis, P. C. (1946). "Recent experiments in statistical sampling", J. Roy. Stat. Soc., Vol. 109, pp. 326-378.

Sukhatme, P. V. and Panse, V. G. (1951). "Crop surveys in India", J. Ind. Soc. Agr. Stat., Vol. 3, No. 2, pp. 98-168.

Aggarwal, O. P. (1968). "On the biases in sampling a plot from a field", Ind. J. Agr. St. (in press).

Horwitz, D. G. and Thompson, D. J. (1952). "A generalisation of sampling without replacement", J. Amer. Stat. Assoc., Vol. 47, pp. 663-685.

I.C.A.R. (1950). Report on the pilot survey on Malabar coast for estimating the catch of marine fish, (unpublished).

Sukhatme, P. V., Panse, V. G. and Sastry, K. V. R. (1958). "Sampling technique for estimating catch of sea fish in India", Biometrics, Vol. 14, No. 1, pp. 78-96.

Table 1

Average size of operational agricultural holding in acres in India in recent surveys

Region	Survey I	Survey II	Survey III
North	3.9	5.3	6.7
North West	11.9	12.6	20.6
East	3.5	4.5	4.8
Central	10.9	12.2	14.6
West	11.1	12.3	14.9
South	3.8	4.5	6.2
All India	6.1	7.5	8.9

Source: V. G. Panse, 1958, I.S.I. Bulletin, Vol. 36, Part 4.

Table 2

Frequency Distribution of the Discrepancies in Patwari's Records Expressed as Percentage of the Acreage Noted by Statistical Staff Lucknow District (U.P.)

% Discrepancy	Frequency
-80 to -100%	3
-60 to - 80%	1
-40 to - 60%	8
-20 to - 40%	7
0 to - 20%	17
0	463
0 to 20%	19
20 to 40%	10
40 to 60%	2
60 to 80%	1
80 to 100%	2
> 100%	3
Total	536

SAMPLING FOR DEMOGRAPHIC VARIABLES

Nathan Keyfitz

University of Chicago

A registration system in which births and deaths are recorded for the entire population cannot be attained quickly by an underdeveloped country. The United States and Canada began to create such systems in the 1920's, and not until the 1930's and 1940's did registration become reasonably complete. Even today we need samples to check on and complement the registration system. Registration requires that everyone in the population be aware of the usefulness of a birth certificate, and such an awareness cannot be much faster in coming than the process of development itself.

Students of population know all this, and have laid emphasis on the need for sampling demographic facts, especially in the underdeveloped countries. But the emphasis has not been accompanied by a corresponding advance in theory for the design and interpretation of such samples. Existing sampling theory is of course available, but the facts which are required for the study of birth and death are interconnected in various ways unknown in labor force, opinion, industrial, or agricultural sampling. Demographers calculate parameters like the expectation of life and intrinsic rates of increase which are complex functions of the original observations. This paper will be concerned with the propagation of error in such functions.

VARIANCE OF LIFE TABLE CHARACTERISTICS

A life table is ordinarily printed in about ten columns, of which one is the probability of dying before he reaches age x + 1 for a person who has just attained his $\underline{x}$th birthday. This column may be written as the vector $\{\underset{\sim}{q}\}$, with braces to remind us that it is vertical:

$$\{\underset{\sim}{q}\} = \begin{Bmatrix} q_0 \\ q_1 \\ \cdot \\ \cdot \\ q_\omega \end{Bmatrix} .$$

Binomial Variates

We will suppose that the interval is one year. Formulas for five years, or some other length of time, may be obtained by modifying the results below in an obvious way.

If q_x is a fraction obtained by the observation of ℓ_x individuals at their $\underline{x}$th birthday, and then noting that d_x of them died during the following 12 months, so that $q_x = d_x/\ell_x$, then q_x is a binomial variable in successive samplings. If the true probability of dying in the following year for a person at exact age $\underline{x}$ is $1 - \lambda_{x+1}/\lambda_x$, then the variance of q_x is

$$\text{Var } q_x = \ell_x\left(1 - \frac{\lambda_{x+1}}{\lambda_x}\right)\left(\frac{\lambda_{x+1}}{\lambda_x}\right) .$$

If the individuals in the sample are drawn independently at random, then the correlation of q_x for different ages will be zero (even though, as Chiang (1960, p.628) shows, the q_x for different ages are not quite independently distributed). Hence (setting $\lambda_0 = 1$) we will have the variance-covariance matrix

$$
\underset{\sim}{V}(\underset{\sim}{q}) = \begin{bmatrix} \ell_0(1-\lambda_1)(\lambda_1) & 0 & 0 & \cdots \\ 0 & \ell_1\left(1-\dfrac{\lambda_2}{\lambda_1}\right)\left(\dfrac{\lambda_2}{\lambda_1}\right) & 0 & \cdots \\ \cdot & & & \cdot \\ \cdot & & & \cdot \\ \cdot & & & \cdot \\ 0 & 0 & 0 & \cdots \end{bmatrix}
$$

whose elements may be called v_{xy}.

Next we have the column for the number of survivors to age x, as obtained from the sample and written ℓ_x. In our notation the ℓ_0 will not be merely an arbitrary radix of the life table, but will be the initial number in the sample from a given cohort, a number of babies born at the same time and observed through life. The number surviving to age $\underline{x}$, ℓ_x in the sample, has the expected value $\ell_0\lambda_x$. Again we are dealing with a binomial variate, and the variance of ℓ_x is

$$
\text{Var } \ell_x = (\ell_0)(\lambda_x)(1 - \lambda_x).
$$

But unlike the q_x, the successive values of ℓ_x cannot be independent of one another. If ℓ_x is high in the amount $\Delta\ell_x = \ell_x - E(\ell_x)$, by the luck of sampling, then ℓ_y, $y > x$, will also tend to be high, one would expect by the amount $\Delta\ell_x(\lambda_y/\lambda_x)$.

The variance of ℓ_y given ℓ_x is binomial and equal to $\ell_x(\lambda_y/\lambda_x)(1 - \lambda_y/\lambda_x)$. The unconditional variance of ℓ_y is $\ell_0(\lambda_y)(1 - \lambda_y)$. The ratio of the conditional to the unconditional variance gives one minus the square of the coefficient of correlation $\underline{r}$. Hence

$$
r^2 = 1 - \frac{\ell_x(\lambda_y/\lambda_x)(1 - \lambda_y/\lambda_x)}{\ell_0(\lambda_y)(1 - \lambda_y)} = \frac{\lambda_y(1 - \lambda_x)}{\lambda_x(1 - \lambda_y)},
$$

where we have approximated by entering λ_x for ℓ_x/ℓ_0, which with samples of useful size cannot be very different. The covariance of ℓ_x and ℓ_y is this multiplied by the product of variances and the square root taken, or $\ell_0\lambda_y(1 - \lambda_x)$, if $Y \geq x$. If $x \geq y$, on the other hand, the covariance is $\ell_0\lambda_x(1 - \lambda\ )$. This makes the variance-covariance matrix symmetrical about its main diagonal:

$$\underset{\sim}{V}(\underset{\sim}{\ell}) = \ell_0 \begin{bmatrix} 0 & 0 & 0 & 0 & 0 & \cdots \\ 0 & \lambda_1(1-\lambda_1) & \lambda_2(1-\lambda_1) & \lambda_3(1-\lambda_1) & \lambda_4(1-\lambda_1) & \cdots \\ 0 & \lambda_2(1-\lambda_1) & \lambda_2(1-\lambda_2) & \lambda_3(1-\lambda_2) & \lambda_4(1-\lambda_2) & \cdots \\ 0 & \lambda_3(1-\lambda_1) & \lambda_3(1-\lambda_2) & \lambda_3(1-\lambda_3) & \lambda_4(1-\lambda_3) & \cdots \\ 0 & \lambda_4(1-\lambda_1) & \lambda_4(1-\lambda_2) & \lambda_4(1-\lambda_3) & \lambda_4(1-\lambda_4) & \cdots \\ \vdots & & & & & \vdots \end{bmatrix}$$

An Example of the Delta Method

The expectation of life is defined as

$$\mathring{e}_x = \frac{T_x}{\ell_x} = \frac{\int_0^{\omega - x} \ell(x+t)\,dt}{\ell_x} ,$$

and we can ascertain its variance by considering how it changes with a small change in q_z, $z > x$. We have identically

$$\mathring{e}_x = \frac{T_x}{\ell_x} = \frac{T_x - T_z + L_z + T_{z+1}}{\ell_x}$$

$$= \frac{T_x - T_z}{\ell_x} + \frac{\ell_z}{\ell_x}.\frac{L_z}{\ell_z} + \frac{\ell_{z+1}}{\ell_x}.\frac{T_{z+1}}{\ell_{z+1}} . \qquad (1)$$

Of the three terms the first, the expected years lived between $\underline{x}$ and $\underline{z}$, is unaffected by a change in q_z. The second is the expected fraction of a year lived between z and z + 1, which may be decomposed into the a_z of a year lived by those dying in the interval and the 1 year by those surviving:

$$\frac{L_z}{\ell_z} = \frac{a_z d_z + \ell_{z+1}}{\ell_z} = a_z q_z + (1 - q_z)$$

$$= 1 - (1 - a_z)q_z \ .$$

The last term of (1) is $(1 - q_z)\mathring{e}_{z+1}$. The variation of e_x in response to a small departure q_z from $e(q_z)$, where $E(\Delta q_z) = 0$, is given by the derivative

$$\frac{\partial \mathring{e}_x}{\partial q_z} = -(1 - a_z + \mathring{e}_{z+1}).$$

We need this for all $z \geq x$:

$$\Delta \mathring{e}_x = \sum_{z \geq x}[-(1 - a_z + e_{z+1})\ \Delta q_z], \qquad (2)$$

and square and average over all samples. If the q_z are independent, then

$$v_x^2 = \text{Var}\ \mathring{e}_x = \sum_{z \geq x}(1 - a_z + \mathring{e}_{z+1})^2\ \text{Var}\ q_z,$$

a result due to Wilson (1938), Irwin (1949), and Chiang (1960).

To complete the variance-covariance matrix for $\mathring{e}_x$ we require $E(\Delta\mathring{e}_x\ \Delta\mathring{e}_y)$. In respect of an age _z_ such that $z \geq y \geq x$, the form of (2) shows that the contribution to covariance is the same as to variance. Hence the typical element of the variance-covariance matrix is

$$v_{x,y} = \sum_{z \geq \max(x,y)} (1 - a_z + \mathring{e}_{z+1})^2 \text{ Var } q_z,$$

the summation being for all z not less than max(x,y), the greater of x and y. In fact,

$$v_{x,y} = v_y^2, \quad x \leq y,$$
$$v_{x,y} = v_x^2, \quad x \geq y.$$

From this the matrix of intercorrelations can be written down. The value of $r_{x,y}$, the correlation of the expectation at age x with that at age y, is

$$r_{x,y} = \frac{\sum_{z \geq \max(x,y)} (1 - a_z + \mathring{e}_{z+1})^2 \text{ Var } q_z}{\sqrt{A\ B}}$$

where

$$A = \sum_{z \geq x} (1 - a_z - \mathring{e}_{z+1})^2 \text{ Var } q_z$$

$$B = \sum_{z \geq y} (1 - a_z + \mathring{e}_{z+1})^2 \text{ Var } q_z$$

which may be expressed in terms of the standard deviations v_x of $\mathring{e}_x$ at individual ages. For any x,y it is

$$r_{x,y} = \frac{v_{\max(x,y)}^2}{v_x v_y} = \frac{v_y}{v_x}, \text{ if } y \geq x;$$

$$r_{x,y} = \frac{v_x}{v_y}, \text{ if } x \geq y.$$

To obtain corresponding results by simulation requires only the programming of a life table in which probabilities are applied to individuals rather than collectively to the entire ℓ_x remaining alive at each age. Starting with 100 males, and working with the male q_x of the United States in 1964, gives the variances and covariances shown in Table 1. Below each figure, in parentheses, is the estimate

$$r_{x,y} = v_y/v_x, \; y \geq x.$$

The ratio, for example 0.727/0.594 for $r_{40,60}$, may be compared with $\chi^2/99$ with 99 degrees of freedom, and has a probability of 0.063 of being exceeded.

TABLE 1. Correlation coefficient r between $\mathring{e}_x$ and $\mathring{e}_y$ resulting from simulation starting with 100 males, and using q_x for males in United States, 1964. Below each figure, in parentheses, is the value of r from $r_{x,y} = v_y/v_x$, $y \geq x$, and $r_{x,y} = v_x/v_y$, $x \geq y$.

	$\mathring{e}_{40}$	$\mathring{e}_{60}$	$\mathring{e}_{80}$
$\mathring{e}_{20}$	0.850 (0.773)	0.575 (0.459)	0.243 (0.169)
$\mathring{e}_{40}$	1.000 (1.000)	0.727 (0.594)	0.229 (0.218)
$\mathring{e}_{60}$	0.727 (0.594)	1.000 (1.000)	0.302 (0.367)

The principal shortcoming of the above results lies in the correlation among q_x that would result from cluster sampling. If we know the design--in particular the arrangement of strata and clusters--we can calculate the variance v_x^2 of each q_x. To obtain the covariance of q_x and q_y we will need the $r_{x,y}$. Finding these from the sample would require complicated procedures, but some shortcuts suggest themselves.

For example, the r's may be established as a function of the distance by which the ages are separated. William Cummings made 83 life tables for developed countries since 1945, and worked out correlations among all pairs of q_x. He found that for ages five years apart the correlation was between 0.94 and 0.97; for ages 10 years apart, between 0.85 and 0.90, etc. Empirically the correlation seems not far from

$$r_{x,y} = 1 - 0.015(y - x),\ y \geq x.$$

This kind of relation could be superimposed on the design effect in the sample to provide rough covariances among ages, and so the preceding results would become applicable to clustered samples.

FUNCTIONS OF FERTILITY AND MORTALITY

The Net Reproduction Rate R_0

To consider the simultaneous effects of birth and death on a population we start with the net reproduction rate R_0, defined as the expected number of girl children by which a girl child just born will be replaced, given the life table for females and the age-specific fertility rates for girl babies. (The definition for males may be set forth in the same way.) If the births of girl children to women of age x last birthday are B_x, and the number of such women is K_x, then the age-specific birth rates are $F_x = B_x/K_x$. The chance that a girl child just born will be alive through the year x to $x + 1$ is L_x/ℓ_0 (strictly the expected fraction of the year she will live). If she does live, the chance of her having a child is F_x. Multiplying these two values and adding through all ages gives the net reproduction rate

$$R_0 = \frac{\sum_x L_x F_x}{\ell_0} .$$

The delta-method again serves for the sampling variation:

$$\text{Var } R_0 = E\left\{\sum_x\left(\frac{L_x}{\ell_0}\Delta F_x + F_x \Delta \frac{L_x}{\ell_0}\right)\right\}^2 .$$

Expanding the square would provide the variance of R_0 in terms of the variances of F_x, L_x/ℓ_0, and the three kinds of possible covariance. If the samples for mortality and fertility have been drawn independently the covariance between F_x and L_x/ℓ_0 would disappear. If the sample for fertility was of individuals drawn separately at random, then the covariance of F_x and F_y would be zero, and correspondingly for mortality. With all three conditions the variance would be

$$\text{Var } R_0 = \sum_x (L_x/\ell_0)^2 \text{Var } F_x + \sum_x F_x^2 \text{ Var}(L_x/\ell_0).$$

If in addition mortality were subject to negligible error the second term would drop out and Koop's (1951) result would appear (though without the finite population correction).

The Intrinsic Rate of Natural Increase

The quantity R_0 may be thought of as the ratio of population in one generation to that in the previous generation. If the rate of increase is r, on an annual basis but compounded momently, and the length of generation is T, then we have

$$e^{rT} = R_0, \text{ or } r = \frac{\ln R_0}{T}.$$

The variation of T would be approximately that of the mean age of childbearing, which in human populations has a standard deviation σ of 5 to 7 years. Hence the standard deviation of T from a sample of size n would be $\sigma/\sqrt{n} \leq 7/\sqrt{n}$. The disturbance of r would be found as

$$\Delta r = \Delta\left(\frac{\ln R_0}{T}\right) = \frac{\Delta R_0}{R_0 T} - \frac{r\Delta T}{T},$$

and the variance of r is the square of this averaged over all samples. If the correlation of R_0 and T among samples is positive but small, the variance is

$$\text{Var } r = \frac{\text{Var } R_0}{(R_0 T)^2} + \frac{r^2}{T^2} \text{Var } T,$$

including a slight overstatement.

Approximations

Quantities such as *r* and T are defined in terms of the life table and the age-specific fertility rates. However, to each corresponds a function of the observed population: to *r* corresponds the annual rate of increase as ascertained from two censuses or from the difference between crude rates of birth and death; to T corresponds the mean age of childbearing in the observed population. We would not want to confuse *r* as the intrinsic rate of natural increase with *r* as the observed increase, but the variances of the two might nonetheless be similar. Experimenting is now going forward to ascertain how close the sampling variance of *r* is to that of the simple population increase, and how close the sampling variance of T is to that of the observed mean age of childbearing.

An alternative approach is to divide the sample into independent portions; calculate the value of the desired parameter from each portion; estimate variance from the several calculations of the parameter. Ways of doing this have been described by W. E. Deming (1956), M. H. Hansen and his staff (United States Bureau of the Census, 1963), N. Keyfitz (1957), L. Kish (1968), and Frederick Mosteller (1968).

VARIANCE REDUCTION BY CURVE FITTING

Without pretending to have completed this discussion, I shall move on to the question of how much can be added to sampling efficiency by parameterization. Standard sampling theory is built on neutral ground; we do not know the form of the curves--for instance of the distribution by age of the unemployed--and samples usually do not suppose such knowledge. Given the published results, anyone can apply whatever curves he wishes, or none at all. But if we could agree on a form of curve with fewer parameters than the number of published age groups, then the same data give us more information.

Age Distribution Fitted by an Exponential Function

Consider as an example the estimation of age distribution for a country or a city. Suppose that the population has been increasing rapidly for some time, and that the death rate is low. This combination of circumstances is to be seen in many parts of the underdeveloped world, and it encourages fitting to the observed age distribution a stable-population model in which the death rate is the same at all ages. The appropriate function is then

$$c(a)da = \lambda e^{-\lambda a} da, \tag{3}$$

where $c(a)da$ is the fraction of the population aged between a and $a + da$, and λ is a constant to be inferred from the data. We are supposing zero migration.

If a random sample of n individuals of one sex has been selected and everyone asked his age, the answers are $a_1, a_2, \ldots, a_n$. From (3) we can say that the likelihood of having obtained these particular results if the true parameter were λ would be

$$L = (\lambda e^{-\lambda a_1} da_1)(\lambda e^{-\lambda a_2} da_2)\ldots(\lambda e^{-\lambda a_n} da_n).$$

The value of λ that maximizes L is the same as that which maximizes the logarithm of L. Thus

$$\frac{\partial \ln L}{\partial \lambda} = \frac{\partial}{\partial \lambda}\left(n \ln \lambda - \lambda \sum_{i=1}^{n} a_i\right) = \frac{n}{\lambda} - \sum_{i=1}^{n} a_i$$

is the quantity to be equated to zero for the desired value of λ. Solving the equation gives

$$\lambda = \frac{n}{\Sigma a}, \tag{4}$$

or the reciprocal of the average age.

The information contained in the estimator (4) is in turn estimated by taking the second derivative of the logarithm of the likelihood with respect to λ, and entering the value of λ from (4):

$$I = -\frac{\partial^2 \ln L}{\partial \lambda^2} = \frac{n}{\lambda^2} = \frac{n}{(n/\Sigma a)^2}$$

The variance of the estimate of λ is the reciprocal of this, or

$$\mathrm{Var}(\lambda) = \frac{1}{I} = \frac{(n/\Sigma a)^2}{n} . \qquad (5)$$

Table 2 illustrates the use of the procedure for fitting to the numbers of Singapore females by age as they were officially published for 1962. The value of the constant λ, the reciprocal of the average age of the population, is

$$\lambda = \frac{1}{22.36} = 0.04472.$$

The fit of the age distribution as calculated by (3) is reasonably good.

TABLE 2.--Age distribution of Singapore females, 1962, showing observed and calculated on simple exponential with $\lambda = 0.04472$

x to x + 4	Observed percentage	Calculated on $\lambda \int_x^{x+5} e^{-\lambda a} da$	Departure of calculated from observed percentage
0-4	18.5	20.0	1.5
5-9	15.4	16.0	0.6
10-14	12.9	12.8	-0.1
15-19	7.9	10.2	2.3
20-24	8.0	8.2	0.2
25-29	7.1	6.5	-0.6
30-34	6.4	5.2	-1.2
35-39	4.8	4.2	-0.6
40-44	4.3	3.3	-1.0
45-49	4.0	2.7	-1.3
50-54	3.4	2.1	-1.3
55-59	2.6	1.7	-0.9
60-64	1.9	1.4	-0.5
65+	2.7	5.6	2.9
			$\Delta = 7.5$

Sampling Variance of Estimates from the Fitting

All this was worked out on a complete census of Singapore. What could have been done with a sample? We see from (5) that a sample of $\underline{n}$ individuals would estimate λ with a standard deviation of $\lambda/\sqrt{n}$. With 100 individuals chosen at a random this would be $\lambda/10 =$ 0.004; a sample of 10,000 would have a standard deviation one tenth of this.

What is the error in the estimate for any particular age group when λ is in error by $\Delta\lambda$? If

$$c(a) = \lambda e^{-\lambda a},$$

then the effect on c(a) of a small perturbation of λ is

$$\Delta c(a) = (e^{-\lambda a})(1 - \lambda a)\Delta\lambda ,$$

and squaring and averaging, gives for any age x

$$\text{Var}[c(a)] = e^{-2\lambda a}(1 - \lambda a)^2 \text{ Var } \lambda ,$$

$$V^2[c(a)] = (1 - \lambda a)^2 V^2(\lambda),$$

where V^2 stands for the rel-variance (Hansen, Hurwitz, and Madow (1953), p. 51) or ratio of the variance to the square of the expected value. The square root of V is the coefficient of variation, and apparently the coefficient of variation of the proportion at age a is $(1 - \lambda a)$ times the coefficient of variation of λ. For the pivotal age $a = 1/\lambda$, this is zero; for ages from about zero to $2/\lambda$, i.e. typically from zero to 40 or 50--about half of the ages of life--the rel-variance of the proportion at age a is less than the rel-variance with which λ is estimated.

The advantage of parameterization is seen in this last statement. For if, per unit observation, the rel-variance of c(a) is about the same as that of λ, then we can make large gains by pooling all ages to estimate λ, and then obtaining the ages from λ by (3). With twenty age groups we can say with only slight exaggeration that for a given sample the parameterization enables us to bring 20 times as much information to bear on each age group.

To discriminate among populations is possible with very small samples. The average age of the 1962 female population of the United States was 32.33, and of Singapore 22.36. Whether one were dealing with an unlabelled sample of individuals from the United States or from Singapore could be decided if the sample had only 25 individuals, provided the exponential curve (3) really fits, and with some practical insurance against the risks of our curve fitting with a sample of 100 individuals.

The dimension of λ is the same as that of an annual rate. If the population is stable, and if it has a death rate μ which is the same for all ages and an intrinsic rate of natural increase $\underline{r}$, then λ = $r + \mu$. Our λ is thus an estimate, made from the age distribution alone, of the intrinsic birth rate. Though we did not set out to calculate the intrinsic birth rate, yet the value of λ is only 0.002 or 5 percent in excess of the intrinsic birth rate of 0.0427, established by the wholly different set of data consisting of registered births and deaths of Singapore in 1962.

REFERENCES

Chiang, C. L., "A stochastic study of the life table and its applications: I. Probability distributions of the biometric functions", Biometrics, XVI (1960), 618-635.

Deming, W. Edwards, "On simplifications of sampling design through replication with equal probabilities and without stages", Journal of the American Statistical Association, LI (1956), 24-53.

Fisher, R. A., "On the mathematical foundations of theoretical statistics", Philosophical Transactions of the Royal Society, CCXXII (1921), 309.

Hansen, M. H., W. N. Hurwitz, and W. G. Madow, Sample Survey Methods and Theory, Vol. 2. New York: John Wiley & Sons, 1953.

Hauser, Philip M., "The use of sampling for vital registration and vital statistics", Bulletin of the World Health Organization, XI (1954), 5-24.

Irwin, J. O., "The standard error of an estimate of expectational life", Journal of Hygiene, XLVII (1949), 188-189.

Keyfitz, N., "Estimates of sampling variance where two units are selected from each stratum", Journal of the American Statistical Association, LII (1957), 503-510.

_______, and William Cummings, "Statistical properties of collections of life tables", 1967, unpublished.

Kish, Leslie, "Standard Errors for Indexes from Complex Samples", Journal of the American Statistical Association, LXIII (1968), 512-529.

Koop, J. C., "Notes on the estimation of gross and net reproduction rates by methods of statistical sampling", Biometrics, VII (1951), 155-166.

Lotka, A. J., "Orphanhood in relation to demographic factors", Metron, IX (1931), 37-109.

Mosteller, Frederick, "Nonsampling errors", International Encyclopedia of the Social Sciences, Vol. 5, 113-131.

Pollard, J. H., "On the use of the direct matrix product in analysing certain stochastic population models", Biometrika, LIII (1966), 397-415.

U. S. Bureau of the Census. The Current Population Survey: A Report on Methodology. Technical Paper No. 7. Washington, D. C.: U. S. Government Printing Office, 1963.

Wilson, E. B., "The standard deviation of sampling for life expectancy", Journal of the American Statistical Association, XXXIII (1938), 705-708.

TESTS OF INDEPENDENCE IN CONTINGENCY TABLES FROM STRATIFIED SAMPLES*

Gad Nathan

Hebrew University and Central Bureau of Statistics

Jerusalem, Israel

1. Introduction and Summary

The standard methods for testing independence between two qualitative variables on the basis of contingency-table sample data (e.g. Pearson's chi-square or Wilks's likelihood ratio test) are based on the assumption that the contingency table is a classification, according to the variables, of a simple random sample. In many practical uses this is far from being the case and in actual fact it is often necessary to infer on independence on the basis of samples of complex design, such as are actually used in sample survey work.

A case in point is that of a stratified sample, with simple random sampling within each stratum and independent selection in different strata. The analysis often carried out is based on adding the relevant frequencies from all strata and treating the resulting frequencies as if obtained from simple random sampling. This is obviously not valid, even with equal sampling fractions in all strata. An alternative approach suggested by Fisher (3) and Kendall (10), is based on statistics with chi-square distributions under the null hypothesis of independence within each stratum. These are added to provide a statistic which can test the null hypothesis of independence between the two characteristics within all strata simultaneously. An alternative approach to testing this hypothesis for a special case is proposed by Armitage (1). This may

* Partially supported by National Center for Health Statistics grant NCHS-IS-1.

well be the hypothesis to be tested when interest is limited to the conditional independence of the two classification characteristics for a given level of the stratification variable.

In many cases, however, the stratification variable is not of primary interest and may only be a technical device used in the design of the sample. The hypothesis of interest is then overall independence of the two characteristics without regard to stratification. It is easy to see that overall independence and conditional independence within strata are different hypotheses, neither of which follow, in general, from the other. Thus, dependence within strata may be in different directions and may cancel out so as to obtain overall independence.

The problem of overall independence can also be viewed as a problem of homogeneity across domains of study in an analytical survey (see, e.g. Hartley (8) and McCarthy(12)). In this case one of the characteristics divides the population into domains of study and the hypothesis states that the conditional distributions with respect to the other characteristic are identical within all domains of study, again without reference to the sample stratification.

In the following it will be shown that by using the exact basic likelihood function of all frequencies, maximum likelihood estimates can be approximated as closely as necessary, by an iterative procedure, for all the parameters involved (i.e. probabilities within each stratum) both in the general and in the hypothesis space. Thus, the likelihood ratio test statistic can be approximated to any required degree of accuracy and used to test the hypothesis of overall independence, its asymptotic distribution being chi-square.

Bhapkar (2) has shown that Neyman's technique of linearization provides a simple expression for the minimum χ_1^2 statistic which can be used to give an approximate test of the nonlinear hypothesis of independence. It is shown below that the values of the parameters at which the χ_1^2 statistic is minimized are exactly the first step of the iterative approximation to the maximum likelihood estimates under the hypothesis. On the other hand, Bhapkar has pointed out that the χ_1^2 statistic thus obtained is equivalent to the large sample test statistic based on the asymptotic normality of unbiased estimates of the linearized hypothesis functions.

Garza-Hernandez and McCarthy (6) and (7) have proposed a test of overall independence, based on a sample ratio estimate of the overall conditional probability of each category of one characteristic, given the category of the other (domain of study). Assuming multivariate normality of these estimates, a maximum likelihood ratio test can be approximated. Again it can be shown that the statistic thus obtained is equivalent to the large-sample test statistic based on estimates of the non-linear hypothesis functions.

Marascuilo and McSweeney (11) use a similar approach to make multiple comparisons in analytical surveys; however in their case the domains of study are defined as strata. In section five it is shown that Gabriel's (4) method of simultaneous inference on all hypotheses of independence between subsets of categories of each characteristic can easily be extended to the case of a stratified random sample. Again this is done by approximating the maximum likelihood ratio for each combination of subsets.

Finally, comparisons of the results of the various methods are made for a numerical example.

2. The Underlying Model, Probability Distributions and Hypotheses

The population of interest is assumed to be divided into t strata and within each stratum classified by a double classification with r and s categories, respectively. Let the joint probability that an item is in stratum t and is classified in the (i,j)-th cell be P_{ijk} (i=1,...,r; j=1,...,s; k=1,...,t). Let the marginal probabilities be:

$$P_{ij.} = \sum_k P_{ijk};\ P_{i.k} = \sum_j P_{ijk};\ P_{.jk} = \sum_i P_{ijk}\ ;$$
$$P_{i..} = \sum_{j,k} P_{ijk};\ P_{.j.} = \sum_{i,k} P_{ijk};\ P_{..k} = \sum_{i,j} P_{ijk}, \tag{1}$$

with:

$$\sum_{i,j,k} P_{ijk} = 1\ . \tag{2}$$

The marginal probabilities of inclusion in strata are assumed known:

$$P_{..k} = P_k \tag{3}$$

In the case of a finite population of size N with strata sizes N_k (k=1,...,t), these would simply be the proportion of elements in each stratum, i.e.:

$$P_k = N_k/N \qquad (k=1,\ldots,t). \tag{4}$$

A stratified random sample of size n is selected with a fixed number of elements, n_k, drawn from the k-th stratum (k=1,...,t). Let n_{ijk} be the number of elements classified in the (i,j)-th cell in the k-th stratum and let the marginal frequencies be denoted, as usual, by $n_{i.k}$, $n_{.jk}$, $n_{ij.}$, $n_{i..}$, $n_{.j.}$, $n_{..k}$, where the marginal strata totals are assumed to be fixed, i.e.,

$$n_{..k} = n_k\ . \tag{5}$$

The t random variable vectors $\{n_{ijk}: i=1,\ldots,r;\ j=1,\ldots,s\}$, k=1,...,t, are distributed independently. If the population in each stratum is infinite or if selection within strata is simple random with replacement, the n_{ijk} are multinomially distributed within each stratum with parameters n_k and probabilities

$$\{\frac{P_{ijk}}{P_k} ;\ i=1,\ldots,r \quad j=1,\ldots,s\},\ \text{i.e.:}$$

$$\Pr\{n_{ijk};\ i=1,\ldots,r;\ j=1,\ldots,s\} = \frac{n_k!}{\prod_{i,j} n_{ijk}!} \prod_{i,j} \left(\frac{P_{ijk}}{P_k}\right)^{n_{ijk}} . \tag{6}$$

The likelihood of the sample is thus given by:

$$L(\{n_{ijk}\}|\{P_{ijk}\}) = \prod_k \left[\frac{n_k!}{\prod_{i,j} n_{ijk}!} \prod_{i,j} \left(\frac{P_{ijk}}{P_k}\right)^{n_{ijk}}\right]. \tag{7}$$

In the case of sampling without replacement from a finite population, let N_{ijk} be the number of elements in cell (i,j) of stratum k (i.e. $P_{ijk} = N_{ijk}/N$). The joint distribution is then hypergeometric, with probability function:

$$\Pr\{n_{ijk};\ i=1,\ldots,r;\ j=1,\ldots,s\} \frac{\prod_{i,j} \binom{N_{ijk}}{n_{ijk}}}{\binom{N_k}{n_k}} . \tag{8}$$

For fixed values of n_k, n_{ijk} and of

$$\frac{N_{ijk}}{N_k} = \frac{P_{ijk}}{P_k} ,$$

(8) tends to (6) as N_k tends to infinity. Thus for small samples from large populations (i.e., when finite factors can be ignored) the likelihood is closely approximated by (7), which can be used even in the case of sampling without replacement.

The hypothesis to be tested of overall independence between the two classifications, without regard to strata, can be formulated as:

$$H_o : P_{ij.} = P_{i..} P_{.j.} \quad (i=1,\ldots,r; 1,\ldots,s). (9)$$

This is to be tested against the alternative of inequality for some pair (i,j). If one of the classifications, say the one denoted by index i, is by domain of study, the null hypothesis (9) is formulated as:

$$H_o: \frac{P_{1j.}}{P_{1..}} = \frac{P_{2j.}}{P_{2..}} = \ldots = \frac{P_{rj.}}{P_{r..}} (= P_{.j.}) \quad (j=1,\ldots,s), (10)$$

i.e., that the conditional probability of having characteristic j (given the domain of study) is constant over all domains of study. This is the formulation used by Garza-Hernandez (6).

It should be noted that the test usually used to test independence in this case as proposed by Fisher (3) or Kendall (10) is based on tests of independence within strata and thus tests the hypothesis:

$$H_o' : \frac{P_{ijk}}{P_k} = \frac{P_{i.k}}{P_k} \cdot \frac{P_{.jk}}{P_k} \quad (i=1,\ldots,r; j=1,\ldots,s; k=1,\ldots,t), \quad (11)$$

or alternatively:

$$H_o': \frac{P_{1jk}}{P_{1.k}} = \frac{P_{2jk}}{P_{2.k}} = \ldots = \frac{P_{rjk}}{P_{r.k}} (=P_{.jk}/P_k) (j=1,\ldots,s; \quad (12)$$
$$k=1,\ldots,t).$$

This is not in general equivalent to H_o[(9) or (10)], except in special cases. It is readily seen that a necessary and sufficient condition for equivalence is

$$P_{ijk} = \frac{P_{ij.} \, P_{i.k} \, P_{.jk}}{P_{i..} \, P_{.j.} \, P_{.k.}}$$

for all i,j,k. Thus, in general, neither does H_o imply H_o' nor does H_o' imply H_o, so that different tests should be used to test these two hypotheses.

The general parameter space is:

$$\Omega = \{P_{ijk} | \sum_{i,j} P_{ijk} = P_k;\ P_{ijk} \geq 0\} \tag{13}$$

of dimension rst-t = (rs-1)t.

The parameter space for the hypothesis H_o, defined by (9) or (10) is:

$$\omega = \{P_{ijk} | \sum_{i,j} P_{ijk} = P_k;\ \sum_k P_{ijk} = (\sum_{x,k} P_{xjk})(\sum_{y,k} P_{iyk});\ P_{ijk} \geq 0\}, \tag{14}$$

which is of dimension rst-t-(r-1)(s-1)=(rs-1)t-(r-1)(s-1).

3. The Maximum Likelihood Estimates and Likelihood Ratio Test Statistic

Maximum likelihood estimates for P_{ijk} in Ω are obtained by solving the equations:

$$\frac{\partial[\ln L - \sum_z \lambda_z(\sum_{x,y} P_{xyz} - P_z)]}{\partial P_{ijk}} \equiv \frac{n_{ijk}}{P_{ijk}} - \lambda_k = 0 \tag{15}$$

$$(i=1,\ldots,r;\ j=1,\ldots,s;\ k=1,\ldots,t)$$

and

$$\sum_{i,j} P_{ijk} = P_k \qquad (k=1,\ldots,t), \tag{16}$$

where λ_k are appropriate Lagrange multipliers. This gives, as might be expected,

$$\hat{P}_{ijk} = n_{ijk} \frac{P_k}{n_k} \quad . \tag{17}$$

To obtain maximum likelihood estimates in ω, under the hypothesis, we set:

$$\frac{\partial}{\partial P_{ijk}} [\ln L - \sum_{z} \lambda_z (\sum_{x,y} P_{xyz} - P_z) -$$

$$\sum_{x,y} \mu_{xy} (\sum_{z} P_{xyz} - \sum_{u,z} P_{uyz} \sum_{v,z} P_{xvz})] = 0 \tag{18}$$

$$(i=1,\ldots,; \ j=1,\ldots,s; \ k=1,\ldots,t)$$

Where λ_k, μ_{ij} are appropriate Lagrange multipliers, with $\mu_{ij} = 0$ for $i=r$ or for $j=s$, as only $(r-1)(s-1)$ of the additional conditions are independent. The resulting equations, to be solved for P_{ijk}, λ_k, μ_{ij}, are as follows:

$$\frac{n_{ijk}}{P_{ijk}} = \lambda_k + \mu_{ij} - \sum_{x,y,z} (\mu_{iy} + \mu_{xj}) P_{xyz}$$

$$(i=1,\ldots,r; \ j=1,\ldots,s; \ k=1,\ldots,t),$$

$$\sum_{i,j} P_{ijk} = P_k \qquad (k=1,\ldots,t)$$

and

$$\sum_{k} P_{ijk} = \sum_{x} P_{xjk} \sum_{y} P_{iyk} (i=1,\ldots,r-1; \ j=1,\ldots,s-1). \tag{19}$$

This set of rst+t+(r-1)(s-1) equations is non-linear but can be solved by numerical methods. Thus the Newton-Raphson method can be employed as follows. Let $\underline{X}$ be the M-dimensional vector of unknowns, where M=rst+t+(r-1)(s-1):

$$\underline{X} = (P_{111},\ldots,P_{rst}, \lambda_1,\ldots,\lambda_t, \mu_{11},\ldots,\mu_{r-1,s-1}) \tag{20}$$

Let $\underline{\Phi}(\underline{X})$ be the M-variate M-dimensional vector function defined by

$$\underline{\Phi} = (F_{111},\ldots,F_{rst}, G_1,\ldots,G_t, H_{11},\ldots,H_{r-1,s-1}) \tag{21}$$

where F_{ijk}, G_k, H_{ij} are single-valued M-variate functions defined by:

$$F_{ijk}(\underline{X}) = \lambda_k + \mu_{ij} - \sum_{x,y,z} (\mu_{iy} + \mu_{xj})P_{xyz} - n_{ijk}/P_{ijk}$$

$$(i=1,\ldots,r;\ j=1,\ldots,s;\ k=1,\ldots,t)$$

$$G_k(\underline{X}) = \sum_{x,y} P_{xyk} - P_k \qquad (k=1,\ldots,t)$$

$$H_{ij}(\underline{X}) = \sum_{z} P_{ijz} - \sum_{y,x} P_{iyz} \sum_{xz} P_{xjz} \tag{22}$$

$$(i=1,\ldots,r-1;\ j=1,\ldots,s-1).$$

Then a solution $\underline{X}^{(r)}$ can be obtained to any desired degree of accuracy from any initial solution $\underline{X}^{(o)}$, by the Newton-Raphson iteration:

$$\underline{X}^{(r+1)} = \underline{X}^{(r)} - \underline{\Phi}(\underline{X}^{(r)})\left(\frac{\partial \underline{\Phi}}{\partial \underline{X}}\right)^{-1}_{\underline{X}=\underline{X}^{(r)}}, \tag{23}$$

where $\left(\frac{\partial \underline{\Phi}}{\partial \underline{X}}\right)^{-1}$ is the inverse of the Jacobian of $\underline{\Phi}$.

As an initial solution, $\underline{X}^{(o)}$, it is convenient to take:

$$P_{ijk} = \frac{n_{ijk}}{n_k} P_k \quad (= \hat{P}_{ijk})$$

$$\lambda_k = \mu_{ij} = 0 \tag{24}$$

A FORTRAN IV programme has been written to perform the necessary iterations and for a typical small size problem (r=2, s=t=3) needs only four iterations to reach 4-digit accuracy with a run time of 7 minutes on the Hebrew University IBM 7040 computer.

Once maximum likelihood estimates are obtained $-\hat{P}_{ijk}$ in ω from (17) and approximations $\hat{\hat{P}}_{ijk}$ in Ω from (23) - the likelihood ratio can be computed

$$\lambda = L(\{n_{ijk}\}|\{\hat{\hat{P}}_{ijk}\})/L(\{n_{ijk}\}|\{\hat{P}_{ijk}\}) \tag{25}$$

and Wilks's log-likelihood ratio (12) defined as:

$$G = -2\ln\lambda = 2 \sum_{i,j,k} n_{ijk} \ln(\hat{P}_{ijk}/\hat{\hat{P}}_{ijk}). \tag{26}$$

Under H_o, G is asymptotically chi-square distributed with (r-1)(s-1) degrees of freedom. Therefore, H_o should be rejected, at level of significance α, if $G > \chi^2_\alpha[(r-1)(s-1)]$, where $\chi^2_\alpha[(r-1)(s-1)]$ is the upper α percentage point of the chi-square distribution with (r-1)(s-1) degrees of freedom.

4. Alternative Approximate Solutions

Bhapkar (2) has shown that, if we consider the χ^2_1 function:

$$\chi_1^2 = \sum_{i,j,k} \frac{\left(n_{ijk} - n_k \frac{P_{ijk}}{P_k}\right)^2}{n_{ijk}} = \sum_k \frac{n_k}{P_k} \sum_{i,j} \frac{(P_{ijk} - \hat{P}_{ijk})^2}{\hat{P}_{ijk}}, \quad (27)$$

where $\hat{P}_{ijk}$ is defined by (17), its minimum over values of P_{ijk}, under a linear hypothesis, is a simple expression which is asymptotically distributed under the hypothesis as chi-square. For testing a non-linear hypothesis, Bhapkar has proposed using Neyman's technique of linearization. Applying this result to the problem of testing overall independence, the non-linear hypothesis:

$$H_o: H_{ij}(\underline{P}) = 0 \qquad (i=1,\ldots,r-1;\ j=1,\ldots,s-1), \quad (28)$$

where $H_{ij}(\underline{P})$ is defined by (22) and $\underline{P}=(P_{111},\ldots,P_{rst})$, is replaced by the linear approximation:

$$H_{oL}: H^*_{ij}(\underline{P}) = H_{ij}(\hat{\underline{P}}) + \sum_{x,y,z} \left[\frac{\partial H_{ij}(\underline{P})}{\partial P_{xyz}}\right]_{\underline{P}=\hat{\underline{P}}} (P_{xyz} - \hat{P}_{xyz}) = 0 \quad (29)$$

$$(i=1,\ldots,r-1,\ j=1,\ldots,s-1)$$

where $\hat{\underline{P}} = (\hat{P}_{111},\ldots,\hat{P}_{rst})$.

Defining:

$$f_{ijxyk} = P_k \left[\frac{\partial H_{ij}(\underline{P})}{\partial P_{xyk}}\right]_{\underline{P}=\hat{\underline{P}}} = P_k [\delta_i^x \delta_j^y - \delta_i^x \hat{P}_{.j.} - \delta_j^y \hat{P}_{i..}] \quad (30)$$

$$b_{ijk} = \sum_{x,y} \frac{f_{ijxyk}}{P_k} \hat{P}_{xyk} \quad (31)$$

$$g_{iji'j'} = \sum_k \frac{P_k}{n_k} \sum_{x,y} \left(\frac{f_{ijxyk} - b_{ijk}}{P_k}\right)\left(\frac{f_{i'j'xyk} - b_{i'j'k}}{P_k}\right) \hat{P}_{xyk} \quad (32)$$

$$c_{ij} = H_{ij}(\hat{\underline{P}}) = \hat{P}_{ij.} - \hat{P}_{i..}\hat{P}_{.j.} \quad (33)$$

$$(i,i'=1,\ldots,r-1;\ j,j'=1,\ldots,s-1;\ x=1,\ldots,r;$$

$$y=1,\ldots,s;\ k=1,\ldots,t),$$

Bhapkar has shown that the minimum of (27) subject to (29) and (16) is given by:

$$\chi_1^2 = \underline{c}\, G^{-1} \underline{c}' \quad , \tag{34}$$

where $\underline{c} = (c_{11},\ldots,c_{r-1,s-1})$

and $G = (g_{iji'j'})$

is an (r-1)(s-1)x(r-1)(s-1) matrix.

In fact, the linear approximation of the hypothesis by (31) is equivalent to the first approximation of the iteration procedure defined in the previous section by (23) and (24). To see this, minimize (29) subject to (29) and (16), be setting:

$$\frac{\partial}{\partial P_{ijk}}\Big[\sum_z \frac{n_z}{P_z}\sum_{x,y}\frac{(P_{ijk}-\hat{P}_{ijk})^2}{\hat{P}_{ijk}} + 2\sum_z \lambda_z' \Big(\sum_{x,y} P_{xyz} - P_z\Big)$$
$$+2\sum_{x,y}\mu_{xy}' H^*_{xy}(\underline{P})\Big] = 0. \tag{35}$$

Thus, the equations to be solved are:

$$\frac{n_k}{P_k}\ \frac{P_{ijk}-\hat{P}_{ijk}}{\hat{P}_{ijk}} + \lambda_k' + \sum_{x,y}\mu_{xy}' f_{xyijk}/P_k = 0$$

$$\sum_{i,j} P_{ijk} = P_k \tag{36}$$

$$H^*_{ij}(\underline{P}) = 0$$

On the other hand, the first iteration of (23), using the initial solution (24), yields the equations

$$-\frac{n_{ijk}}{\hat{P}_{ijk}} + \frac{n_{ijk}}{\hat{P}^2_{ijk}}(P_{ijk}-\hat{P}_{ijk}) + \lambda_k + \sum_{x,y} \mu_{xy} \frac{f_{xyijk}}{P_k} = 0$$

$$\sum_{i,j} P_{ijk} = P_k$$

$$H_{ij}(\hat{\underline{P}}) + \sum_{x,y,z} \frac{f_{ijxyz}}{P_z}(P_{xyz} - \hat{P}_{xyz}) = 0. \quad (37)$$

But (36) is equivalent to (37) if we set:

$$\lambda'_k = \lambda_k - \frac{n_k}{P_k} \; ; \; \mu'_{ij} = \mu_{ij} \quad (38)$$

Thus the values of P_{ijk} which minimize the χ_1^2 statistic, (solutions of (36)) are first approximations (solutions of (37)) to the values of P_{ijk} which minimize the likelihood ratio statistic, according to the iteration procedure of the previous section.

Bhapkar (2) has pointed out that the minimal χ_1^2 statistic, as given by (34), is exactly equivalent to the large sample test statistic based on the asymptotic normality of the unbiased estimates of the hypothesis functions (29), whose variance-covariance matrix is estimated by the sample variance-covariance matrix. The method proposed by Garza-Hernandez and McCarthy (6) and (7), leads essentially to the same large sample statistic, except that it is based on biased ratio-estimates of the exact hypothesis function (30). Denoting:

$$q_{ij} = \frac{P_{ij}}{P_{i..}} \; (i=1,\ldots,r; j=1,\ldots,s-1); \; \underline{q}=(q_{11},\ldots,q_{r,s-1}), \quad (39)$$

ratio estimates of q_{ij} are:

$$\hat{q}_{ij} = \frac{\hat{P}_{ij}}{\hat{P}_{i..}} \; ; \; \hat{\underline{q}} = (q_{11}, \ldots, q_{r,s-1}) \tag{40}$$

$$(i=1,\ldots,r;\; j=1,\ldots,s-1).$$

The elements of the sample variance-covariance matrix of these estimates,

$$\sum = (\sigma_{ij,i'j'}),$$

are given approximately by:

$$\sigma_{ij,i'j'} = \frac{1}{\hat{P}_{i..}\hat{P}_{i'..}} \sum_k \frac{a_k}{n_k} \{\delta_i^{i'} P_k[\delta_j^{j'} \hat{P}_{ijk} - \frac{\hat{P}_{ijk}\hat{P}_{i'j'.}}{\hat{P}_{i..}} - \frac{\hat{P}_{i'j'k}\hat{P}_{ij.}}{\hat{P}_{i'..}} + \frac{\hat{P}_{i.k}\hat{P}_{ij.}\hat{P}_{i'j'.}}{\hat{P}_{i..}\hat{P}_{i'..}}]$$

$$-[\hat{P}_{ijk}\hat{P}_{i'j'k} \frac{\hat{P}_{ijk}\hat{P}_{i'.k}\hat{P}_{i'j'.}}{\hat{P}_{i'..}} - \frac{\hat{P}_{i'j'k}\hat{P}_{i.k}\hat{P}_{ij.}}{\hat{P}_{i..}}$$

$$+ \frac{\hat{P}_{i.k}\hat{P}_{i'.k}\hat{P}_{ij.}\hat{P}_{i'j'.}}{\hat{P}_{i..}\hat{P}_{i'..}}]\} , \tag{41}$$

where

$$a_k = \frac{N_k - n_k}{N_k - 1}$$

is a finite factor which may be ignored if the sampling fractions are small. Denoting:

$$\gamma_{i'j'}^{(ij)} = \begin{cases} 1 \; ; \; j'=j; \; i'=i \\ -1 \; ; \; j'=j; \; i'=i+1 \\ 0 \; ; \; \text{otherwise} \end{cases}$$

$$i=1,\ldots,r-1;\ i'=1,\ldots,r;\ j,j'=1,\ldots,s-1$$

$$\Gamma'_{ij} = (\gamma_{11}^{(ij)},\ldots,\gamma_{r,s-1}^{(ij)})$$

$$\Gamma = (\Gamma_{11},\ldots,\Gamma_{r-1,s-1}) \, , \tag{42}$$

the hypothesis (28) is equivalent to:

$$\underline{q}\,\Gamma = 0 \tag{43}$$

Garza-Hernandex (4), has shown that, assuming multivariate normality of the estimates $\hat{q}_{ij}$ with variance-covariance matrix, Σ, maximum likelihood estimates, $\hat{\hat{q}}_{ij}$, subject to (43), are obtained as the solutions of the equations:

$$\Sigma^{-1}\,(\underline{q}-\underline{\hat{q}})' + \Gamma\underline{\mu}' = 0$$

$$\underline{q}\,\Gamma = 0 \tag{44}$$

where $\mu = (\mu_{11},\ldots,\mu_{r-1,s-1})$ is a vector of Lagrange multipliers. The solutions are:

$$\underline{\hat{\hat{q}}}=\underline{\hat{q}}-\underline{\hat{q}}\,\Gamma(\Gamma'\Sigma\Gamma)^{-1}\,\Gamma'\Sigma$$

$$\underline{\mu} = \underline{\hat{q}}\,\Gamma\,(\Gamma'\Sigma\Gamma)^{-1} \tag{45}$$

Substituting these values in the likelihood ratio statistic gives:

$$(\underline{\hat{q}}-\underline{\hat{\hat{q}}})\,\Sigma^{-1}\,(\underline{\hat{q}}-\underline{\hat{\hat{q}}})' = (\underline{\hat{q}}\,\Gamma)(\Gamma'\Sigma\Gamma)^{-1}(\underline{\hat{q}}\Gamma)' \tag{46}$$

This is exactly the large sample test statistic based on the estimates, $\underline{\hat{q}}$, of the hypothesis functions (43), with the sample variance-covariance matrix of $\underline{\hat{q}}\,\Gamma$, which is

$$\Gamma'\Sigma\Gamma.$$

Thus, both the methods of Bhapkar and of Garza-Hernandez and McCarthy are essentially based on large-sample test statistics which, under the hypothesis, are asymptotically distributed chi-squares, as is the likelihood-ratio test statistic computed iteratively in the previous section. Although the methods of this section involve far less computation, the likelihood ratio test of the previous section may be considerably more asymptotically powerful at many points of the alternative space, as shown by Hoeffding (9), for the case where error probabilities tend to zero; when the error probabilities are bounded away from zero, certain asymptotically optimal properties have been shown to hold by Wald (13).

5. Extension to Simultaneous Inference

The rejection of the hypothesis (9) does not provide any insight on the nature of the dependence, if any, between the two characteristics. Thus it is possible that there is independence between sub-groups of the categories of each characteristic.

If $V(\subset R = \{1,\ldots,r\})$ and $W(\subset S = \{1,\ldots,s\})$ are subsets of the sets of categories, define:

$$\begin{aligned} P_{Vj.} &= \sum_{i\in V} P_{ij.} \\ P_{iW.} &= \sum_{j\in W} P_{ij.} \\ P_{VW.} &= \sum_{\substack{e\in V \\ j\in W}} P_{ij.} \quad . \end{aligned} \tag{47}$$

Then independence between the categories of V and W is equivalent to:

$$H_{OVW}: P_{ij.}\, P_{VW.} = P_{Vj.}\, P_{iW.};\ i\in V,\ j\in W\ . \tag{48}$$

Gabriel (4) has shown that, for a simple random sample (i.e. k=1), simultaneous tests of the hypotheses (48) for all V R and W S are provided by rejecting those hypotheses, H_{OVW}, for which the maximum likelihood ratio test statistic, computed over the subsets V,W, is greater than the critical value of the chi-square distribution with (r-1) (s-1) degrees of freedom.

This simultaneous test procedure can obviously be extended to the case of stratified random sampling. The maximum likelihood ratio statistic (26) can be computed for any subsets V,W, by the method of section three, if n_k is replaced by

$$n_{VWk} = \underset{i\epsilon V}{\quad} \underset{j\epsilon W}{\quad} n_{ijk}$$

and the indexes i,j limited to the subsets V,W respectively. Let G(V,W) be the statistic thus computed. Then a simultaneous test of the hypotheses (48) is to reject H_{OVW} if:

$$G(V,W) > \chi^2_\alpha \, [(r-1)(s-1)]. \qquad (49)$$

This simultaneous test procedure has the following properties:

(I) The probability of not rejecting hypothesis H_{OVW} for any subsets V,W for which H_{OVW} is true is at least $1 - \alpha$;

(II) The test decisions are "Coherent", in the sense that if H_{OVW} is "rejected" for any V,W then it is also rejected for all $H_{OV'W'}$ such that V' V, W' W. These properties have been established generally by Gabriel (5) for all simultaneous test procedures usi g likelihood ratio statistics.

6. Numerical Example

Garza-Hernandez (4) has used the following data, which classifies responses from a stratified group survey conducted in 1952 in a Canadian Maritime Province. The responses are classified by groups of "Occupational Disadvantage" (r-2 groups) and by incidence of psychiatric disorders (s=3 classes), within (t=3) geographic- and social-area strata. Using the notation introduced above, the data is given in Table 1.

Table 1: Observed Frequencies

k	i / j	n_{ijk} Total	1	2	P_k
All Strata	Total	282	93	189	1.0000
	1	85	31	54	
	2	151	44	107	
	3	46	18	28	
k = 1	Total	92	18	74	0.1708
	1	28	10	18	
	2	52	6	46	
	3	12	2	10	
k = 2	Total	112	54	58	0.3465
	1	35	17	18	
	2	60	29	31	
	3	17	8	9	
k = 3	Total	78	21	57	0.4827
	1	22	4	18	
	2	39	9	30	
	3	17	8	9	

For stratum K independence within the strata can be tested by means of the statistic:

$$G_k = 2\sum_{i,j} n_{ijk} \ln n_{ijk} - 2\sum_{i} n_{i.k} \ln n_{i.k}$$

$$- 2\sum_{j} n_{.jk} \ln n_{.jk} + 2n_k \ln n_k \quad .$$

The values obtained are:

k	G_k
1	6.449
2	0.011
3	4.363
Total	10.823

Comparing these values with critical values of the chi-square distribution with two degrees of freedom, it is seen that the hypothesis of independence within strata (11) is rejected for the first stratum at the .05 level of significance (but not at the .01 level) but it is not rejected in the second or third stratum (even at a .10 level of significance). The overall test made by comparing the total with the critical chi-square value with 6 degrees of freedom is also not significant even at the .1 level of significance.

Only four iterations are necessary to reach four-digit accurate maximum likelihood estimates under the hypothesis of general independence. The resulting estimates of the marginal probabilities, $P_{.j.}$, iteration estimates used in Bhapkar's χ_1^2 statistics, and the Garza-Hernandez maximum likelihood estimates are given in Table 2. together with the final test statistics.

Although this example shows that the approximate test statistics of section four may differ considerably from the maximum likelihood ratio test statistic, it should be pointed out that the hypothesis would not be rejected at the .05 level by any of the methods proposed, as none of the statistics exceed the critical .05 level chi-square value (2 degrees of freedom) of 5.991.

For this example no further computation is required for the simultaneous test procedure of section five, since the overall null hypothesis is not rejected. Had it been rejected three further tests on the 2x2x2 submatrices would be carried out to determine whether psychiatric-disorder incidence in any pair of classes is independent of occupational disadvantage.

7. Acknowledgements

This paper has benefited from several helpful suggestions made by Prof. K. R. Gabriel and by Dr. E. Peritz. Thanks are also due to Ruth Sheshinski and to Israel Adler for their programming.

Table 2: Estimates of Marginal Probabilities, Under the Null Hypothesis, and Test Statistics

Method	Marginal Probabilities-$P_{.j.}$			Test Statistic
	j=1	j=2	j=3	
Maximum Likelihood (Sec. 3)	.2965	.5236	.1799	2.288
Bhapkar (Minimal χ_1^2)	.2925	.5389	.1686	2.130
Garza-Hernandez	.2983	.5246	.1771	1.929

REFERENCES

(1) Armitage, P., "The Chi-square Test for Heterogeneity of Proportions, after Adjustment for Stratification", Journal of the Royal Statistical Society (B), 28 (1966), 150-163.

(2) Bhapkar, V. P., "Some Tests for Categorical Data", Annals of Mathematical Statistics, 32 (1961), 72-83.

(3) Fisher, R. A., Statistical Methods for Research Workers, Oliver and Boyd, London, 1946.

(4) Gabriel, K. R.,"Simultaneous Test Procedures of Multiple Comparisons of Categorical Data", Journal of the American Statistical Association, 61 (1966), 1081-1096.

(5) Gabriel, K. R., "Simultaneous Test Procedures - Some Theory of Multiple Comparisons", University of North Carolina, Institute of Statistics, Mimeo Series No. 536 (1967).

(6) Garza-Hernandez, T., "An Approximate Test of Homogeneity on the Basis of a Stratified Random Sample", M. Sc. Thesis, Cornell University, 1961.

(7) Garza-Hernandez, T. and McCarthy, P. J., "A Test of Homogeneity for a Stratified Sample", American Statistical Association, Proceedings of the Social Statistics Section, 1962, pp. 200-202.

(8) Hartley, H. O., "Analytical Studies of Survey Data", American Statistical Association, Proceedings of the Social Statistics Section, 1954, pp. 146-154.

(9) Hoeffding, W., "Asymptotically Optimal Tests for Multinomial Distributions", Annals of Mathematical Statistics, 36 (1965), pp. 369-401.

(10) Kendall, M. G., The Advanced Theory of Statistics, Griffin, London, 1945.

(11) Marascuilo, L. A. and McSweeney, M., "Multiple Contrast Methods for Analytical Surveys", American Statistical Association, Proceedings of the Social Statistics Section, 1967.

(12) McCarthy, P. J., "Replication - An Approach to the Analysis of Data from Complex Surveys", National Center of Health Statistics, Ser. 2, No. 14, Washington, 1966.

(13) Wald, A., "Tests of Statistical Hypotheses Concerning Several Parameters when the Number of Observations is Large", Transactions of the American Mathematical Society, 54 (1943), pp. 426-482.

(14) Wilks, S. S., Mathematical Statistics, Wiley, N. Y., 1962.

IMPACT OF DESIGN AND ESTIMATION COMPONENTS ON INFERENCE

Walt R. Simmons and Judy Ann Bean

National Center for Health Statistics, Washington, D.C.

1. Introduction

The central function of a sample is to provide estimates of parameters of the population which it represents. Probability samples accomplish this function in a manner that permits the precision of the estimates to be determined from the sample data themselves--one of the most striking phenomena to be found anywhere in scientific method, as others have noted. It has become common in survey work to say that the sample is designed to produce minimum variance of the estimators for a fixed budget, given the prevailing environment. In many situations we ask further that the estimators be essentially or nearly unbiased.

Theoretical work has in the main moved along two lines. One of these may perhaps be called *classical analysis*. It encompasses a considerable part of the entire body of statistical theory. Typically it contemplates securing n independent observations of one or more variables from a specified distribution by a simple random process, and the calculation of expected value, variance, and perhaps other attributes of sample statistics derived from the n observations, in terms of population parameters, which in turn can be similarly estimated from the sample data. This approach underlies the dominant part of our arsenal of devices for making both interval estimates and tests of significance. Its delineation is to be found in scores of textbooks, and perhaps outstandingly in Kendall and Stuart, *The Advanced Theory of Statistics*, particularly Vol. 2.

The second principal avenue of theoretical development is the design of techniques for survey sampling of finite universes. This second area is much the newer of the two, but has had extensive growth in the last three decades, as is evident from the journals and such textbooks as those by Yates; Deming; Hansen, Hurwitz, and Madow; Sukhatme; Cochran; and Kish. (1/, 2/, 4/, 6/, 11/, 12/)

Today it is common that a survey of social, economic or biological phenomena includes in its structure such complexities as controlled selection, imputation for nonresponse, ratio or regression estimation, two or more stages of sampling, stratification with unequal sampling fractions, and post-stratification. A variety of questions arises. One very general question is how can the statistics from a complex survey be properly analyzed? What methods and instruments from classical analysis are appropriate? At the heart of this problem is the fact that classical analysis rests in part on assumptions that the sample observations are mutually independent, and have been drawn by a simple random procedure; while the data from the social survey have neither of these characteristics. Also of importance is the circumstance that usually operational survey data contain substantial and confounded measurement errors, whereas classical models assume either no observational error, or simply distributed errors.

There is, too, the matter of the relationship between sample data and the universe which those data are presumed to represent. When the sample units are drawn with known probability from an identifiable finite population, the sample results certainly represent that finite population. In what sense, if any, do they represent some more general universe? For example, suppose a large probability sample of adult males is drawn from the resident population of North Carolina as of July 1, 1967. Data from that sample yield a simple correlation coefficient of 0.80 between height and knee height. What inference may one draw on this evidence about height and knee-height correlations for female adults in North Carolina, or for other age groups in North Carolina, or for Carolinians in 1776, or for Chinese? And if any inferences are drawn for these universes, should the estimation process be different than it was for making estimates for adult males in North Carolina in 1967?

We and our colleagues at the National Center for Health Statistics have been much concerned for several years with such questions as these, and how best to convert sample data from complex surveys of finite universes to useful inferences. Many hours of discussion on the subject have taken place. One aspect was treated in 1962 in a paper by Simmons, McDowell, and Gordon,[9] which dealt with the use of collateral data; another aspect of appropriate methods of analysis was presented in 1965 in a paper by McCarthy, Simmons and Losee.[7] McCarthy, under a contract with the Center, wrote a more extensive report in 1965 on analytic methods, emphasizing pseudo-replication techniques.[8] His paper presented at this symposium is a further elaboration of that earlier work. The Center also has financed recent related work by Kish in the use of pseudo-replication to study correlation and regression in survey data.

We do not here summarize that work, but offer two observations:

A. We are convinced that much classical analysis is inappropriate when applied to data from complex surveys, in the sense that usual measures of precision and tests of significance are not valid, or at best are crude approximations.

B. Many difficulties are encountered in attempting to treat data from the complex survey as though they were derived from a simple random sample, but the most serious impact arises from lack of independence, brought about by clustering in the sample design.

2. Strategy of a Study

After having raised these major issues and identified several matters fundamental to statistical analysis, this paper treats only one facet of the problem, and that in a rather pedestrian manner. The emphasis is on a method of analysis. The objective is to determine the contribution to or reduction in sampling variance and mean square error which can be credited to each of several components of the design and estimation procedure. The relevance of results bears on the inference problem and incidentally on the design problem. The technique employed is an empirical one, based on the pseudo-replication half-sample computational device.

For many analytic purposes the difference between two designs or estimating procedures can be described adequately in terms of the expected value, variances, and mean square errors of statistics derived from the two processes. Accordingly, it becomes desirable to estimate expected values, variances and mean square errors for alternative procedures, and to try to quantify the extent to which these values are affected by different design and estimating factors. A by-product of the effort is an evaluation of the utility--or disutility--of specific components of design or estimation.

Primary emphasis is placed on comparison of variances for two procedures, since expected values do not differ much from one another for most of the procedures under study; but in some of the situations it is necessary that the comparison be in terms of mean square error. Comparisons could be attempted with the aid of models and theoretical analysis of the effect that should be observed from such factors as unequal sampling fractions, non-response adjustment, ratio estimation, clustering, post-stratification, and all of these combined. An earlier version of the present paper contained a section which studied the impact of various components of design and estimation, using simplified models and theoretical analysis. For example, the loss in efficiency caused by unequal sampling fractions can be analyzed as follows:

While there are, of course, distinct advantages in varying the sampling fraction from one stratum to another, it is also true in household sampling that selection with equal-probability ordinarily is most efficient within a stratum. For example, compare for a binomial variable with mean P the variance of estimated P' from an equal-weighted sample of n of N units, with that of P" obtained also from a sample of n units, but with the λ*th* part of the sample secured at a sampling rate of 1-in-r, and $(1-\lambda)$*th*-part of the sample secured at a rate of 1-in-kr--i.e., "subsampled" at a rate of 1-in-k. The ratio of variance of the latter to the former is

$$R = (k\lambda+1-\lambda)\frac{\lambda+k-\lambda k}{k} \quad .$$

This rather simple rule can serve as a guide to the loss to be expected when one is forced to accept unequal weighting within the stratum. The relative loss can be expressed as

$$L = R-1.$$

As an illustration, suppose $\lambda = 0.65$ and $k = 2$--i.e., 35 percent of the sample is further subsampled at a 1-in-2 rate: $R = 1.12$, and the loss is 12 percent. Other notes treated the theoretical impacts of (1) stratification, (2) Ratio Estimation, (3) Post-stratification, (4) Non-response, and (5) Clustering. These analyses and the resulting guidelines formulae are, however, of but limited help: in the more complex designs, factors of interest are confounded, and even more importantly, it is very difficult to obtain values for parameters needed to give quantitative character to the formulae. Because of space limitations, most of that analysis has not been included here. Portions of it are suggested by references 1/, 3/, and 10/. In the material presented, we have adopted an empirical scheme for comparing two procedures, using pseudo-replication[8/] as the instrument of analysis. This technique by no means resolves all the difficulties. For example, we were unable to separate the effect of stratification in measuring the impact of clustering. And some of the calculations are approximations. But we hope others will agree that the general approach has merit as a scheme for analysis of design and estimating techniques.

3. Methodology

Description of Health Examination. The Health Examination Survey is a highly stratified multi-stage sample of the civilian noninstitutional population of the conterminous United States. The design is most easily understood when described in terms of a set of defined building blocks which are associated with the different stages of sample selection. The first of these is the Primary Sampling Unit. For use in the Health Interview Survey, the geographical territory of the mainland United States has been divided into 1,900 areas. Each area is a county or a small group of contiguous counties. With minor modifications, these areas became the PSU's of the Health Examination Survey. The PSU was divided geographically into segments, each containing an expected six households. From a listing of households within the segment, a random sampling procedure created a subsegment of approximately four households, each of which was interviewed. Every alternate person in the sample household who was an eligible adult (civilian, in the age range 18-79) became a sample person for inclusion in the HES panel. Thus the successive building blocks are person, household, subsegment, segment, and PSU, and the sampling process was applied to successive stages of these blocks in reverse order to that just listed. Table 1 presents description statistics for the design.

The wide target of the design was selection of persons with equal probability. To a substantial degree this objective was met, but other requirements caused some relaxation. The final distribution of sample persons by relative weights was

Relative Weight	Percent of Sample
All Weights	100.0%
1/2	13.8
1	58.6
2	16.4
3	9.3
4-7	1.9

In addition to certain editing or laundering of data, estimation included four principal operations.

A. Inflation by the reciprocal of the probability selection. This probability of selection is the product of the probabilities of selection from each step of selection: PSU, segment, household, and person.

B. First Stage Ratio Adjustment. Primary sampling units are ratio adjusted to 1960 population with eight geographic and population concentration classes. (See Appendix Table 2.)

C. Nonresponse adjustment carried out in 294 age-sex-PSU cells. The nonresponse adjustment is a multiplication factor; the numerator is the designed sample number in an age-sex class within a primary sampling unit, and the denominator is the examined number of persons in that cell.

D. A post-stratification by 12 age-sex cells. [The control for each cell is an independent estimate of population as of the survey period, prepared by the Census Bureau.] (See Appendix Table 4.)

Method of Analysis. The object of the analysis is to determine the effect of different design and estimating components upon bias and variance. The components include unequal sampling fractions, clustering, stratification, ratio estimation, nonresponse adjustment, and post-stratification--alone and in combination. Two separate sets of comparisons were made. In the first, ratios of variances were computed where the numerator is the variance of one of the estimates or estimating processes, and the denominator is the variance of another of the estimates--often a synthetic estimate calculated under the assumption of simple random sampling, but in some situations, one of the other estimates.

That first set of ratios does not take bias into account. When measuring the accuracy of a biased estimate, a satisfactory criterion is the mean square error. Accordingly, a second set of ratios is examined: The ratios of the mean square errors of one estimate to the mean square error of another-again frequently the simple random statistic.

The major part of this analysis utilizes data from the Health Examination Survey for four items. Each of the items is a binomial variate, the proportion of the population or of a subclass of the population which has a specific health condition. The four were selected to represent a wide range of proportions or P-values. The item hypertensive heart disease (HHD) reflects relatively high prevalence (9.5 percent of the civilian non-institutional population; age 18-79 years), myocardial infarction (MI) and angina pectoris (AP) appear relatively infrequently (1.3 percent and 1.4 percent of the civilian non-institutional population, age 18-79 years), and syphilis (STS) an intermediate value (4.0 percent of the civilian non-institutional population, 18-79 years). Calculations were made separately for all persons, for males, for females, and for subcategories of total persons, males and females cross-classified by income, by marital status, or by education.

For each of the items appropriate formulae were applied to the HES data to produce estimates of the P-values, and estimates of the sampling variance of those P-values. The objective, as noted above, was to produce sets of estimates which are the basis for determining impact of different design and estimation components. For convenience in reference the various estimates have been given letter designations as follows:

A. PQ-Model. These estimates are computed using unweighted sample data in conformity with a binomial model.

B. Equal Weighting. Each examined person received a weight of 10,000, but otherwise calculations used data which reflected all features of the actual design. (But not more elaborate estimation.)

C. Basic Weighting. Same as B except that each person received a weight equal to the reciprocal of the probability of selection.

D. Nonresponse Adjusted. Same as C except that each person's basic weight is multiplied by the nonresponse adjustment.

E. Adjusted for nonresponse and first-stage ratio. Same as D except that each person's nonresponse adjusted weight is multiplied by the first stage ratio adjustment factor.

F. Final estimates. Full design and estimation procedure is reflected. Same as E except post-stratified by age and sex.

The key formula for calculation of the ratio of two variances, when the denominator is the variance of the synthetic simple random model is

$$R_{ijk} = \frac{s^2_{ijk}}{s^2_{jk,\ SRS}} = \frac{\sum_{\alpha=1}^{16}(x'_{ijk\alpha}-x'_{ijk})^2/16}{P_{jk}Q_{jk}/n_{jk}}$$

where

s^2_{ijk} = estimated half-sample pseudo-replication variance for the ith estimate (i = method B, C, D, E or F) for the jkth subgroup (j = males, females, or totals; k = income class, marital status, education group or total of all classes)

x'_{ijk} = estimate using the ith method for the jkth subgroup in the total sample

$x'_{ijk\alpha}$ = estimate using the ith method for the jkth subgroup in the αth half-sample

$s^2_{jk,\ SRS}$ = estimated variance for the estimate using unweighted data for the jkth subgroup

P_{jk} = unweighted sample estimate for jkth subgroup

Q_{jk} = $1-P_{jk}$

n_{jk} = number of examined persons in jk*th* subgroup

For a similar ratio of Mean Square Errors, where Mean Square Error is defined as the sum of variance and the square of bias, the formula used is the following. Note that bias is estimated as the difference between the estimate secured in the full HES design and procedure, and that obtained for the method in the comparison.

$$R^*_{ijk} = \frac{MSE_{ijk}}{MSE_{jk,\ SRS}}$$

where

$$MSE_{ijk} = s^2_{ijk,} + (x'_{ijk} - x'_{fjk})^2$$

$$MSE_{jk,\ SRS} = s^2_{jk,\ SRS} + (P_{jk} - x'_{fjk})^2$$

x'_{fjk} = estimate using final weights for the jk*th* subgroup in total sample--which is assumed for this purpose to be the criterion value.

Both sets of ratios--ratio of variances and ratio of mean square error--were computed for each of the four statistics for all persons, for males, for females, and for all the subcategories.

For larger sets of variables than the four--those in the Table on page 616-the ratios were calculated only for the full design and estimating procedure, thus yielding estimates of the overall design effect.

For measuring the impact of components, the average R-value and R*-value for the four statistics for all persons were used. For situations for which calculations were made for larger numbers of variables, medians of R-values and R*-values were utilized in order to lessen the influence of extreme values, some of which occurred because of high sampling error in small cells.

Since there are reasonable alternatives which might be advanced for utilizing the basic data to estimate impact of some of the components, it will be helpful to indicate explicitly the interpretation chosen in the present report. The capital letters in the following tabular arrangement refer to the designation given the methods earlier in this section on methodology.

Relationship of Methods Computations	Interpreted as Yielding
F/A	Overall Design Effect.
B/A	Combined Impact of Clustering and Stratification.
C/A	Combined Impact of all Design Components.
D/C	Relative Impact of Nonresponse Adjustment.
E/D	Relative Impact of First-Stage Ratio.
F/E	Relative Impact of Post-Stratification.
C/B	Relative Impact of Unequal Weighting.
$\frac{C}{A} - \frac{B}{A}$	Impact of unequal weighting on Base A.
F/C	Impact of combined estimating technique improvements over a linear unbiased estimator.

4. Summary and Conclusions

The analysis of impact of components, as distinguished from some other aspects, of our investigation, rests on data for only four statistics, all binomial, and all having reference to cardiovascular conditions. They have P-values--i.e., proportion of the population with specified conditions--over the range 1 to 10 percent. Conclusions are based on those statistics, but may be substantively correct for a much wider range of statistics. The analytic processes themselves are subject to sampling variability, and that fact means that numerical findings must be interpreted as indicators rather than definitive measures.

For these four statistics the average overall design effect (the ratio of estimated "true" variance for the full design with all the complicating steps of estimation to a hypothetical simple random variance PQ/n, where n is the number of persons examined) is 1.86. That is, the true variance is nearly double the simple random variance.

This overall design effect is the resultant of deviations in design from simple random, and deviation in estimation from simple inflation. On the design side, the pseudo-replication procedure detected an average increase in variance of 27 percent over a simple random model, caused by variation in the probability of selection. The net increase from clustering, dampened by stratification is 110 percent of simple random. The combined effect of unequal weighting, clustering, and stratification--i.e., all <u>design</u> elements--yields a factor of 2.37, or a 137 percent increase in variance, over a simple random design.

The difference between the 2.37 factor and the overall 1.86 factor, represents the improvement in precision brought about by estimation components--a decrease in variance of 51 percentage points measured on the scale of simple random variance, or 21.6 percent over the unbiased linear estimate. The components of this decrease are summarized in the table, which also shows percentage decreases in mean square error. The evidence from these data is that the first-stage ratio adjustment had little effect on either variance or bias. Contrastingly, the reductions in variance and mean square error are substantial for both the nonresponse adjustment and post-stratification. Collectively, the estimating procedures reduced mean square

Impact of Estimation Refinements

Estimating Step or Refinement	Average Decrease over Linear Unbiased Estimator			
	In Variance		In Mean Square Error	
	Amount in Percent	Cumulative Percent	Amount in Percent	Cumulative Percent
First-stage ratio Estimate	2.1%	2.1%	1.8%	1.8%
Nonresponse Adjustment	12.2	14.3	44.1	45.9
Post-Stratification By Age and Sex	7.2	21.5	20.2	64.3

error by over 60%--three times the variance reduction alone. [There is a suggestion of possible overstatement of improvement arising from a particularly large estimated bias in one disease category.] These improvements are the consequence of a variety of factors, but derive especially from the effect of steps which make the estimate more closely representative of the age-sex structure of the civilian non-institutional U.S. population.

Since some of the elements under discussion are confounded with others, there is a certain degree of risk in combining the above findings. Further, all the estimates are subject to considerable sampling error, and to computational approximations. They should be interpreted as indicators of magnitude rather than definitive measures. The general outline of the picture is clear, however, and is brought out by focusing on these approximate figures.

A. Net bias of an unweighted estimate	5 percent of the estimate
B. Impact on Variances	Percent of PQ/n
(1) Increase due to unequal sampling factions	27
(2) Net increase from clustering and stratification	110
(3) Reduction from first-stage ratio adjustment	5
(4) Reduction from Nonresponse adjustment	29
(5) Reduction from Post-stratification by Age and Sex	17
C. Ratio of true variance to PQ/n	1.86

Data for the larger sets of variables, as expected, show variation in the design effect--part of the variation undoubtedly reflecting different characteristics of the different statistics, and part of it being the consequence of sampling error in the experiment. The table following shows a summary of the sets of variables studied. Medians and quartiles are used here to avoid distortion from a few atypical observations.

Description of Statistics	Design Effect	
	Median	Upper Quartile
35 Demographic Subgroups with Hypertension (PQ Type)	1.3	1.7
6 Sex-Region Groups with Corrected Distance Visual Acuity of 20/15 (PQ Type)	1.8	4.0
8 Sex Groups with Syphilis, Hypertensive Heart Disease, Myocardial Infarction, or Angina Pectoris (PQ Type)	1.5	1.9
116 Socio-Economic Groups of Males with one of the same four diseases (PQ Type)	1.3	2.0
116 Socio-Economic Groups of Females with one of the same four diseases (PQ Type)	1.4	1.9
16 Mean Body Characteristics for Adult Males	3.6	4.7

All of the above statistics are estimated mean values, and most of them are binomial variates. We have only very limited analyzed experience thus far with other types of statistics. But from multivariate analysis of the set of 16 body measurements--each with three independent variables--we do have results from pseudo-replication calculations carried out for NCHS by the Survey Research Center of the University of Michigan. These show a trivial bias of less than one percent in simple unweighted calculation of the multiple correlation coefficient. The average Index or Ratio of properly computed variance using pseudo-replication, to the variance under an assumption of a simple random model differs notably from unity. The mean Index for the 16 measurements is 2.2. The median index is also 2.2 and the upper quartile value is 3.2

Some Speculative Conclusions

Both on *a priori* grounds and on empirical evidence, the following conclusions may be warranted, although on the empirical side, the possibility of sampling variation should be kept in mind.

The overall Design Factor--i.e., the ratio of variance for the multistage design with full ratio and post-stratified estimation to the simple PQ/n variance is:

A. Greater for broader categories of persons included and smaller for smaller domains or subclasses of population. In the former case, the Design Factor for many statistics will run as high as 2.0 or 3.0 and occasionally higher. For the latter, the Design Factor often may be of the order of 1.3, and in some cases is near unity.

B. Greater for estimates of attributes which are common in the population and smaller for estimates of attributes which are rare in the population.

Both of these tendencies are largely a consequence of the effective cluster, with "effective" having a rather special definition. For a smaller domain, the cluster will contain fewer individuals of the specified category, and so the effective cluster size moves toward unity and simple random sampling. For low frequency attributes, a cluster of not more than moderate size is likely to have not more than one or only a few individuals with the attributes of interest, and so again the clustering effect is minimized.

There is a countervailing influence which might be noted. For any statistic bounded by age-sex or PSU lines, much of the effect of the more elaborate estimating technique--which tends to lower the true variance--is nullified, and so for such statistics the Design Factor is likely to be <u>increased</u> inasmuch as clustering impact remains but estimating efficiencies are lost.

5. Discussion

As noted in the Summary, the methods of analysis employed in this study are in some instances only approximations to conceptual targets, and often the estimates themselves are subject to substantial sampling error. Therefore most individual pieces of evidence in the study must be interpreted with caution. Nevertheless, attention may be called to a few features of the data which are suggestive of special characteristics probably present in the parent universe.

While there are exceptions, the common finding is that none of the design or estimating steps introduced important biases in statistics, whether those statistics are proportions, means, or regression coefficients. Indeed the only significantly biasing element is differential nonresponse by age groups and geography, and that influence has been considerably dampened by the nonresponse adjustment. It ought to be noted further, however, that if weighting appropriate to the design and adjustment for nonresponse had not been introduced some of the biases would have been substantial.

Just as clearly, it is apparent that both design and estimating steps do have nontrivial impacts on variance. As observed earlier, the typical overall design effect for the HES appears to lie in the range 1.3-3.6 for the statistics investigated. Except for the factors mentioned in the Summary in Section 4, most of the variation among statistics in size of overall design effect quite probably is simply the consequence of sampling error. For example, we are unable to advance a plausible substantive reason for the rather marked observed difference in design effect between angina pectoris and myocardial infarction. Scanning of more detailed tables than those reproduced here also fails to adduce any attractive explanation.

The situation is different for the mean value data on body measurements which yield design effects that are consistently higher than those for the proportions of persons with specified morbidity. In fact, 14 of the 16 body measurements show a higher calculated design effect than the median of all the prevalence proportions, and the median body measurement design effect was roughly double the median for the proportions. This is more than an accidental or random result. Whereas one may expect not more than minor positive intraclass correlations for the morbidity proportion, it is easily believable that the intraclass correlations--both for segments of neighboring households, and for PSU's--will run higher for the body measurements that tend to be correlated with ethnic and economic characteristics which in turn are clustered.

It ought not be concluded, however, that all observed high Indexes, even of the morbidity proportions, are the consequence of sampling error. Some are. But some undoubtedly reflect real situations. Consider for instance the data for subcategories of the population for STS (seriological test for syphilis). The condition is one which tends to be concentrated in localities. For a particular subcategory, a high prevalence yields a comparatively low rel-variance for the simple random model. But the very existence of the high prevalence implies likely existence of pockets of the condition, and consequent clustering for the true design. Thus the calculated ratio of true variance to simple random variance is properly high for the subcategory.

The matter of cost has not been treated in the present paper, although cost is the other side of the coin in evaluating the efficiency of any design or procedure. Two general comments may be in order. The design factor which increased variance most is clustering. This choice more than doubled variance. But for the Health Examination Survey a nonclustered sample would have been totally impractical, and if it could have been handled operationally would have multiplied costs several times. Thus clustering was a desirable step despite its impact on variance.

The other aspect of cost which should be mentioned bears on the substantive significance of modest reductions in variance. Our analysis suggests that some of the estimating refinements reduced sampling variance by amounts that are not large. A first impression may be that a, say, 10 percent reduction in variance is not worth the additional difficulty which these refinements entail in more complicated calculations. But when it is recalled that a 10 percent variance reduction can be converted to an approximately 10 percent reduction in cost, the importance of techniques for minimizing variance is reestablished. This is valuable information quite aside from its significance in drawing inferences from the evidence of complex surveys.

References and Footnotes

1/ Cochran, W. G.: Sampling Techniques, ed. 2. New York. John Wiley and Sons, Inc., 1963.

2/ Deming, W. E.: Some Theory of Sampling. New York. John Wiley and Sons, Inc., 1950.

3/ Durbin, J.: Sampling theory for estimates based on fewer individuals than the number selected. Bulletin of the International Statistical Institute, 36: 113-119, 1958.

4/ Hansen, M. H.; Hurwitz, W. N.; and Madow, W. G.: Sample Survey Methods and Theory, Vols. I and II. New York. John Wiley and Sons, Inc., 1953.

5/ Kendall, M. G.; and Stuart, G.: The Advanced Theory of Statistics, Vol. II, Inference and Relationship. London. Griffin and Co., 1961.

6/ Kish, L.: Survey Sampling. New York. John Wiley and Sons, Inc., 1965.

7/ McCarthy, P. J.; Simmons, W. R.; and Losee, G. J.: Replication techniques for estimating variances from complex surveys. Epidemiology and Statistics Sections, American Public Health Association, October 1965.

8/ National Center for Health Statistics: Replication: An Approach to the Analysis of Data from Complex Surveys. Vital and Health Statistics. PHS Pub. No. 1000-Series 2-No. 14. Public Health Service, Washington, D. C. U.S. Government Printing Office, April 1966.

9/ Simmons, W. R.; McDowell, A. J.; and Gordon, Tavia: Some Statistical Features of the Health Examination Survey. Proceedings of the Social Statistics Section of the American Statistical Association, 1962, pp. 126-137.

10/ Stephan, F. F.: The expected value and variance of the reciprocal and other negative powers of a positive Bernoullian variate. Annals of Mathematical Statistics, 16: 50-51, 1945.

11/ Sukhatme, P. V.: Sampling Theory of Surveys with Applications. Ames, Iowa. Iowa State College Press, 1954.

12/ Yates, F.: Sampling Methods for Censuses and Surveys, ed. 3. New York. Hafner Publishing Co., 1960.

Appendix B--Tables

Table 1. Summary Statistics on Design and Estimation Characteristics of the Health Examination Survey of the U.S. Civilian Non-Institutional Adult Population, 1960-62.

Item	Statistic
Number of Primary Sampling Units (PSU's)	42
Number of Segments	2,174
Number of Sample Persons (Design)	7,710
Number of Sample Persons (Examined)	6,672
Typical no. of sample persons per segment	4
Typical no. of sample persons per PSU	160
Number of Physicians	62
Number of First Stage Ratio Categories	9
Number of Non-Response Adjusted Cells	294
Number of Post-Strata	12
Number of Pseudo-Replications	16

Table 2. First-Stage Ratio Adjustment Factors: Health Examination Survey, United States, 1960-62.

Geographic Location	Self-Representing Areas[1]	Non-Selfrepresenting Areas	
		Standard Metropolitan Statistical Areas	Other Areas
Northeast	1.00	0.97	0.98
South	--	1.09	0.88
West	1.00	0.88	1.04

[1]New York, Chicago, Los Angeles, Philadelphia, Detroit, and Boston sample areas represented only themselves.

Table 3. Frequency Distribution of Multipliers for Nonresponse Adjustment, Health Examination Survey, United States, 1960-62

Size of Adjustment Factor	Number of PSU-AGE-Sex Cells in which Factor was used			
	Total	Males Under 65 yrs.	Females Under 65 yrs.	Over 65 years
All Cells	294	126	126	42
1.00	39	23	14	2
1.01-1.09	86	40	38	8
1.10-1.19	60	27	27	6
1.20-1.29	49	19	20	10
1.30-1.39	25	5	22	8
1.40-1.49	9	4	3	2
1.50-1.59	12	1	7	4
1.60-1.69	3	1	1	1
1.70-1.79	2	-	2	-
1.80-1.89	4	4	-	-
1.90-1.99	-	-	-	-
2.00	5	2	2	1

Table 4. Poststratifying Age-Sex Adjustment Factors: Health Examination Survey, United States, 1960-62

Age	Multiplier Factor	
	Male	Female
18-24 years	1.15	1.05
25-34 years	0.97	0.96
35-44 years	1.00	0.97
45-54 years	1.16	0.95
55-64 years	1.08	1.11
65-79 years	1.14	1.25

Table 5. Estimates of the Percent of the U.S. Adult Population with Specified Health Characteristics and Estimates of Variance and Bias for Several Design and Estimation Procedures

Health Condition and Design-Estimation Procedures a/	Estimate of Percent with Characteristic	Estimate of Relative Variance	Estimate of Mean Square Error	Estimate of Bias
HHD				
A. Simple Random Model	9.25	.00147	.1658	-0.20
C. Basic Weighting	8.37	.00447	1.4689	-1.08
D. Non-Response Adjusted	8.82	.00366	.6778	-0.63
E. Non-Response and Ratio-Adjusted	8.85	.00386	.6516	-0.60
F. Final Estimate	9.45	.00271	.2401	--
STS				
A. Simple Random Model	4.15	.00355	.0936	0.18
C. Basic Weighting	3.96	.01068	.1601	-0.01
D. Non-Response Adjusted	3.94	.00959	.1453	-0.03
E. Non-Response and Ratio-Adjusted	3.96	.00955	.1445	-0.01
F. Final Estimate	3.97	.00989	.1521	--

(CONTINUED)

Table 5 (CONTINUED)

Health Condition and Design-Estimation Procedures a/	Estimate of Percent with Characteristic	Estimate of Relative Variance	Estimate of Mean Square Error	Estimate of Bias
AP				
A. Simple Random Model	1.35	.01096	.0216	-0.04
C. Basic Weighting	1.26	.02853	.0610	-0.13
D. Non-Response Adjusted	1.28	.02606	.0521	-0.11
E. Non-Response and Ratio-Adjusted	1.29	.02512	.0500	-0.10
F. Final Estimate	1.39	.02345	.0441	--
MI				
A. Simple Random Model	1.11	.01366	.0421	-0.16
C. Basic Weighting	1.09	.01162	.0445	-0.18
D. Non-Response Adjusted	1.18	.01020	.0202	-0.09
E. Non-Response and Ratio-Adjusted	1.19	.00904	.0185	-0.08
F. Final Estimate	1.27	.00924	.0144	--

a/ See Section 2 for description of Design-Estimation Procedure

Table 6. Index or Ratio of Variance for Selected Design and Estimation Procedures to Variance for Simple Random Model for Each of 4 Disease Categories, and the Average of these Ratios

Method which is Numerator of Ratio	The Index				
	Average Index	Disease Category			
		HHD	STS	AP	MI
A. Simple Random Model	1.00	1.00	1.00	1.00	1.00
B. Equal Weighted	2.10	2.24	1.99	3.45	0.72
C. Basic Weighted	2.37	3.04	3.01	2.60	0.85
D. Non-Response Adjusted	2.08	2.49	2.70	2.38	0.75
E. Non-Response and Ratio Adjusted	2.03	2.63	2.69	2.29	0.66
F. Final Estimates	1.86	1.84	2.79	2.14	0.68

Table 7. Index or Ratio of Mean Square Error for Selected Design and Estimation Procedures to Mean Square Error for Simple Random Model for Each of 4 Disease Categories, and the Average of these Ratios.

Method which is Numerator of Ratio	The Index				
	Average Index	Disease Category			
		HHD	STS	AP	MI
A. Simple Random Model	1.00	1.00	1.00	1.00	1.00
B. Equal Weighted	1.93	1.96	1.58	3.25	0.92
C. Basic Weighted	3.81	8.86	1.71	2.82	1.06
D. Non-Response Adjusted	2.13	4.09	1.55	2.41	0.48
E. Non-Response and Ratio-Adjusted	2.06	3.93	1.54	2.31	0.44
F. Final Estimates	1.36	1.45	1.63	2.04	0.34

INFERENTIAL ASPECTS OF THE RANDOMNESS OF SAMPLE SIZE IN SURVEY SAMPLING

J. Durbin

London School of Economics and Political Science
England

1. Introduction

Estimates derived from sample surveys of human populations are almost invariably based on samples of random size. The basic reason for this is that we usually require estimates for sub-populations, e.g. the various age/sex groups. Even if an estimate is required for the whole population the number of individuals actually observed is usually less than the number selected by a random amount depending on such factors as death, removal or refusal to cooperate.

The purpose of this paper is to consider the consequences for statistical inference of this randomness of sample size. Since this symposium is concerned with foundations I make no apology for confining myself to the simplest situations capable of bringing out the essential ideas. In particular I will deal mainly with the case where the sample size n is large with a known distribution and where the observations are independent normal variables with known variance which, without loss of generality, can be taken as unity. In fact to a first approximation this is not a completely unrealistic situation since in surveys, samples are usually large, and the distribution of sample size can often be approximated to a reasonable degree of accuracy by a binomial distribution with a parameter whose value is known from sources outside the survey, e.g. census data.

On these assumptions sample size is an ancillary statistic and this brings us up against the question, which will be our main concern in this paper, of whether the inference should be made conditionally given n or unconditionally over variations in n. Although the question is introduced in the context of the survey situation the problem is of general importance and the treatment is therefore presented in general terms where this seems appropriate.

Throughout the paper the discussion is conducted mainly in terms of testing hypotheses although I am well aware that from a practical point of view estimation, which I take to include interval estimation, is much more important than testing. However, I believe that most of the results obtained have direct and obvious analogues for interval estimation.

The problem is first considered from the standpoint of Neyman-Pearson theory. The powers of the most powerful, the conditional and the unconditional one-sided tests on the mean are derived to $O(1/N)$ where $N = E(n)$. It is found that to this order neither the conditional nor the unconditional test is uniformly more powerful than the other, and that each is most powerful against particular alternatives.

The fact that neither the conditional nor the unconditional test is uniformly more powerful than the other is particularly interesting in the light of Welch's (1939) example of a situation in which an unconditional test is found which is uniformly more powerful than the conditional test given the ancillary. This raises the question whether an example can be found of the opposite situation in which the conditional test is uniformly more powerful than the appropriate unconditional test. A simple example of this kind is given in Section 3.

After referring in Section 4 to the case where the distribution of sample size depends on unknown parameters a brief treatment along Bayesian lines is given in Section 5.

The final section consists of some general remarks on inference in the presence of ancillary statistics. My basic position is that conditional tests and intervals seem more appealing than unconditional procedures on grounds of common sense, but that it seems very disturbing that after over forty years of the theory of statistical inference a satisfactory non-Bayesian theory justifying this position has still not been developed.

2. Neyman-Pearson treatment.

Suppose sample size n has known density p(n) and that observations are independent $N(\mu,1)$. The likelihood is

$$p(n)\ (2\pi)^{-n/2} e^{-\frac{1}{2}\sum_1^n (x_i-\mu)^2},$$

so the likelihood-ratio criterion for a test of the hypothesis that μ takes a specified value, which without loss of generality may be taken as zero, against $\mu = \mu_1 > o$ is

$$\ell = e^{-\frac{1}{2}\sum_1^n x_i^2 + \frac{1}{2}\sum_1^n (x_i - \mu_1)^2}$$

$$= e^{-n\bar{x}\mu_1 + \frac{1}{2} n \mu_1^2}$$

It should be noted that the density p(n) has disappeared on taking the ratio. The most powerful test is therefore given by the critical region

$$-n\,\bar{x}\,\mu_1 + \frac{1}{2}\,n\,\mu_1^2 < \log k, \text{ i.e.}$$

$$\sqrt{n}\,\bar{x} > \frac{1}{2}\sqrt{n}\,\mu_1 - \frac{\log k}{\sqrt{n}\mu_1},$$

where $\sqrt{n}\,\bar{x}$ is $N(0,1)$. Since this region depends on μ_1 there is no uniformly most powerful test.

In order to develop a suitable asymptotic theory let us consider a family of alternatives $\mu_1 = \gamma/\sqrt{N}$ where $N = E(n)$ and γ is a positive constant.

Since p(n) is known N is known. The region is then

$$\sqrt{n}\ x > \frac{1}{2}\sqrt{\frac{n}{N}}\,\gamma \quad \sqrt{\frac{N}{n}}\,\lambda \tag{1}$$

where $\lambda = -\log k/(\sqrt{N}\ \mu)$ and is determined so that the probability that (1) holds is α, the significance level. We therefore have

$$\alpha = \Sigma\ p(n) \int_{\frac{1}{2}\sqrt{\frac{n}{N}}\gamma + \sqrt{\frac{N}{n}}\lambda}^{\infty} (2\pi)^{-\frac{1}{2}} e^{-\frac{1}{2}z^2}\, dz \tag{2}$$

where Σ denotes summation over the range of n.

Assume that the cumulants of n are O(N) as is the case with binomial and Poisson sampling and let

$$b = \lim_{N\to\infty} V(n)/N.$$

Expanding (2) about $\frac{1}{2}\gamma + \lambda$ we have

$$\alpha=\Sigma p(n)\Bigg[\int_{\frac{1}{2}\gamma+\lambda}^{\infty} (2\pi)^{-\frac{1}{2}} e^{-\frac{1}{2}z^2} dz - \left(\frac{1}{2}\sqrt{\frac{n}{N}}\gamma+\sqrt{\frac{N}{n}}\lambda-\frac{1}{2}\gamma-\lambda\right)(2\pi)^{-\frac{1}{2}} e^{-\frac{1}{2}(\frac{1}{2}\gamma+\lambda}$$

$$+\frac{1}{2}\left(\frac{1}{2}\sqrt{\frac{n}{N}}\gamma+\sqrt{\frac{N}{n}}\lambda-\frac{1}{2}\gamma-\lambda\right)^2\left(\frac{1}{2}\gamma+\lambda\right)(2\pi)^{-\frac{1}{2}} e^{-\frac{1}{2}(\frac{1}{2}\gamma+\lambda)^2} + 0(N^{-3/2})\Bigg] \tag{3}$$

Putting $dn = n - N$ and neglecting terms of smaller order than N^{-1} we have

$$E\left(\frac{1}{2}\sqrt{\frac{n}{N}}\gamma+\sqrt{\frac{N}{n}}\lambda-\frac{1}{2}\gamma-\lambda\right)$$

$$= E\left[\frac{1}{2}\left\{\frac{1}{2}\frac{dn}{N} - \frac{1}{8}\left(\frac{dn}{N}\right)^2\right\}\gamma + \left\{-\frac{1}{2}\frac{dn}{N} + \frac{3}{8}\left(\frac{dn}{N}\right)^2\right\}\gamma\right]$$

$$= \left(-\frac{1}{16}\gamma + \frac{3}{8}\lambda\right)^2 bN^{-1}.$$

To the same order we have

$$E\left(\frac{1}{2}\sqrt{\frac{n}{N}}\,\gamma + \sqrt{\frac{N}{n}}\,\lambda - \frac{1}{2}\,\gamma-\lambda\right)^2 = E\left(\frac{1}{4}\,\frac{dn}{N}\,\gamma - \frac{1}{2}\,\frac{dn}{N}\,\lambda\right)^2$$

$$= \frac{1}{4}\left(\frac{1}{2}\gamma-\lambda\right)^2 bN^{-1} \quad .$$

Let a be defined by $\alpha = \int_a^\infty (2\pi)^{-\frac{1}{2}} e^{-\frac{1}{2}z^2} dz$. Expanding terms in $\frac{1}{2}\gamma+\lambda$ in (3) about a and taking only terms of order N^{-1} we find that λ is determined by the relation

$$0 = -\left\{\frac{1}{2}\gamma+\lambda-a+\left(-\frac{1}{16}\gamma + \frac{3}{8}\lambda\right) bN^{-1}\right\} + \frac{1}{8}\left(\frac{1}{2}\gamma-\lambda\right)^2 ab\ N^{-1} \quad . \quad (4)$$

Thus to a first approximation $\lambda = a - \frac{1}{2}\gamma$.

On the alternative hypothesis $E(\sqrt{n}\ \bar{x}|n) = \sqrt{\frac{n}{N}}\gamma$. The power β_{mp} is the probability that $\sqrt{n}\bar{x} > \frac{1}{2}\sqrt{\frac{n}{N}}\,\gamma+\sqrt{\frac{N}{n}}\lambda$, i.e. $n\ \bar{x} - \sqrt{\frac{n}{N}}\,\gamma > -\frac{1}{2}\sqrt{\frac{n}{N}}\,\gamma + \sqrt{\frac{N}{n}}\,\lambda$, and is therefore

$$\beta_{mp} = \Sigma\ p(n) \int_{-\frac{1}{2}\sqrt{\frac{n}{N}}\gamma + \sqrt{\frac{N}{n}}\lambda}^{\infty} (2\pi)^{-\frac{1}{2}} e^{-\frac{1}{2}z^2} dz \ .$$

Expanding about $a - \gamma$ and ignoring terms of smaller order than N^{-1} gives

$$\beta_{mp} = \Sigma\ p(n)\ \Big[\int_{a-\gamma}^{\infty} (2\pi)^{-\frac{1}{2}} e^{-\frac{1}{2}z^2} z$$

$$-\left(-\frac{1}{2}\sqrt{\frac{n}{N}}\,\gamma + \sqrt{\frac{N}{n}}\,\lambda-a + \gamma\right)(2\pi)^{-\frac{1}{2}} e^{-\frac{1}{2}(a-\gamma)^2}$$

$$+ \frac{1}{2}\left(-\frac{1}{2}\sqrt{\frac{n}{N}}\gamma + \sqrt{\frac{N}{n}}\lambda - a + \gamma\right)^2 (a-\gamma)(2\pi)^{-\frac{1}{2}} e^{-\frac{1}{2}(a-\gamma)^2}]$$

$$= \beta_f + [-\{\frac{1}{2}\gamma + \lambda - a + (\frac{1}{16}\gamma + \frac{3}{8}\lambda) bN^{-1}\}$$

$$+ \frac{1}{8}(\frac{1}{2}\gamma + \lambda)^2 (a-\gamma) bN^{-1}]\ (2\pi)^{-\frac{1}{2}} e^{-\frac{1}{2}(a-\gamma)^2}$$

on expanding as before, where $\beta_f = \int_{a-\gamma}^{\infty} (2\pi)^{-\frac{1}{2}} e^{-\frac{1}{2}z^2} dz.$

Note that β_f is the power of the fixed sample size test with n = N. On substituting from (4) the expression in square brackets reduces to

$$[-\{\frac{1}{8}\gamma + \frac{1}{4}\gamma\lambda\ a - \frac{1}{8}\gamma(\frac{1}{2}\gamma + \lambda)^2\} bN^{-1}] \quad .$$

Substituting the first approximation to λ, namely $a - \frac{1}{2}\gamma$, we obtain finally

$$\beta_{mp} = \beta_f - \frac{1}{8}\gamma(1 + a\gamma - a^2) b \quad (2\pi)^{-\frac{1}{2}} e^{-\frac{1}{2}(a-\gamma)^2} N^{-1}, \text{ i.e.}$$

$$\beta_{mp} = \beta_f - (1 + a\gamma - a^2) cN^{-1} \tag{5}$$

to $O(N^{-1})$, where

$$c = \frac{\gamma b}{8\sqrt{2\pi}} \quad e^{-\frac{1}{2}(a-\gamma)^2} \quad .$$

It is interesting to note that $\beta_{mp} > \beta_f$ when $\gamma < a - a^{-1}$. This means that for γ in this range the most powerful random sample size test is asymptotically more powerful than the corresponding most powerful fixed sample size test with $n = N$. The possibility of obtaining greater power by randomization of sample size has been systematically studied by Cohen (1958).

Of course in practice one is rarely in a position where one wishes to discriminate against a particular value of γ. In order to use the test one has to select arbitrarily a particular value of γ, say γ_1, and one hopes that the test which is most powerful against this particular value has sufficiently high power against other values of γ in the range of interest. Let us therefore compute power as a function of γ when the test is used in this way.

The rejection region is $\sqrt{n}\,\bar{x} \geq \frac{1}{2}\sqrt{\frac{n}{N}}\,\gamma_1 + \frac{N}{n}\lambda$ where λ is obtained from (4) with $\gamma = \gamma_1$. On a general alternative $E(\sqrt{n}\,\bar{x}) = \sqrt{\frac{n}{N}}\,\gamma$. The power is therefore

$$\beta_1 = \Sigma\, p(n) \int_{\sqrt{\frac{n}{N}}(\frac{1}{2}\gamma_1 - \gamma) + \sqrt{\frac{N}{n}}\lambda}^{\infty} \sqrt{2\pi}\; e^{-\frac{1}{2}z^2}\, dz$$

Using the same techniques as before we obtain

$$\beta_1 = \beta_f + [-\{\tfrac{1}{2}\gamma_1 + \lambda - a + (-\tfrac{1}{16}\gamma_1 + \tfrac{1}{8}\gamma + \tfrac{3}{8}\lambda)bN^{-1}\}$$

$$+ \tfrac{1}{8}(-\tfrac{1}{2}\gamma_1 + \gamma + \lambda)^2 (a-\gamma)bN^{-1}]\ (2\pi)^{-\frac{1}{2}} e^{-\frac{1}{2}(a-\gamma)^2}$$

which, substituting from (4) with $\gamma = \gamma_1$ and then putting $\lambda = a - \frac{1}{2}\gamma_1$, gives

$$\beta_1 = \beta_{mp} - (\gamma-\gamma_1)^2 cN^{-1} . \tag{6}$$

to $O(N^{-1})$. The term in $(\gamma-\gamma_1)^2$ represents the power lost by constructing the test to give maximum power against γ_1 instead of against the true value γ.

Next we consider the conditional test. This is constructed by treating the observed sample size n as fixed, i.e. rejecting when $\sqrt{n}\,\bar{x} > a$ where $\alpha = \int_a^\infty (2\pi)^{-\frac{1}{2}} e^{-\frac{1}{2}z^2} dz$ as before. On the alternative $E(\sqrt{n}\,\bar{x}) = \sqrt{\frac{n}{N}}\gamma$ so the power is

$$\beta_c = \Sigma\, p(n) \int_{a-\sqrt{\frac{n}{N}}\gamma}^\infty (2\pi)^{-\frac{1}{2}} e^{-\frac{1}{2}z^2} dz$$

$$= \Sigma\, p(n)\Big[\int_{a-\gamma}^\infty (2\pi)^{-\frac{1}{2}} e^{-\frac{1}{2}z^2} dz - \Big(a-\sqrt{\frac{n}{N}}\gamma-a+\gamma\Big) 2\pi e^{-\frac{1}{2}(a-\gamma)^2}$$

$$+ \frac{1}{2}\Big(-\sqrt{\frac{n}{N}}\gamma+\gamma\Big)^2 (a-\gamma)(2\pi)^{-\frac{1}{2}} e^{-\frac{1}{2}(a-\gamma)^2}\Big] .$$

$$= \beta_f + E\Big[-\Big(1-\sqrt{\frac{n}{N}}\Big)\gamma + \frac{1}{2}\Big(1-\sqrt{\frac{n}{N}}\Big)^2\gamma^2 (a-\gamma)\Big](2\pi)^{-\frac{1}{2}} e^{-\frac{1}{2}(a-\gamma)^2}$$

$$= \beta_f - \frac{1}{8}\gamma(1-a\gamma+\gamma^2)b\ (2\pi)^{-\frac{1}{2}} e^{-\frac{1}{2}(a-\gamma)^2}\ N^{-1} \tag{7}$$

$$= \beta_{mp} - (a-\gamma)^2\ cN^{-1} \tag{8}$$

to $O(N^{-1})$ using the same techniques as previously.

From (7), we note the unexpected fact that if $a > \gamma + \gamma^{-1}$ the conditional test is more powerful than the fixed sample size test with $n = N$.

(8) shows that the conditional test is most powerful to $O(N^{-1})$ against an alternative with $\gamma = a$, i.e. one with symptotic power $\beta_c = 0.5$. Moreover the test has the same power to $O(N^{-1})$ as a test constructed to have maximum power against the alternative $\gamma = a$.

Finally we consider the unconditional test constructed from the marginal distribution of the maximum likelihood estimator $\bar{x}$ of μ. The best critical region based on this distribution is $\bar{x}$ > constant where the constant can be taken as $N^{-\frac{1}{2}}\nu$. The value of ν is determined by the condition $\alpha = \Pr(\sqrt{n}\ \bar{x} > \sqrt{\frac{n}{N}}\ \nu)$, i.e.

$$\alpha = \Sigma\ p(n) \int_{\sqrt{\frac{n}{N}}\nu}^{\infty} (2\pi)^{-\frac{1}{2}} e^{-\frac{1}{2}z^2} dz$$

$$= \Sigma\ p(n) [\int_a^{\infty} (2\pi)^{-\frac{1}{2}} e^{-\frac{1}{2}z^2} dz - (\sqrt{\frac{n}{N}}\ \nu - a)(2\pi)^{-\frac{1}{2}} e^{-\frac{1}{2}a^2}$$

$$+ \frac{1}{2}(\sqrt{\frac{n}{N}}\nu - a)^2 a (2\pi)^{-\frac{1}{2}} e^{-\frac{1}{2}a^2}]$$

$$= \alpha + E\{-\sqrt{\tfrac{n}{N}}\nu + a + \tfrac{1}{2}(\sqrt{\tfrac{n}{N}}\nu - a)^2\ a\}\ \ (2\pi)^{-\frac{1}{2}}e^{-\frac{1}{2}a^2}$$

to $O(N^{-1})$. This gives

$$-\nu + a + \tfrac{1}{8}(\nu + \nu^2\ a)\ bN^{-1} = 0. \tag{9}$$

The power of the test is $\Pr\{\sqrt{n}\,\bar{x} - \sqrt{\tfrac{n}{N}}\gamma > \sqrt{\tfrac{n}{N}}(\nu-\gamma)\}$ given the alternative, i.e.

$$\beta_u = \Sigma\ p(n) \int_{\sqrt{\frac{n}{N}}(\nu-\gamma)}^{\infty} (2\pi)^{-\frac{1}{2}}e^{-\frac{1}{2}z^2}\,dz$$

$$= \beta_f + [-\nu+a+ \tfrac{1}{8}\{\nu-\gamma+(\nu-\gamma)^2(a-\gamma)\}]$$

$$\cdot\ b(2\pi)^{-\frac{1}{2}}e^{-\frac{1}{2}(a-\gamma)^2}\ N^{-1} .$$

Using (9) this gives

$$\beta_u = \beta_{mp} - (2a-\gamma)^2\ cN^{-1} . \tag{10}$$

It follows that to $O(N^{-1})$ the unconditional test is most powerful against an alternative for which $\gamma = 2a$. To the same order it has the same power as a test constructed to have maximum power against the alternative $\gamma = 2a$.

More important is the comparison with the conditional test. We find

$$\beta_c - \beta_u = a(3a-2\gamma)\ cN^{-1}\ . \tag{11}$$

This implies that to $O(N^{-1})$ the conditional test is more powerful than the unconditional test for $\gamma < \frac{3}{2}$ a and vice versa for $\gamma > \frac{3}{2}$ a. In other words, for relatively low power the conditional test is better and for relatively high power the unconditional test is better. For example, for a one-sided test at the 5% level the conditional test is better if the asymptotic power is less than approximately 0.80.

3. An example of a conditional test which is uniformly more powerful than the corresponding unconditional test.

One of the chief points of interest in the results of the last section is the fact that for the particular problem under discussion neither the conditional nor the unconditional test is asymptotically uniformly more powerful than the other. There is already available in the literature an example of a situation in which a conditional test given an ancillary statistic is uniformly less powerful than a particular unconditional test. This is the problem of testing for the location parameter of a uniform distribution which was studied by Welch (1939). We give in this section an example of the converse situation where the conditional test given the ancillary is uniformly more powerful than obvious unconditional test.

Suppose that the sample size n is n_1 with probability $\frac{1}{2}$ and n_2 with probability $\frac{1}{2}$, $n_2 > n_1$. For given n the observations are independent values from the uniform distribution on the interval $(0,\theta)$. Let x_n denote the largest observation. It is easy to show that the pair of values n, x_n is minimal sufficient for θ. n is therefore an ancillary statistic.

We have $\Pr(x_n \leq a|n) = (a/\theta)^n$. The most powerful conditional test of $\theta = 1$ against a particular value of $\theta > 1$ at significance level α is to reject when $x_n > a_n$ when $a_n^n = 1 - \alpha$. The power of the test is

$$\beta_c = \frac{1}{2}\{1 - (a_{n_1}/\theta)^{n_1} + 1 - (a_{n_2}/\theta)^{n_2}\}$$

$$= 1 - \frac{1}{2}(1 - \alpha)\{\theta^{-n_1} + \theta^{-n_2}\} .$$

Note that this test has no more power than the randomised test: reject with probability α if $x_n \leq 1$ and with probability one if $x_n > 1$. This fact does not affect the point at issue, however.

The maximum-likelihood estimator of θ is x_n so the obvious unconditional test is based on the marginal distribution of x_n. The most powerful test based on this distribution is to reject when $x_n > b$ where $1 - \alpha = \frac{1}{2}(b^{n_1} + b^{n_2})$. The power is

$$\beta_u = \frac{1}{2}\{1 - (\frac{b}{\theta})^{n_1} + 1 - (\frac{b}{\theta})^{n_2}\} .$$

Thus $\beta_c > \beta_u$ if

$$(\frac{b}{\theta})^{n_1} + (\frac{b}{\theta})^{n_2} > (1 - \alpha)\ \{(\frac{1}{\theta})^{n_1} + (\frac{1}{\theta})^{n_2}\}$$

$$> \frac{1}{2}(b^{n_1} + b^{n_2})\ \ \{(\frac{1}{\theta})^{n_1} + (\frac{1}{\theta})^{n_2}\} ,$$

$$\text{i.e. } \frac{1}{2}\left(\frac{b}{\theta}\right)^{n_1} + \frac{1}{2}\left(\frac{b}{\theta}\right)^{n_2} - \frac{1}{2}\, b^{n_1}\left(\frac{1}{\theta}\right)^{n_2} - \frac{1}{2}\, b^{n_2}\left(\frac{1}{\theta}\right)^{n_1} > 0,$$

$$\text{i.e. } \frac{1}{2}\left(b^{n_1} - b^{n_2}\right) \left\{\left(\frac{1}{\theta}\right)^{n_1} - \left(\frac{1}{\theta}\right)^{n_2}\right\} > 0.$$

Since $b < 1$, $\theta > 1$ and $n_2 > n_1$ we have $b^{n_1} > 0$ and $\left(\frac{1}{\theta}\right)^{n_1} - \left(\frac{1}{\theta}\right)^{n_2} > 0$. The conditional test is therefore uniformly more powerful than the unconditional test for all θ.

4. Unknown distribution of sample size.

Suppose the distribution of sample size depends on one or more unknown parameters θ but does not depend on the parameter of interest. Returning to the problem considered in section 2 the likelihood would now be written

$$p(n,\theta)\ (2\pi)^{-n/2} e^{-\frac{1}{2}\sum_{1}^{n} (x_i - \mu)^2} \tag{11}$$

so the likelihood ratio for a test of $\mu = 0$ against $\mu = \mu_1 > 0$ would remain as before; however, its distribution depends on the unknown nuisance parameters θ. In this situation the Neyman-Pearson theory requires us to construct a most powerful similar test by obtaining the most powerful test conditional on n fixed. See for example Kendall and Stuart (1961), chapter 23. We therefore have the curious situation in which if the distribution of n is completely known the theory calls for an unconditional test, whereas if the distribution is unknown even only to a very slight extent a conditional test is required.

This seems to be a very unsatisfactory feature of the theory. Consider for example the common situation in which n is determined by a binomial distribution. It will often be the case in practice that the binomial parameter is known almost exactly, e.g. that the population of males in a given population of adults is very nearly equal to a half, whereas it would be incorrect to claim that the value is known with absolute precision. It seems quite wrong that by following the theory literally, this case would be treated on equal terms with the case in which nothing whatever is known about the binomial parameter.

5. Bayesian treatment.

Let us consider the Bayesian treatment of the general problem in which the distribution of n depends on a set of unknown parameters θ. Suppose the joint prior distribution of θ and μ is $\pi(\theta,\mu)$. Multiplying by the likelihood (11) we obtain the joint posterior distribution

$$\pi(\theta,\mu|x_1,..,x_n) \propto \pi(\theta,\mu)\ p(n,\theta)e^{-\frac{n}{2}(\bar{x}-\mu)^2} . \tag{12}$$

If, as will usually be justifiable, we can assume that θ and μ have independent prior distributions we can integrate out θ from (12) to give

$$\pi(\mu|x_1,..,x_n) \propto \pi(\mu)\ e^{-\frac{1}{2}n(\bar{x}-\mu)^2} \tag{13}$$

Consequently we obtain exactly the same analysis whether p(n) depends on unknown parameters or not. It is in fact the same as the conditional analysis derived by treating n as if it were fixed.

Where $\pi(\mu)$ can be taken as uniform on the range $(-\infty,\infty)$ the posterior distribution becomes $N(\bar{x},\frac{1}{n})$. The The resulting Bayesian tests and confidence intervals are then equivalent to those given by the conditional tests considered in section 2.

6. Some general remarks.

(a) The common-sense arguments in favor of using conditional tests and confidence intervals where the sample size is random seem to me to be extremely strong. In any given situation the information available on the quantities under study comes only from the sample actually observed. If the sample size is determined by a random mechanism and one happens to get a large sample one knows perfectly well that the quantities of interest are measured more accurately than they would have been if the sample size had happened to be small. It seems self-evident that one should use the information available on sample size in the interpretation of the results. To average over variations in sample size which might have occurred but did not occur, when in fact the sample size is exactly known, seems quite wrong from the standpoint of the analysis of the data actually observed. Yet when the distribution of sample size is known this is what seems to be required by the Neyman-Pearson theory. On the other hand if the distribution of sample size depends on unknown parameters even to the minutest degree, the theory calls for a conditional analysis with sample size held fixed. Since the two analyses can be very different there is a discontinuity here which seems to be very undesirable.

(b) There seems to be a common impression that a conditional test can be generally expected to be less powerful than the corresponding unconditional test based on the marginal distribution of the maximum likelihood estimator of the parameter of interest. This is perhaps attributable to the influence of the examples discussed by Bartlett (1939) and Cox (1958, p. 360). In fact, as we have seen, no clear picture is found. It seems to be possible for either the conditional test or the unconditional test or neither to be uniformly more powerful than the other.

(c) Numerous attempts have been made from Fisher onwards to construct a satisfactory theory of conditional tests given the ancillary. Most workers in the field would agree that these attempts have failed. The main difficulties seem to be the following:

(i) The ancillary is not always unique. Where it is not, different conditional analyses can be constructed which lead to different conclusions. There seems to be no way of choosing between the different possibilities. A simple example is discussed by Basu (1964, Section 3)

(ii) There appears to be no general algorithm for discovering ancillary statistics when they exist. There is no guarantee that inspection of the minimal sufficient statistic will automatically reveal any ancillary that is present.

To meet the first of these difficulties Basu (1964) attempted to distinguish between "performable" and "non-performable" experiments, and suggested that conditional inference should be confined to the former. This, however, seems to bring in considerations from outside the theory that may not be applicable in all cases and therefore cannot be regarded as giving a satisfactory solution of the theoretical difficulties.

(d) In all the examples considered so far the conditioning variable has been an ancillary statistic in the strict sense, i.e. a component of the minimal sufficient statistic whose distribution is known and does not depend on the parameter under study. Examples can be constructed, however, where a conditional analysis seems to be called for on common sense grounds yet over most of the range of the observations the conditioning variable is not part of the minimal sufficient statistic.

Consider for example a situation where with probability 1/2 a random number of X_2 of trials is determined by a known distribution with density $p(x_2)$ and then x_2 binomial trials are performed with unknown probability θ giving x_3 successes; similarly, with probability 1/2 a random integer X_3 is determined by a known distribution with density $q(x_3)$ and binomial trials are performed with probability θ until x_3 successes are observed, the number of trials necessary being x_2.

Thus the experiment consists of a mixture of direct and inverse binomial sampling with probabilities one half each.

Let $x_1 = 0$ if direct sampling is used and $x_1 = 1$ if inverse sampling is used. The probability of the observed values x_1, x_2, x_3 is therefore

$$\frac{1}{2}(1-x_1)p(x_2)\binom{x_2}{x_3}\theta^{x_3}(1-\theta)^{x_2-x_3}+\frac{1}{2}x_1 q(x_3)\binom{x_2-1}{x_3-1}\theta^{x_3}(1-\theta)^{x_2-x_3}, \tag{14}$$

$$x_1 = 0;\ x_2 = 1, 2, ..;\ x_3 = 0, 1, .., x_2$$

<u>or</u> $$x_1 = 1;\ x_2 = x_3, x_3+1, ..;\ x_3 = 1,2, ...$$

The conditional probability that $x_1 = 0$ given x_2, x_3 is

$$= \frac{p(x_2)\binom{x_2}{x_3}}{p(x_2)\binom{x_2}{x_3} + q(x_3)\binom{x_2-1}{x_3-1}} \qquad (x_3 \neq 0) \tag{15}$$

$$= 1 \qquad (x_3 = 0).$$

which does not depend on θ; consequently (x_2, x_3) is a sufficient statistic. The theory of sufficiency therefore requires us to ignore the observed value of x_1 and base the analysis only on the observed values of x_2, x_3. It is clear from (14) that the marginal distribution of (x_2, x_3) would be rather complicated so the optimal analysis based on it would be a complex affair.

But in practice one would know the value of x_1 so one would know whether direct or inverse sampling had been used. Moreover, in the direct sampling case one would know the number of trials x_2 and in the inverse case one would know the number of successes x_3. Since these values would have been determined independently of the unknown θ, common sense would seem to require that the analysis is carried out conditionally given $x_1 = 0$ and x_2 or $x_1 = 1$ and x_3 as the case may be. The case for doing so seems intuitively almost as strong as in the situation where the conditioning variable is a component of the minimal sufficient statistic. The fact that this procedure would be in conflict with the theory of sufficiency should perhaps serve as a warning that one should proceed somewhat cautiously in adopting conditional analyses purely on intuitive grounds.

It is noteworthy that the Bayesian analysis is consistent with the conditional analysis just described, e.g. if $x = 0$ is observed the analysis is the same as if the experiment had consisted solely of a fixed number x_2 of binomial trials.

(e) A theory of conditional inference which has attracted considerable interest is that of Birnbaum (1962). Starting with the concept of the evidential meaning of an experimental outcome together with a pair of axioms called by him the principles of sufficiency and conditionality, Birnbaum proves that evidential meaning depends only on the likelihood function. This implies that the sampling distribution of the observations is irrelevant so far as evidential meaning is concerned. At first sight this would seem to suggest that the search for a system of conditional inference based on conditional sampling distributions is pointless. It turns out, however, that if attention is restricted to conditional inferences based only on the value of the minimal sufficient statistic, Birnbaum's proof fails (Durbin, 1968). Since this restriction will often, if not always, be appropriate, it cannot be maintained that Birnbaum's theorem establishes the irrelevance of sampling distributions to conditional inference, even if the basic ideas on which Birnbaum's theory is based are otherwise acceptable.

(f) We have seen that the concept of power and its analogue for confidence intervals do not provide satisfactory criteria for evaluating conditional procedures. The advantage of the conditional analysis is that when the value of the ancillary statistic indicates that the estimate of the parameter of interest is relatively accurate, or inaccurate as the case may be, this information is directly utilized in the analysis. This seems so sensible that it is surprising that no satisfactory objective criterion appears to have been put forward which indicates an advantage in favor of the conditional analysis. One feels that the user of the results of the analysis ought to be "better off" in the long run in some simple way for being told which in a sequence of estimates is relatively accurate and which is relatively inaccurate.

The nearest to such a criterion known to me is an idea of Buehler (1959) based on betting. Buehler considers two players, Peter and Paul, the first of whom sets confidence limits at level α, thus indicating his readiness to accept bets that the true parameter lies inside the interval with odds in favor of $1-\alpha:\alpha$, while Peter is free to bet against Paul for or against inclusion or alternatively to refrain from placing a bet. It is clear that for the situation considered in Section 2, if Peter bets on unconditional confidence intervals and Paul bets on conditional confidence intervals, Paul will gain in the long run. We are supposing that the players will operate in the following fashion. If A's interval is longer than B's, A will conclude that B's is "too short" and will therefore bet against inclusion while B will bet that A's interval includes the parameter.

Apart from general objections to criteria based on betting it is clear that this idea is not altogether satisfactory since, as pointed out by Buehler, it does not even work completely satisfactorily for intervals for the normal mean based on the t distribution. Buehler's point is as follows. Let a be an arbitrary positive number. If one adopts the strategy of betting against inclusion where the sampling variance $s^2 < a$ and for inclusion when $s^2 > a$ one achieves a positive expected return.

(g) The Bayesian treatment of the problem is internally consistent and appears to give sensible results. Although, so far as I am concerned, this does not lead inexorably to the adoption of a firm Bayesian position, it cannot be denied that the considerations discussed in this paper lend strong support to the Bayesian viewpoint.

(h) Bartholomew (1967) has recently given a general review of problems of inference in the presence of random variation in sample size, but his results overlap very little with the treatment given here. Indeed some of Bartholomew's conclusions are in flat contradiction with my own. For example, in the Summary (p. 53) the following is stated

> "In the Neyman-Pearson theory of hypothesis testing it is customary to calculate significance levels and power functions on the assumption that the sample size is fixed. The main purpose of this paper is to propagate the view that the restriction of the theory to reference sets in which the sample size is constant is neither necessary nor desirable. The paper falls into two parts. In the first we consider the test of a simple null hypothesis versus a simple alternative. In this case it is pointed out that the optimum test requires the rejection of the null hypothesis if the likelihood ratio is less than a constant, K, which is independent of n, the sample size. In this respect it is identical with Bayesian and likelihood methods of dealing with the same problem."

However, it is clear from the results of sections 2 and 5 above for a composite test against a general alternative that the Bayesian analysis is equivalent to a conditional analysis given n fixed and not to an optimum likelihood-ratio procedure in which n is permitted to vary. Consequently, any inference that might be drawn from Bartholemew's treatment of the test of a simple hypothesis or from parts of the printed discussion of the paper, that there is a close affinity between Bayesian methods and methods based on averaging out random variations in sample size, seems to me unjustified.

On the following page (54) Bartholomew says the following:

"The question of fixing sample size is a special case of a more general situation in which other random variables associated with the experiment, but not depending on the parameter under test, may be treated as constant. Such quantities are called ancillary statistics. The argument in support of this practice is usually expressed as follows; see, for example, Cox (1958). Random variables, whose distributions do not involve the parameter of interest, cannot tell us anything about that parameter. If, therefore, we use the reference set in which they are allowed to vary, the values of α and β will depend upon these distributions. Since the inference is based on α and, to a lesser extent, on β the inference is made to depend, in part, on 'irrelevant' features of the experiment.

In this paper we propose to explore the consequences of abandoning the principle of conditioning when the ancillary statistic in question is the sample size. This means that α and β will be determined by reference to repetitions of the experiment as it was actually performed. The essence of our reply to the criticism of the preceding paragraph will be that although α and β do depend on the outcome of an irrelevant chance event the <u>inference to be drawn from them should not</u>."

These remarks are in conflict with my general view that on the whole the conditional inference is to be preferred. Fortunately, the practical differences between us are less than might appear at first sight. There are two reasons for this. First, for a number of the particular procedures considered by Professor Bartholomew the probability of Type I error α_n is either constant or approximately constant (c.f. his remarks at the end of Section 5 on p. 66), and a constant value for α_n is equivalent to the use of a conditional test. Secondly, I believe that a number of his remarks on the importance of relating α_n and the Type II error rate β_n to n apply quite independently of whether or not n has been determined at random.

References

Bartholomew, D. J. (1967). Hypothesis testing when the sample size is treated as a random variable. J. Roy. Statist. Soc. B., 29, 53.

Basu, D. (1964). Recovery of ancillary information. Sankhya, A., 26, 3.

Bartlett, M. S. (1939). A note on the interpretation of quasi-sufficiency. Biometrika, 31, 391.

Birnbaum, A. (1962). On the foundations of statistical inference. J. Amer. Statist. Assoc., 57, 269.

Buehler, R. J. (1959). Some validity criteria for statistical inferences. Ann. Math. Statist., 30, 845.

Cohen, L. (1958). On mixed single sample experiments. Ann. Math. Statist. 29, 947.

Cox, D. R. (1958). Some problems connected with statistical inference. Ann. Math. Statist. 29, 357.

Durbin, J. (1968). On Birnbaum's theorem on the relation between sufficiency, conditionality and likelihood. (To be published.)

Kendall, M. G. and Stuart, A. (1961). The Advanced Theory of Statistics, Vol. 2, Griffin, London.

Welch, B. L. (1939). On confidence limits and sufficiency, with particular reference to parameters of location. Ann. Math. Statist., 10, 58.

BOUNDARIES OF STATISTICAL INFERENCE

W. Edwards Deming

Consultant in Statistical Surveys

WASHINGTON

Purpose. The purpose of this paper is to point out what I believe to be insufficiency of some of the theory or models used in the design of studies carried out for analytic purposes, and to point out need of modification of teaching and of textbooks on statistics.

There are three kinds of statisticians: (1) mathematical statisticians, (2) theoretical statisticians, (3) practical statisticians. The mathematical statistician engages himself in extension of man's knowledge through new mathematical theory. It is his work that circumscribes the boundaries of knowledge, providing a foundation for more efficient collection and use of statistical information in the future, and for long-range improvement of industrial processes and of complex equipment, military and civilian. His interest is mathematical. His work is not glamorous except to mathematicians, and he receives little recognition outside a close group of specialists.

Second, there is the theoretical statistician. The theoretical statistician is engaged in consultation with the aid of theory: he guides his work with the aid of the theory of probability. The theoretical statistician is the servant of experts in other fields, such as traffic, marketing, medicine, engineering, production, chemistry, agriculture, labor force, Census methods. A theoretical statistician has mastered enough theory of statistics to use it safely in sampling, design of experiments, and in a variety of related work. He engages the help of a mathematical statistician from time to time to achieve his goals of delivering maximum information per unit cost.

Third, there is the practical statistician, described years ago by the great Thomas Henry Huxley who said, "The practical man practices the errors of his forefathers." This kind of practical man can be a real hazard in any field.

It is the needs of the student who intends to join the ranks of theoretical statisticians that I am concerned about here.

The contribution of the statistician has its seat in the theory of probability. It is the statistician's knowledge of the theory of probability that distinguishes him from other experts. Knowledge of the theory of probability furnishes a guidepost to the power of statistical theory and also to its limitations. I write from the standpoint of the professional statistician. It is the limitations of statistical inference that this paper is mainly addressed to, and to suggestions that recognition of these limitations lead to for modifications in the teaching of statistics.

It was Shewhart who pointed out first, I believe, that practice demands more care if not also greater depth of knowledge than is required for pure research or for teaching. Philosophy, theory of knowledge, questions of what constitutes establishment of a basis for action, all suddenly loom up as important when one goes into practice. The results of an investigation and the statistician's inference based thereon will be subjected to test. If the statistician is wrong, his error will be discovered, often without much delay. The statistician in practice must expect to face cross-examination by a board of directors or by a lawyer whose aim is to find flaws.

The statistician is the architect of a study. He does not just "draw samples" and write complicated formulas for variances. He designs studies, including controls to detect departures from the prescribed procedure and other nonsampling errors. He extracts information from results. The statistician, in the planning stages, protects the people that he works with from getting too much precision in a study, and from getting so little precision that the study is nigh useless. He helps his client to decide how much to spend for a study. He may recommend no study at all. His chief aim is to do his work in a way that will bring honor to his profession. He knows about the various uncertainties in data, the difficulties of investigations of various kinds, the difficulties of coding, errors in tabulation, the extra variance caused by influence of the investigator, variation attributable to the season of the year, differences attributable to a new set of instructions, even though there be no change in content. This knowledge and experience help the statistician to protect a project from procuring meaningless information. The statistician protects a project from unwarranted conclusions. Too often the statistician's client* sees in data only what he wishes to see to establish a point: this trait is not confined to business research. The statistician, in protecting his client, protects also the client's competition, fellow scientists, and the public.

Naturally, to act as architect of a study, the statistician must work with the substantive experts and must learn superficially enough about the subject-matter to work in it and be helpful. The interchange of ideas between statistician and the substantive experts may extend over a period of weeks or over many months, before formulation of the problem in statistical terms is anywhere near the final stage. This foundation is usually the hardest part of the statistician's participation.

What are facts ? Some people talk about facts as if there were a definite number of people unemployed today, or a definite number of agricultural workers, or of people engaged in the production of food, or of the number of people that ever heard of some specified brand of oven cleaner, or the number of automobiles in need of repair. The paper by Hansen and Tepping in this conference gives adequate illustration, if any were needed. The statistician in practice, knowing that somebody will base a business decision on his inferences, must under-

---

*The word client will denote the expert or group of experts in a substantive field (medicine, engineering, production, marketing, psychology, agriculture, or other) who are responsible to the man that will pay the bill for the study.

stand the meaning of figures. He does not speak of the true value of anything. Instead, he speaks of the result of applying an operational definition.

One of the important lessons in a course in management, I am told, is to get THE FACTS. But what facts ? What are facts ? It might be better to teach students of management to enquire what facts are, and to enquire into the kinds of information that would be relevant to a problem: also to teach something about statistical studies and about operational definitions.

Some people suppose that there is a definite value for the velocity of light, or for the ratio e/m, or for the atomic weight of carbon C_{12}. The statistician thinks in terms of an operational definition and knows that this is not so. My colleague Dr. W. J. Youden showed in the Scientific American for April 1961 a chart on which he had plotted the velocity of light on the vertical axis and the ordinal number of the experiment on the horizontal. The determinations of the velocity of light had continually decreased from the earliest experiments of 100 years ago. Not a single experiment gave a result within the probable error of the preceding result.

Intuitive arguments have no place in the statistician's analysis, because he can not define intuition.

All data come from marks on paper, put there with a purpose, whether by man or by machine. These are the results of investigation. Coding and summarization produce tables. Any data are the result of a long series of operations. The velocity of light is no exception, nor e/m, nor an atomic weight. If you change interviewers in a monthly or quarterly survey, or change only 15% of them, you may expect to see a rise or fall, even if you interview in the same households. Change households, and you get still different figures. History itself is only the events that somebody thought would be worth while to record, and how he elected to record them.

These dreary truths do not mean that no data are any good, nor that the situation is hopeless, nor that there is no use to study history. Rather, they mean that, to use data, one must think in terms of an operational definition and must understand the chain of operations that produces figures.

Even concepts that we apply daily without much thought require operational definitions. To see the truth of this statement, one need only try to communicate to someone else what he means by green, tired, round, unemployed, unemployable, a farm, good qualtiy, uniform quality, accurate, satisfactory, hard, strong, beautiful music, cured. It can only be done by means of an operational definition -- i.e., by empirical tests which applied to material or to a sample thereof will come through with an answer, yes, this is green, no it is not, etc.* There is nothing new about these axioms, but it was Shewhart who saw clearly how important they are for the statistician.

We are now in position to think about the question, What are facts ? They are the results of studies that apply a specified operational definition. If one changes the definition, he gets a new figure. To say, for example, that 4.7% of the labor force is unemployed is only to say that the result of applying certain operations embodied in a questionnaire and the answers thereto gave 4.7%. Any economist knows that this number is sensitive to the questionnaire: a simple change in one question may produce a change of half a million in the number unemployed.

Statistical theory will not of itself originate a substantive problem. Statistical theory can not generate an operational definition. Thus, statistical theory can not of itself generate ideas for a questionnaire, nor for a test of hardness, nor specify tests that might decide what would be acceptable colors of dishes or of a carpet; nor does statistical theory originate ways to teach field-workers or inspectors how to do their work properly, nor what characteristics of workmanship constitute a meaningful measure of performance. This is so in spite of the fact that statistical theory is essential for reliable and economical comparisons of proposed alternative questionnaires and tests (i.e., alternative operational definitions).

---

*Walter A. Shewhart, STATISTICAL METHOD FROM THE VIEWPOINT OF QUALITY CONTROL (The Graduate School, Department of Agriculture, 1939), page 130.

It is a good time now, I believe, in this day of the marvels of computers, to return to the decade of enthusiasm of 1945-55 for improvement of survey-techniques, and concern for input. EDP does not produce input (though we must of course not overlook the benefits of machine-editing for the detection of blunders in input). Machine-sheets that show regression analyses and discriminant functions carried out correctly to 10 places are not by themselves analysis of data. Machines are aids, and can be important aids, but nothing can take the place of quality of input, nor take the place of humdrum search with dogged perseverance for ideas that might explain the results that come out of an experiment.

The frame, the universe, and the limitations of statistical inference. A statistical study proceeds by investigation of the material in a frame.* The frame is an aggregate of tangible units of some kind, any or all of which may be selected and investigated. The frame may be lists of people, areas, establishments, materials, manufactured parts, or other units that would yield useful results if the whole content were investigated. The frame must also in many kinds of tests specify conditions such as the season, climates, rainfall, levels, ranges of concentration, dosages, pressures, temperatures, speeds, voltages, or other stresses that the material or product will be subjected to in service. Experimentation on a few convenient levels of stress may be inconclusive. Tests of varieties of wheat would specify ranges of climate and rainfall, as a minimum. Tests of a medical treatment, to be useful, would specify ranges of dosage, severities and other characteristics of illness. Consumer research on some products is nigh meaningless without reference to the season and climate -- e.g., soft drinks, analgesics.

How does man establish a scientific law ? By observing that the law still holds when one restriction after another is removed.

---

*First defined by without use of any specific term by F.F. Stephan, American Sociological Review, vol. 1, 1936: pp. 569-580.

The frame is the limitation of statistical inference. The results of a study, and any statistical inference (estimates, probabilities, likelihood, confidence limits, etc.) calculated by statistical theory from the results relate only to the material, product, people, business establishments, etc., that the frame was in effect drawn from, and only to the range of environment specified for the study, such as the date, weather, rates of flow, levels of concentration used in tests of a production-process, range of voltage, maximum gust or other stress specified for the tests.*

Knowledge of statistical theory discloses to the statistician the limitations of statistical inference. The statistician has an obligation, as architect of a study, to help his client in advance to preceive the limitations of any study that is contemplated, and to alter the design to cover the range. Only the statistician, with his knowledge of theory, can perceive the limitations of the results that will come out of a study.

My favorite question when someone proposes a study is this: what will the results refer to? How do you propose to use them?

Enumerative and analytic aims of studies. The aim of a statistical study is to provide a basis for action. In an enumerative study, action will be taken on material termed the universe, which is covered sufficiently well by the frame to be studied. Only substantive knowledge (engineering, psychology, medicine, agriculture, consumer research) can bridge the gap between the frame and the universe when they are not identical. In an analytic study, action will be taken on a cause-system or process with the hope of changing product of the future (which might be people, manufactured product, agricultural product).

Briefly, an enumerative aim is to count or evaluate something. The uses to be made of an enumerative study depend only on the evaluation of a specified frame. Examples:

---

*William G. Madow at the meeting of the American Statistical Association in Washington, December 1967, treated the problem optimum placement of experiments within specified ranges of stress. (Not published at this writing).

1. An inventory of materials to assess their total value in dollars. This evaluation may determine the selling price, or evaluation for an auditor's report, or for tax-purposes.
2. Examination of equipment to estimate cost of maintenance.
3. The Census of the U.S. for the purpose of Congressional representation.
4. Census of a city taken in an attempt to justify greater financial benefit from the State.
5. Evaluation of accounts receivable.

When there is an earthquake or a flood, a vital question is how many people, adults, infants, and infirm are there in need of the necessaries of life. The aim is not to try to find out why so many people lived there, nor how to foretell the coming of an earthquake.

In contrast, in an analytic study the aim is to seek causes. The Census of any country, aside from enumerative uses, is used also by economists and sociologists to test theories of migration, fertility, growth of population, aging of the population, to understand better the changes that take place in education, income, employment, occupation and industry, urbanization. The aim is to predict population by sex and age, and amongst other aims to alter the causes of poverty and malnutrition.

A firm advertises to change people, to change their tastes and habits. The firm conducts consumer research to learn how effective a scheme of promotion was last month, the aim being to acquire ideas on how to advertise more effectively in the future. Why do people buy this brand and not some other? A study of accounts receivable, primarily for an enumerative purpose, may also yield information by which errors can be reduced in the future. Why are there so many errors of this or that type in the accounts receivable? What brought so many people to this area, or why did so many leave? What are some of the special causes of variation of a particular dimension? What are some of the common causes?

As a current example of an analytic study, we could mention that there are before the Interstate Commerce Commission for consideration at this very moment a number of studies of rail-traffic that purport to show how the business of one railway will suffer or will gain if certain other railways merge. The frame for such studies is records of last year's business (interline abstracts):

the universe is railway business in the year 1975, 1980, and later. What the Interstate Commerce Commission really needs is knowledge about future business. Unfortunately, we can only study history as recorded on last year's interline abstracts.

Studies of last year's business are dependent on the judgment of experts in traffic. These judgments are predictions, as they relate to future business under specified conditions. The results of such studies appear to be highly dependent on the rules and criteria laid down as a basis for judgment on the possible amount of business that may be lost because of a merger. These judgments are also allegedly at times dependent to some extent on the background and auspices of the expert. Extreme precision in a result only means that the same expert would derive substantially the same figure were he to study under the same rules the whole frame (which might be a million abstracts).

In the planning stages, the statistician will therefore aim at some moderate degree of precision, which might mean a standard error of 10 or 15 per cent. He should explain to his client, and later in testimony before the Interstate Commerce Commission, the futility of extreme precision, because of the long range predictive nature of the results.

In acceptance sampling, tests of a sample from a lot determine whether to accept the lot or to reject it (re-work it, down-grade it, throw it out). A definite rule of acceptance, applied to every lot in a succession of lots, guarantees an A.O.Q.L. or better. (A.O.Q.L. is an expected quality in the material accepted and re-worked.) The action is disposition of lots, an enumerative purpose. The aim in acceptance sampling is not to find reasons why so many lots are rejected. In contrast, in the use of a Shewhart control chart or in some other test, the aim is analytic, to detect the existence of a special cause of variation, elimination of which (action on the process) will reduce the number of lots rejected in the future.

A manufacturing process, brought by analytic methods to a state of statistical control produces hour after hour, product with quality-characteristics that vary from one item to another like ideal random drawings from a bowl. One can then safely predict, by statistical methods the characteristics of the product of tomorrow or of next week. This is possible because, in a state of statistical control, the cause-system remains constant, stable, unless something drastic happens to the process. There are also of course processes in nature that exhibit statistical control, for example, radioactive disintegration. One may apply with confidence the theory of probability to the half life and to statistical fluctuations in the decay of small quantities of radio-active material.

We may generalize to say that in a enumerative study, the aim is to evaluate some numerical characteristic of a frame, not to study the causes that made the frame what it is. The action to be taken will be disposition of the frame. In contrast, in an analytic study, the aim is to discover causes to find out what made the frame what it is, and how to modify the process that will produce frames that we desire to see in the future.

In enumerative studies, formulas for variances contain the finite reduction factor (N-n)/(N-1). In analytic studies, there is no such factor.* A complete census is for analytic purposes still a sample. It is a sample of all the product or of all the people that the cause-system has produced in the past and will produce in the future.* If 1200 cases of a reportable disease are reported in one year for the whole country, classes tabulated for that year by color, income, type of residence, etc., will for analytic purposes be small samples, despite the fact that the reported numbers are presumably complete. There is no factor (N-n)/(N-1) that reduces the sampling error to 0 just because n=N.

Hints on the teaching of statistical inference. One can only learn theory. One can only teach theory.

---

*W. Edwards Deming, SOME THEORY OF SAMPLING (Wiley 1950, Dover 1966), Chapter 7.

W. Edwards Deming, "On the distinction between enumerative and analytic surveys," Journal of the American Statistical Association, vol. 48, 1953: pp. 244-255.

*On the interpretation of censuses as samples, by F. F. Stephan and W. Edwards Deming, Journal of the American Statistical Association, vol. 36, 1941: pp.45-49.

The question is, what theory ? What modifications of theory now in the books and in the classroom are desirable in the teaching of the design of experiments and surveys ?

First, I shall mention the testing of hypotheses. This is beautiful theory, with errors of the 1st and 2nd kind, and power of a test. What the books fail to point out is that this theory applies only to enumerative studies, the evaluation of a frame, where action is to be taken on the material in the frame. Moreover, with a little thought, face to face with a problem, the reader will perhaps agree that even for an enumerative aim, the aim is much more usefully stated and interpreted as a problem in estimation than as test of a hypothesis. The question is how many ? What proportion ? How big is the difference ?

To illustrate, we may take an evaluation of accounts receivable. Students learn to formulate the hypothesis that the fraction defective $p = 0$, with the alternative hypothesis that $p > 0$. The hypothesis that the fraction p defective is equal to 0 is not worth testing: we know without making any examination at all that p is bigger than 0. It might be important to know if $p < .04$. An error of the 1st kind would be to conclude that $p < .04$ when a bigger sample processed with the same care would convince us that $p < .04$. An error of the 2nd kind would be to conclude that $p < .04$ when a bigger sample would convince us that $p > .04$.

No such beauty of theory exists, however, I fear, in an analytic study. People certainly make errors of the 2nd kind and other wrong decisions in business, law, agriculture, medicine, manufacturing processes. However, that it is rarely possible to evaluate from comparisons of Method A and Method B the probability of being wrong in actually adopting Method A over Method B. It is only in the ideal circumstance of a project that embraces in scope a sample of all the samples that could be conducted on the materials produced by the processes under test, and covering the ranges and stresses specified, that we may evaluate by the theory of probability the risk of a wrong decision.

How big is the difference between the outputs of two machines or two manufacturing processes, or the difference between the rates of rejection of what they produce ? The aim, wherever the experiment be carried out, is not to learn whether Treatments A and B will produce the same yield: in symbols, the problem is not whether $\mu_a = \mu_b$. One may state categorically without spending a nickel that $\mu_a \neq \mu_b$. We don't need an experiment for that. The statical question to be answered is what is the magnitude of $\mu_a - \mu_b$ under specified ranges of conditions, and does the difference warrant the expense of a change ? What is the cost of change ? How sure are we that the difference is big enough to warrant this expense ? The problem is estimation and economics, not a matter of testing a hypothesis.

Suppose that Treatment A turns out to be superior superior to Treatment B in an experiment conducted in a certain hospital. The difference is "highly significant". What do the results mean? What is the meaning of the comparative adjective "superior"? Let us suppose that the circumstances are ideal for the experiment, and that the patients tested were drawn at random from the eligible patients in the hospital. The aim of the experiment if to learn which is the better treatment for patients <u>not yet treated</u>. Unfortunately, we can not evaluate from the given experiment the probability of being wrong in choosing Treatment A over Treatment B for patients yet to be treated, even if they are to be treated in the same hospital, and still less in another hospital. The trouble is that for patients to be treated in the future, the conditions will be different from those that existed in our one experiment.

Only substantive knowledge can bridge the gap between (a) the frame and conditions studied and (b) the cause-system that will produce other frames (industrial product, people, patients in hospitals, crops) under other conditions. A doctor of medicine or a chemist may, over his own signature, assert by substantive knowledge that a test carried out on patients in Chicago would give the same results in Denver. No statistical theory can make such a generalization. To bridge the gap between experimentation on animals and the effect of a drug on human beings is for the doctor of medicine: statistical theory can not help.

Two varieties of wheat tested at Rothamsted may show that Variety A delivers under certain conditions much greater yield than Variety B. In other words, the results are statistically significant. But does this result tell you which variety would do better on your farm in Illinois? Can you evaluate from the experiment at Rothamsted the probability of going wrong in adopting Variety A in Illinois? The answer is no. But is this question in the books? in classes?

It is only when Treatment A appears to be better than Treatment B in a great many places and under a wide variety of conditions that we can begin to feel confident in asserting that tests seem to indicate the superiority of A over B. Only substantive knowledge can form a judgment and decision, unless the tests cover all possible conditions likely to be met in the immediate future. Statistical theory ends with inference about the frame and the conditions studied.

To the statistician, difficulties with new drugs that come on to the market well tooted at first, only to be taken off the market later, and (e.g.) innovations in the preservation of foods, only to be brought under suspicion after preparations for introduction to the market are well underway, all represent cases of failure of the expert in the subject-matter to specify in advance the conditions under which the product must perform. Perhaps the fault really lies with the statistician who fails to compel the expert in the subject-matter to state these conditions and to appreciate the consequences of failure to think the matter through in advance. Again, what universe will the results of the experiment refer to?

What are the statistican's obligations? His job is (1) to explain to his client the boundaries of statistical inference; (2) to explain the client's obligation to specify the ranges of stress, voltage, temperature, duration of treatment, etc., within which the product may have to perform; (3) to design experiments within the specified conditions that will meet the client's needs at lowest cost and which will comply with any administrative conditions imposed; (4) to interpret the results with care; (5) to have no part of a study that is apparently doomed to failure.

Now another point. A possible weakness of textbooks and of teaching is to place too much reliance in the standard error in the analysis of data. Speculations on the standard error to be expected from alternative designs are rightly the foundation of the statistician in the planning stages. After the results are in, however, and the statisticain faces the problems of analysis and interpretation, there are many statistical techniques at his disposal. Standard errors, t-tests, chi-square, etc., are all useful, but they may be in actual use woefully inefficient, in spite of teaching to the contrary. The trouble is that these tests are symmetrical functions of the numerical data, and that they therefore bury all the information that is contained in the order of the data. In my own work, I rely heavily on simple tools like distributions, correlation diagrams, runs, patterns and comparisons, sign-tests, control charts, which do not bury the information that is contained in time- and geographic-order.

We learn in the books that r/n is a valid and efficient estimate of p, r being the number of red balls in n random drawings, and p the statistical limit of r/n. How many students, given 7 heads running followed by 3 tails, would form the ratio 7/10 for an estimate of p? I wouldn't. The estimate r/n buries the information in the order of occurence of heads. What does an unbiased estimate mean to a student? The most valuable outcome from the experiment might arise from an investigation into the reasons for 7 consecutive heads. Where will he learn to investigate the run of 7 heads before he proceeds?

We learn from statistical theory that for a simple random sample with replacement drawn from a normal frame, the mean of the sample and its standard deviation, together with the number of degrees of freedom in the estimate thereof, convey all the information that there is with respect to the bowl; $\bar{x}$, s, and n are sufficient statistics. This is a powerful theorem that every statistician knows.

However, the most important thing about the bowl the sample can not tell us--<u>who created the bowl?</u> The results of a survey are no good to us if we don't know anything about the organization that carried out the field-work. We must have in hand the questionnaire or the method of test. We must know something about the training, supervision of the investigators, and something about the variance between investigators. These points must all be included and evaluated in a report of an investigation, or reasons given for not including them, along with a warning of possible uncertainties not yet evaluated.

The statistician's duty in a legal case or in fact in any report is to present data and inferences therefrom with a full report on the procedures, what went wrong, and the possible effects of these departures. The rule is that the statistician's report on the statistical reliability of a figure should give the reader the same chance to agree or to disagree with the conclusions as the writer had.

Statistical teaching centres might well include in the curriculum some small amount of time on preparation of procedures. How many students with doctor's degrees have had the experience of writing a set of procedures for a study, complete with statistical controls to detect departures from the procedure prescribed, and of learning the hard way that he may have to write these instructions twelve times before he learns that if there is any possibl way to misconstrue an instruction, someone will do it?

Our statistical teaching centres might well also give some attention to analysis of actual (dirty) data, and by this I don't mean more theory in estimation and analysis of variance. How many students have had the experience of trying to dig out information from the results of a study and to write a report in intelligible words to say what the results mean and what they don't mean, with a full disclosure of what went wrong? Under what circumstances might the conclusions be expected to hold? What training do students get in the evaluation of differences between investigators, nonsampling errors, and departures from procedures, and their possible effects on the conclusions? How many students know that a piece of scientific work is often more noted for its evaluation of the uncertainties in the results than for the results themselves? The following quotation is worth noting:

> The outcome of the inherent difficulties of the method is that, although Cornu's discussion of his experiments is a model in the care taken to determine so far as practicable every source of error, his definitive result is shown by other determinations to have been too great by about one-thousandth part of its whole amount.--Encyclopaedia Britannica, Velocity of Light, edition of 1952, vol. 35, page 36.

The statistician is today a powerful figure in society. Some regulatory agencies in the U.S. now require that quantitative evidence be based on studies designed and interpreted by statistical theory. Our statistical teaching centres could possibly use an extra seminar or special lectures to impart to students some of the obligations of the statistician to his profession and to society. This could be arranged, I believe, without stealing any appreciable time from the teaching of theory. If some chiseling from theory is necessary, then I submit that consideration should be given to the problem of overall optimization of the student's career, through the first five years beyond his degree. The legal and medical schools do not turn their students loose without guidance.

We learn in the theory of sampling that the variance of the mean of a simple random sample is σ /n, where σ is the variance between the sampling units of the frame and n is the size of the sample. The variance σ is defined on the assumption that every sampling unit has a numerical value x_i for Unit i. It is as if we put N perfect poker chips into a bowl, numbered them serially 1, 2, 3, and onward to N, with the number x_i, on Chip i, and drew with random numbers a sample of size n. We could read the number x_i on each chip in the sample and we could compute the mean, variance, skewness or any other property of the sample.

The fact is, however, that for many types of investigation, a specified characteristic of a sampling unit may not have a definite numerical value like x_i. Ask any man in a rate bureau if the freight charges on a shipment and the divisions of revenue between carriers thereon have definite values. The answer is yes, they ought to, but he will also tell you that these values depend a lot on what division-clerk worked it out. The formulas for freight charges are complex. Ask a chemist if a sample of naphthalene has a determinable purity. He will tell you that he can analyze it and get a value for it, but that what he gets will depend on the method of analysis that he uses.

When we come to dealing with households, people, the problem is equally complex, maybe more so. A common joke in the Census in 1940, where the questions on employment were of prime importance, was that if you asked a woman three times if she were seeking employment, the answer would be yes at the 3d trial, no matter how she answered the first two. The results of any study depend on the interviewer, supervisor, crew leader, etc.* Re-phrasing of a question will change the answers. The answers to any question will depend on what other questions precede it.

Advanced courses in statistical theory and elementary courses for people that are studying business, psychology, medicine, engineering, or whatever line should, I believe, acquaint the students with some of the complexities of a statistical study, and the hazards of doing a good job on part of it without looking at the whole. There are also the hazards of failing to fix in advance the division of responsibilities. In my own work, every engagement is subject to a code of professional conduct.*

To find problems is the responsibility of management, delegated perhaps to expert knowledge of specific subject-matter such as chemistry, engineering, economics, marketing, medicine. The screening of problems, to decide which ones have possible solution, and their formalization, is largely statistical. Once formulated, the statistical portion of a problem is dependent purely on statistical theory. Statistical problems are not solved by chemistry, administration, engineering, medicine, nor by anything but statistical theory. Statistical work is not a sideline. The teaching of statistics should emphasize this point.

---

*A limited number of references will have to suffice here.

1. Herbert Hyman, Survey Design and Analysis (The Free Press, 1955).

2. Morris H. Hanse, William N. Hurwitz, and Max. A. Bershad, Bulletin of the International Statistical Institute, vol. 38, 1960; pp. 359-374.

3. Joseph Steinberg et al., The reinterview program in the current population survey, Technical Paper No. 6, Bureau of the Census, 1963.

4. Barbara Bailar, Effects of interviews and crew leaders, Technical Paper No.7, Bureau of the Census,1968.

*Principles of professional statistical practice, Annalls of Mathematical Statistical, vol. 36, No. 6, 1965: pp. 1883-1900; Code of professional conduct, Sankhya, Series B, vol. 28, 1966: pp. 11-18.

Another point to bring out that I believe requires change in practice as well as change in teaching is the matter of interpretation of a confidence interval. We learn in the books that if one were to carry out 1000 experiments and for every one make calculations like the one at hand, then only 50 of them would fall below the lower 5% limit calculated for this experiment. This is small confort to a client. He probably burst his budget or badly bent it to do this one experiment: he is not going to do another one, at least not for a while. What the statistician can tell him is that in his own experience, he (statistician) will be wrong only 1 time in 20 (or some other fraction adjusted to the requirements) by repeatedly carrying out this type of inference.

There is the beautiful theory of outliers for deciding "whether an unusual observation is worth investigation." This is sometimes wrongly described as theory for the rejection of observations, though one should never reject an observation on statistical grounds alone.* Actually, this theory is really a theory for deciding whether, when applied routinely to all sets of observations, the minimum or the maximum of a given set of observations should set off an investigation to try to find a special cause of nonuniformity.

The unfortunate fact is that, in practice, the only time when a chemist would use such a test and the accompanying tables, if ever, is when his eye-ball test tells him that something may be wrong. The probabilities associated with the theory are then badly distorted, as the application is highly selective. But do the textbooks or class-room teaching give any such warning?

Along with the limitations of statistical theory, the statistician must of course not forget its power. Theory enables him to make objective inferences when the theory is applicable.

Statistical theory does more than may be inferred from the foregoing. For example, application of statistical theory to production enables the engineer not only

---

* Sir Ronald Fisher, "On the mathematical foundations of theoretical statistics," Phil.Trans., vol. 222A, 1922: pp. 309-368.

to detect the existence of conditions that need correction for improvement of a process (as to decrease the variability in dimensions or in performance of product); the same techniques indicate which administrative level (local operator or foreman, or upper management) is responsible to identify the cause and to correct it. A point out of control on a Shewhart control chart indicates a condition that the operator or his foreman is responsible to correct. On the other hand, too much variability in a state of statistical control represents a condition that be changed only by upper management. Examples: poor raw material, poor light, wrong humidity, lack of a quality-program. Whether it would be economically feasible for management to correct a condition that only management can correct is a decision of management--a matter of engineering and economics.*

The practice of statistics and the teaching thereof are intimately bound together. What one learns in practice should be taught. Practice can not go beyond the bounds of theory. One does not dabble in statistics either in practice or in teaching. No lustre of personality in the teaching of statistics can atone for teaching error instead of truth.

I am accordingly deeply interested in the summarization of data and how to make the most helpful inferences from data. I am sure that if our theoreticians put their minds to it, the next few years will see contributions and changes in the teaching of statistics and in textbooks, with reverberations in practice, greater than we have seen in any like period of the past.

---

*W. Edwards Deming, "What happened in Japan?" INDUSTRIAL QUALITY CONTROL, vol. 24, No. 2, August 1967: pp. 89-93.

SOME REMARKS ON STATISTICAL INFERENCE IN FINITE SAMPLING

Oscar Kempthorne
Iowa State University
Ames, Iowa

Introduction

The problem of forming opinions about a finite population by examining only a part of it seems to me to be the basic problem of statistical inference. If as a profession we are unable to make sense of this problem, I do not think we can claim any real understanding of inductive inference. I am sure I am not at all alone in feeling considerable dissatisfaction with the state of knowledge of inductive inference. The occurrence of widely antithetic overall views on the general problem exhibits the fact that as a profession we seem to be able to agree on essentially nothing.

It is, however, highly relevant to note that practitioners in sampling and in design and analysis of comparative experiments have to a considerable extent settled down to modes of operation which are fairly standard. It appears, therefore, that these modes of operation are found to be useful. What is obscure is why they are useful. Whether this question can ever be answered in a completely convincing way is I think a moot point. The development of knowledge is a peculiar process which one would hope that philosophers would help us understand. But the lesson of the past 2500 years seems to be that philosophers are not merely of no help but are actually an hindrance. We do not insist on our

---

Prepared in connection with Contract No. AF33(615)1737, supported by the United States Air Force, monitored by the Aerospace Research Laboratories, Wright-Patterson Air Force Base, Ohio.

science students spending a lot of time studying the philosophy of knowledge, and this happens, I suppose, because such study has not been found useful. Part of the reason for this is that the philosophers have rarely tackled real problems like the development of knowledge of a population by looking at a part of it.

I am reminded in this context of some remarks by Oliver Heaviside:

> "Now, the prevalent idea of mathematical works is that you must understand the reason why first, before you proceed to practice. This is fudge and fiddlesticks. I speak with confidence in this matter, not merely from experience as a boy myself, and from knowledge of other boys, but as a grown man who has had some practice in applications of mathematics. I know mathematical processes, that I have used with success for a very long time, of which neither I nor any one else understands the scholastic logic. I have grown into them, and so understand them that way. Facts are facts, even though you do not see your way to a complete theory of them. And no complete theory is possible."
>
> "I seem to be running down logic. I do not mean to. But there is logic and logic. There is narrow-minded logic confined within narrow limits, rather conceited, and professing to be very exact, with absolutely certain premisses. And there is a broader sort of logic, more common-sensical, wider in its premisses, with less pretension to exactness, and more allowance for human error, and more room for growth."

While there is much to be said for Heaviside's viewpoint, it clearly goes too far. I am very glad that this conference is being held, and hope that understanding of the logic of sampling will be increased.

Deficiencies of Present State of Knowledge

I have to state that I find the presently existing array of textbooks on sampling theory highly unsatisfactory. In order to present my views on the whole matter, I have to state why I think this.

The overall problem of sampling from a finite population is to develop ways of forming opinions about the whole population from examination of a subset chosen in some way. The textbooks fall down almost totally on this because they are directed almost completely, if not completely, at the development of opinion about the population mean. A course in sampling theory can easily degenerate into a lot of fairly elementary but tedious algebra of calculating expected values and variances of estimators of population mean with inadequate attention to inferential processes.

It is necessary in this context to differentiate somewhat between design of surveys and the analysis of surveys. The basic ideas of stratification seem to be unassailable, and much of what is presented with regard to design of surveys seems reasonable providing one ignores the matter of analysis. It is really in the area of analysis that the deficiency lies, and this must have implications with regard to design, because the two aspects are really inseparable in the last resort.

What then do the books tell us about analysis? And to cut through all the complexities, it is adequate to discuss the case of a single unstructured finite population. The most primitive problem is the following. I have, say, 100 slips of paper. On each piece of paper I write a number. I then insert each of these slips of paper in an envelope, and number the envelopes from 1 to 100. The primitive problem is that <u>you</u> are permitted to examine, say, 10 of these envelopes at your choice, and, on seeing the numbers inside the envelopes, you are required to state opinions about some aspects of the set of 100 numbers. This is a problem which was formulated many decades ago. The labelling of the envelopes is <u>not</u> a new idea but was present, I think, from the beginning, i.e. at least 40 years ago. The recipe for design of the survey that is universally favored, except for some neo-Bayesians, is to draw a random sample. I do not recall any of the neo-Bayesians, except Dr. Ericson in this volume, saying what precisely they would do, and I wish they would come out into the open.

What now is the recipe for analysis? If we look at the books on sampling they concentrate totally on the population mean. They tell us to compute $\bar{x}$ and SE($\bar{x}$), and leave us high and dry. I was pleased to note that Cochran gives at least a brief discussion of what follows. Without any real discussion, we are led to the idea that $(\bar{x}-\mu)/SE(\bar{x})$ is distributed approximately as a N(0,1) variable or as a Student 't' variable. We are then told to make the usual inversion so as to reach an interval $\{\bar{x}-tSE(\bar{x}),\ \bar{x}+tSE(\bar{x})\}$, which is intended to convey our opinion that the population mean lies in this interval with some degree of confidence.

One often hears the cliche that in sample surveys we are not interested in tests of significance or tests of hypotheses but in estimation. I think this type of remark is completely misleading. Supposing that we are interested in forming opinions about a parameter θ. We need an "estimate" $\hat{\theta}$ and a standard error of estimate SE($\hat{\theta}$), such that the joint distribution of $\hat{\theta}$ and SE($\hat{\theta}$) is known. Otherwise the knowledge of SE($\hat{\theta}$) is useless. If we have this information we can make a test of significance. Alternatively if we can write down a system of successively nested intervals which have some "confidence" properties, we can look at them from another angle as giving us a test of significance.

The practice of analysis then appears to use without questioning that $(\hat{\theta}-\theta)/SE(\hat{\theta})$ is nearly normally distributed. The function $(\hat{\theta}-\theta)/XE(\hat{\theta})$ is assumed to be a pivotal quantity with known distribution.

It is common to buttress this idea by calling some sort of Central Limit Theorem, but this merely says that with increasing sample size and increasing population size, the mean will tend to be normally distributed. But this says nothing about the distribution of the sample mean for particular sample and population sizes. Also it says nothing about the joint distribution of the error of the estimate and the standard error of estimate. There is some theoretical work on this for infinite populations (by Gayen and others), and limit theorem by Hajek. But I find these results singularly unconvincing for an actual situation at hand. It is not at all difficult to specify a population of numbers such that the frequency with which the interval

$$\bar{x} - 3SE(\bar{x}),\ \bar{x} + 3SE(\bar{x})$$

contains the true population mean is zero. All one has to do is to suppose that the population of x values consists of 99 values near zero say, and one value of 10^6. Very little thought shows that the 3S.E. interval from a sample of 10 does not contain the true mean with probability 1. What appears to be going on is that the practitioners of sampling are using without saying so, additional ideas of the nature of the population they are sampling. So they are really adjoining something else to the idealized problem of x-values on slips of paper in numbered envelopes. The whole question is the robustness of the 't' distribution with finite populations. It appears to me that there is remarkably little reported research on this matter.

The issue of joint distribution of actual error of estimate and standard error of estimate seems much more difficult in the case of unequal probability sampling. Horvitz and Thompson, for instance, produced an estimate of variance which could take negative values. I do not wish to denigrate their work by saying this. How would one construct a system of intervals of uncertainty with a negative variance estimate?

It is essential to differentiate between the problems of the governmental agencies which are continuously performing surveys on populations which are changing only slowly, and the problems of the individual research worker who wishes to examine a unique finite population. The samplers in the governmental agency obtain empirical information on the distribution of $(\bar{x}-\bar{X})/SE(\bar{x})$, and use vague but useful Bayesian ideas that the present population is "like" previous populations.

The logic of their intellectual processes is obscure, and I would wish that they paid more attention to the actual inferential process. I shall discuss the unique unknown population problem, which is surely relevant to the other, but ignores information available in that case.

Inference

Any general discussion of the sort I am following requires some specification of what one wishes to mean by the term "inference". The great bulk of problems considered under inference are artificial. We assume that we have a random sample from one of a class of distributions and wish to say something about the parameters of the particular distribution from which the observations came. But in fact we rarely, if ever, have a random sample. The idea that we have a random sample from some mathematical distribution is a pure conceptual construct, the main purpose of which is to enable a condensed representation of the data, and to give us an idea of what we might observe if we could take indefinitely more observations "like" the ones we have.

The interesting aspect of random sampling from a finite population is that there is a real population from which we are sampling at random. That is the reason that I find the problem so interesting and elemental.

Ideas of inference seem to be broadly classifiable into 3 types of conclusion:

(a) The decision theory conclusion which requires specification in some way of a loss function, and then the nomination of a point estimator of some property of the population.

(b) The reaching of a posterior distribution for a parameter.
and

(c) The specification in some pre-specified ways of the tenability of various possible values for the parameters without using a prior distribution.

There are some who appear to think that the type of conclusion (b) is the only one worth having. On the contrary, I have never felt a real need for such a conclusion.

The theory of sampling appears to be dominated strongly by the decision theory conclusion--the providing of the best (in some sense) point estimate, without any real attention to the assessment of how good the estimate is.

My interests are predominantly on the type of conclusion (c), which attempts to assess the tenability of models and parameter values.

Before turning to this, however, I would like to make a few remarks about admissibility. It seems to me that admissibility ideas have been relatively sterile in the case of parametric inference. Also it seems to me that the idea of admissibility has been totally sterile in the case of finite population theory.

Optimality

For some years there has been concern, particularly from Godambe (1955) and people who have worked with him, that the sample mean has no optimality properties. Quoting from Royall (1967), who was quoting Godambe, "It (the sample mean) is not the UMV unbiassed linear estimator." Royall quotes a small example with equal probable samples of two from a population of three, and the estimators of population mean:

$$t_1(s_1,\underline{x}) = \tfrac{1}{2}x_1 + \tfrac{1}{2}x_2$$

$$t_1(s_2,\underline{x}) = \tfrac{1}{2}x_1 + \frac{2}{3}x_3$$

$$t_1(s_3,\underline{x}) = \tfrac{1}{2}x_2 + \frac{1}{3}x_3.$$

He states that this estimator has variance less than the sample mean if $x_3(3x_2-3x_1-x_3)$ is greater than zero. This little example exhibits the sterility of the underlying question, because one does not know the x's, and it seems highly doubtful if one would have prior knowledge of the type that $x_3(3x_2-3x_1-x_3) > 0$. Nor will one have any knowledge of this _after_ drawing a sample.

A more reasonable basis for considering optimality seems to be the following. We suppose that there exists an underlying unknown set of N numbers, say $Z_1, Z_2, \ldots, Z_N$ and that these are associated with the labels of the units in an unknown way. Suppose samples are drawn without replacement with equal probability. Let the structure of sample i be given by

$$\delta_{i1}, \delta_{i2}, \ldots, \delta_{iN}$$

in which $\delta_{ij}=1$ if unit j is in sample i and is zero otherwise. Let the estimate of $\bar{x}$ from i be

$$\sum_{j=1}^{N} \alpha_{ij}\delta_{ij}x_j. \tag{1}$$

The estimator depends on the sample number and the particular members in the sample. Now impose two requirements:

(a) _unbiassedness regardless of the N elements_

$$\frac{1}{M}\sum_{i=1}^{M} \alpha_{ij}\delta_{ij} = \frac{1}{N}, \tag{2}$$

M = number of possible samples;

(b) _origin invariance_

if each x value is increased by a constant ν then the estimate given by each sample is to be increased by ν, so that

$$\sum_{j=1}^{N} \alpha_{ij}\delta_{ij} = 1. \tag{3}$$

The expected square of the estimate (1) is

$$\frac{1}{M}\sum_{i=1}^{M}\left\{\sum_{j=1}^{N}\alpha_{ij}\delta_{ij}x_j\right\}^2$$

$$= \frac{1}{M}\sum_{i=1}^{M}\sum_{j=1}^{N}\alpha_{ij}^2\delta_{ij}^2x_j^2 + \frac{1}{M}\sum_{i=1}^{M}\sum_{j,j'}^{\neq}\alpha_{ij}\delta_{ij}\alpha_{ij'}\delta_{ij'}x_jx_{j'} \tag{4}$$

Now incorporate the ideas that the values $x_1, x_2, \ldots, x_N$ are a random permutation of $Z_1, Z_2, \ldots, Z_N$. Then the expected value of the variance is

$$\frac{S_2}{MN}\sum_{i=1}^{M}\sum_{j=1}^{N}\alpha_{ij}^2\delta_{ij}^2 + \frac{S_{11}}{MN(N-1)}\sum_{i=1}^{M}\sum_{j,j'}^{\neq}\alpha_{ij}\delta_{ij}\alpha_{ij'}\delta_{ij'} \tag{5}$$

where

$$S_2 = \Sigma Z_i^2, \quad S_{11} = \sum_{i,i'}^{\neq} Z_i Z_{i'}.$$

But (5) equals

$$\frac{S_2}{MN}\sum_{i=1}^{M}\sum_{j=1}^{N}\alpha_{ij}^2\delta_{ij}^2 \frac{S_{11}}{MN(N-1)}\sum_{i=1}^{M}\left\{\left(\sum_{j=1}^{N}\alpha_{ij}\delta_{ij}\right)^2 - \sum_{j=1}^{N}\alpha_{ij}^2\delta_{ij}^2\right\}$$

$$= \frac{1}{MN}\left(S_2 - \frac{S_{11}}{N-1}\right)\sum_{i=1}^{M}\sum_{j=1}^{N}\alpha_{ij}^2\delta_{ij}^2 + \frac{S_{11}}{N(N-1)} \quad . \tag{6}$$

It is natural to minimize the expectation of (6) subject to (2) and (3). The minimum value of (6) is achieved by $\alpha_{ij} = \frac{1}{n}$.

This seems to convince me that in a certain sense the sample mean has an optimal property. The use of a random association of the $\{Z_i\}$ with the labelling seems reasonable. Certainly if one had knowledge that there is some system in the association, one would try to modify the sampling plan, e.g., by stratification.

I suppose some will take the view that I have slipped in a Bayesian argument. There are, of course, many ways of doing this and some seem more innocuous than others. This version seems plausible to me. Also note that the argument is used to justify a mode of estimation but the value of the estimate is not to be judged by a Bayesian argument.

This type of argument, however, does raise a curious question. It is clear that in order to be able to draw a random sample one must have a labelling of members of the population or a physical random shuffling apparatus. Suppose then that one sees some evidence for a relation between the x value of a unit and its labelling number. To go to an extreme case suppose we see that in our sample x_i is very close to $\alpha + \beta i$ for some value of α and β. Should we then use this suggested relation in order to estimate the mean of the population? The simple theory of random sampling seems to require that one not look for or use any such association.

I wonder seriously about the use of standard error intervals in the case of unequal probability sampling. Is there any published work on this?

Likelihood

In recent years, there have been various attempts to produce compelling arguments for something called "the likelihood principle". I have tried to understand what this principle is but have failed.

In the case of a random sample from some parametrically defined population which may have any degree of complexity, e.g., intricately correlated variables and so on, the likelihood appears to me to have a useful purpose as a condensation of the data, and the way the unknown parameters enter into the probability of the actual data. In a certain obscure sense the likelihood function appears to tell us all that the data can tell us about the values of the parameters. But I do not think this type of loose statement can be given compelling force. I am very glad that Professor Barnard is here to "clue us in on"these matters. By obscurities about the likelihood function arise from at least two related sources:

(i) The likelihood function is a random function under the model of repeated sampling. So I am compelled to ask "What is the variabliity of this function under repetitions of sampling?". This seems a very difficult question. In some cases one can pick out a particular aspect of the likelihood function such as its point of maximum $\hat{\theta}$, say, and then one can develop some pivotal quantity involving θ and θ, and some other aspects of the data which has a known distribution. Or one may feel some confidence in using some asymptotic theory. But in thinking about the whole of the likelihood function, I have to think about how this whole function will vary in repetitions. I find that I do not have the necessary mathematical language to do this. Nor do I find the matter treated in the literature.

(ii) Related to these remarks, is the fact that I do not know in general how to look at a realized likelihood function. In certain simple cases such as the exponential family I can develop a little feel by considering the distribution of the sufficient statistic. But in general I am at a loss. Is a difference of log-likelihood equal to 2 big or small? The matter is somewhat like looking at a graph of points and a fitted line. It is not enough to say "look at it". I need yardsticks to develop some idea of what are big deviations and what are small.

The role of likelihood if one has a valid prior distribution is, of course, quite clear. But I have to state my view that I reject in general Bayesian distributions of parameter values. It seems to me that in general when one uses a Bayesian argument, one accepts errors for the particular case at hand, because if one indeed had a population of situations of the hypotheized (by the prior) type, one would do better on the average. I find no comfort at all in this type of balancing. The question at issue is not what opinions may reasonably be formed about a population of populations, so that in some average sense one does well for the popluation of populations. The question is what opinions can be formed about the particular unknown population at hand, which has a totally unambiguous existence in the finite population case.

Notwithstanding these remarks, however, I would like to comment on the likelihood as applied to a random sample from a finite population. It seems necessary to distinguish between two cases which are at opposite ends of a spectrum.

Case I. There is a small number of multinomial classes with proportions $p_1, p_2, \ldots, p_k$, say with $p_1+p_2+\ldots+p_k$ equal to 1. In this case with n_i, $i=1,2,\ldots,k$, observed in the classes the likelihood is equal to

$$\prod_{i=1}^{k} \binom{Np_i}{n_i} \Big/ \binom{N}{n}$$

and one can examine the behavior of this function restricting the Np_i to be integral. This has been done by Royall (1967), and by Hartley and Rao (1967). Presumably one can also try to develop a variance-covariance matrix of the maximum likelihood estimate $Np_1, Np_2, \ldots, Np_k$. Hopefully, one can also develop pivotal function of the n_i and the Np_i which has a known distribution at least asymptotically. If one could do this, then one could hope to develop a pivotal quantity $[\Sigma x_i \hat{p}_i - \Sigma x_i p_i]/s$ which has a known distribution asymptotically.

It also seems worth noting that in this case one can use the likelihood function to order possible populations. A simple example is the following:

3 classes: $n_1=1$, $n_2=1$, $n_3=2$, and $N=7$.

The possible population structures with relative likelihoods in parentheses are:

1/1/5(10); 4/1/2(4); 1/4/2(4); 2/1/4(8); 1/2/4(8);

3/2/2(6); 2/3/2(6); 2/2/3(12); 3/1/3(9); 1/3/3(9).

Also in this case one can test goodness of fit of a defined population structure, in that one can order the possible samples by their probabilities and get a probability ordering goodness of fit. One could also use this goodness of fit test to specify the populations which are consonant with the sample at various levels of probability, and hence get a confidence interval for $\Sigma x_i p_i$ with bounded coefficients.

Case II. Suppose the population x-values to be $x_1, x_2, \ldots x_N$. Several authors have written down very nasty looking mathematical expressions which are supposed to represent the likelihood of $x_1, x_2, \ldots, x_N$, given a sample denoted by $x_{\alpha 1}, x_{\alpha 2}, \ldots, x_{\alpha n}$. What these amount to, I think, is the statement that the likelihood of any set $x_1, x_2, \ldots, x_N$ is 1/P(sample) for those sets which contain the x-values of the sample and zero for sets which do not. So if we had a population of two elements x_1, x_2 and we draw one at random we shall have one of two likelihood functions:

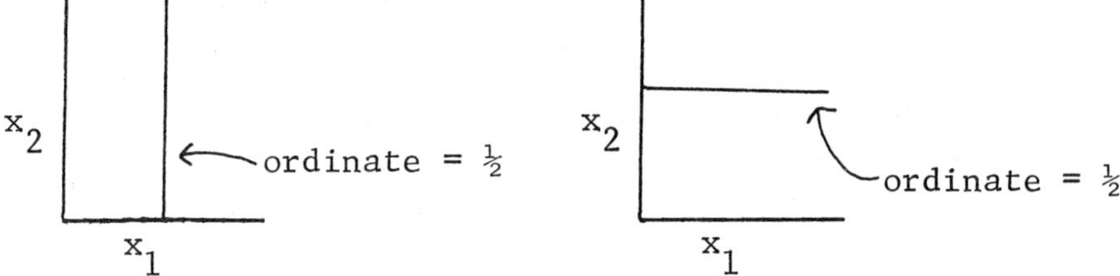

these likelihood functions "tell" me nothing about the unobserved x-values.

I am extremely skeptical of this whole approach. It seems to me that likelihood has meaning only as the probability prior to observation of the actual data, and that in the case of random sampling from a finite population, the probability of the whole set of data is not at all dependent on the x-values of the sample. I hope that Professor Barnard will comment on these matters.

If one accepts the above definition of likelihood, and if one accepts the idea that in some vague way, the ordinates of the likelihood function tell one something about the tenability of parameter values, it seems that the likelihood function merely tells one that all sets of population values which contains the sample values are equally tenable. If this is indeed the case, it seems that the likelihood function is completely uninformative. If also the only way to go from the sample to the population is by way of the likelihood function, then it would appear that there is no way to form opinions about the population. But this is patently false.

Perhaps the situation is that there is no unique way of doing so.

The problem I referred to above of how one is to look at the likelihood function seems to become acute in the case of a sample of n distinct x-values. Suppose that the jack-up factor $\frac{N}{n}$ is 20, and that we observe the sample

2, 3, 7, 11, 18.

The population structure which has maximum likelihood, at least if the likelihood is viewed as Hartley and Rao (1967) view it, is that in which there are 20 units with an x-value of 2, 20 units with an x-value of 3, and so on. The likelihood of this population relative to a population in which there are N different x-values containing the values 2, 3, 7, 11, 18 is 20^5. This is surely a big ratio. But I am certainly not inclined to the view that this maximum likelihood population structure is at all "likely". In this respect, the case of a finite population is no different from the classical parametric situations, in my opinion. The maximum-likelihood estimate is of no interest to me unless I have some invertible significance tests associated with it.

I also note that if we take the model of N different x-values and try a probability ordering goodness of fit test, then any population containing the sample values has perfect goodness of fit. This remark may, however, lead to other reactions:

(i) That probability ordering goodness of fit tests are useless in general.

or

(ii) That probability ordering goodness of fit tests cannot be applied in this situation.

If, however, I disregard some aspects of the sample, by making a small number of multinomial classes, I can make a goodness of fit test. It is this aspect of the matter that really confuses me. It seems as though one must not look at the sample too finely or one has nothing.

This raises the question of definition of parameters of a distribution. Perhaps I am being stupid, but I wonder if the values $x_1, x_2, \ldots, x_N$ are parameters. They certainly do not enter into the probability of the sample. Or perhaps one should take the view that there is just <u>one</u> population parameter, the whole set of x-values.

The "likelihood" approach seems sterile because it is obvious that a random sample tells us something about the population. Suppose, for instance, that we ask about the members in the population below and above some designated value x_0, say, $N(x_0)$ and $N-N(x_0)$. Then if $n(x_0)$ and $n-n(x_0)$ are the sample values, we can write an informative probability

$$\mathrm{Prob}[n(x_0)] \binom{N(x_0)}{n(x_0)} \binom{N-N(x_0)}{n-n(x_0)} \Big/ \binom{N}{n} .$$

On the basis of this probability we can obtain a measure of goodness of fit of $n(x_0)$ to $N(x_0)$, and we can delimit those values of $N(x_0)$ which have goodness of fit above some probability value, say, of 0.05. This assures me that the sample does tell me something about the population.

I can nominate two values, say, x_L and x_U and can form opinions from the sample of the number of units with x-values less than x_L and with x-values greater than x_U, and so on.

But if I nominate a very large number of x-values and the x-values in the population are all different, I run out of maneuvering space, so to speak.

These aspects tend to support the view that we can obtain opinions about the population only if we are prepared to ignore some aspects of the sample. I do not know that I can give this sort of idea any precision. But such an idea seems to underly all statistical inference. It is only by disregarding some aspects of the data, so as to render them non-unique, that one can embed them in a population of repetitions, and then perform probability calculations of one sort or another.

The fact that one can say something about the population given a random sample is evidenced by the results given by Wilks (1962), under the assumption that the population x-values are all different, which does not seem an unreasonable assumption with measurement data. Wilks in Section 8.8 presents the distribution of the population rank of the k-th order statistic in the sample. Also in section 11.4, he presents confidence intervals for quantities in finite populations.

I have done a little initial work on the distribution in repetitions of

$$D = \int (F-\hat{F})^2 dF$$

where F is the population c.d.f. and $\hat{F}$ is the sample c.d.f This is probably closely related to some of the infinite population non-parametric work. It appears that the moments of D are calculable, so that in a certain sense D is a pivotal quantity with respect tothe whole distribution function F. It is possible to form opinions about what c.d.f.'s F are tenable on the basis of the sample c.d.f.

The problem, therefore, seems to lie not with random sampling. We can form opinions from a random sample, but not from the full likelihood of the sample. We seem to be in the position of Heaviside that I quoted at the beginning. We know that random sampling works, but not why.

It is perhaps worthwhile to express the view that the problem illustrates the point that there is no one unique way of going from the sample to the population. We can look at the sample in various ways. We can form an opinion about the population c.d.f., about the population median, and many other aspects, but we cannot boil all these down into a single "most powerful" inference.

Some General Questions

1. Unbiassedness. Why is there so much attention to bias in the theory of sample surveys? Even those who say they are not interested in bias, say they do not mind bias providing it is small or can be estimated. The literature on mathematical statistics is also remarkably vague on this point. Unbiassedness is introduced and worked on with no discussion of why it is relevant. Is the fact that an estimator is UMVU of any real relevance, in the absence of additional properties?

2. Consistency. This seems to be a purely mathematical property of no interest in inferential problems except as a mathematical approximation which has been proved to be a good approximation for a finite situation. Or am I wrong?

3. Analytical Surveys. What questions are the analysers of "analytical surveys" trying to answer? I have listened to them for many years without receiving answers. What are the populations aimed at by the inferences? Is it not likely that the errors of estimation for most populations which have plausible conceptual reality are strongly underestimated?

The Case of Unequal Probability Sampling

It seems intuitively clear that if one takes a random sample with equal probabilities, the sample c.d.f. is a reasonable "estimate" of the population c.d.f. Here I use the word "estimate" in a loose way. I do not know exactly what I wish the word to mean. And it seems entirely clear that aspects of the population c.d.f. are testable by means of the sample c.d.f. I now wish to put a question to the group here, and I will do so by making an assertion:

Assertion

With unequal probability sampling, aspects of the population c.d.f. are not testable from the sample c.d.f.

I state it loosely, and mathematical work would require

tightening. Obviously if one takes (N-1) from a population of N, one knows a lot about the population. But can one test in any way what one does not know? In the case of equiprobable samples of size 2 from a population of 3, the probability is $\frac{1}{3}$ that the median is between the smallest and the largest.[3] Can one make any such statement with arbitrary probabilities for the 3 possible samples? My intuition on this is bolstered by existence of results such as those of Wilks, that I have quoted. It appears to me that equiprobability of samples is a critical aspect of the development. I hope someone in the field will be stimulated to examine the assertion.

The Case of Randomization in Experiments

There are several close analogues between sampling from finite populations and randomization in the choice of an experimental plan in comparative experiments. These analogies which I and others have used in talking about the randomization analyses of comparative experiments should not suggest, however, that the two situations are essentially identical.

The case which is simplest to describe so as to illustrate the difference is the paired experiment. If we are comparing a control and a treatment, with n pairs of experimental units, the randomization procedure tells us to pick at random one of the 2^n possible assignments of the treatments to the experimental units. The model is assumed that the control treatment on unit ij will give a yield of, say, x_{ij}, and the treatment on that unit will give a yield of $x_{ij} + \theta$. The purpose of the whole exercise is to form opinions about θ, and <u>not</u> about the x_{ij}. This is made possible by the existence under the procedure of pivotal quantities with a known distribution. If we denote by d_i, the observed difference (Treatment minus Control) in the i-th pair, then we can use the Wilcoxon signed rank pivotal quantity

$$\sum_{i=1}^{n} r\{|d_i-\theta|\}\text{sgn}(d_i-\theta)$$

where $r\{|d_i-\theta|\}$ is the rank of $|d_i-\theta|$ in the set of n quantities $|d_i - \theta|$. This pivotal quantity has a distribution independent of the x_i and enables us to form opinions about tenable values of θ.

Alternatively, and better, in my opinion, one can look at the rank of

$$\sum_i (d_i - \theta)$$

in the assemblage of 2^n sums

$$\sum_i \pm(d_i - \theta)$$

and this rank has a distribution independent of the differences of units within pairs, $x_{i1}-x_{i2}$. This rank is again a pivotal quantity which allows us to make tests of the tenability of various values of θ, and to specify values for θ which are "consonant" with the data at different levels of probability. Incidentally, inversion of either of these pivotal quantities gives results very close, except at the tails, to what we would get by inverting the normal theory t-pivotal, which is convenient but misleading.

The purpose in the randomized experiment is <u>not</u> then to form opinions about the x_{ij}, but to form opinions about tenable values of a shift parameter.

The virtue of the procedure is that it must carry some force to anyone. It is robust with regard to whatever opinions the experimenter and the data interpreter may have. If one permits oneself to look at the data more closely, one will reach personal opinions which may have force for oneself, but may or may not have force for the general public. This is not to say that one should not examine data from many points of view. If data inspection suggests a model which appears to explain the data very well, one would be foolish to ignore it. The point, I think, is that we are again "bugged" by the optimality idea--the hope that there is one analysis of the data which is all powerful and on which all should agree. I see no objection at all to analysing data in several ways, and even reporting more than one analysis. The real difficulty with data analyses which are highly suggested by the data is that we cannot develop ways of monitoring our thinking. I am inclined to the view that this monitoring is the basic role of statistics in the accumulation of knowledge.

The Use of Concomitant Variation in Sampling

I do not claim to have any real understanding of ratio and regression estimates in finite populations, and I pass over this subject with just a restatement of my overall general question--What sort of probability inferences, of tenability of population mean, or of confidence in interval statements about the mean, are available I conclude from my previous thinking expressed above that probability statements are possible really, only about quantities of the population. I see no way of introducing concomitant variation into this type of approach.

The use of concomitant variation in stratification presents no additional problems, of course.

Approaches to Sampling on a Line or on a Plane

I wish to use my remaining time to describe some preliminary work on sampling on a line, which can be extended to a plane with no particular difficulty. I became interested in this from the point of view of experimental design, but it seems to me that it has real relevance to sampling existent populations on a line or a plane, as in forestry sampling.

The ideas behind this work are:

(i) in these cases, one is surely not only intereste in the population mean, but in a map of the population;

(ii) a sampling procedure should therefore aim not only at estimating the mean but also a map;

(iii) if the aim of the exercise is to produce a map, one should also make provisions for estimating the precision of the map;

(iv) these considerations suggest that a sample should consist of two parts: one part would be used to develop a map, by whatever mapping rule the actual data may suggest, and the second part should be used to estimate the accuracy of the map and any quantities derived from it;

(v) estimation of the population mean can be made by first making a map, and then computing a map mean; this leads to unusual methods of estimating the population mean, not the simple ratio or regression estimates.

The accompanying appendix has a summary report of some Monte Carlo work. We compare two methods:

(i) the use of 20 observations with 10 strata each having two observations,

and

(ii) the use of 2 end-points and a randomly started systematic sample with an additional 6 points at random to assess the precision of conclusions.

In the second case, the mapping rule used was merely to form a polygonal line. One could certainly use more complex mapping rules. The conclusion is rather obvious, for me. One can do much better in all respects by procedure (ii) than by procedure (i), not only with regard to the population mean but also with respect to the map of the population. Note that in procedure (ii) the precision is estimated only from the 6 observations that are <u>not</u> used in estimating the map. (They can, however, be used to adjust the map mean.)

Any mapping rule leads to an estimation procedure. One could fit a 15th degree polynomial to all the 16 points if one wished. One could go through a variety of preliminary tests of significance to choose a mapping rule. The precision of the procedure is estimated externally, so to speak, by the additional 6 points.

The extension to two dimensions is clear, though the variety of possible mapping rules is greater.

I am firmly of the opinion that this type of procedure should be used in surveys of forests, and possibly other populations fixed in space, in which one may reasonably expect a relationship of attributes of interest to position.

This line of thought suggests that the actual precision of ratio and regression estimators can be assayed with a supplementary sample which is not used in developing the estimator. I am inclined to the view that this is in fact the general case.

Concluding Remarks

I hope that my critical remarks about some aspects of sampling theory are not interpreted as criticism of most sampling practice. I would often do what they do even though I do not know completely why. But I believe inadequate attention has been given to the basic inferential questions, and I hope this conference will advance that cause.

Appendix

The Examination of an Unknown Function Over an Interval

O. Kempthorne and Leon Jordan-Filho

The population lies on a line segment. The problems are to estimate the mean and to estimate a map of the population. In the case of forestry, for example, one would not wish only to determine how much timber there is in the line segment but also where the timber is. Two schemes of observation are compared:

(a) simple random stratified sampling (SRS)

and

(b) randomly placed systematic sample (SYS) with additional random points and the two end points.

To estimate the mean one can use the ordinary SRS mean, a map mean obtained by using the values for the end points and for the points in the random sample. For the systematic sample one can draw a linear map using only the systematic and end points. One then uses the additional random points to (a) adjust the linear map mean, and (b) to estimate the error of the map given by this linear mapping. The latter is obtained as the average over the additional points of (actual value-linear map value)^2.

The graphs presented below exhibit what happened with 10 strata of two points for the random stratified sample, with 12 points in the systematic sample plus 2 end points plus 6 random points. Histograms are presented for the following attributes:

EMI - error of usual SRS mean
ETI - error of linear map mean from SRS (with end points)
D4 - error of linear map mean from SYS corrected
T1 - EM1/S.E. (EM1)

TT - D4/S.E. (D4)
SY2 - true map error of SYS linear map with end points
ST2 - true map error of linear SRS map with end points
ST4 - true map error of SRS bar map
E5 - estimate of SYS map error
P - SY2/E5.

Obviously what happens depends critically on the nature of the true relationship of the attribute to its position on the line segment. Two cases were considered; a polynomial and an approximate realization of a Gaussian stationary process. These are also illustrated. It is quite clear that if the function has considerable near-regularity, the systematic procedure does much better than the stratified random sample. If on the other hand there were no near-regularity (speaking loosely) one would lose by the systematic procedure, but one can readily surmise one would lose little. Also one can be sure that existent forest populations do have real trends, so the question of what happens with a purely random pattern of values is irrelevant.

This line of thought suggests that much of the theoretical work on systematic sampling has been ill-directed and that the use of a systematic sample with a few random points to assess accuracy solves the practical problem.

Examples of f(x) Considered

(1) Polynomial

$$f(x) = 1 + 132x - 640x^2 + 1024x^3 - 512x^4$$

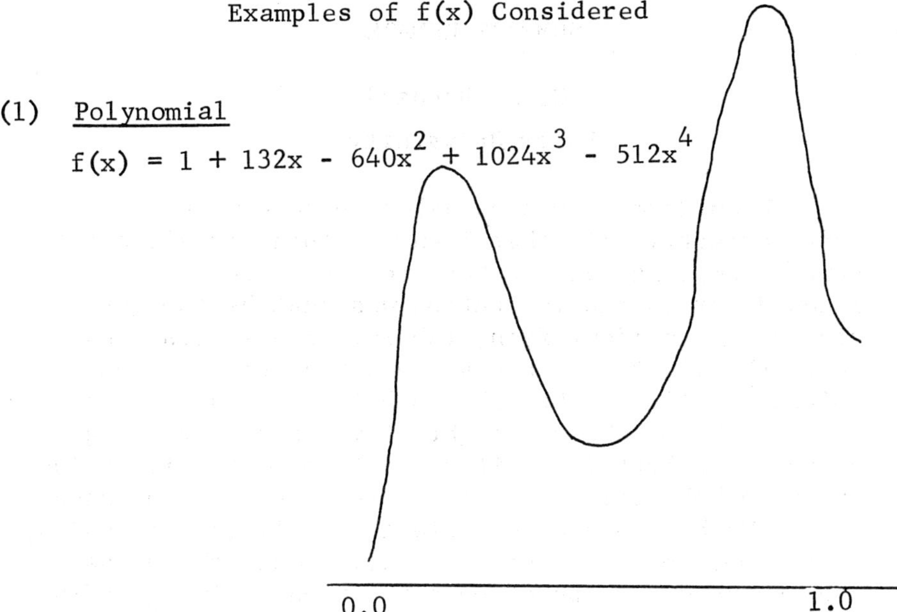

(2) Stochastic Function

An approximate realization of a Gaussian stationary process with covariance kernel

$$K(s,t) = e^{-\lambda|t-s|}$$

with $\lambda = 0.4$. (This based on work of Matern).

SUMMARY REMARKS

G. A. Barnard
Essex University

When Norman Johnson asked me to come here he knew perfectly well that I know nothing at all about survey sampling, and perhaps the best construction I could put on his invitation was that he thought discussion on identifying labels, likelihoods, and so forth, might generate so much heat than an uninvolved outsider might help to hold the ring. Alternatively, he may have thought of me as a disturbing element, or butt for criticism, to promote discussion. Whichever objective he had, it would have been unnecessary. We have had an excellent, civilized discussion, on a number of controversial topics; and there has been no need for promotion nor for any calming influence.

However, I was very glad to accept his invitation, because I have been anxious to learn about survey sampling and analysis. And, so far as one can learn without actually doing in this area, I think no better circumstances than have obtained this week could be imagined. Furthermore, although Norman did not know it at the time, I was and am involved in a major survey directed towards studying the transition from secondary school to University or other form of tertiary education so that I can even lay claim to a limited amount of practical experience - not to mention smaller studies of transport behaviour, skilled manpower studies, and so on, with which I have been concerned.

The history of this secondary to tertiary education study serves to illustrate a number of points, most of which have been made by others here, but which perhaps deserve reemphasis. The Royal Statistical Society has been concerned with the teaching of statistics in schools, and from this point of view some fifteen years ago a study by Egon Pearson disclosed a threat of a drastic shortage of school mathematics teachers, which seemed likely to become most acute in about ten years from that time. The Society drew attention to this, but perhaps not so forcefully as, with hindsight, one might have wished; and as a result we in England are facing a crisis in science education. However, about four years ago the Society saw that the statistical information, available and in prospect, on Higher Education, stood in need of improvement, and it set up a committee to look into the situation. We were able to note studies which Claus Moser was putting in train, concerned with the transition from tertiary education to industrial and other employment; and a number of other studies, some of which we helped to develop. But what came out most clearly was that no provision was being made to cover the gap in the chain of data between the secondary and tertiary stage. So the Society proposed to our Department of Education and Science that a study in this area should be undertaken. The Society was willing, if necessary, to sponsor the study, though it would have preferred that some University or other body could be persuaded to undertake the exercise. In the end, the Society did sponsor it, though mainly for technical reasons the workers have been attached to the University of Essex. It took two years to get final clearance for the study from the D.E.S.; and almost from that very time the D.E.S. has been breathing down our necks, to know when our results might be available - so serious has the situation patently become.

I do not wish hereby to imply criticism of the D.E.S., who have really been most helpful, almost uniformly. But the story is, I think, worth putting on record as illustrating what Frank Yates said on Wednesday about the duty of the statistician to call attention to gaps in the information available for social decisions, and to urge the adoption of steps to improve the situation when appropriate. We do not do as much as we should in this direction.

The D.E.S. survey also illustrates another point which perhaps may be of local relevance only, though perhaps others may find it useful. As planned, it certainly was concerned with subclass comparisons, etc., very much of the type discussed by Leslie Kish. And I originally considered adopting a rather elaborate sampling plan, designed to maximize the precision of the relevant contrasts. I owe to John Tukey the advice, which I took, to make it basically a pps survey with schools as psu's; and I am glad to think that we used Bill Madow's method for drawing the sample.

Why am I so glad? Because as it is, the limitations on available survey analysis facilities are such, at present, that we are held up for the preliminary results, beyond rather crude total and some cross tabulations; and had a more elaborate design been used this would have put still more strain on our rather limited computing manpower. (Perhaps I should make clear that our University has only been in existence for three years. We <u>shall</u>, I am sure, eventually have excellent facilities for survey analysis, because the whole University is determined that we should and I hope we can get Frank Yates' programme to run on our machine. But meanwhile things are a little delayed!)

For all that we designed as for a descriptive survey, there is no doubt that ours is an instance which supports Leslie Kisch's argument, and what here seems to be something of a consensus, that surveys of human populations are becoming more and more analytical, rather than descriptive - a development which corresponds to a welcome rise in the level of sophistication in the use by administrators and politicians of the information provided for them.

Mention of the sophistication of our clients brings me to the main point I would like to make. It seems to me that the time has come when we can consider abandoning the practice of expressing the results of our investigations in terms simply of "point estimates", with or without "standard errors", or in terms of acceptance or rejection of hypotheses.

For some time now the nuclear physicists have been in the habit of expressing the information they have about the various constants of nuclear physics - say, the difference δ in mass between the K_1 and K_2 mesons in what they call an ideogram like Fig. 1 (See, for example, Reviews of Modern Physics, 39, (1967), p. 11).

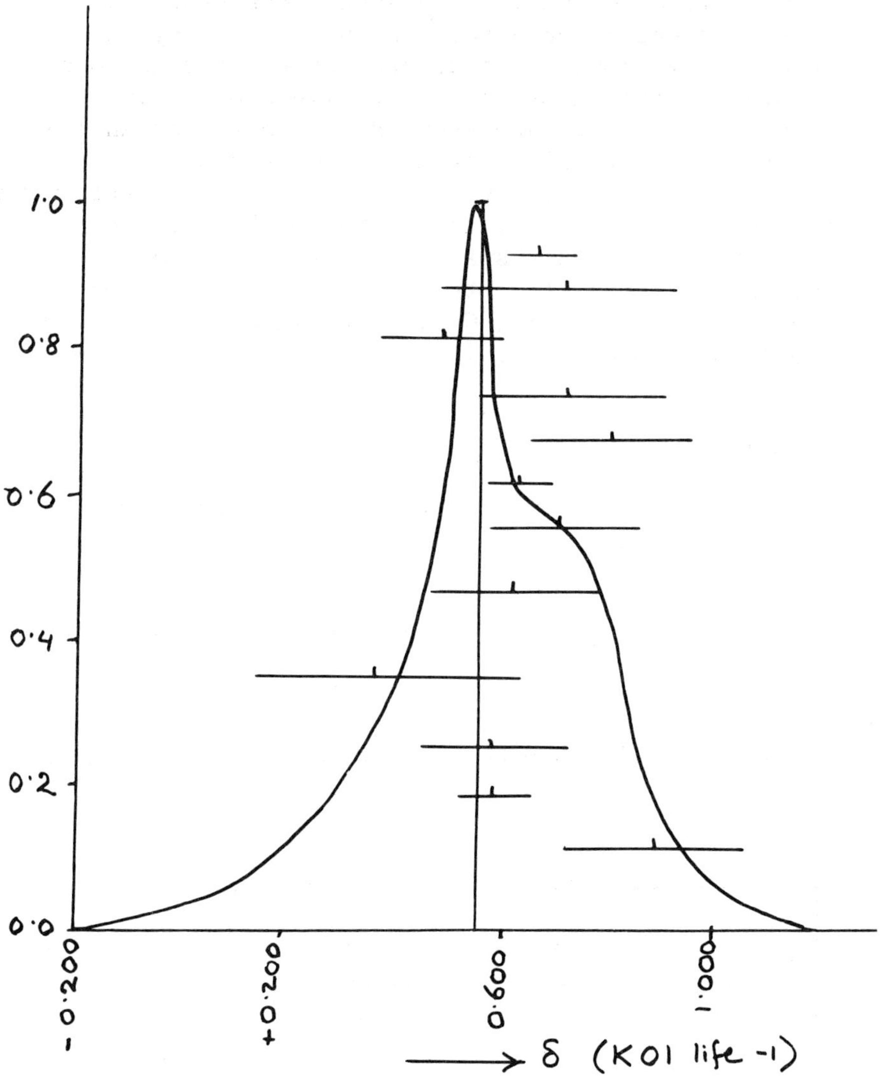

Figure 1. Ideogram

The ordinate is a form of credibility, very close to a likelihood on the available data. Indeed, in some publications the physicists and geneticists too, express their information directly in terms of likelihoods. Would the heavens fall if we started doing the same? I think not; and in support I would refer you to the Journal of our British Consumer Association, called "Which", which about four years ago carried a report on a test on the life of nylon stockings (I commend it to you for reflection to determine whether in sampling stockings of a given brand and price we are sampling a finite, or an infinite, population). There is a very wide scatter in the life of these things (corresponding, in fact, more to the habits of the wearer than to the brand of stocking) and as a result, although quite extensive trials were conducted, the results were not conclusive. Because of this, Chris Winsten suggested they should be expressed as in Fig. 2.

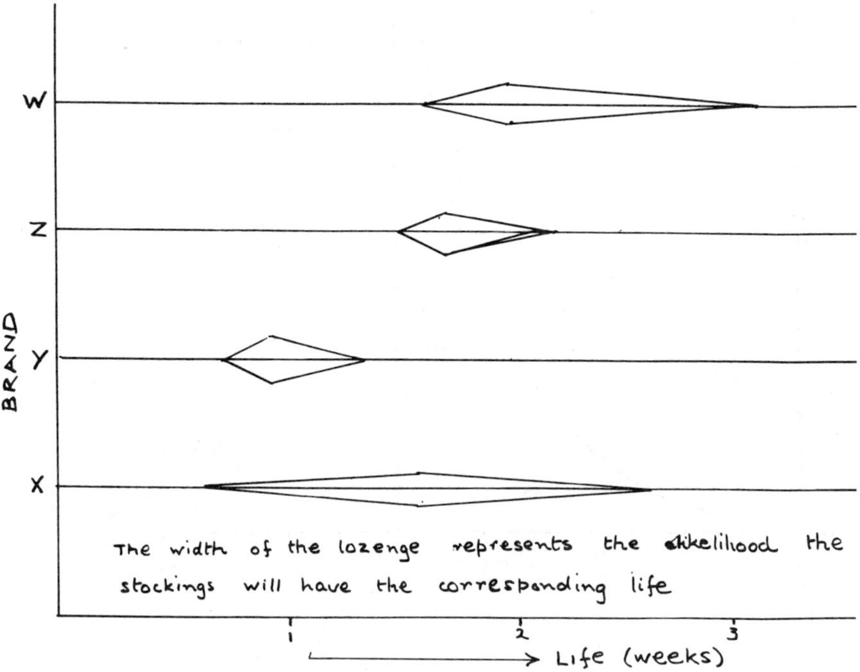

Figure 2.

and I can report that although the journal in question has a mass circulation, there were no complaints that the way in which the results were expressed made them incomprehensible.

If I may refer, for example, to Dr. Robson's paper on capture-recapture methods, in the presence of mortality and other parameters, one can't help feeling that, as Derek Hudson suggested, the expression of the message in his data would be made most clearly in terms of a plot like Fig. 3.

One is often asked, and Prof. Kempthorne repeats this, what such a likelihood plot means. It can be interpreted in many ways, but perhaps the simplest is to determine, by the usual test procedures, an interval or set I for θ such that, unless an event of probability (say) 10^{-3} has occurred, the true value of θ must lie in I. Then, it would seem sensible to imagine the value of θ as <u>possibly</u> having been picked out at random (ie. uniformly distributed) in I.† If it had been, then on the basis of the given data one could say for instance, that the odds, that the true value lay in the set AB $\cup$ CD would be the ratio of the double shaded area to the total shaded area (ie. about 5/6, as drawn).

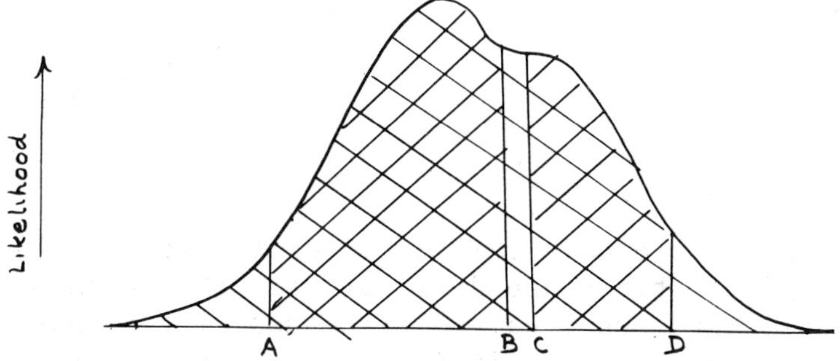

Figure 3.

† Thus one might say loosely a likelihood man is a non-exclusive Bayesian who doesn't believe in his priors. A non-exclusive Bayesian is one who accepts the need for classical tests of significance.

All these statements, I would have you note, are rigorously demonstrable no matter what theory of probability we adopt. And I hope this mode of expression will not be ignored although it may seem strange at first.

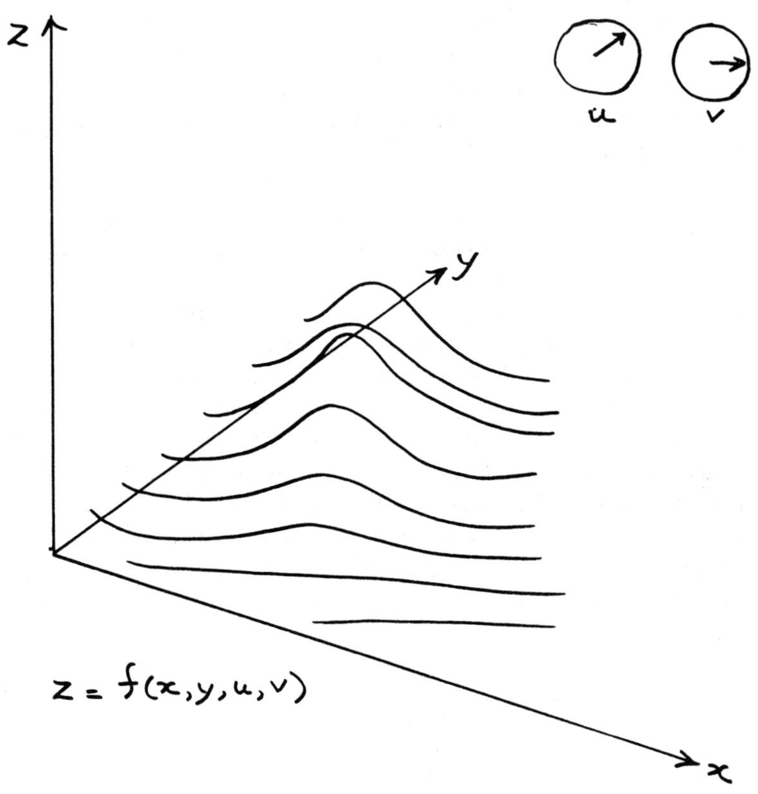

Figure 4.

Many people have raised the difficulty that such likelihood functions are difficult to take in when many parameters are involved. The use of too many parameters carries with it its own dangers (just as does the use of too few); but I would like to mention the way in which devices like the Stromberg Carlson 4020 graphical output machine may change matters in this respect. This machine produces a frame of 35 or 16 m.m. film, with graphical data such as, for example Fig. 3, at the rate of 4 frames a second. So we could represent a function of, say, four variables, as in Fig. 4., with the clock representing u rotating rapidly in a sequence of frames, and the clock representing v rotating slowly. The surface will then both quiver (by u) and heave (by v).

However we need to be very careful to avoid introducing too many parameters into our description of any situation. Perhaps the least serious dangers that can arise is illustrated by the likelihood used by Godambe and Ericson, which effectively expresses the fact that populations which could not have contained the observed sample are impossible, and while any population which could contain the observed sample is possible. The apparent triviality of this likelihood function arises from the unwillingness, expressed in Ericson's very general class of prior distributions, to make any assumption beyond those relating to the exchangeability of the sample elements.

If I may interpolate here, I would remark that the dependence of the likelihood function on the model is something which would repay further study. It would seem at first sight that we should be able to treat it very simply, by the consideration that choice of one model among many involves nothing more than a restriction of the parameter space from Ω, say, to a subset Ω_1, and would therefore always be equivalent to setting the function zero in the complement Ω_1.

For example if we take a simple dose-response experiment, in which dose x is related to the probability p of response by the logistic law

$$\log \frac{p}{q} = \alpha + \beta x$$

and we have a three point assay with doses equally spaced at x = -1, 0, +1, the assumption of the logistic law has the effect of confining the parameter space, which in general consists of the cube

$$0 < p_1 < 1$$
$$0 < p_2 < 1$$
$$0 < p_3 < 1 ,$$

to the surface within the cube whose equation is

$$\left(\frac{p_1}{q_1}\right) \cdot \left(\frac{p_3}{q_3}\right) = \left(\frac{p_2}{q_2}\right)^2$$

But if we consider another example, originating a long time ago with Prof. Neyman, in which x_{1i}, x_{2i} are supposed independent $N(\mu_i, \sigma^2)$, and the parameter of interest is σ^2, the likelihood function has its maximum where $\sigma^2 = \sum_i (x_{1i} - x_{2i})^2/4n$

just half the value we could expect. But if we say that the data are equivalent to

$$m_i = (x_{1i} + x_{2i})/2$$

$$d_i = (x_i - y_i)$$

and then say that the m_i values are irrelevant, then the model for the problem becomes

$$d_i \text{ distance as } N(0, 2\sigma^2)$$

and the likelihood for σ^2 now has its maximum at $\frac{1}{2}\Sigma\, d_i^2/n$, just where it ought to be

To obtain the second likelihood from the first, in Bayesian terminology we have to integrate out the nuisance parameters, using as a "prior" the invariant Haar measure of the group of transformations corresponding to our concept of irrelevance.

Another example, - and one more directly relevant - to survey sampling practice, - I would refer to the work of Charles Stein and Alan James on the multivariate normal distribution. If we have a sample $\underline{x}_i$ $i = 1, 2, \ldots\ldots n$ from a multivariate population

$$N(\underline{\mu}, \underline{\sigma}'\underline{\rho}\,\underline{\sigma})$$

where $\underline{\mu}$, $\underline{\sigma}$ are K- rowed vectors and ρ is K×K with $\rho_{ii} = 1$, we often want to estimate $\underline{\rho}$, and are not concerned with $\underline{\mu}$ and $\underline{\sigma}$. Stein shows that in spite of theorems guaranteeing good behaviour of the maximum likelihood estimates when $n \to \infty$, with K fixed and finite we can have really serious trouble if $n \to \infty$ and $K \to \infty$ as well, with $K = O(\sqrt{n})$. And considering that we are sometimes required to discuss the factor analysis, for example, of a 30 - rowed vector, on the basis of 1000 observations, it will be seen that the dangers Stein points to are quite real and of practical importance. What Alan James has shown is, that if we take the usual sample estimate $\underline{r}$ of $\underline{\rho}$, whose density function depends only on $\underline{\rho}$, and use this to generate our likelihood function, the changes which Stein draws attention to are removed.

Before leaving this matter I would like to cite another example - an amplification of one which I venture to put forward as suitable for meditation: how to form an estimate of the probability that in the U. S. a Catholic will marry a Protestant. This problem has so many aspects that I feel one might use it as a basis for a whole course on survey sampling, much as I remember the philosopher Wiltgenstein who gave a course of lectures (which I attended, happy days!) each 3 hours long, 3 times a week, through two semesters, devoted to discussing what is the king of chess. But let us be very simple minded and suppose that we can use the simple multinomial distribution

θ_1 = probability that a man is a Catholic

θ_2 = probability, given a man is a Catholic, that he marries a Protestant

Then if a simple random sample of n men splits into

n_1 non-Catholics

$m \begin{cases} n_2 & \text{Catholics not married to Protestants} \\ n_3 & \text{Catholics married to Protestants} \end{cases}$

the simple-minded approach would condition m fixed, and work in terms of the binomial distribution for n_3, generated by

$$[\overline{1 - \theta_2} + \theta_2]^m$$

or the likelihood function proportional to

$$L(\theta_2) = \theta_2^{n_3} (1 - \theta_2)^{n_2}$$

We are taking the n_1 non-Catholics as irrelevant.

The complete likelihood is, however proportional to

$$(1 - \theta_1)^{n_1} [\theta_1 (1 - \theta_2)]^{n_2} [\theta_1 \theta_2]^{n_3}$$

which factorises into a function of θ_1 and a function of θ_2,

$$(1 - \theta_1)^{n_1} \theta_1^{n_2+n_3} \cdot L(\theta_2).$$

Now a Bayesian would require to have a prior $P(\theta_1, \theta_2)$ to this experiment, and it is clear that if, and only if

$$P(\theta_1, \theta_2) = P_1(\theta_1) P_2(\theta_2)$$

he would effectively take $L(\theta_2)$ as the likelihood for θ_2. Now the factorisation of $P(\theta_1, \theta_2)$ corresponds to saying that a priori, the proportion of Catholics in the population at large is irrelevant to the estimation of θ_2.

Now this θ_1 is obviously not irrelevant. If we knew that θ_1 were very large, along with the proportion of Protestants among the women, then our knowledge of human nature would leave us to suppose that θ_2 would be likely to depart somewhat from zero. Whereas if θ_1 is small, along with the proportion of Protestants among women, then θ_2 is likely to be small. However, with moderate θ_1, as, I suppose is so in the U.S. and the U.K., it is not unreasonable to suppose that θ_1 is irrelevant. Thus the judgement of irrelevance is not purely logical; it has empirical content also.

I hope these examples will make clear my attitude to Prof. Ericson, and other $\frac{\text{HIS}}{\text{WFFD}}$ Bayesians like him. HIS stands for Highly intelligent
WFFD subjective Bayesians with a feeling for data.

The attitude to data appears to me to be very similar to the approach to epistemology which was popular in

Cambridge in the 1920's in which one was taught to think of a table as a bundle of sense-data. Then in the 1930's Wiltgenstein persuaded us that to replace the word "table" whenever it occurred by "bundle of sense data ordinarily called a table" really served only to lengthen the time it would take to express our thoughts. The painstaking analysis that Prof. Ericson has made of stratification is a good example of how, by a thorough Bayesian analysis, we are led back to do what we did before, just as we come back to calling a table a table.

I hope my brief reference to Catholics and Protestants will indicate that I do not think such excursions as Prof. Ericson has made as in any sense a waste of time. But the main point to be emphasised, I think, is the identity of practical treatment of which we are led, in nearly all cases.

Next let me come to Prof. Godambe. I hope he will forgive me if I say that if I feel a short answer to him and Dr. Hanurav (not to mention Prof. Ericson in this connection) might be said to be "round objects". I refer here to a favourite story of the late Edgar Fieller, who told of a Civil Service colleague who felt like expressing himself in vulgar terms on the merits of a proposal, but felt impelled to avoid the use of the five letter word, and so wrote "round objects" on the file. When this came to the Minister, he wrote "Who is Round and why does he object?".

The reason why round objects provide an answer to Dr. Godambe is, that by putting unlabelled round objects, or balls, into a sufficiently large urn, and agitating it sufficiently vigorously, we can then pick out a sample of any size less than the population size which is random, although it is unlabelled.

Prof. Godambe, I think, suggests that this is based on intuitive reasoning, not mathematical. A distinguishing feature of some of the populations which are dealt with in survey sampling is, that they have a real existence (I don't here wish to exclude referring them to a superpopulation in the way in which Prof. Sprott and Mr. Kalbfleisch do). So when we talk of a random sample of such a real population we have in mind a <u>physical</u> process, like that of drawing balls from an urn. And we surely do accept that "random" is an atribute which can meaningfully be attached to a physical process.

We need not be discussing within this universe of discourse, however, and Prof. Godambe and Prof. Ericson, and Dr. Hanurav, are, at least part of their time, talking within a strictly mathematical universe of discourse. In this connection, I think the objection that might be made to their mathematical analysis is, that they take a random sequence as given, in some sense axiomatically. Whatever may be said about the need for this up till a few years ago, the work of Martin-Löb and Kolmogoroff on the definition of a random sequence now enables us to go somewhat deeper.

Since neither M-L nor Kolmogoroff has done more than indicate verbally what their approval is, it may be helpful if I say that, roughly speaking, they say that a sequence

0110111001......

of 0's and 1's is random if the set of rules defining the sequence has to form a sentence as long as the sequence itself.

A sequence like

010101......(to, say, 10 million terms)

is not random, because it can be described by saying simply

$$f(1) = 0$$

$$f(n+1) \neq f(n)$$

I conjecture (I have no proof) that when a proper analysis of randomness is incorporated into the mathematical theory, it will turn out that invariance under permutations of labels is, in one way or another required.

While in this area, may I say that I accept Prof. Godambe's analysis of the fiducial distribution of the population mean, in the situation specified as he does it. And although Fisher did restrict his fiducial arguments to the continuous case he did in conversation with me agree that cases could arise in which discrete fiducial distributions could be found. I think this is one such. We need, however, to be clear on the precise specification of conditions for its appliability; and I must admit I find it difficult to see how Dr. Godambe's conditions could arise in practice.

If I might go on, I am led to wonder whether Prof. Godambe's name means, in Maharastrian, "Gadfly". I mean to convey the idea that Dr. Godambe serves to fulfill the very useful function of keeping us prodded, so that we do not fall victims, as we are prone to do, to the intellectual laziness embodied in loose modes of expression. But in the case of Prof. Hartley and Dr. Rao I wonder whether the gadfly hasn't stung the horse so sharply as to make it bolt. The treatment of an observation in terms of scaling points was something which Fisher did very frequently, with good effect (see, for example, his 1934 RSS paper); and it does help to view samples from this point of view. But my feeling is that Dr. Godambe is now satisfied that the point he was making is well taken and we need not pursue the matter further.

Finally let me come to my last main point. This could be summed up by saying that we might do well, in sample surveys, to make more frequent use of models like those put forward by Prof. Sprott and Mr. Kalbfleisch - in which we regard our given sample, and the remainder of the finite population, as two samples from an underlying super population about which distributional form assumptions are in order. For example I might refer to the data quoted by Dr. Sedransk in the discussion on Dr. Rao's paper on ratio estimators, on dimensions of timber stands: my recollection was that in Dr. Bliss's book there is some data on tree stands which suggests that the joint distribution of, for instance, tree volume and diameter at breast height, is log-normal-and I am grateful to Dr. Sedransk for confirming that this is so, provided the stands are of uniform age. If we make use of this information we can, of course, apply the standard likelihood procedures for comparing alternative estimators, so that the approach via the theory of ratio estimators becomes irrelevant.

It is noteworthy that the only discussion we have had as to population form has, I think, been given by Dr. Patrick - though I venture to suggest that if the distributions she gave really were truncated normal, rather than censored logarithmic, I would be very much surprised. But surely survey samplers should follow the example of biometricians and put on record the histograms of their data, so that a body of knowledge about distributional forms can be built up.

Of course I'm not suggesting we should take our distributional models over seriously. Never. As in other statistical fields we must have regard to robustness, etc. But in adopting, I think by tradition, an almost entirely non-parametric approach (though this would better be called an over-parametric approach since it involves too many parameters rather than too few) we surely have been lacking somewhat in efficiency.

If this point is taken we may look forward to a greater integration of survey sample theory with other areas of statistical inference. This process has already gone a long way, and I must say that, while I still feel ignorant, I do not feel a stranger in present company. The development of half-sample replication offers a lot in this direction - One value of this approach lies in the prospect it offers of accumulating empirical data about clustering, sampling, and response errors - such as we had from Dr. Neter, for example, about forecasts of car buying.

But I would not wish to sit down without referring to Dr. Joshi's work on the admissibility of the sample mean. He said, when challenged as to the value of his work that his proof that the sample mean was admissible was to be evaluated in the light of the possibility of the opposite conclusion - but one should in this connection consider what has been the bearing on practical statistics of Stein's result about the inadmissibility of the sample mean when we are estimating 3 or more means at one time. It has, I think been virtually ignored - wrongly, in respect of situations like factorial analysis, where a natural origin of measurements presents itself. But in other cases, where we have translation invariance, the fact that we have continued to use the translation invariant estimate, in spite of its inadmissibility, shows that the invariance requirement takes precedence.

Dr. Joshi, like Dr. Ericson and so many other contributors to this symposium, has based his analytical work on the rigorous mathematical model for finite sampling provided by Dr. Godambe in his fundamental papers (especially in J.R.S.S. (B) 1955 and 1966). I hope the light-hearted tone of some of my comments on Dr. Godambe will not lead anyone to suppose that I underestimate the value of his pioneering work on the mathematical formulation of the foundations of the subject. And indeed, we are, I am sure, all of us most grateful to him as being the initiator of the idea of such a conference as this. I am expressing a general consensus in saying that few, if any of us, can recall a conference which has been so rewarding as this one. To Godambe, and of course to Norman Johnson and all those here at North Carolina who have been responsible for the practical arrangements and organization, we are most grateful for this.

To answer the question with which I began, I think I was put up at this point to enable everyone to have a last fling or discussion. I hope that in doing so, we may focus on the very great extent to which we are all agreed.

BIAS IN SURVEYS DUE TO NONRESPONSE*

Closing Address

Jerzy Neyman

University of California, Berkeley

0. I feel very honored by the invitation of Professor Norman L. Johnson to take an active part in his Symposium on Sampling Surveys and, more particularly, by his suggestion that I deliver the closing address. The pleasure of doing so is enhanced by the ties between this Symposium and the period of my life in London 1934-1938, the period which predetermined the pattern of my subsequent scholarly activity.

First I shall mention the personal ties. In the middle 1930's two important participants of this Symposium were Junior Members of the University College, London, working on their Ph.D. theses and were frequent visitors in my office there. One of these individuals is Norman L. Johnson himself, currently Professor at one of the most illustrious centers of statistical research and instruction and the organizer of the present Symposium. The other individual is Dr. P. V. Sukhatme, an eminent leader in agricultural sampling surveys in developing nations and head of statistics in the F.A.O. in Rome. I enjoy the memories of our contacts in London and I enjoy our present encounter!

The ideological ties, which to me are also emotional ties, between this Symposium and my London period stem from the fact that the subject of this Symposium was also the subject of a paper [1] which in 1934 I delivered before a meeting of the Royal Statistical Society. This particular paper contained certain ideas which at the time were highly unconventional. As a result, I became involved in disputes, on several fronts at once, and the exchange of arguments and counterarguments influenced greatly my subsequent research activities.

*This paper was prepared with partial support from the United States Army Research Office - Durham, Grant No. DA ARO(D) 31-124-G816.

One of the subjects of disputes is specifically mentioned in the title of the paper: "On the two different aspects of the representative method: the method of stratified (random) sampling and the method of (nonrandom) purposive selection." In formulating the paper I expressed the hope that "as a result... the general confidence ... in the method of purposive selection will be somewhat diminished."

The title of the paper and the quotation reflect on the contemporary thinking of the leading statisticians in many countries. That was the epoch when the advisability of randomness at any stage of a survey was questioned and when "purposive" selection was frequently preferred. My own paper, originating from a sampling survey conducted in about 1930 by the Institute for Social Problems in Warsaw, represented a call for a serious change in attitudes. As reflected in the various publications [2,3], currently this particular battle appears to have been very definitely won. The battle for randomization in surveys of human populations must be considered as a part of a more general battle for randomization of experiments with variable material. This more general battle was initiated in the 1920's by R. A. Fisher and continued by Frank Yates and their followers, and also must be considered won. In the United States the credit for leadership in this particular domain belongs to G. Snedecor, Gertrude Cox, W. G. Cochran and Alexander Brownlee. It is true that from time to time we continue to witness publications which, while paying lip service to randomization, advocate reliance on nonrandomized experiments. However, such cases are rare and one is uncertain whether they are manifestations of ignorance or of intentions to mislead.

The other battles in which I became involved through my 1934 paper stemmed from an effort I made to incorporate the sampling surveys into a comprehensive system of ideas which, to my mind, constitute the theory of statistics. The descriptive term I invented (in 1937) to describe this theory was the theory of "inductive behavior." Subsequently, in the 1940's, Abraham Wald invented a better term: theory of "statistical decision making." Characteristically, questions of this general category crop up more or less spontaneously at every serious attempt to discuss empirical statistical research. This

Symposium is no exception and it was a pleasure to listen to papers by George Barnard, J. Durbin and Oscar Kempthorne, all dealing with these particular matters. As is frequently the case in philosophical discussions, each of these papers contained points with which I agree and also some other which are not consistent with my own thinking.

As far as the sampling of human population is concerned, I made just two attempts to contribute to the theory. In both cases I was concerned with optimum stratification and optimum allocation. As years went by, I learned that my results of 1934 were anticipated by Tschuprov [4]. My other effort [5] was stimulated in 1937 by certain questions regarding an interesting method of double sampling invented by Sydney Wilcox and Milton Friedman. Thereafter, sampling of human populations disappeared from my horizon and was replaced by interests in other domains. As a result, my contribution to the present Symposium will be concerned not specifically with what one ordinarily calls "sampling surveys" relating to human populations, but with two aspects of a very general problem occurring in practically all domains of empirical statistical studies. Regretfully, while essentially the same, this problem is described in varying terms depending upon the domain of study. Im sampling human populations it is called the problem of nonresponse. The importance of this problem is illustrated by the fact that in this Symposium it was mentioned in a very large number of papers, beginning with the opening address of Morris H. Hansen and Benjamin J. Tepping. Subsequently, essentially the same problem was treated by D. H. Cox with reference to textile fibres, by P. V. and B. V. Sukhatme with reference to farms, by Ruth Patricle dealing with aquatic communities, by Walt R. Simmons and Judy Ann Bean with reference to health statistics, and by Lester R. Frankel in market research. To this great variety I propose to add two somewhat extreme domains: the problem of "nonresponse" in studies of galaxies and in studies of carcinogenesis.*

---

*I am grateful to Professor Tore Dalenius for calling my attention to the so-called Politz-plan in dealing with the problem of "not at home" as described in [3]

The general nature of the problem of nonresponse is in all these domains the same: the intention is to study the distribution of a variable X characterizing the members of a certain population Π; however, some of the members of this population fail to "respond" and the frequency of "nonresponse" is, or may be, correlated with X. Unfortunately, the details of this problem differ from one domain to the next and, thus far, no general methodology seems to exist. Most of the results reported below were obtained jointly with my colleague in Berkeley, Elizabeth L. Scott.

I. "Nonresponse" problem in galactic research"

1. Substantive background. The studies to be reported concern galaxies [6], [7]. According to current views, galaxies occur in agglomerations, called clusters, with membership varying between very broad limits. For some particular clusters the membership estimates run into thousands. Some other clusters occasionally called "small groups" have estimated memberships below 10. Also, there are single so called "field" galaxies which, for purposes of uniformity, we consider as clusters of one member only.

In the theory we were working on, it is assumed that the actual clustering of galaxies may be of some nth order, where n has not been estimated. However, the work reported here is based on the tentative assumption of first order clustering. This amounts to postulating that the centers of clusters are Poissonwise distributed in space, that to each cluster center there correspond a random variable ν, representing the number of cluster members, and a nonnegative function f, representing the probability density of the location $\tilde{x} = (x_1,x_2,x_3)$ of a cluster member with respect to the location of the cluster center, say at $\tilde{u} = (u_1,u_2,u_3)$. It is assumed that the function f depends only on the Euclidean distance between $\tilde{x}$ and $\tilde{u}$ and, for typographical simplicity, we write $f(\tilde{x}-\tilde{u})$. It is postulated that the locations of the ν cluster members are mutually independent.

The subjects of study of galaxies include the measure of the galaxy's faintness denoted by M and described as the absolute magnitude of the galaxy. The larger the value of M, the fainter the particular object. In order to measure M directly, it is necessary to approach the galaxy to a specified distance, which is impossible. From the Earth all that can be measured is the "apparent magnitude," which is usually denoted by m. Apart from obscurations, and some other details which we shall not discuss here, the relationship between the apparent and the absolute magnitudes is given by the formula

$$m = M - 5 + a \log \xi \tag{1}$$

where $a = 5 \log_{10} e$, and ξ stands for the distance of the object measured in appropriate units. As is intuitively clear, for a galaxy of fixed absolute magnitude M, its apparent faintness m will vary as a function of the distance ξ.

As is well known, galaxies differ in their appearance: some look like spirals of a varying degree of regularity and development. Some others appear as smooth ellipses, etc. For present purposes it must suffice that astronomers distinguish several morphological types of galaxies, that there are several systems of classification and certain theories of evolution of galaxies. In this connection it is important to know the proportions of galaxies that belong to this or that morphological type. These proportions, called abundances, are natural subjects of study.

In our own work we treat the type of galaxy as a random variable with a probability, say Λ_t, that a galaxy will be of type number $t = 1,2,\ldots$. Further, we consider that the type of one galaxy is independent of that of another and of all other variables of the system. Also, we postulate that each type t of galaxies is characterized by its own distribution of the absolute magnitude with a probability density which I shall denote by $\phi_t(M)$. The absolute magnitude of one galaxy is again assumed to be independent of that of another and of all other variables of the system. The function $\phi_t(M)$ are also obvious subjects of study. In the astronomical literature they are called "luminosity functions."

2. <u>Problem of selection bias</u>. The obvious sources of information regarding the abundances Λ_t and of luminosity functions ϕ_t are surveys of the sky and their results compiled in catalogues of galaxies. The simplest description of the process of compiling a catalogue is as follows.

An astronomer embarking on the project selects a region in the sky, say R, convenient for his observations. A series of photographs are taken over R. The astronomer inspects these photographs and marks those galaxies which, in his opinion, will be accessible to the available instruments without excessive trouble. The objects marked are numbered and detailed observations begin starting with no. 1, no. 2, ... etc. These observations include the following items: whether the galaxy is a field galaxy or a member of a group or cluster, the morphological type of the galaxy, the galaxy's apparent magnitude and its spectrum. The analysis of the latter provides the so called redshift and, eventually, the distance ξ of the galaxy and its absolute magnitude M.

The reader must realize that this description is grossly oversimplified. For example, quite frequently, the apparent magnitude of a galaxy is measured by one astronomer in one observatory and its redshift by another astronomer in a different observatory. Also, ordinarily, a substantial catalogue represents a combination of several "subcatalogues," each compiled independently from the others, etc. However, for the present purposes the above schematic description must suffice.

Consider, then, the catalogue of galaxies compiled more or less as described and concentrate on those galaxies of the various types that are marked as field galaxies. Let N denote the total number of these galaxies and N_t that of those of the tth type, $t = 1,2, \ldots$. The quotient $n_t = N_t/N$ is described as the catalogue abundance of field galaxies of the type considered. Not unnaturally, one is tempted to consider the catalogue abundance n_t as an estimate of the abundance Λ_t which we will now describe as the space abundance. However, it is easy to perceive that, considered as an estimate of space abundance Λ_t, the catalogue abundance is likely to be biased.

Using the observed values m and ξ for each of the catalogued field galaxies of the type t , one can easily compute their absolute magnitude M and construct the empirical distribution of these values. At the present moment the numbers N_t of field galaxies of the same morphological types are rather small, but one can visualize an improvement in this respect so that we are led to considering the possibility of an empirical estimate of the distribution of absolute magnitude of those galaxies of type t that are in the catalogue. The density of this distribution will be called the catalogue luminosity function of type t galaxies and denoted by ϕ_t^*. Again one may be tempted to consider that an empirical ϕ_t^* is an estimate of ϕ_t , the "space luminosity function" of the particular type of objects. Here again it is easy to perceive that, as an estimate of ϕ_t , the function ϕ_t^* is likely to be biased.

The manner in which Miss Scott and I approached this situation, the manner which may perhaps be useful in other domains of study, is as follows.

3. Selection probabilities and space luminosity functions.

We assumed that the morphological classification used is so fine, that the probability that a particular galaxy of type t located in R will be included in the catalogue depends on this galaxy's apparent faintness m and on nothing else. Denote by $\Phi_t(m)$ this probability and call it the selection probability. A priori, little is known about the function $\Phi_t(m)$ except that for very bright galaxies, those with small m perhaps of the order of 10, the functions Φ_t must be close to unity, that the functions must be monotone nonincreasing and must tend to zero as m grows.

Our problem was to see whether the selection probabilities $\Phi_t(m)$ could be directly estimated from the data in the catalogue and whether the estimates so obtained can be used in order to obtain both the space abundances Λ_t and the space luminosity functions ϕ_t . A priori the problem seemed hopeless and I personally was convinced that, in due course, it will be necessary for us to postulate some particular parametric form of $\Phi_t(m)$ and also of ϕ_t which we could fit jointly to the observations with a perpetual uncertainty as to whether either of the two fitted functions has anything to do with reality. Incidentally, this is exactly the path followed by the famous Swedish astronomer-statistician K. G. Malmquist who, in 1922, studied the luminosity functions of stars [8]. After postulating that the space luminosity function ϕ is a normal density with unknown mean and variance, he established its relation to the catalogue density ϕ^* It was a pleasure to find that this relation is correct and can be asserted without the assumption of normality, in fact, without any postulate regarding ϕ whatsoever (except convergence of certain integrals).

To persons not accustomed to concepts of absolute and of apparent magnitudes, of clusters and field galaxies, etc., the above description may seem a bit messy. However, this is a case where the problem of selection bias appears to me unique in its simplicity and in the elegance of the final result.

Following the easy lines of first order clustering theory and assuming that field galaxies are unambiguously identified, it is easy to deduce that the catalogue joint probability density of Ξ (the random distance of a galaxy) of a specified type t and of its apparent magnitude μ, is given by, say,

$$p_{\Xi,\mu}(\xi,m|t) = C\xi^2\Phi_t(m)\phi_t(m+5-a\log\xi) , \tag{2}$$

where C is the usual norming factor. In order to obtain the catalogue density of the apparent magnitude μ, it is sufficient to integrate (2) for ξ from 0 to $+\infty$. The substitution $m+5-a\log\xi=\tau$ leads to the following result

$$p_\mu(m|t) = Ca^{-1}\Phi_t(m)\ e^{3(m+5)/a}\int_{-\infty}^{+\infty}e^{-3\tau/a}\phi_t(\tau)d\tau\ . \tag{3}$$

In writing this formula we have to make the unique assumption limiting the arbitrariness of the density ϕ_t. This is that the integral in the right side must converge. Granting this, formula (3) can be rewritten in its final form

$$p_\mu(m|t) = C_1\Phi_t(m)\ e^{3m/a}, \tag{4}$$

which is a very interesting and important result. Let us look at it. In the left side we have the catalog density of the apparent magnitude of type t galaxies. Given a substantial number of observations, this density can be estimated empirically, without making any arbitrary assumptions. In the right side we have the norming factor C_1, and the product of the exponential $\exp\{3m/a\}^1$ by the unknown (and I used to think unknowable) selection probability $\Phi_t(m)$! It follows that this selection probability can be calculated from (4) yielding

$$\Phi_t(m) = e^{-3m/a} p_\mu(m|t)/C_1 \tag{5}$$

with just one little difficulty. This is connected with the norming factor C_1 which remains unknown. However, given that the catalogue of galaxies is compiled with an effort not to miss galaxies that are readily available for observation, which I think is a very plausible assumption, we can take it for granted that for small values of m (that is, for bright galaxies) the probability $\Phi_t(m)$ must be close to unity. It will be seen that this extra assumption determines the selection probability function $\Phi_t(m)$ uniquely.

I wish anything like this situation existed in other domains of research where we meet with selection bias!

The result just described is not the only surprising result implied by formula (2). As easily seen, this formula provides the possibility of calculating the joint catalogue probability density of μ, the galaxy's apparent magnitude, and of this same galaxy's absolute magnitude. In order to obtain this result, we note that the variable τ I used above is nothing but the absolute magnitude of interest. Thus the joint catalogue probability density is given by, say,

$$p_{\mu,\tau}(m,M\ t) = C_2 a^{-1}[\Phi_t(m)e^{3m/a}][\phi_t(M)\ e^{-3M/a}] , \tag{5}$$

where, in the right side we find the product of two functions, one depending upon the apparent magnitude m only and the other upon the absolute magnitude M only. It follows that in the catalogue, the two magnitudes of galaxies are mutually independent and that the catalogue luminosity function ϕ_t^* of type t galaxies is connected with its space counterpart ϕ_t by the simple relation

$$\phi_t{*}(M) = C_3\phi_t(M)\ e^{-3M/a} \tag{6}$$

This provides the solution of the fundamental problem of estimating the space luminosity function ϕ_t as follows

$$\phi_t(M) = C_4\phi_t{*}(M)\ e^{3M/a} \tag{7}$$

where, as usual, C_4 stands for an appropriate norming factor.

The results obtained can be summarized as follows: granting that field galaxies are Poisson distributed in space, and granting that for the particular morphological types of galaxies the selection probabilities Φ_t depend upon the apparent magnitude m and on nothing else, in order to estimate the space luminosity functions ϕ_t it is sufficient to estimate their catalogue counterparts $\phi_t{*}$, to multiply by $\exp\{3M/a\}$ and to norm. No other arbitrary assumptions are necessary! I wish something of this sort existed in demographic, in economic or in health studies of our society!

Attractive as these results may look, we must not close our eyes on the difficulties that do exist. Are the two basic assumptions of Poissoneness and of dependence of Φ on the apparent magnitude only truly realistic? One way of checking is through the implications of formula (5): given a catalogue of type t field galaxies, is it really true that the apparent and the absolute magnitudes of these galaxies are independent? This, of course, is a subject worthy of study. In particular, the results of this study may indicate that the classification of morphological types is not fine enough and that a particular type should really be subdivided into certain subtypes t' and t''. However, there is also another way of checking how far the estimates of Φ_t and of ϕ_t are realistic. Before proceeding to this subject I should point out that formula (7) represents a generalization of the earlier result of Malmquist, already mentioned. Formula (7) in its present generality was found by us when we tried to answer the questions of two astronomers, N. U. Mayall and C. D. Shane, about the validity of Malmquist's result which, somehow, took root in astronomical literature, for example in the works of the pioneer researcher of the realm of galaxies E. Hubble, while the deduction of the formula was forgotten.

4. <u>Abundances of morphological types of galaxies in clusters.</u>

At the outset I mentioned that currently it is considered that, generally, galaxies occur in clusters and that field galaxies are treated as clusters of just one member. In thinking about the possibility of empirical validation of formulas (5) and (7) it occured to us that they could be used to calculate the abundances of the various types of galaxies in clusters. The extra assumption we had to adopt was that the space abundances of the various types is the same, whether we consider field galaxies, members of small groups, or members of known giant clusters.

As a matter of fact, in the astronomical literature this latter assumption is a matter of dispute or even of disbelief. The point is that as astronomers study the more and more distant clusters of galaxies, they find increasing difficulties in identifying the spirals which, at relatively small distances, are quite frequent. Because the time that the light emitted by galaxies takes to reach our telescopes is measured in millions and in hundreds of millions of years, when looking at a very distant cluster we are looking also into the past. Thus, the scarcity of spiral galaxies in the very distant clusters is occasionally interpreted as a result of evolution of galaxies. Be that as it may, we conceived the idea of the following calculations.

(i) Use the estimates of Φ_t and of ϕ_t obtained from formulas (5) and (7) applied to field galaxies found in the famous Humason-Mayall-Sandage catalogue (HMS, for short) in order to estimate the space abundances of types Λ_t.

(ii) Use the same formulas (5) and (7) and certain deductions thereof in order to calculate the expectations of the numbers of galaxies of the various types to be found in the various clusters recorded in the same catalogue, on the basic assumption that the space abundances of these types in clusters are the same as in the field.

(iii) Compare the expectations so obtained with the actual numbers of the various types of galaxies in the particular clusters.

Let me state this problem a little differently. The total number of galaxies belonging to the Coma Cluster listed in the HMS catalogue is 46. <u>Suppose that the Coma</u>

Cluster galaxies have the same abundances of types and the same luminosity functions as the field galaxies. Suppose further (very likely not true) that the selection probabilities for cluster galaxies are the same as for those in the field. On these assumptions, how many spirals should one expect in the HMS catalogue listed as members of the Coma Cluster? How many ellipticals, how many galaxies of type S0, etc.?

The following table, reproduced from the Astronomical Journal [9] Vol. 67(1962) pp. 582-3, summarizes our findings, obtained in cooperation with Prof. W. Zonn.

The outstanding feature of this table is the parallelism between the systematic changes in the expected and the actual frequencies of the various morphological types as one proceeds from left to right in the table, or as one proceeds from less distant to more distant clusters. In particular, at small distances one expects, and one observes, many spiral galaxies of the several subtypes symbolized by Sa, Sb and Sc, etc. On the contrary, at near distances the frequency of elliptical galaxies, symbolized by E0, E1, etc., is rather small. Our calculations of space abundance, see table, show that, among field galaxies at least, ellipticals are scarce and spirals frequent. The same calculations show also that, on the average, elliptical galaxies and also S0 galaxies, are much brighter than spirals. As a result, as we move from the immediate vicinity of our galaxy to farther and farther clusters, the relative frequency of catalogued spirals steadily decreases and that of catalogued ellipticals steadily grows.

Whether there is anything to the belief of some astronomers that distant clusters have practically no spiral members, we do not know and we do not dispute the assertion. However, there is the empirical fact, that even if in reality these distant clusters contained spirals and ellipticals in exactly the same proportions as they are in the field, the differences in the brightness of these morphological types would cause the apparent depletion of spirals and the apparent increase in the frequency of ellipticals, similar to those observed.

Looking at the table more closely, one notices certain systematic errors in the expectations. These are due to the fact that all estimates of frequencies of types are based on the same estimates of space abundances, of selection probabilities and of luminosity functions. For example, one of the space abundances is overestimated, then this overestimate will be felt all along the line in the table that corresponds to the particular morphological type.

So much for "nonresponse" in galactic research. As I mentioned before, this is probably the only case in which the problem can be solved without adopting assumptions extraneous to the background theory. In the next section I wish to describe certain "nonresponse" incidents which occur in our studies of carcinogenesis. Here, the situation is much less satisfactory.

II. "Nonresponse" problems in studies of carcinogenesis.

5. Background. In the study of the origin of cancer we distinguish between one hit and multihit hypotheses. Another distinction is between one stage and multistage assumptions. Commonly, it is assumed that the initial event leading to cancer occurs in a single normal cell when it is "hit" by carcinogen, either a chemical or radiation. If one hit is sufficient to initiate the process, we speak of one hit mechanism. However, some evidence points to the possibility that more than one hit is necessary. Whatever the case may be, the initial event, perhaps mutation, is supposed to generate a clone of modified cells. If this clone is already cancer, we speak of one stage process. However, the clone generated by the mutation in a normal cell may be benign, except that each of its live cells may be exposed to the risk of further mutation which would be cancer. If this is so, then the mechanism is called a two stage mechanism, etc. Both the number of hits and the number of stages are the subjects of studies and, occasionally, beliefs.

On the empirical side there is a variety of things that one can do to make the experimental animals. One possibility is to inject groups of animals, usually mice, with varying doses of a chemical carcinogen which is known to remain in the bodies only for a very short time (less than 24 hours) so that all the "initial events" leading to tumors could be effectively considered as simultaneous. Next, after some 10 weeks or more, at consecutive times $t_1, t_2, \ldots, t_s$ groups of mice are killed and their organs examined. Two kinds of variables are usually considered. One is the number of tumors, say in the lungs. The other is the number of cells per tumor.

This is the basic form of an experiment and various combinations of such experiments are used. For example, rather than give the animals just one injection of the carcinogen, several injections are made at preassigned intervals of time, etc. In the following I shall be concerned with the simplest experiment of just one injection and with the simplest observation representing the number of cells in a single tumor identified at time t since the injection of the carcinogen. More details will be found in the joint paper [10], recently published.

6. Cancer clone as a sample of the birth and death process.

In the simplest possible model of carcinogenesis it is assumed that the appearance at time $t = 0$ of a single mutated cell is followed by the realization of an age independent time homogeneous birth and death process with rates λ and $\mu < \lambda$, respectively. Thus the process is supercritical and, if it grows indefinitely, it will kill the organism. Let $X(t)$ be the number of live cells in this growth. As is well known, the distribution of $X(t)$ is determined by

$$P\{X(t) = n\} = P_n(t) = \psi(t)\, R^n(t) \quad \text{for } n>0 \tag{8}$$

where

$$\psi(t) = \frac{q^2 e^{qt}}{\lambda(e^{qt}-1)(\lambda e^{qt}-\mu)}, \tag{9}$$

$$R(t) = \frac{\lambda(e^{qt}-1)}{\lambda e^{qt}-1}, \tag{10}$$

with

$$q = \lambda - \mu > 0 \tag{11}$$

One might be inclined to think that a group of animals killed at some time t would provide a set of observations on the variable $X(t)$ which, when compared to formula (8), would lead to estimation of the two unknown constants λ and μ. If such groups of animals are killed and examined at several moments $t_1, t_2, \ldots t_s$, then the resulting material would be sufficient for validation, or otherwise, of the tentative model considered.

Unfortunately, the situation is much more complex. In fact, it appears that tumors are simply not noticeable unless they are composed of a very substantial number of cells. Here, then, we are again confronted with the ubiquitous problem of nonresponse and, as in the case of galactic research, are forced to introduce selection probabilities.

7. Number of cells in a noticed tumor. In parallel with the variable $X(t)$ defined above we shall now consider another variable $X^*(t)$ defined as the number of cells in a tumor aged t that has been noticed by the experimenter. Let $\Phi(n)$ stand for the probability (the selection probability) that a tumor with exactly n cells will be noticed. We visualize that $\Phi(0) = 0$ and that, as n is increased, $\Phi(n)$ is monotone nondecreasing and tends to unity. It will be convenient to define $\Phi(n)$ in terms of an infinite sequence of nonnegative numbers $a_0, a_1, a_2, \ldots, a_n \ldots$ subject to the restriction that their sum is equal to one. Then $\Phi(n) = \sum_0^{n-1} a_i$ (12)

$$\text{Let} \qquad g(u) = \Sigma_0^\infty a_n u^n \tag{13}$$

be the generating function of the numbers $\{a_n\}$. Also, let $G(u|t)$ stand for the probability generating function of $X^*(t)$. Obviously

$$G(u|t) = \frac{\Sigma_1^\infty P_n(t)\Phi(n)u^n}{\Sigma_1^\infty P_n(t)\Phi(n)}$$

$$= \frac{\Sigma_1^\infty (Ru)^n \Phi(n)}{\Sigma_1^\infty R^n \, \Phi(n)} \quad . \tag{14}$$

Thus, in order to complete the calculation, we have to evaluate sums of the type $\Sigma x^n \Phi(n)$. We have

$$\Sigma\, x^n \Phi(n) = \Sigma_1^\infty\, x^n\, \Sigma_0^{n-1} a_i$$
$$= \Sigma_0^\infty \frac{a_i x^{i+1}}{1-x} \tag{15}$$
$$= \frac{x}{1-x}\, g(x)\ ,$$

and it follows

$$G(u|t) = u\, \frac{1-R}{1-Ru}\, \frac{g(Ru)}{g(R)}\ . \tag{16}$$

This appears as a compact formula having a degree of elegance. However, it is not very useful in actual research because of the presence of the unknown function g . The best that one can do is to substitute for g a function depending upon several parameters and use the observations in order to estimate not only the parameters already involved in the model λ and μ, but also those appearing in g . My own favorite for g is the negative binomial

$$g(u|\alpha,\beta) = \{1 + \beta - \beta u\}^{-\alpha} \tag{17}$$

Formula (16) can be used in order to evaluate the expectation of $X^*(t)$. Differentiating $G(u|t)$ with respect to u and substituting $u = 1$, we obtain

$$E\, X^*(t) = \frac{1}{1-R} + R\, \frac{g'(R)}{g(R)}\ . \tag{18}$$

If (17) is used as a tentative approximation of the real g , then we have

$$E\, X^*(t) = \frac{1}{1-R} + \frac{\alpha\beta R}{1+\beta(1-R)}\ . \tag{19}$$

Ordinary experiments yield mean numbers of cells in tumors of consecutive ages $t_1 < t_2 < \ldots < t_s$. These means can be used in connection with the last formula in order to estimate λ, μ, α and β . The multiplicity of these parameters is likely to result in an excellent fit, which however, will not necessarily mean a validation of the model. Some further experiments and crosschecks will be needed and the particular difficulty is that the counts of tumors and cells should be made by the same individual (with the same g) and also the same population of mice must be used in order to ensure the sameness of the parameters λ and μ . In practice this means that the several different experiments should be conducted simultaneously. The difficulties are formidable, but we continue to struggle with them.

References

[1] J. Neyman, "On the two different aspects of the representative method: the method of stratified sampling and the method of purposive selection." J. Royal Sta. Soc., 97, pp. 558-625, 1934.

[2] Morris H. Hansen, William N. Hurwitz and William G. Madow, "Sampling Survey Methods and Theory." Two volumes. John Wiley and Sons, Inc., New York, 1953.

[3] William G. Cochran, "Sampling Techniques." John Wiley and Sons, Inc., New York, 1953.

[4] A. A. Tschuprow, "On the Mathematical Expectations of the Moments of Frequency Distributions in the case of Correlated Observations." Metron, Vol. 2 (1923) pp. 646-680.

[5] J. Neyman, "Contribution to the theory of sampling human populations," J. Am. Stat. Assoc., Vol. 33 (1938), pp. 101-116.

[6] J. Neyman and E. L. Scott, "A Theory of the Spatial Distribution of Galaxies," Astrophysical Jour., Vol. 116 (1952), pp. 144-163.

[7] J. Neyman and E. L. Scott, "Field galaxies: luminosity, redshift, and abundance of types. Part I. Theory," Proc. Fourth Berkeley Symposium Math. Stat. and Prob., Vol. 3, pp. 261-276, Univ. of California Press, Berkeley and Los Angeles, 1961.

[8] K. G. Malmquist, "On some relations in stellar statistics," Lund Medd. I, No. 100 (1922) Ark. Mat. Astro. Fys., Vol. 16 (1922), pp. 1-52.

[9] J. Neyman, E. L. Scott and W. Zonn, "Abundances of morphological types among galaxies in clusters and in the field," (Abstract) Astronomical Jour., Vol. 67 (1962), pp. 582-583.

[10] J. Neyman and E. L. Scott, "Statistical aspect of the problem of carcinogenesis," Proc. Fifth Berkeley Symposium Math. Stat. and Prob., Vol. 4, pp. 745-776, Univ. of California Press, Berkeley and Los Angeles, 1967.

Percentage of galaxies of

Morphological type	Field HMS n	galaxies Space λ %	Near groups n = 32 Exp	Near groups n = 32 Obs	Virgo n=80 Exp	Virgo n=80 Obs
E0-E3	83	3.7	7.0	6.2	10.6	12.5
E4-E7,Ep	28	7.6	5.3	9.4	5.0	15.0
SB0,SBa	21	2.2	4.1	9.4	5.6	10.0
SBb	26	1.9	3.4	3.1	5.0	2.5
S0,S0p	66	11.5	12.5	6.2	13.7	15.0
Sa,Sap,Sab	51	3.3	7.1	12.5	11.2	8.8
Sb,Sbc	77	20.8	21.8	12.5	20.6	12.5
Sc,Scp,SBc	94	49.0	38.8	40.6	28.5	23.8
Correlation coefficient			0.90		0.73	

different morphological types.

Intermediate groups n = 72		Coma n=46		"Far" Clusters n = 48	
Exp	Obs	Exp	Obs	Exp	Obs
18.1	22.2	34.8	34.8	44.1	47.9
6.2	9.7	8.8	19.6	12.3	8.3
4.6	13.9	2.3	0.0	1.1	2.1
6.3	4.2	7.8	2.2	5.8	0.0
16.1	23.6	20.5	28.3	21.0	14.6
11.6	11.1	8.8	8.7	4.3	18.7
17.5	8.3	11.0	2.2	8.0	6.2
19.5	6.9	5.9	4.3	3.3	2.1
0.32		0.88		0.90